Stress Physiology of Woody Plants

Stress Physiology of Woody Plants

Edited by
Wenhao Dai

CRC Press is an imprint of the
Taylor & Francis Group, an **informa** business

CRC Press
Taylor & Francis Group
6000 Broken Sound Parkway NW, Suite 300
Boca Raton, FL 33487-2742

First issued in paperback 2021

ISBN 13: 978-1-03-209277-5 (pbk)
ISBN 13: 978-1-4987-4608-3 (hbk)

Library of Congress Cataloging-in-Publication Data

Names: Dai, Wenhao, editor.
Title: Stress physiology of woody plants / edited by Wenhao Dai.
Description: Boca Raton, Florida : CRC Press, 2019. | Includes
bibliographical references and index.
Identifiers: LCCN 2018057268| ISBN 9781498746083 (hardback : alk. paper) |
ISBN 9780429190476 (e-book)
Subjects: LCSH: Woody plants--Effect of stress on.
Classification: LCC QK754 .S767 2019 | DDC 582.1/8--dc23
LC record available at https://lccn.loc.gov/2018057268

Visit the Taylor & Francis Web site at
http://www.taylorandfrancis.com

and the CRC Press Web site at
http://www.crcpress.com

Contents

Preface

Woody plants are a group of the most fascinating living organisms on the planet. They produce food, fiber, and fuel as well as help to beautify our living environment and conserve natural resources. Woody plants have distinct growth and development habits: tree vs. shrub, deciduous vs. evergreen, broadleaved vs. coniferous, perennial vs. biannual, temperate vs. subtropical vs. tropical, forest vs. fruit vs. landscape, and so on.

Woody species, particularly tree species, have a large and deep root system that helps them survive some extreme environmental conditions, such as extreme temperatures and drought. However, the perennial nature, large size, and long life expectancy make woody species prone to repeated and frequent challenges from different stresses during their lifetime. For example, a woody plant may face the stresses of freezing, heat, drought, water logging, and cold in one year. The same plant may also face other stresses, such as nutrient deficiency, diseases, and insects in certain periods. Although different plant species share the same basic physiological processes such as photosynthesis and respiration, the specific characteristics of woody plants make them unique in certain physiological processes including their responses to stresses.

Recently accumulated advances in physiology and molecular biology in plants have produced a large body of knowledge on how plants respond to different stresses and how plants evolve resistance and tolerance to these stresses. Existing books focus on either general physiology in woody plants or on stress physiology in herbaceous species. With the increasing importance of woody plants in production of food, fiber, and bioenergy, and in protection of the environment, understanding the mechanisms involved in the responses of woody species to different stresses is necessary.

The main objective of this book is to provide a comprehensive review of physiological and molecular aspects of woody plants in response to major environmental stresses. Some chapters specifically address the mechanisms involved in the development and regulation of resistance to the individual or combined stresses and the molecular and biotechnological approaches to improve stress resistance or tolerance of woody plants. This book comprises 11 chapters. Chapters 1 through 4 cover the basics of plant physiology and discuss uniqueness of plant structure, plant growth and development, photosynthesis and respiration, and plant growth regulation in woody species. Chapters 5 through 10 cover plant responses to the stresses including drought, nutrient deficiency, salinity, low temperature, oxidative stress, and heavy metal toxicity. Chapter 11 describes how plants respond to multiple environmental stresses.

This book is intended for use as a text or reference for students and scientists who already have knowledge of general plant physiology and basic woody plant materials in the field of horticulture, forestry, and plant molecular biology. It provides a basic background of stress physiology in woody plants and the essential insights in how plants respond and regulate different stresses to protect themselves, as well as an overview of some successful studies in woody plant stress physiology. It aims to bridge gaps in research between herbaceous and woody species and between general physiology and stress physiology in woody species, leading to the goal of improving resistance to environmental stresses in woody species.

This book is truly a group effort made by chapter contributors, reviewers, and editors, and I am indebted to them for their considerable patience and excellent work. I would like to thank Ms. Randy Brehm, who suggested that I should write this book for her help and advice throughout this lengthy endeavor was invaluable. Finally, I am extremely grateful to my wife, Qi Zhang, my daughter, Ann, and my son, Andrew, for their encouragement and patience during the long course of this effort.

About the Editor

Dr. Wenhao Dai received his PhD in plant science, specializing in woody plant physiology and biotechnology, from North Dakota State University in 2001. He was a Postdoctoral Research Associate in the AgBiotech Center at Rutgers University from 2001–2002. Currently, he is a professor in the Department of Plant Sciences at North Dakota State University, leading the Woody Plant Physiology and Biotechnology program. Dr. Dai has more than 30 years of experience in woody plant research. His research areas cover woody plant physiology, biotechnology, molecular genetics, propagation, and production. Dr. Dai has taught several horticultural courses including Introductory Horticulture, Horticultural Science Laboratory, Plant Propagation, Fruit Crop Breeding, Fruit Crop Production, Nursery Production and Management, and Plant Tissue Culture and Biotechnology. Dr. Dai has published 39 research papers in peer-reviewed journals.

Contributors

Nir Carmi
Institute of Plant Sciences
Agricultural Research Organization-Volcani Center
Bet Dagan, Israel

Wenhao Dai
Department of Plant Sciences
North Dakota State University
Fargo, North Dakota

Rakefet David-Schwartz
Institute of Plant Sciences
Agricultural Research Organization-Volcani Center
Bet Dagan, Israel

David Granot
Institute of Plant Sciences
Agricultural Research Organization-Volcani Center
Bet Dagan, Israel

Yu-Jin Hao
National Key Laboratory of Crop Biology
National Research Center for Apple Engineering and Technology
College of Horticulture Science and Engineering
Shandong Agricultural University
Tai-An, Shandong, China

Rolston St. Hilaire
Department of Plant Environmental Sciences
New Mexico State University
Las Cruces, New Mexico

Da-Gang Hu
National Key Laboratory of Crop Biology
National Research Center for Apple Engineering and Technology
College of Horticulture Science and Engineering
Shandong Agricultural University
Tai-An, Shandong, China

Danqiong Huang
Department of Plant Sciences
North Dakota State University
Fargo, North Dakota

Byoung Ryong Jeong
Division of Applied Life Science (BK21 + Program)
Graduate School
Gyeongsang National University
Jinju, Republic of Korea

Tamir Klein
Department of Plant and Environmental Sciences
Weizmann Institute of Science
Rehovot, Israel

Abinaya Manivannan
Division of Applied Life Science (BK21 + Program)
Graduate School Gyeongsang National University
Jinju, Republic of Korea

and

Vegetable Research Division
National Institute of Horticultural and Herbal Science
Rural Development Administration
Jeonju, Republic of Korea

Renae E. Moran
School of Food and Agriculture
University of Maine
Orono, Maine

Yoo Gyeong Park
Institute of Agriculture and Life Science
Gyeongsang National University
Jinju, Korea

Bryan J. Peterson
School of Food and Agriculture
University of Maine
Orono, Maine

Eran Raveh
Institute of Plant Sciences
Agricultural Research Organization-Volcani Center
Bet Dagan, Israel

Raj Narayan Roy
Dr. Bhupendra Nath Dutta Smriti Mahavidyalaya
Hatgobindapur, Burdwan
West Bengal, India

Bidyut Saha
University of Burdwan
West Bengal, India

Guo-qing Song
Plant Biotechnology Resource and Outreach Center
Department of Horticulture
Michigan State University
East Lansing, Michigan

Prabhakaran Soundararajan
Division of Applied Life Science (BK21 + Program)
Graduate School Gyeongsang National University
Jinju, Republic of Korea

and

Genomics Division
Department of Agricultural Bioresources
National Academy of Agricultural Science
Rural Development Administration (RDA)
Wansan-gu
Jeonju, Republic of Korea

Hanan Stein
Institute of Plant Sciences
Agricultural Research Organization-Volcani Center
Bet Dagan, Israel

Todd P. West
Department of Plant Sciences
North Dakota State University
Fargo, North Dakota

Qi Zhang
Department of Plant Sciences
North Dakota State University
Fargo, North Dakota

1 Woody Plant Structure

Todd P. West

CONTENTS

1.1 INTRODUCTION

A woody plant is defined as a plant having secondary growth (wood). This secondary growth provides structure allowing for perennial growth resulting in increased aboveground size. Woody plants generally consist of six primary organs: leaves, stems, roots, flowers, fruits and seeds. These six can be classified as vegetative (leaves, stems and roots) and reproductive (flowers, fruit and seeds). The different organs are composed of various types of tissues.

1.2 WOODY PLANT CLASSIFICATIONS

Woody plants are classified by growth habit and include three categories: tree, shrub and liana (vine). A tree is defined as a woody perennial plant, typically having one dominant vertical trunk with height greater than 5m (15ft). A shrub is defined as a woody perennial plant with multiple vertical or semi-upright branches originating at or near the ground with a height less than 5m (15ft). A liana (vine) is defined as a woody perennial plant with a slender, flexible stem requiring support. Woody vines can climb with tendrils or by coiling their stems around other plants (often trees) or objects.

1.3 LEAVES

Leaves are the primary organ in vascular plants where photosynthesis occurs, converting light energy into chemical energy. Leaves are composed mainly of primary tissues including the epidermis, ground tissue (parenchyma, collenchyma and sclerenchyma) and vascular tissue (xylem and phloem). Leaves differ structurally depending on classification, angiosperm (seeds borne within a pericarp) or gymnosperm (seeds not borne within a pericarp). Leaf structure and shape vary considerably from species to species depending on adaptations resulting from climate including light, nutrient and water availability.

1.3.1 Angiosperm Leaves

Angiosperm leaves are composed of a leaf blade (lamina) and petiole for support and attachment to the stem. The lamina is usually broad and flat and consists of two epidermal layers (upper and lower) with a tightly packed palisade (adaxial, upper layer) and loosely arranged mesophyll layer (abaxial, lower layer) (Figure 1.1). These layers consist of parenchyma tissue with photosynthetic capabilities. Sclerenchyma or collenchyma tissue may be present to provide structure and support. Veins (vascular tissue) are present Leaves may be simple or compound depending on location of the bud (compressed embryonic tissues that develops into a new stem) with respect to petiole attachment to the stem. If the bud is located at the base of a single leaf, then this is termed a "simple" leaf, if

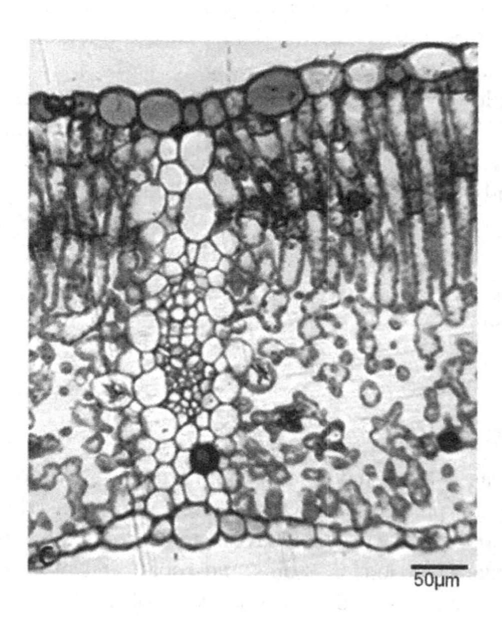

FIGURE 1.1 Cross section of angiosperm leaf (*Hibiscus* spp.). Epidermis layers on top and bottom with tight palisade layers on the adaxial side of the leaf blade and spongy mesophyll in the abaxial layer.

the bud is located at the base of more than one leaf (leaflets) then this is termed a "compound" leaf (Figure 1.2). Compound leaves can be either palmately compound, even or odd pinnately compound depending on the arrangement of the leaflets (Figure 1.3). Leaves, simple and compound have one of three types of venation within the lamina, pinnate, palmate or dichotomous branching (Figure 1.4). Leaves are attached to the stem at nodes and are arranged alternately (singly), oppositely (paired), suboppositely (singly) or whorled (three or more) (Figure 1.5). The arrangement of leaves on the axis is called phyllotaxy. Leaf shape including apices, bases and margins vary considerably from species to species depending on adaptations (Figure 1.6).

1.3.2 Gymnosperm Leaves

Gymnosperm leaves are often modified as a needle, scale or awl form (Figure 1.7). These modified leaves contain essentially the same tissues (epidermis, ground tissue and vascular tissue) as

FIGURE 1.2 Simple (left, *Ulmus* spp., elm) and compound (right, *Fraxinus americana*, green ash) leaves. Vegetative bud positioning determines if the leaf is simple or compound, the bud is located in the axil of the petiole and the connecting stem as seen in both leaf types at the base of the petiole.

FIGURE 1.3 Compound leaf types including palmate and pinnate compound. Palmately compound consisting of each leaflet attachment at single point (left, *Aesculus glabra*, Ohio buckeye), odd pinnate compound consisting of each leaflet attached opposite that terminates in a single leaflet (center, *F. americana*, green ash), even pinnate compound consisting of each leaflet attached oppositely that terminated with two leaflets (right, *G. triacanthos*, honeylocust).

FIGURE 1.4 Leaf venation patterns including pinnate, one main vein or midrib starting at petiole and extending to the apex of the leaf blade with several secondary veins branches off from the midrib (left, *Quercus macrocarpa*, bur oak), palmate, several main veins starting at the petiole and extending nearly equally to the apices of each leaf lobe (center, *Acer saccharum*, sugar maple), and dichotomous, several parallel veins starting at the petiole and dividing as they extend toward the leaf margin (right, *G. biloba*, gingko).

FIGURE 1.5 Leaf stem arrangements including alternate leaf/bud (vegetative) arrangement with one leaf/bud attached at a single node and subsequent attachments alternately along the stem (top left, *Ulmus* spp., elm), opposite leaf/bud arrangement with two leaves/buds attached at the nodes along the stem (top right, *Fraxinus* spp., ash), whorled leaf/bud attachment with three or more leaves/buds attached at the nodes along the stem (bottom left, *Catalpa* spp., catalpa), and subopposite leaf/bud attachment with two leaves/buds attached along the stem that are not perfectly opposite or spaced far enough to be considered alternate arrangement (bottom right, *Cercidiphyllum* spp., katsuratree).

FIGURE 1.6 Common woody plant leaf shapes, ovate (top left, *Betula papyrifera*, paper birch), lanceolate (top center, Salix pentandra, bay willow), cordate (top right, *Tilia americana*, American linden), elliptical (bottom left, *Malus* spp., crabapple), linear (bottom center, *Eleaeagnus angustifolia*, Russian-olive), and obovate (bottom right, *Ulmus davidiana* var. *japonica*, Japanese elm).

FIGURE 1.7 Different gymnosperm leaf types: scale-like (left, *Thuja occidentalis,* American arborvitae) consisting of overlapping flattened leaves and generally a soft feel, awl-like (center, *Juniperus communis,* common juniper) consisting of leaves that are attached to the stem singly and are generally sharp feel, needle-like (right, *Pinus sylvestris,* Scotch pine) consisting of slender elongated leaves that are either attached singly to the stem (*Abies* spp., *Cedrus* spp., *Picea* spp., *Pseudotsuga* spp. and *Taxus* spp.) or fastened together in bundles of 2, 3 or 5 (*Pinus* spp.).

angiosperm leaves. They also perform similar photosynthetic functions. Except for a few genera (several exceptions including *Ginkgo biloba*, *Larix* spp., *Metasequoia glyptostroboides* and *Taxodium* spp.), most gymnosperm have evergreen leaves. These evergreen species usually have leaves that typically persist three or more years.

1.3.3 Epicuticular Waxes

Many angiosperms and gymnosperms leaves have a waxy coating (cuticle). The cuticle covers the epidermal layer and may extend into the stomatal openings to decrease moisture loss. Epicuticular waxes vary in amount among species and may affect the visual color of the leaf to the human eye often giving a white to blue-white appearance. These waxes assist the plant with maintaining water balance, ultraviolet light reflection and assist in reduction of pathogen/insect incidences.

1.4 STEMS

1.4.1 Stem Structure and Functions

Stems have four primary functions: (1) support the crown, (2) conduct water and mineral elements upwards from the roots, (3) translocation of photosynthate and hormones from synthesis sources to site of use or storage and (4) storage of nutrients.

The stem is the main organ, typically supporting leaves and buds (vegetative and floral), responsible for the above ground structure of plant. Structure of the stem consists of nodes (point of leaf and/or bud attachment), internodes (space between adjacent nodes) (Figure 1.8). A bud is an unexpanded vegetative stem or flower (reproductive), and depending on the species, may have protective coverings know as bud scales (comprise of modified leaves). Buds are typically located at the stem apex (terminal) or lateral (axillary). The apical bud controls the level of apical dominance, which is the growth inhibition of the lateral buds. Woody plants undergo primary growth through rapidly dividing cells in the apical meristems within the stem apex to increase length.

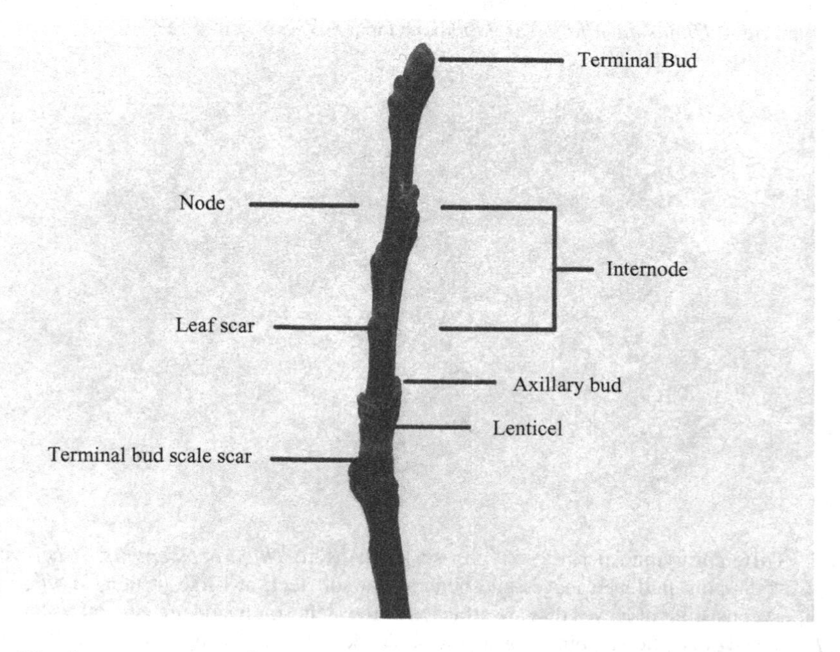

FIGURE 1.8 Woody stem anatomy. Current (one year of growth) season growth from terminal bud to terminal bud scale scar with subsequent year growth from terminal bud scale scar to terminal bud scale scar.

1.4.2 Secondary Growth in Stems

Secondary growth occurs in woody plants through lateral meristems, including both the vascular cambium and cork cambium to increase stem thickness (Figure 1.9). During secondary growth, more xylem than phloem is produced. Secondary growth in young woody plants, especially with trees, generally have strong apical control resulting in excurrent growth (strong terminal growth with weak lateral growth) and may or may not maintain this growth as it matures (Figure 1.10, left). Tree species that lose this strong apical control will have a more rounded form that is termed decurrent growth (weak terminal grown with strong lateral growth) (Figure 1.10, right).

1.4.3 Suckers vs. Watersprouts

Stems may arise in woody plants as suckers or watersprouts. Suckers are adventitious (in an unusual anatomical position) stems that arise from root tissue (e.g. *Rhus* spp., sumac). Watersprouts are shoots that often arise from latent buds (preformed buds, usually located under the bark of woody plants) within stem tissue. The latent buds are generally released from dormancy as a result of apical dominance occurring because of loss of apical bud through herbivory, damage or pruning.

1.4.4 Stem Modifications

Stems can be modified to facilitate different adaptations for flowering, support and defense. A spur is a compact stem that is modified for flower and fruit production (e.g. *Malus* spp., apple and *Pyrus* spp., pear). Stems can be modified to form tendrils for support (e.g. *Vitis* spp., grape). Tendrils have the capability to curve toward contact and have subsequent sclerenchyma development for weight support of the plants. For protection, stems may be modified for thorns or prickles. Thorns (e.g. *Gleditsia triacanthos*, honeylocust) are a pointed modified stem and prickles (e.g. *Rosa* spp., rose) are sharp outgrowths of the epidermis or bark. Leaves also may be modified for protection by forming spines. Spines are a modified leaf subtending a bud that is a sharp pointed structure (e.g. *Robinia pseudoacacia*, black locust).

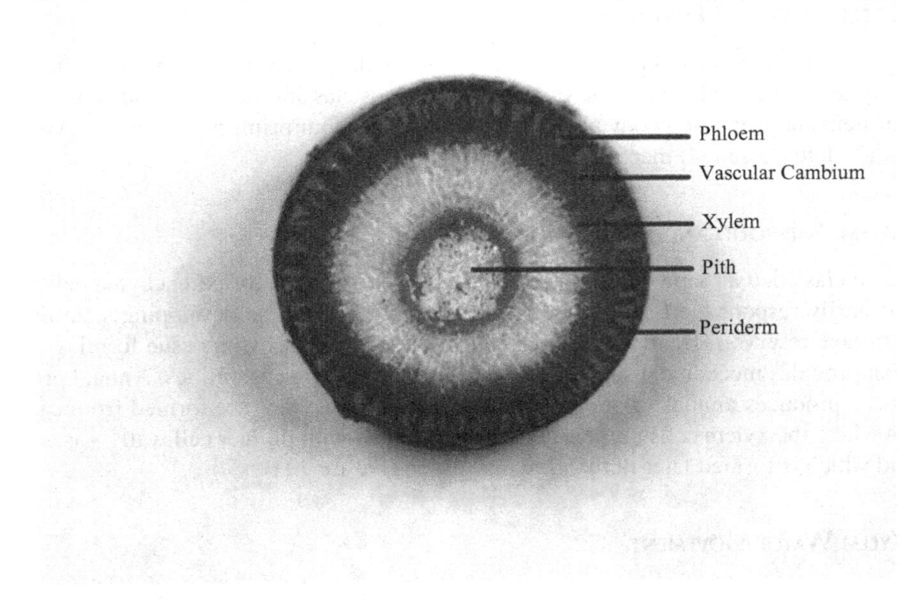

FIGURE 1.9 Cross section of woody stem.

FIGURE 1.10 Excurrent vs. decurrent growth differences in mature form (left, excurrent growth of *T. americana* 'Redmond', Redmond linden vs. right, decurrent growth of *U. americana*, American elm).

1.5 WOOD FORMATION

1.5.1 Xylem Tissue and Function

In woody plants, stems and roots are strengthened by secondary growth resulting from the vascular cambium. Xylem tissue produced by the vascular cambium is classified as wood and comprises most of the main stem and trunk of woody plants. Xylem tissue has four primary functions: (1) transport, (2) storage, (3) defense and (4) mechanical support.

1.5.2 Xylem: Sapwood and Heartwood

Xylem can be classified as sapwood or heartwood. Sapwood consists of parenchyma cells (living) that are primarily responsible for conducting water, strengthening the developing stem and serve as a food storage reservoir. Heartwood is formed from dead parenchyma tissue forming a central cylinder that provides mechanical support and infection restricting capabilities. Annual production of xylem cells produces annual xylem incremental rings. These rings are formed from earlywood production where the xylem cells are less dense – larger cells with thinner cell walls – as compared to latewood which is formed later in the growing season (Figure 1.11).

1.5.3 Xylem Water Movement

Water movement is conducted through xylem tracheids (gymnosperms) and vessels (angiosperms) primarily within the first few annual rings. In hardwoods (angiosperms), the wood is categorized

FIGURE 1.11 Annual growth representing incremental annual xylem ring with earlywood (wider banded rings) and latewood (narrow banded rings).

based on the vessel orientation. If the vessels are distributed uniformly throughout each annual ring, this is termed diffuse porous. In contrast, if the vessels are concentrated within the earlywood of each annual ring, this is termed ring-porous.

1.5.4 Rays

Secondary xylem will produce laterally oriented parenchyma cells forming rays. Rays formed in the sapwood are primarily responsible making the pathway for food movement to the phloem tissue and act as the primary storage sites of food storage in the main trunk and branches in conjunction with the longitudinal parenchyma cells.

1.5.5 Reaction Wood

Reaction wood will be produced when a main stem grows on a lean or in branches that need to counteract the increasing weight of the structure. There are two types of reaction wood: (1) tension wood which develops on the upper surfaces of angiosperm branches and leaning main stems (trunks) and (2) compression wood which develops on the underside of branches and leaning trunks of gymnosperms. Reaction wood results from xylem cells with greater amounts of elongation resulting in compression force. This compression force tends to force the trunk upright or in branches, provides more structural support for increasing weight, as well as assisting with restoring or maintaining proper growth angles.

1.6 BARK

Bark (periderm) has two primary functions: (1) protection from water loss and (2) protection from damage by external agents. In many woody species, there is another layer of expanding cells outside of the phloem tissue. This meristematic cell layer is called the cork cambium (phelloderm) and gives rise to the protective layer of cork cells (periderm). Cork cells lack a protoplast at maturity and its walls are thickened with a substance called suberin that repels water. These necrotic suberized cork cells lack intercellular spaces and form two common configurations in woody plants, either a wide, thin-walled cell arrangement or a flat, thick-walled cell arrangement. These configurations vary among species and give rise to the different bark characteristics and patterns seen within the range of woody plants forming smooth bark, rough bark, fibrous bark and exfoliating bark species. Since suberized cork cells are impermeable to water and gases, a portion of the periderm will have air spaces between the cells to assist with gas exchange. These air spaces are called lenticels and may appear as raised rounded, oval or elongated structures on the periderm. Some woody species have highly pronounced lenticels (e.g. *Betula* spp., birch and *Chionanthus virginicus*, fringe tree).

1.7 ROOTS

Roots have four primary functions: (1) anchorage, (2) absorption of water and mineral elements, (3) food reserve storage and (4) organic synthesis of certain organic materials. The genetic make-up of the different plant species dictates the development, size, form and function of the root system and is directly influenced by the plant's environment and management.

1.7.1 Root Development

Roots consist of three different zones. These zones include the meristematic region or zone, zone of elongation and zone of maturation (Figure 1.12). The root apex consists of a root cap (calyptra) and apical meristem. The root cap performs several functions for the developing root tip, and these include: protection of the root tip, lubrication of the soil by cells that have been sloughed off and gravitropic perception (presence of statocytes in the root cap). The zone of elongation is the section where root cells elongate to facilitate root lengthening. The zone of maturation includes the section where root hairs and subsequent lateral root branching occurs. Cell differentiation also occurs within this zone with cells differentiating into xylem, vascular cambium, phloem, endodermis, pericycle (thin layer of cells between the phloem and endodermis) and the cortex. The root hairs develop from epidermal cells and are primarily responsible for water and nutrient update by the root. Root hairs do not develop into lateral roots, but they may persist for a few weeks to more than a year. Lateral root branching differs from stem branching with new lateral roots developing from the pericycle within the zone of maturation, behind where root hairs have developed, instead

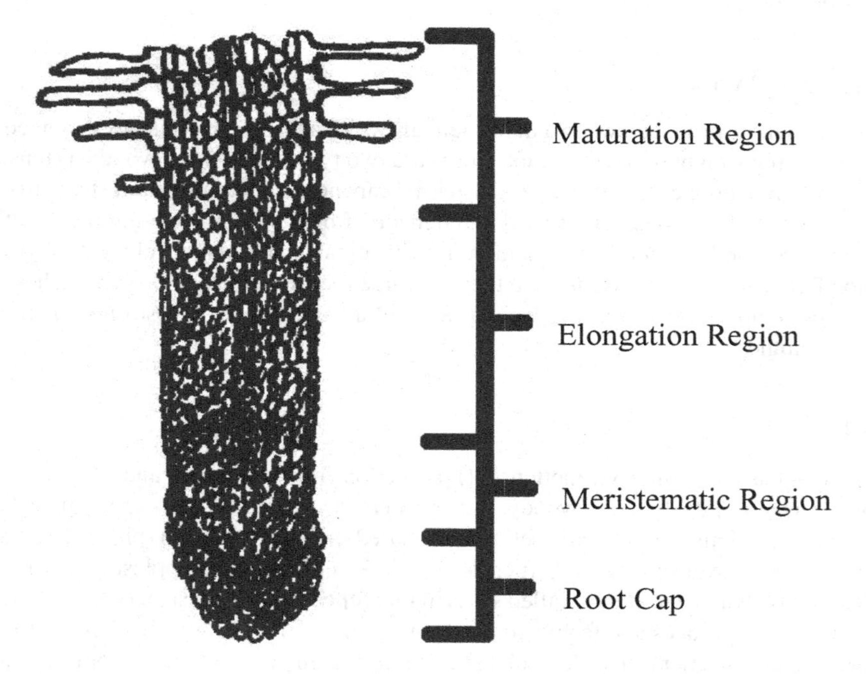

FIGURE 1.12 Longitudinal section of root tip showing the three main regions (maturation, elongation and meristematic) with root cap. The maturation region allows for rapid water/nutrient absorption, while the elongation region has limited water/nutrient absorption and the meristematic region has little to no water/nutrient absorption.

of at nodes as in stems. Primary and lateral roots will enlarge in size as a result of cell expansion and the production of secondary tissues.

1.7.1.1 Adventitious Roots

Adventitious roots are roots that arise from plant parts other than roots, such as stems and leaves. Adventitious roots are produced from parenchyma cells and are often a result of injury but also occur naturally on some plants (e.g. *Ficus benghalensis*, banyan tree). Adventitious roots are of two types: (1) preformed or latent root initials or (2) wound-induced roots. Adventitious roots are essential for commercial clonal propagation of many woody plants using cutting propagation or plant tissue culture.

1.7.2 Roots and Soil Organisms

Woody plants roots are capable of creating a symbiotic association (mycorrhiza) with soil fungi. This mycorrhizal association can have a significant benefit to woody plants including increased uptake of water and nutrients, and increased resistance to drought, pathogens, and soil salinity. These symbiotic associations are between nonpathogenic or weakly pathogenic fungi and the living cells of roots. Mycorrhizal associations are divided into two broad groups – ectotrophic (these do not penetrate root cells and area associated outside and inside of the root structure) and endotrophic (these penetrate root cell walls). Endotrophic associations do not extend into the endodermis tissues of the root. Ectotrophic mycorrhizae are commonly associated with woody plants in Betulaceae, Fagaceae Juglandaceae, Pinaceae, and Salicaceae. Endotrophic mycorrhizae are associated with woody plants in Altingiaceae (*Liquidambar styraciflua*, American sweetgum), Ericaceae, Juglandaceae, Magnoliaceae (*Liriodendron tulipifera*, tuliptree), Sapindaceae (*Acer* spp.) and Ulmaceae. When conditions support high fertility, mycorrhizae may not be necessary for successful growth, but mycorrhizal associations have a significant role in tree physiology when conditions are not ideal.

1.8 FLOWERS

Flowers are the sexual reproductive organs facilitating the union of gametes (egg and sperm). Flowers of woody plants differ greatly between angiosperms and gymnosperms.

1.8.1 Angiosperm Flowers

A typical angiosperm flower is composed of reproductive structures consisting of sepals, petals, stamens and pistils and is considered to be a complete flower (Figure 1.13). Sepals are found in a ring at the base of the flower and enclose the structure (bloom) to protect the developing flower bud from environmental stresses such as dehydration. They are typically green and collectively referred to as the calyx. In some woody species, the sepals are not distinguishable from the petals and are referred collectively as tepals (e.g. *Magnolia* spp., magnolia). Petals are attached to the floral structure above the calyx and collectively referred to as the corolla. The petals often assist with pollinator attraction and may be brightly colored or have a specific form. Wind pollinated species may have obscure petals with very little ornamental value. Collectively the calyx and corolla are referred as the perianth. The stamens are the male reproductive organs located above the petals. They are typically found on an elongated stalk (filament) that supports the anther which produces pollen (male gamete). Pollen is shed by the splitting (dehiscence) of the anthers. In the center of the flower structure is the pistil which is the female reproductive organ. The pistil is generally flask shaped with the swollen ovary at the base with a stalk-like portion (style) and terminates with the stigma where pollen (sperm) is deposited for fertilization. Within the ovary, there is an ovule which is the female gamete. A complete flower has all the reproductive structures present – perianth, stamens and pistil. Incomplete flowers lack one or more of these reproductive structures. Flowers

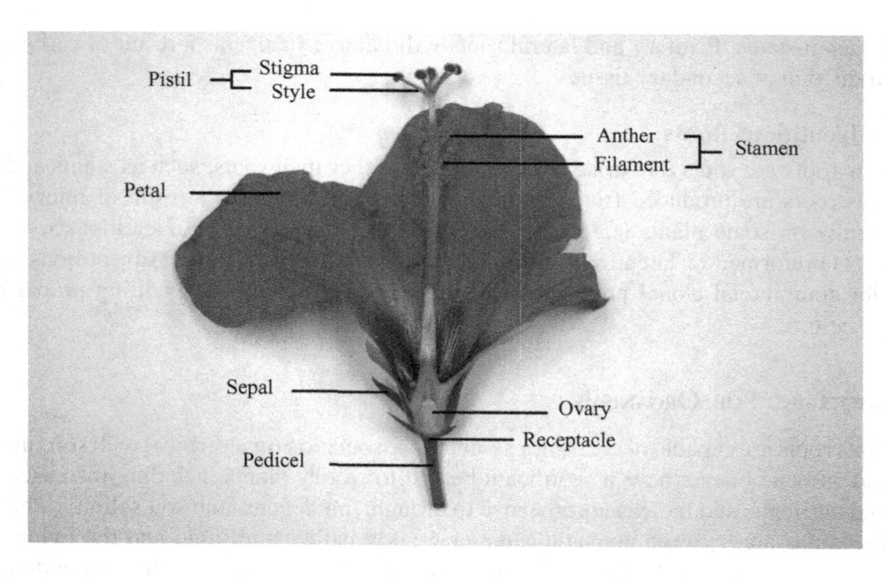

FIGURE 1.13 Typical angiosperm flower (*Hibiscus rosa-sinensis,* Chinese hibiscus).

that contain both sexual reproductive structures – stamens and pistil – are considered to be perfect. Some woody angiosperm species have staminate (male, stamens only) and pistillate (female, pistil only) in separate structures on the same tree (e.g. *Betula* spp., birch) and are considered to be imperfect and monoecious. Other species have staminate and pistillate flowers on separate plants which is dioecious (e.g. *Gymnocladus dioicus,* Kentucky coffeetree and *Salix* spp., willow). With dioecious trees, many commercial nurseries select "male" trees to have fruitless trees.

1.8.2 Gymnosperm Flowers

Gymnosperms are classified as a group of plants that produce naked seeds outside of an ovary. Flowers of gymnosperms are classified as imperfect and can be either monoecious (e.g. *Picea* spp., spruce and *Pinus* spp., pine) or dioecious (e.g. *G. biloba,* ginkgo and *Taxus* spp., yew). Gymnosperms do not produce elaborate flower parts similar to angiosperms. They typically have flowers that are structured as cones with exceptions (e.g. *G. biloba,* ginkgo and *Taxus* spp., yew) which have naked seeds that are not protected within a cone structure. These cones are either pollen (staminate) cones or seed (ovulate) cones. The pollen cones on average are smaller (10–12mm in length) than seed cones (25–500mm in length) and are generally produced in multiple clusters at the lower portion of the tree. Seed cones are generally borne singly or in doubles at the top of the tree. This placement of the pollen and seed cones ensures cross-pollination because conifers are wind pollinated.

1.8.3 Inflorescences

Flower arrangement on a stem varies greatly from species to species. Flowers may be borne singly (solitary) or in groups or clusters termed inflorescences. There are many different forms of inflorescences found in woody plant species which include several recognizable forms such as solitary (e.g. *Magnolia* spp., magnolia), spike, catkin (modified spike; e.g. *Betula* spp., birch; *Morus* spp., mulberry; *Quercus* spp., oak; and *Salix* spp., willow), raceme (e.g. *Wisteria* spp., wisteria), corymb (e.g. *Malus* spp., apple and *Prunus* spp., cherry), umbel (e.g. *Hedera helix,* English ivy), cyme (e.g. *Cornus* spp., dogwood and *Viburnum* spp., viburnum) and panicle (e.g. *Pieris* spp., pieris and *Syringa* spp., lilac) (Figure 1.14). The main stalk (stem) of an inflorescence is a peduncle with

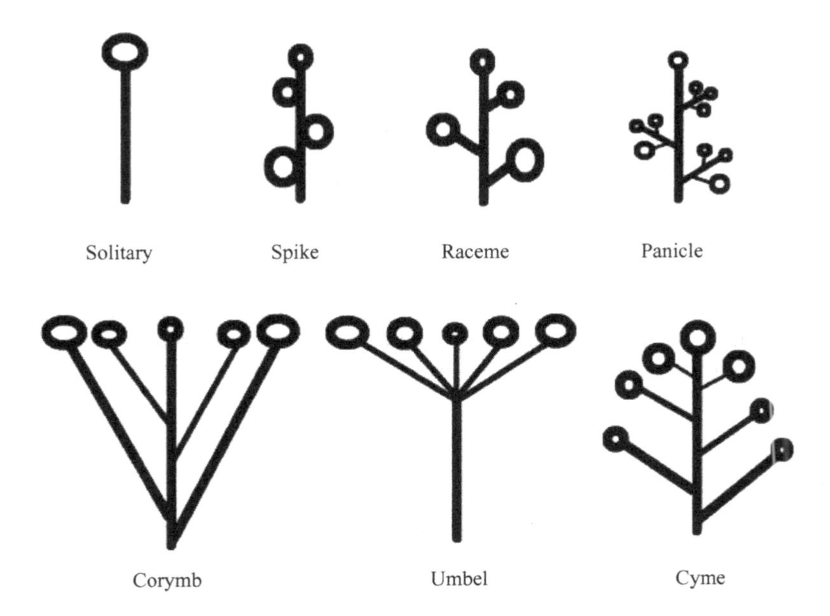

FIGURE 1.14 Common inflorescence types found in woody plants.

individual flower stalks on divided inflorescence types termed pedicels. These inflorescences are often consistent for a genus or even a family, and are often useful in plant identification.

1.8.4 Pollination and Fertilization

Pollination is necessary for initiating seed formation. Pollination is the transfer of pollen from a single flower (self-pollination) or from separate flowers (cross-pollination) to a receptive stigma. The process of fertilization is the fusion of haploid (1n) male and female gametes to form the viable zygote (2n). Double fertilization occurs in angiosperms with one sperm nucleus and egg to form the zygote and a second sperm nucleus fusing with two polar nuclei to form the endosperm (3n). Fertilization differs in gymnosperms because there is not a true stigma. Instead there is an opening or a sugary pollination exudate at the opening of the ovule within the cone structure to collect wind-borne pollen. Double fertilization does not occur in gymnosperms, and results in an endosperm that is a haploid female gametophyte tissue surrounding the embryo.

1.9 FRUITS

The result of fertilization is the development of a seed (embryo) and a fruit. A fruit is classified as a ripened ovary or group of ovaries along with any associated parts. Fruit can be formed without fertilization and occurs through the process of parthenocarpy. In some woody species such as apple (*Malus* spp.), the developing seed will produce hormones which aid in fruit development. As the fruit ripens, the ovary wall (pericarp) often swells and consists of three parts: (1) exocarp (outside layer), (2) mesocarp (middle layer) and (3) endocarp (inner layer). Angiosperm fruits are classified by three main categories: (1) simple, (2) aggregate and (3) multiple. Simple fruits are formed from a single pistil or ovary. Simple fruits are further classified as either dry or fleshy. There are two categories of simple, dry fruits: (1) dehiscent, where the pericarp splits open at maturity exposing the mature seed (legume, follicle and capsule) and (2) indehiscent, where the pericarp does not split open (achene, caryopsis, samara, nut and nutlet). There are three types of simple fleshy fruits: (1) berry (e.g. *Lonicera* spp., honeysuckle), (2) drupe (e.g. *Prunus* spp., cherry) and (3) pome (e.g.

Malus spp., apple). Multiple fruits are formed by the development of several flowers on a compact inflorescence that fuses during ripening and often has a common receptacle (enlarged area at the apex of the flower stem) (e.g. *Morus* spp., mulberry). Aggregate fruits are formed from clusters of individual fruits that develop from several pistils of a single flower and have a common receptacle (e.g. *Rubus* spp., raspberry and blackberry). Depending on the species, usually a plant will set more fruit than can mature, resulting in fruit drop shortly after fertilization.

1.9.1 Gymnosperm Fruits

Gymnosperm fruit is the mature seed cone. The maturation of cones and dissemination of seeds is similar to that of angiosperms. Cones may take one or two years to mature depending on the species. Some cones require specialized conditions to release seed, such as with many pine (*Pinus* spp.) species where high heat treatment is required.

1.10 SEED

A seed is a matured ovule containing an embryo generally resulting from sexual fertilization. Seeds of woody plants vary in color, shape, size and structure. Most fruit contains one or more seeds. In some species, limited viable (capable of germinating) seed may be produced. For example, *Acer triflorum*, threeflower maple, will produce fruit but a very low percentage of the fruit are filled out with a viable embryo with endosperm tissue. Many woody plants also have a dormancy requirement in order to germinate.

Viable seed in many woody plant species will not germinate even when fully mature because of dormancy. Dormant seed will not germinate even when the environment is suitable. Seed dormancy prevents immediate germination that is regulated by time and germination conditions. There are two major categories of seed dormancy: (1) primary and (2) secondary dormancy. Dormancy conditions can be overcome through several practices such as stratification (cool moist treatment) and/or scarification (physical weakening or opening of the seed coat).

BIBLIOGRAPHY

Kramer, P.J. and T.T. Kozlowski. 1979. *Physiology of Woody Plants*. 1st ed. Academic Press, Inc., New York, NY.

MacAdam, J.W. 2009. *Structure & Function of Plants*. 1st ed. Wiley-Blackwell, Ames, IA.

Preece, J.E. and P.E. Read. 2005. *The Biology of Horticulture: An Introductory Textbook*. 2nd ed. John Wiley & Sons Inc., Edison, NJ.

Shigo, A. 1986. *A New Tree Biology: Facts, Photos, and Philosophies on Trees and Their Problems and Care*. 1st ed. Shigo and Trees, Associates, Durham, NC.

Zimmermann, M.H. and C.L. Brown. 1974. *Trees: Structure and Function*. 1st ed. Springer-Verlag New York Inc., New York, NY.

2 Plant Growth and Development

Bryan J. Peterson and Renae E. Moran

CONTENTS

2.1 INTRODUCTION

The means by which woody plants develop and grow and modify their forms and functions in response to a diverse range of environmental conditions, is a broad topic. It would be impossible to describe the mechanisms underlying all these aspects of woody plant growth, development, and adaptation in full detail. Instead, this chapter provides a brief summary of some of the essential concepts in plant growth and development and discussion of woody plant form and function as it relates to the environment. In many respects, woody plants are similar to their herbaceous relatives, which should not be surprising when one considers that many families comprise both woody and herbaceous members. Clearly, the evolutionary switch from an herbaceous to a woody nature (or back again) has occurred independently in many different lineages. Thus, many of the mechanisms central to growth and development in herbaceous plants are shared with woody plants. Even secondary growth, a phenomenon generally associated with woody plants, occurs to a limited extent in the model species *Arabidopsis thaliana* and other herbaceous taxa.

Lignification (the development of secondary cell wall thickening) and secondary growth (the capacity for roots and stems to grow in diameter each season) are key evolutionary developments in the origins of woody plants. Plant cells are characterized by the formation of a primary cell wall comprising celluloses and hemicelluloses, and, in many cases, by additional thickening of the cell wall through the accumulation of lignin. The presence of lignin is crucial to the rise of woody plants, both literally and figuratively. Lignification gives woody plants the mechanical strength and durability to persist as enduring, and sometimes towering, features of the landscape through sequential additions of primary growth. Likewise, the evolution of the vascular cambium, a ring of meristematic cells within stems and roots, enables the progressive enlargement of these organs over time by secondary growth. Together, these characteristics and various adaptations to common developmental themes have facilitated the evolution, competitive success, and diversification of woody plants across the globe.

This chapter covers key themes in plant growth and development presented with a focus on the diverse adaptive characteristics important to woody plants. Topics covered include seed development and dormancy, root growth and development, shoot growth and development, control of bud dormancy and bud outgrowth, leaf growth and development, flower initiation and floral development, and fruit development and ripening. The chapter ends with a brief discussion of adaptation, which highlights the relevance of adaptation to horticulture, agriculture, forestry, and restoration ecology, as well as challenges associated with the study of adaptation.

2.2 SEED DEVELOPMENT AND DORMANCY

2.2.1 The Seed Contains the Embryo

Seed development begins with the union of gametes to create a zygote with a unique genetic blueprint. This zygote, the first structure of the sporophytic generation, begins to grow by mitosis to produce a multicellular embryo. Development of the embryo is characterized by the differentiation of cells along both a root-shoot axis and a radial axis (Su et al. 2015). Radially, the embryo consists of layers of cells forming the epidermis, cortex, endodermis, pericycle, and vascular tissues. The root-shoot axis comprises an embryonic root with a root apical meristem (RAM), an embryonic shoot with a shoot apical meristem (SAM) and cotyledons, as well as a portion of stem joining them, the hypocotyl. New tissues added to the plant by primary growth after germination are descended from the original RAM and SAM of the embryo.

The embryo develops within the ovule, which is located inside the ovary of an angiosperm flower or on the scale of a female cone produced by a gymnosperm. The ovule, plus all its contents, develops into a mature seed. It is through the funiculus, the connection between the ovule and maternal tissues, that the developing ovule obtains the water and other substances necessary to sustain the developing embryo. Eventually, as the embryo matures, the connection between the maternal plant and the ovule is lost and seed abscission occurs. The mature seed comprises a seed coat (testa) derived from integuments of the ovule, whereas all the contents internal to the seed coat are derived from megagametophytic tissue and the fertilized egg. Following maturation, seeds are dispersed into the environment by vectors such as gravity, water, wind, or animals, or by ballistic dispersal.

2.2.2 Hormonal Influences on Embryo Development

The development of a sporophyte from a single-celled zygote to a mature embryo with differentiated tissues depends on complex networks of hormonal signaling, changes in gene expression, and modifications to cellular activities (Weijers and Jürgens 2005). For example, the plant hormone auxin that plays a role in many developmental processes is essential in establishing the polarity of the embryo. Auxin gradients develop from the polar movement of auxin through transport proteins localized within cell walls of the developing embryo. In its protonated form, auxin is a weak, nonpolar acid that can diffuse from the acidic apoplast through the cell membrane. However, when in the less acidic environment of the cytoplasm, the auxin is deprotonated, rendering it polar and trapping it within the cytoplasm. Active movement out of the cytoplasm is restricted to locations in which members of the PIN-FORMED (PIN) family of auxin efflux proteins span the polar membrane and facilitate auxin movement from the cytoplasm to the apoplast. The asymmetric distribution of PIN proteins within and among cells directs the movement of auxin during embryogenesis (Möller and Weijers 2009). Because of both its interactions with auxin response factors in the regulation of gene transcription and its non-transcriptional regulation of cellular activities (e.g. loosening of cell walls to permit cell elongation), variations in auxin concentration play many roles in the growth and development of plants throughout their entire life cycles (Paque and Weijers 2016).

In the embryo, gradients in auxin concentration are necessary for the patterning of cell identities from the two-celled stage through the development of the mature embryo (Weijers and Jürgens 2005; Möller and Weijers 2009). With the first cell division, the proembryo comprises an apical cell, which will develop into the structures of the embryo proper, attached to a basal cell that develops into the suspensor. Already at this stage, PIN proteins are localized at the apical end of the suspensor cell, directing the movement of auxin to initiate the patterning of cell fates. During development of the proembryo following subsequent cell divisions, highly ordered patterns of PIN protein localizations are evident and lead to the establishment of position-dependent developmental effects as cells obtain informational cues about their positions relative to one another during embryogenesis (Souter and Lindsey 2000; Weijers and Jürgens 2005). These positional effects based on the localized transport of auxin culminate in the development of an embryo consisting of a highly structured apical-to-basal axis with a RAM, SAM, and cotyledon(s), as well as a structured radial axis with an epidermis, ground tissues, and vascular tissues.

2.2.3 Seed Maturation

Seed development occurs in several phases, including the histodifferentiation phase, the seed filling stage, and the maturation drying stage (Hartmann et al. 2011). Histodifferentiation is characterized by the development and growth of the structured embryo. In the seed filling stage, the seed grows rapidly in size and mass as cells expand and fill with proteins, lipids, and carbohydrates essential to support maintenance and growth of the embryo during germination. An increase in endogenous levels of abscisic acid (ABA) inhibits precocious germination and alters expression of ABA-responsive genes that function in embryo maturation (Thomas 1993). In orthodox seeds, the tissues of the seed also undergo desiccation during maturation drying, which slows metabolic activity and further prevents precocious germination. Under conditions favorable to germination, non-dormant orthodox seeds will germinate once they imbibe sufficient water; dormant seeds will either fail to imbibe water, or will imbibe but fail to germinate until dormancy is released. Recalcitrant seeds, such as those of buckeye (*Aesculus* spp.) and willow (*Salix* spp.), do not undergo maturation drying or exhibit dormancy; instead, such seeds must germinate soon after they mature or they will die.

2.2.4 Maintenance and Release of Seed Dormancy

Primary seed dormancy is characterized by the failure of viable seeds to germinate in an environment typically considered favorable to germination, due to the presence of germination-inhibiting factors that developed as the seed matured. Such dormancy might be exogenous, imposed by physical or chemical factors external to the embryo, or endogenous, imposed by factors internal to the embryo (Hartmann et al. 2011). Primary dormancy delays germination of mature seeds until the onset of environmental conditions suitable to support germinant survival and growth. Examples of conditions that might cause the release of seeds from primary dormancy include exposure to prolonged periods of cold or warmth, freeze-thaw cycles, or being ingested into animal digestive systems. The delay in germination afforded by these mechanisms presumably ensures that seedlings begin their growth at the start of a suitable season, and not at the end of the season in which they mature.

Seeds with physical dormancy, such as those of honeylocust (*Gleditsia triacanthos*) and many other members of the Fabaceae, will not germinate following dispersal because their hard seed coats prevent the seed from imbibing water (Fordham 1965; Baskin and Baskin 2004). With this type of dormancy, accompanied by slow rates of respiration, seeds of some plants in the Fabaceae can remain viable in dry storage for more than a century. In contrast, germination of seeds with chemical dormancy is prevented by growth-inhibiting factors originating from the pericarp of the fruit, which promote the maintenance of a quiescent state (i.e. not true dormancy) within the seed (Baskin and Baskin 2004).

Endogenous dormancy, imposed by factors internal to the embryo, includes various morphological and physiological conditions (Baskin and Baskin 2004). Seeds with morphological dormancy contain embryos that have not matured sufficiently for germination and typically require a period of exposure to warm, moist conditions (warm stratification) for the embryo to enlarge before germination can take place. For example, seeds of *Lonicera caerulea* require a period of warm stratification between dispersal in the summer and germination in the fall (Phartyal et al. 2009). Seeds with physiological dormancy are those in which physiological aspects of the embryos inhibit germination. When seeds are physiologically dormant, the growth potential of the embryo is insufficient to overcome the resistance imposed by the seed coat or other surroundings, such as remnants of the endosperm or pericarp tissues. This particularly common form of dormancy can usually be overcome by exposure of the seeds to cool, moist conditions (cold stratification) to increase the growth potential of the radicle and/or decrease the resistance offered by the seed coverings (Baskin and Baskin 2004). Seeds of some species, such as elderberry (*Sambucus* spp.) are characterized by morphophysiological dormancy, in which both morphological dormancy and physiological dormancy work in concert to delay germination (Hidayati et al. 2000).

Combinatorial dormancy is dormancy in which seeds have both physical dormancy imposed by an impermeable seed coat, as well as physiological dormancy. To germinate, such seeds require the release of physical dormancy, after which they can imbibe water and commence the release of physiological dormancy over time. For example, the seeds of eastern redbud (*Cercis canadensis*) require scarification like those of many other members of the Fabaceae, but also require a period of cold stratification to relieve hormonal inhibition of germination (Baskin and Baskin 2004).

Under favorable conditions, a seed may germinate following the release of primary dormancy; if conditions are unfavorable, the seed may become dormant again. This secondary dormancy develops in a previously non-dormant seed as a strategy to prevent germination in environments unlikely to support seedling growth. A seed may cycle through dormancy and non-dormancy as environmental conditions such as moisture, temperature, nitrate availability, and light change over the seasons (Finch-Savage and Leubner-Metzger 2006). Although secondary dormancy is well studied in herbaceous plants, its investigation in woody plants seems to be limited to the recent documentation of the phenomenon in several shrubs (Cao et al. 2014; Copete et al. 2015).

2.2.5 Hormonal Control of Seed Dormancy

Primary physiological dormancy, an especially common characteristic of the seeds of temperate woody plants, is maintained by hormonal balances within the seed. ABA biosynthesized within the seed during maturation imparts dormancy on seeds of taxa exhibiting this type of dormancy (Finch-Savage and Leubner-Metzger 2006; Finkelstein et al. 2008). Moreover, new ABA biosynthesized as seeds imbibe water maintains this dormancy even as the seed is exposed to environments that favor germination of non-dormant seeds. The biosynthesis and persistence of ABA within a seed maintains high ratios of ABA to gibberellic acid (GA), a ratio critical to the maintenance of seed dormancy (Finch-Savage and Leubner-Metzger 2006).

Several factors that are implicated in the release of dormancy and the promotion of germination include the degradation of ABA and changes in sensitivities of the embryo to ABA and GA signals. For instance, the degradation of ABA decreases the relative strength of the ABA signal relative to the GA signal. Within the embryo, a decrease in sensitivity to ABA and an increase in sensitivity to GA alters the perception of hormone signals (Finkelstein et al. 2008). Although the picture is not complete, induced changes in expression of genes for hormone signaling, and changes in the abundance of hormone receptor proteins, can alter cellular sensitivity to hormone signals (Hauvermale et al. 2015). When dormancy is released, GA acts as a positive promoter of germination by both weakening seed coverings, and increasing the growth potential of the radicle. At this point, the embryo is more sensitive to cues from the surrounding environment and will respond to favorable conditions by germinating (Finch-Savage and Leubner-Metzger 2006).

Although ABA is the hormone credited for maintaining seed dormancy, crosstalk with other phytohormones is important to dormancy processes. For example, auxin synthesis and signaling are essential for the maintenance of dormancy in *Arabidopsis*, probably through an auxin-mediated increase in ABA sensitivity initiated by auxin response factors (Liu et al. 2013). Likewise, crosstalk between ethylene and ABA within the seed may regulate ABA-induced dormancy, with high concentrations of ethylene countering the influence of ABA and permitting germination even in its presence (Arc et al. 2013).

2.2.6 Seed Germination

Germination is a three-phase process starting with rapid imbibition of the seed by water (Phase 1), followed by an increase in metabolic activity within the embryo that prepares it for germination (Phase 2), and finally the emergence of the radicle (embryonic root) through the seed coat or other seed coverings (Phase 3), after which seedling growth continues rapidly (Nonogaki et al. 2010). As a morphologically mature and viable seed becomes hydrated in Phase 1, it will begin catabolizing stored reserves to supply energy necessary for the completion of germination, unless it is physiologically dormant. ABA in dormant seeds remains present at high concentrations, because ABA biosynthesis occurs more quickly than its catabolism unless environmental cues sensed and integrated by the seed have triggered a change in dormancy status (Endo et al. 2014). In non-dormant seeds, catabolization of stored reserves coupled with translation of both stored and newly synthesized mRNAs lead to the growth of the embryo and the emergence of the radicle from the seed at the start of Phase 3. From radicle emergence to the development of photosynthetic competence in the seedling, the plant further mobilizes and utilizes stored reserves for additional growth of roots and shoots.

2.2.7 Commercial Methods to Release Seed Dormancy

Seeds play an important role in the commercial propagation of many woody crops grown for forestry, ornamental horticulture, and fruit production. Among forestry crops and Christmas tree farms, many trees are propagated using seeds collected from seed orchards. These are managed populations of elite genotypes selected as seedling parents because of their superior phenotypes or strong performance (Hartmann et al. 2011). Although nurseries in the ornamental horticulture industry may sometimes produce landscape-ready trees propagated from seed, seedlings play a much more important role as rootstocks onto which many shade, ornamental, nut, and fruit tree cultivars are propagated by grafting (Hartmann et al. 2011). In many cases, the commercial seed propagation of woody taxa for landscape use or for use as rootstocks depends on reliable methods to overcome seed dormancy that may be present.

Depending on the mechanisms imposing seed dormancy, propagators may subject seeds of woody plants to pre-germination treatments of stratification, scarification, or sometimes both. In nature, the need for stratification is more common than the need for scarification. When cold stratification is necessary, such as for many seeds of temperate woody plants (e.g. *Quercus*, *Malus*, *Tsuga*, and *Corylus*), seeds are cleansed of fruit tissue and placed into a moist but well-aerated substrate such as sphagnum moss, vermiculite, sand, or bark and stored at cold, nonfreezing temperatures. For seeds of species with immature embryos, warm stratification at temperatures sufficient to encourage growth of the embryo is conducted prior to cold stratification (Hartmann et al. 2011). The duration of stratification treatments depends upon the taxon and may vary from approximately one to six months.

Scarification can be used to quickly release dormancy that is caused by the presence of a seed coat preventing imbibition with water. For example, members of the Fabaceae (e.g. *Gleditsia*), Anacardiaceae (e.g. *Cotinus*), and Rosaceae (e.g. *Rubus*) may require such treatment. Common methods of scarification include treatment by sanding or filing seeds, placing seeds in concentrated sulfuric acid for durations ranging from 30 minutes to several hours, or placing seeds in several volumes of water hot enough to damage the seed coat without killing the embryo. Regardless of the scarification treatment, the successful disruption of a hard seed coat is evident by the swelling of a seed when it is placed in water.

Some taxa, such as *Cercis* and *Tilia* require a two-step seed pretreatment because of their exogenous and endogenous dormancies. In such taxa, scarification will overcome the effects of the impermeable seed coat, and a period of cold stratification will overcome the physiological dormancy within the seed. Conversely, seeds of other species require no special treatments, because they lack dormancy altogether. Examples include both orthodox seeds such as those of *Catalpa*, and recalcitrant seeds, such as those of *Aesculus* and *Salix*.

2.3 ROOT GROWTH

2.3.1 Root Structure and Development

The radicle, the first structure to emerge from a seed during germination, develops into a large, branching root system through the accumulation and enlargement of cells laid down by the RAM. The RAM, protected from mechanical damage by a root cap, pushes through the soil as new cells are produced and accumulate behind it. At the heart of the RAM is the "stem cell" niche composed of the quiescent center, a group of slowly dividing cells that signal multi-potent stem cells around it to remain undifferentiated (Perilli et al. 2012). These undifferentiated stem cells undergo continuous cell division, with one daughter cell remaining in the stem cell niche and another entering the proximal meristem, where it divides several times. These new cells next enter the transition zone, where their differentiation is initiated by the activity of cytokinin (Ioio et al. 2008), and finally the elongation/differentiation zone, where they elongate and complete their differentiation into cells of one of the primary meristems (Perilli et al. 2012). Cell-to-cell communication and modified gene expression determine the patterning of cells specified within the root, with various phytohormones involved in the process. For example, auxin produced in plant shoots is transported acropetally within the stele to the root tip, where it is redistributed and transported basipetally away from the root tip through the lateral root cap and cells of the epidermis (Jones et al. 2008). Such transport establishes local maxima and minima of auxin that are, in part, responsible for the different zones of activity observed during primary root growth. The root primary meristems specified during primary growth include the protoderm, which will develop into the epidermis; the ground meristem, which will develop into the cortex and the endodermis; and the protostele, which will develop into the stele (vascular cylinder).

As the epidermis develops from the protoderm within the zone of differentiation, two types of cells are specified: those with no root hair, and those with a thin root hair that functions to increase the absorptive surface of the root for the uptake of water and nutrients. Cells of the epidermis rely on positional information such that those straddling two underlying cells of the cortex may adopt a root-hair fate, whereas those simply stacked over a single cell of the cortex will not (Jones et al. 2008). A surprising number of genes and competitive interactions among transcription factors are involved in this seemingly simple decision about cell fate, with feedback loops that push some epidermal cells to the root-hair fate while inhibiting that fate in others (Schiefelbein et al. 2014). Basipetal transport of auxin from the root tip is involved in this process, with a tenfold higher concentration of auxin accumulating in non-hair cells because of the positions of auxin influx carriers localized in them. An environmental component to epidermal cell specification is also evident, with factors like nutrient availability modifying the regulation of cell fate to produce longer root hairs and/or more cells with root hairs (Giehl and von Wirén 2014).

During development of the ground meristem from cells in the RAM, periclinal division induced by transcription factors gives rise to a two-layered ground meristem, the innermost layer of which differentiates into the endodermis (Miyashima and Nakajima 2011). In addition to functioning for selective uptake of mineral nutrients, the endodermis is a major regulator of hormonal balance within different layers of root tissue, inducing phototropic and gravitropic responses. For example, PIN proteins within cells of the endodermis are located by default on both sides of the cell to transport auxin toward both the stele and the cortex, but light or gravity signals cause localization of PIN proteins to only one side of the endodermis, directing auxin flow in a response-specific pattern (Dinneny 2014). Moreover, auxin regulation by the endodermis is implicated in lateral root initiation from root founder cells of the pericycle, with auxin directed toward the founder cell fated to develop a lateral root initial (Marhavy et al. 2013).

The stele develops during primary growth of the root from stem cells near the root tip. Following primary growth, periclinal divisions of procambial cells establish the vascular cambium as a secondary meristem, with cytokinin produced in the root cap playing a key role (Mähönen et al. 2006). Several lines of evidence indicate that the fates of procambial cells as meristematic cells vs. differentiated vascular cells are determined by patterns of cytokinin and auxin signaling (Mähönen et al. 2006; Aloni et al. 2006).

Signaling by cytokinin, produced primarily by cells of the root cap, is essential for various growth and development outcomes within not only roots, but also shoots. After biosynthesis, cytokinin travels through plasmodesmata of adjacent root cells from the meristematic zone to the zone of differentiation, at which point its transportation shootward is by bulk flow in the xylem (Aloni et al. 2006). By this mechanism, cytokinin is delivered disproportionately to shoots and leaves supporting high rates of transpiration where cytokinin manifests responses such as the development of leaves suited to sun vs. shade. Interestingly, cytokinins produced in the roots promote growth and development of shoots but generally constrain growth and development in roots (Aloni et al. 2006).

2.3.2 Factors Affecting Root Growth and Root System Architecture

A key question in plant growth and development is how plants balance the production of roots and shoots to match the needs of the plant in a given environment. Certainly, environmental conditions like water and nutrient availability are the primary regulators of root growth. In the case of water stress, the view has long been that roots in dry soils produce ABA, which is transported through the xylem to shoots, where it causes stomatal closure in response to drought (e.g. Fort et al. 1998). Recent evidence demonstrates that ABA is produced primarily in shoots in response to water stress sensed in the shoots and is sent via phloem to the roots, where its presence promotes primary root growth while inhibiting development of lateral roots (McAdam et al. 2016). The regulatory network responsible for plant root responses is complex but involves, at a minimum, crosstalk between ABA, cytokinin, ethylene, and auxin (Rowe et al. 2016). In roots, the ratio of auxin to cytokinin determines growth by affecting the size of the RAM and the rates of cell division and differentiation (Ioio et al. 2008; Rowe et al. 2016).

Nutrient availability also affects root growth. Within the soil, nutrient availability is heterogeneous across both space and time, and plant nutrient acquisition may deplete available nutrients at the interface between roots and soil. Roots have evolved to be exquisitely capable of responding to variation in nutrient concentration, to forage efficiently, and to maximize the uptake of required nutrients when they are available. For example, plants under nutrient deficiency may alter root system architecture and root hair development, or exude compounds that mobilize nutrients (Giehl and von Wirén 2014). Since nutrients are often found more abundantly in upper layers of the soil, it should come as no surprise that trees and shrubs typically produce shallow root systems with many lateral roots and/or root hairs to increase efficiency of nutrient foraging.

Plant root architecture is particularly responsive to nutrient availability, with pronounced responses demonstrated for nitrogen, phosphate, and iron (López-Bucio et al. 2003). For example, experiments with *Arabidopsis* show that lateral root formation is promoted in soils low in available phosphorus or nitrogen, and that plants produce more, and longer, root hairs when iron or phosphorus are limited (Linkohr et al. 2002; López-Bucio et al. 2003; Giehl and von Wirén 2014). Root system architecture may be controlled either systemically by sensing of whole-plant nutritional status or locally by sensing of nutrient availability in the environment around the root (Giehl and von Wirén 2014). In fact, the combined influence of both systemic nutritional status and localized sensing mediate patterns of lateral root initiation and elongation, as plants forage for nutrients among soil patches heterogeneous in nutrient availability (Giehl and von Wirén 2014; Hu et al. 2014). Formation of symbioses with microbes can act as an additional modification to root system architecture, to produce increased nutrient uptake. For example, root hairs serve as sites of nodule formation in plant taxa that can be colonized by N-fixing bacteria in the genus *Rhizobium* (Libault et al. 2010). Likewise, mycorrhizal fungi infect roots of many woody species, where they extend the effective reach of the root system through networks of fungal hyphae in the soil. Infection of plant roots with mycorrhizae is also known to modify root system architecture through the promotion of lateral root and fine root development (Yao et al. 2009).

Another factor that controls root system architecture is exposure to mechanical strain, a stress that is particularly important in shaping the architecture of large woody plants. When intense forces are exerted on the root system as the tree canopy moves in the wind, the tree develops a large and rigid soil-root plate to avoid windthrow. In fact, trees exposed to wind and mechanical strain respond with a disproportionate rate of secondary thickening among structural roots relative to stems (Nicoll and Ray 1996). Carbon and nitrogen are allocated to the parts of the plant under the greatest strain, which allows for differential growth that matches the conditions of the environment

(Read and Stokes 2006). Consequently, a tree produces greater structural root mass on its leeward side than its windward side, presumably as a means to prevent bending or breaking of roots as stress from the movement of the canopy is transferred to the leeward side of the root plate (Nicoll and Ray 1996). In contrast, roots on the windward side, which are under tension with wind stress, produce greater branching to increase the holding capability of the roots in the soil (Read and Stokes 2006).

2.4 SHOOT GROWTH

Woody plants develop from small, morphologically simple embryos found in seeds into the staggering diversity of forms and sizes found throughout the world's biomes. The contrast in woody plant forms is represented well by subshrubs of dwarf willow (*Salix herbacea*), which reaches heights of 3–6 centimeters in subarctic regions of the northern hemisphere, and the coast redwood (*Sequoia sempervirens*), which may reach heights exceeding 100 meters where they grow in coastal portions of northern California and southern Oregon. Indeed, one of the most visually striking aspects of woody plant diversity is the variation in aboveground plant biomass caused by different rates and patterns of shoot growth. Such growth not only shapes the forms of individual plants but also ultimately determines the structure and function of entire plant communities through the impacts of plants on their environments and their neighbors.

2.4.1 Primary Growth of Shoots

The primary growth of woody plant shoots is generated by the activity of the SAM, a population of actively dividing cells that produce new stem tissues or lateral organs such as leaves and vegetative or floral buds (Murray et al. 2012). Both the rate of cell division within the meristem and the patterned differentiation of cells characteristic of primary growth are specified within the SAM by hormonal gradients and crosstalk that alter local patterns of gene expression within the meristem. For example, auxin is critical for the differentiation of leaf primordia from cells in the peripheral zone of the SAM. Developing leaf primordia act as sinks for auxin, causing localized depletion in auxin in the adjacent portions of the meristem. Consequently, whichever region in the peripheral zone is furthest from developing primordia, and therefore highest in auxin, becomes the next site of leaf primordium formation on the shoot (Fleming 2005). These auxin maxima and minima develop dynamically within the peripheral zone of the SAM during primary growth and establish the regular, patterned positioning of leaf primordia around the stem (Murray et al. 2012).

During primary growth, the SAM generates the basic "shoot module" comprising an internode and a node with a leaf and an axillary meristem. These shoot modules, termed phytomers, are produced successively during primary shoot growth and give the shoot its characteristic architecture, which may differ markedly among species (Sanchez et al. 2012). In this way, a basic rhythm of primordium formation determined by the SAM gives rise to a taxon-specific arrangement of leaves with axillary buds in their axils. The arrangement of leaves on a stem, called phyllotaxy, may be whorled with three or more leaves per node, opposite with two leaves per node, or alternate with one leaf per node. Axillary buds themselves are formed by the enclosure of the axillary meristems in new leaf primordia that have developed into bud scales. Depending on the developmental phase of the shoot, among other contributing factors, these axillary buds may be competent to develop into vegetative or reproductive lateral shoots, and may or may not be dormant upon their developmental completion.

2.4.2 Origins and Functions of Shoot Primary Meristems

Populations of cells within the SAM that give rise to plant organs may seem at first glance to be relatively unpatterned, but the meristem actually comprises distinct lineages of cells occupying different layers. Although the number of layers may vary by taxon, the typical dicot pattern of L1, L2, and L3 layers is a useful model. L1 cells contribute to the formation of the epidermis, and L2 cells contribute to the formation of sub-epidermal tissues. L3 cells generally contribute to the formation of vascular tissues, as well as the pith and cortex of the shoot (Meyerowitz 1997). Interestingly, the specification of different tissue identities from L1-, L2-, and L3-derived cells is not determined by the clonal lineage of the mother cells that gave rise to the tissues, but directly from positional information communicated cell to cell during tissue development. Therefore, if a cell from the L2 layer shifts into the L1 layer of the meristem, the developmental fate of its descendent cells will be modified to that of L1-derived tissues.

The epidermis is generated through the action of the SAM when cells in the L1 layer divide anticlinally to form a continuous sheet of cells around the stem or other plant organ (Javelle et al. 2010). Cells in the L1 layer receive hormonal and other chemical signals, which specify the maintenance and/or modification of epidermal cell fates. Within the epidermis, pavement cells are most abundant and constitute the bulk of a plant's surface area, typically developing a waxy cuticle by the accumulation of extracellular compounds that protect the plant from water loss and other abiotic and biotic stressors (Javelle et al. 2010). Other specialized epidermal cell types include trichomes, guard cells of leaf stomata, and lenticels.

Formation of the vascular tissues in the stem begins with the development of a continuous procambial strand among cells in the newly produced shoot tissues. This process of procambial development seems to be self-organizing and a consequence of polar auxin transport down the shoot generating continuous columns of cells high in auxin (Carlsbecker and Helariutta 2005). In fact, the localization of PIN auxin efflux proteins along the route of auxin transport in the shoot is an early sign of procambium initiation (Sanchez et al. 2012). Such procambial strands also vascularize developing primordial organs, which themselves produce a strong auxin signal transported to the stem. As the shoot tissue further develops, the procambial strands mature into vascular bundles with locations of primary xylem and phloem cells specified based on positional cues (Carlsbecker and Helariutta 2005). Within each vascular bundle, xylem cells are specified adaxially (toward the inside of the stem), and phloem cells are specified abaxially (toward the outside of the stem), with procambial cells between them. To date, several dozen genes have been identified that function in the activity and differentiation of vascular stem cells, with auxin, cytokinins, gibberellins, ethylene, jasmonates, and brassinosteroids, as well as various non-hormonal signaling molecules, implicated in a complex network that regulates vascular tissue development (Miyashima et al. 2013).

2.4.3 Secondary Growth of Shoots

Secondary growth, the radial thickening of plant organs (such as stems and roots), allows plants like giant sequoias (*Sequoiadendron giganteum*) to develop trunks of extraordinary diameter with age. In stems, secondary growth is accomplished when fascicular cambia (cambia within vascular bundles that are derived from the procambium) expand laterally to generate interfascicular cambia connecting the secondary meristematic regions of each vascular bundle (Sanchez et al. 2012). This meristematic population of stem cells, now forming a continuous ring of vascular cambium, produces xylem cells toward the inside of the shoot, and phloem cells toward the outside. Within the cambium, anticlinal divisions maintain the population of meristematic cells as the stem increases in circumference, and periclinal divisions generate new xylem and phloem cells. A continued supply of auxin from the shoot apex to cambial cells is necessary to maintain pleuriopotent stem cell populations available to differentiate into xylem and phloem (Sanchez et al. 2012).

Xylem tissues comprise the vast majority of secondary growth within a woody plant stem. For many species, the periodicity of cambial activity in response to favorable and unfavorable environmental cycles leads to the development of xylem in visually discernable rings. These rings are characterized by larger-diameter vessels produced in the spring contrasted with narrow vessels produced later in the season. In species for which the rings are visible, these rings are a record of the number of secondary growth cycles of a stem or limb.

In woody plant stems, xylem tissues provide not only a means of transporting water and dissolved solutes but also provide mechanical support for the shoot. This support allows a growing plant to bear the increasing weight of distal shoots and fruits as the tree accumulates primary and secondary growth in its organs. For some taxa, this means that the tiny, fragile stem of a seedling will radically transform itself through secondary growth into a structure capable of supporting tens to hundreds of metric tons of biomass! Ironically, given our interest in woody plants throughout this book, most of our understanding of the molecular aspects of secondary growth has been obtained in work with the tiny, herbaceous model species *Arabidopsis thaliana* (Barra-Jiménez and Ragni 2017).

Another form of secondary growth in woody plant shoots is the production of periderm by the cork cambium. This ring of meristematic cells does not originate during primary growth, but arises from a layer of cells in the cortex, or less frequently, the epidermis or phloem, after vascularization of the stem has been completed (Pereira 2011). When the cork cambium initially produces periderm (bark) toward the outside of the shoot, it displaces the epidermis to become the protective covering of the stem. From a molecular perspective, regulatory pathways in periderm formation are poorly understood. In one of the few investigations conducted on this topic, the activity of a transcription factor regulating periderm production in cork oak (*Quercus suber*) was shown to be influenced, at least in part, by abiotic stressors of heat and drought (Almeida et al. 2013). From an evolutionary perspective, the development of periderm plays important roles in plant adaptation to environmental conditions. For example, the development of thicker bark among some trees of savannahs and forests around the world probably represents adaptation to seasonally dry or fire-prone habitats, where thick bark limits moisture loss or damage by the heat of fire (Pellegrini et al. 2017).

2.4.4 Photomorphogenesis

Photomorphogenesis is the adaptive growth and development of plant shoots or other organs in response to the light environment. Central to photomorphogenesis are photoreceptors, which are proteins that change conformation in different light environments to alter gene expression and, consequently, growth and development. Several classes of photoreceptors have been described, including phytochromes, cryptochromes, phototropin, and UVR-8 (Briggs and Olney 2001; Huché-Thélier et al. 2016). Their regulation of biosynthesis and signaling of hormones, including auxin, gibberellins, cytokinin, brassinosteroids, and ethylene (Arsovski et al. 2012; Liscum et al. 2014), give these proteins a profound influence on plant growth and development in response to the light environment.

Phytochromes are a family of proteins that alter conformation in a reversible fashion between active and inactive forms in response to the ratio of red to far-red wavelengths of light detected (Briggs and Olney 2001). Such changes mediate wavelength-specific responses that, among other things, are responsible for the classic shade-avoidance of plants competing for light (Franklin and Whitelam 2005). Under light competition, phytochrome receives a high ratio of far-red to red light, as red light is disproportionately absorbed by neighboring plants or by leaves of a forest canopy, whereas a greater proportion of far-red light bounces off or passes through leaves (Weinig 2000; Franklin and Whitelam 2005). The red-light absorbing form, Pr, is the inactive form of the protein produced when far-red wavelengths dominate. When red wavelengths dominate, Pr converts to Pfr, the biologically active and far-red-light absorbing form. Through this mechanism, a plant is superbly capable of sensing its light environment and responding adaptively through downstream signaling pathways and transcriptional modification mediated by active and inactive forms of phytochrome and other photoreceptors (Franklin and Whitelam 2005; Briggs and Olney 2001). Considering the classic shade-avoidance syndrome, the inactivation of phytochrome by high far-red to red light causes plants to develop adaptive responses such as longer internodes and hyponastic leaves, until shoots encounter a more suitable light environment. Phytochrome has also been implicated in the regulation of resource allocation, metabolism, biomass accumulation, and several stress responses, indicating that the roles of phytochrome are diverse and probably not yet fully understood (Yang et al. 2016).

Cryptochromes are photoreceptors that detect blue light and UVA, and have important roles in the regulation of shoot and internode length, branching, and many other aspects of plant development (Huché-Thélier et al. 2016). Cryptochromes are known to play a role in the shade-avoidance response by contributing essential information about the quantity of light, complementing the red:far-red information obtained by phytochrome (Franklin and Whitelam 2005). Phototropin, another blue-light sensitive photoreceptor, is crucial to phototropic responses of shoots (e.g. hypocotyls of seedlings or shoots produced by the SAM) to light direction. Through the impact of conformational changes in phototropin on the accumulation and transport of auxin within the shoot, a shoot exposed to directional blue light develops an auxin gradient that leads to the elongation of cells on the side of the shoot opposite the origin of light (Liscum et al. 2014). Clearly, the varied forms of shoot growth and canopy architecture manifested by a single genotype when exposed to different environments are the products of interactions among several photoreceptors receiving and integrating information from the light environment.

2.4.5 Shoot Resistance to Mechanical Stresses

Stems and branches must not only support the weight of distal organs against gravity, but also tolerate the various mechanical stresses to which they are subjected. In large woody plants, these include dynamic stresses such as those imposed by exposure of the shoot system to wind, snow loads, and negative pressures within the xylem (Pitterman et al. 2006; Read and Stokes 2006). The bulk of woody plant trunks and branches comprise lignified cells of the secondary xylem that provide both the rigidity and sufficient diameter necessary to support the canopy. In the presence of mechanical stresses like wind loading, radial growth of the stem is increased and the rate of stem elongation is decreased, a phenomenon known as thigmomorphogenesis (Nicoll and Ray 1996). At the bases of the trunk and major limbs, where mechanical stresses are the greatest, secondary thickening produces pronounced stem tapering that provides additional structural stability against stress imposed by wind (Read and Stokes 2006). Crown growth is also influenced by gravity, with specialized reaction wood produced in branches and leaning trunks to correct stems that are not in a vertical orientation. In angiosperms, cellulose-rich tension wood forms on the upper side of the branch or trunk, enabling new stem tissues to curve up in the direction of tension (Plomion et al. 2001). In gymnosperms, very dense and lignin-rich compression wood forms on the lower side of the branch or trunk, enabling the stem to curve up and away from the side under compression.

Wood density within the trunk and branches may be less related to the need for mechanical support of the shoots against gravity and more a function of the mechanical resistance necessary to withstand the negative pressures in the xylem (Read and Stokes 2006). Xylem conduits must resist implosion when faced with the negative pressure of the water column, which may span tens of meters against the force of gravity, and the water column must avoid cavitation and air embolisms (Hacke et al. 2001; Pitterman et al. 2006; Lens et al. 2011). The tendency for a woody plant to suffer from such problems is presumed to be a function of the anatomy of tracheids and/or vessel elements, including their lumen diameters and cell wall thicknesses. Species adapted to drier conditions are expected to have narrower xylem conduits with thicker cell walls, presumably to avoid cavitation and embolism. A general tradeoff between mechanical reinforcement of vessel conduits and hydraulic conductivity or efficiency suggests that the wide range in wood density and specific xylem characteristics among woody plants represents adaptation to environments that differ in water availability (Pitterman et al. 2006; Hacke et al. 2001).

2.4.6 Bud Outgrowth and Lateral Shoot Growth Determine Canopy Architecture

The unique architecture of a plant canopy is the product of the radial arrangement and longitudinal spacing of axillary buds produced during primary growth, the spatial and temporal distribution of outgrowth from these buds, and the length and girth of lateral shoots that develop from them. Two distinct mechanisms, apical dominance and apical control, modify canopy architecture by regulating initiation and growth of axillary shoots, respectively. Apical dominance is the suppression of axillary bud outgrowth by regulatory signals originating in the SAM which determines whether or not a bud begins to develop into a shoot (Dun et al. 2006). Apical control, in contrast, describes the suppression of primary (and secondary) growth on an existing lateral shoot by a shoot or shoots located more distally (Cline and Harrington 2007). The strength of, and interactions between, these two influences explain much of the variation in branching patterns and relationships of branch size found among taxa.

Two general patterns of crown architecture are well represented by the narrow and upright canopy of Norfolk Island pine (*Auraucaria heterophylla*), which has an excurrent form, and the wide and spreading canopy of white oak (*Quercus alba*), which has a decurrent form. There has been some confusion regarding the roles of apical dominance and apical control in producing these patterns. Although many have attributed an excurrent growth habit to the influence of strong apical dominance and a decurrent growth habit to comparatively weak apical dominance, excurrent and decurrent canopy forms are actually the product of strong vs. weak apical control, respectively. When apical control is strong, distal branches suppress the growth of more proximal branches to produce a pyramidal canopy with a straight central stem. Conversely, in taxa with weak apical control, growth of lateral shoots is less strongly inhibited, which results in a spreading canopy that develops multiple codominant leaders. Another crucial aspect of apical control is the determination of branch angle. In plants with strong apical control, lateral branch angles are nearly 90° from vertical. In plants with weaker apical control, lateral branches typically develop on an angle closer to vertical. Interestingly, lateral branches of most, but not all, plants with strong apical control will respond to decapitation of the primary shoot by growing more upright over time (Wilson 2000). In contrast to apical control, the degree of apical dominance varies widely among taxa regardless of overall canopy architecture. For example, many woody plants that produce excurrent growth also exhibit rather prolific development of lateral shoots from axillary buds, and many woody plants that produce decurrent growth are comparatively resistant to lateral bud outgrowth (Brown et al. 1967).

Canopy architecture and lateral bud outgrowth, as it varies with cultivar, is well studied in apple. Lauri and Lespinasse (1993) and Lauri et al. (1995) classified four types of growth habits that vary in the degree of inhibition of lateral bud outgrowth. When lateral bud outgrowth is strongly inhibited, the resulting tree shape is columnar and the tree forms one primary vertical branch with very short lateral shoots called spurs. A pyramidal tree canopy arises from strong, erect branches that have a predominance of lateral spurs and some weak shoots. Trees with a globose-shaped canopy develop in genotypes that have stronger lateral shoots and proportionately fewer spurs.

2.4.7 Hormonal Basis for Apical Dominance and Apical Control

In the classical hypothesis for apical dominance, auxin originating from the shoot tip travels down the stem by polar transport and suppresses lateral bud outgrowth by inhibiting cytokinin biosynthesis by stem tissues. As long as the SAM remains intact, endogenous cytokinin will be insufficient to induce axillary bud outgrowth (Dun et al. 2006). Evidence for this hypothesis comes from a variety of sources, including decapitation studies in which removal of the shoot tip increases cytokinin biosynthesis and the outgrowth of lateral buds but removal of the shoot tip followed by exogenous application of auxin maintains apical dominance (Wilson 2000; Dun et al. 2006). Finally, the application of cytokinin to lateral buds may induce outgrowth of the buds, confirming the causal role of cytokinin in the process (Cline and Harrington 2007). Several alternative hypotheses have been advanced to account for observed responses of plants to auxin and cytokinin, plus potential roles for other signaling molecules and the dynamics of sugar transport (Shimizu-Sato and Mori 2001; Horvath et al. 2003; Dun et al. 2006; Mason et al. 2014; Barbier et al. 2015). It is also unclear what role, if any, cytokinin signals originating from the roots play in apical dominance (Müller and Leyser 2011). In short, the physiology of apical dominance is not settled.

The physiology responsible for apical control is even less settled than that of apical dominance. Auxin seems to play a role, as applying auxin to a decapitated primary shoot may be sufficient to suppress both primary and secondary growth of lateral shoots (Wilson 2000), although not all species respond in this way (Cline et al. 2009). It is unclear precisely how auxin is involved in apical control, and the hormones cytokinin, gibberellin, and ethylene have also been implicated. Other hypotheses pertain to differential transport of water and mineral nutrients, rates of photosynthesis, allocation of assimilates, and a role for gravimorphism (Wilson 2000; Cline et al. 2009).

2.4.8 Types of Bud Dormancy and Their Control

Bud dormancy is a key adaptive trait in the evolution of perennial woody plants. As a necessary consequence of seasonality and environmental limits on plant growth, SAMs of woody plants do not grow unchecked through all seasons. The cessation of shoot primary growth and the formation of buds enclosing the shoot meristems are mechanisms for protection from stressors such as desiccation or freezing damage. By regulating the dormancy status of buds, a plant can fine-tune its response to environmental conditions to favor survival and productivity.

Three broad categories of dormancy have been described, including endodormancy imposed by factors within the bud, paradormancy imposed by signals received by the bud from elsewhere in the plant, and ecodormancy imposed by external factors in the environment (Lang 1987; Horvath et al. 2003). Endodormancy can be released only if conditions within the bud are met, such as the accumulation of sufficient chilling hours. A release from paradormancy requires physiological changes within the plant, such as the removal of apical dominance. Ecodormancy is released readily if environmental conditions favor plant growth.

2.4.9 Endodormancy and Cold-Acclimation Are Distinct but Overlapping Processes

Of particular interest in woody plants is the endodormancy that prepares them for seasonal changes in environmental conditions. For example, many woody plants of temperate and boreal regions enter a state of endodormancy in the fall induced by the short-day photoperiod preceding the onset of the winter season (Olsen 2010). In some taxa, such as many members of the Rosaceae, it is temperature alone, and not photoperiod, that induces the onset of dormancy (Heide and Prestrud 2005). In this state of endodormancy, a bud will not resume growth, even if environmental conditions (e.g. temperature) could sustain growth. Instead, dormancy is released only when an internal inhibitor, such as the requirement for sufficient chill-units sensed by the bud, is satisfied (Arora et al. 2003). Such a mechanism prevents endodormant buds from initiating outgrowth during transient periods of favorable weather within a season otherwise hostile to plant growth, such as might be encountered with fluctuating temperatures in winter (Horvath et al. 2003; Kalberer et al. 2006; Arora and Taulavuori 2016).

For many taxa of the temperate zone, seasonal endodormancy is accompanied by cold acclimation, the process by which exposure to short photoperiods and/or decreasing temperatures in the fall cues a plant to prepare for the onset of seasonally low temperatures and cellular desiccation (Kalberer et al. 2006). For example, within several hours after exposure of *Arabidopsis* to low temperatures, radical modifications to the transcriptome are apparent, with the CBF family of transcription factors altering the expression of numerous cold-responsive genes (Thomashow 2010). However, regulatory mechanisms independent of CBF are also involved in plant responses to low temperature (Penfield 2008). Interestingly, light sensing by phytochrome seems to play a role during the induction of cold-acclimation, with both photoperiod and quality influencing expression of cold-responsive genes (Kim et al. 2002; Franklin and Whitelam 2007). The crosstalk between light and temperature during the induction of cold-acclimation, and their integration by responsive tissues, is not well characterized (Catalá et al. 2011). In all, the literature related to cold-acclimation is vast, with various mechanisms implicated in the induction and maintenance of cold acclimation. Among these are the synthesis, transport, and accumulation of hormones (e.g. ABA), proteins such as dehydrins that protect cells from freezing injury and dehydration stress, and cryoprotectant sugars, sugar alcohols, and nitrogenous compounds of low molecular weight (Janská et al. 2010; Thomashow 2010).

Although endodormancy and cold acclimation are distinct phenomena, the temporal overlap between the two has made it difficult for researchers to tease apart factors crucial to dormancy from those factors crucial to the cold acclimation process (Arora et al. 2003). Arora et al. (2003) review the progress made in disentangling the molecular and physiological mechanisms for each of these processes. What is clear is that low temperatures and photoperiod can act to induce both endodormancy and cold tolerance within plants. However, it is also clear that cold acclimation can persist even after buds transition from endodormancy to ecodormancy following the perception of sufficient chill-units over the winter. In such plants, low temperatures alone maintain cold acclimation and prevent bud outgrowth. Buds of ecodormant, cold-acclimated plants can begin growing soon after temperatures increase, a process that is accompanied by rapid de-acclimation to cold (Arora and Taulavuori 2016). The survival of plants in cold climates depends, therefore, on the timescales and magnitudes of temperature fluctuations and the capacity of plants, once released from endodormancy, to rapidly re-acclimate to cold upon re-exposure to low temperatures (Kalberer et al. 2006; Arora and Taulavuori 2016).

2.4.10 Ecodormancy Is Imposed by the Environment

Ecodormancy of buds may be induced in woody plants by exposure to environmental conditions such as water stress, nutrient deprivation, and high temperatures. Unlike the endodormancy induced by low temperatures and/or short photoperiods, ecodormant buds can resume outgrowth as soon as conditions become favorable, because such buds are highly responsive to the external factors that restrict bud growth. For example, many growers of woody plants for horticulture or forestry have witnessed ecodormancy caused by nutrient stress when crops set terminal buds early in the season due to limited nutrient availability in their substrates. Fortunately, fertilizer application releases plants from this dormancy and promotes the resumption of growth. Likewise, seasonal dormancy of plants in regions with a distinct dry season is a form of ecodormancy that is rapidly overcome when soil moisture increases to a level that favors plant growth. Although the molecular basis for ecodormancy under various environmental conditions is not well understood, ABA accumulation under drought conditions might impose dormancy by preventing cell division within the meristem (Horvath et al. 2003). It is clear that more research is needed to understand the regulatory mechanisms that underlie ecodormancy.

2.5 LEAF GROWTH

2.5.1 Leaf Structure and Development

In evolutionary time, the development of leaves is a recent occurrence in vascular land plants. Early land plants consisted of photosynthetic, branching stems with a relatively low density of stomata in their epidermal layer (Steemans et al. 2009; Kidner and Timmermans 2010). To maintain rates of carbon fixation as CO_2 levels dropped during the late Paleozoic, multiple distinct lineages evolved flat lamina with high densities of stomata. Leaves as we know them today are extensively diverse and evolutionarily successful in a wide range of habitats so that one can scarcely imagine a world without them.

The development of a leaf begins with the initiation of leaf-specific gene expression in cells of the peripheral zone of the meristem. During development of a leaf primordium, these populations of cells take on identities corresponding to the abaxial/adaxial polarity typical of a leaf. Signals from the meristem specify the development of an "adaxial" cell population proximal to the SAM, whereas an "abaxial" population develops on the side away from the SAM (Kidner and Timmermans 2010). At this stage, patterns of gene expression commit these cell populations to their distinct developmental fates, and the boundary between adaxial and abaxial populations determines the location of laminar outgrowth to generate the leaf blade. Intercellular signaling polarizes leaf primordia and maintains polarity in developing leaves that are beyond the influence of the SAM.

Once initiated, a leaf undergoes both primary and secondary morphogenesis. During primary morphogenesis, the basic structure of the leaf develops, including the initiation of the lamina and specification of its anatomy and morphology. During secondary morphogenesis, the leaf expands to attain its mature size, and cells complete their differentiation to produce structures of the mature leaf (Bar and Ori 2014). A strand of high auxin concentration is established between the developing leaf primordium and the provascular strand of the developing shoot, forming a vascular connection between the shoot and the leaf (Nelson and Dengler 1997).

During leaf development, meristematic cells located at the vicinity of the junction between the leaf blade and the petiole are responsible for the growth of a leaf by cell proliferation. Cells produced to the distal side of the meristemic region develop into tissues of the leaf blade, whereas cells produced to the basal side of the meristematic region contribute to the petiole (Ichihashi and Tsukaya 2015). The process continues as the leaf undergoes development, with leaf maturation beginning with the arrest of the cell cycle among the oldest tissues, i.e. the cells furthest from the meristematic region (Kümpers et al. 2017). Eventually, cells within the proliferative zone of the leaf cease their meristematic activity, as the leaf approaches the end of its determinate growth (Ichihashi and Tsukaya 2015). Following arrest of the cell cycle, final differentiation is accompanied by modifications to the transcriptome, until cell fate (size, shape, and function) is fully realized (Efroni et al. 2010).

Asymmetric tissue differentiation is a hallmark of a developed leaf. For example, the upper and lower epidermises differ in the shape of pavement cells, as well as the number of guard cell and trichomes (Fukushima and Hasebe 2013). Internally, the layer of cells under the epidermis on the adaxial side of the blade develops into palisade mesophyll, where chloroplasts are well-positioned to effectively harvest light energy. Spongy mesophyll, in contrast, forms within the abaxial portion of the leaf, where the bulk of the stomata are located. Finally, the vascular bundles in leaf veins are polarized, with xylem to the adaxial side and phloem to the abaxial side (Fukushima and Hasebe 2013). This patterning is the result of complex signaling and developmental pathways that provide positional information to cells that determines their fates during organogenesis. Work with mutant plants shows that interruptions of such signals may result, for example, in a fully abaxialized, cylindrical leaf with radial symmetry (Fukushima and Hasebe 2013).

Hormones play a primary role in leaf primordium initiation, morphogenesis, differentiation and maturation. For example, auxin signaling is a positive regulator of leaf primordium initiation, as evidenced by the role of auxin maxima in the specification of incipient leaf primordia. It seems that cytokinin also plays a role in the process, but its specific roles are complex and might be context-dependent (Bar and Ori 2014). However, it is clear that fine-tuning of auxin:cytokinin in the vicinity of the SAM is important for both the maintenance of the meristem and the initiation of leaf primordia. Finally, the hormones GA and cytokinin are known to play roles in cell proliferation, expansion, and differentiation associated with leaf morphogenesis and maturation. High concentrations of GA, for example, can induce premature arrest of cell division and leaf maturation. In contrast, cytokinin promotes cell division, delaying the onset of final differentiation and leaf maturation (Bar and Ori 2014).

2.5.2 Leaf Development Responds to the Light Environment

During development and growth of leaves, the light environment influences final leaf morphology and anatomy. Cytokinin, which is translocated to the shoot through the xylem from its primary site of synthesis in the root, has emerged as a central regulator of leaf developmental responses to light. The rate of transpiration through leaves developing in the sun is greater than that of leaves in the shade, leading to differential import of cytokinin into developing leaves based on their light environment (Boonman et al. 2007). Consequently, substantial variation in cytokinin delivery to developing leaves can occur. Cytokinin-responsive pathways promote a greater leaf mass and photosynthetic capacity per unit area among leaves developing in the sun, presumably to better capitalize on the available light at that location in the canopy (Boonman et al. 2007). Such leaves may be characterized by more layers of palisade parenchyma, sun-adapted chloroplasts well-suited to rapid photosynthesis, and an increase in stomatal density on the abaxial surface (Kim et al. 2005; Sarijeva et al. 2007). It is understandable that shade leaves might be reduced in their development, as the maintenance of overbuilt and underutilized photosynthetic tissues could be costly. Although the development of leaves in sun and shade seems to depend heavily on differential transport of cytokinin based on local light environments, it remains unclear what roles other signaling mechanisms might play (Kim et al. 2005). For example, work with various photoreceptor mutants shows that normal photoreceptor function is not necessary to produce sun and shade leaves in response to the light environment (Terashima et al. 2005).

2.5.3 Leaf Senescence and Abscission

Senescence and abscission of leaves or reproductive structures is an important aspect of the life cycle of plants. During senescence, cellular contents of the leaf are dismantled for the orderly resorption of nutrients (e.g. N, P, and K) which can be used for construction of new shoots and leaves the following year (Marchin et al. 2010; February and Higgins 2016). A complex transcriptional network, for which thousands of senescence-associated genes have been identified (Guo 2013), reprograms gene expression to make the leaf competent to senesce and then initiate the process of senescence itself. Ethylene, ABA, salicylic acid, jasmonic acid, brassinosteroids, cytokinins, auxin, sugars, and other signals have been implicated in the processes of leaf aging and senescence (Guo 2013; Schippers 2015). With sufficient age-related changes in the leaf, senescence can be induced by environmental signals such as drought or nutrient stress, or may occur over the course of time as an age-related developmental process. Chlorophyll, proteins, nucleic acids, and lipids are degraded, and their nutrients resorbed into the shoot (Guo 2013). Evidence indicates that mitochondria, in contrast to chloroplasts, maintain their function until late in the process of leaf senescence to provide energy needed for the catabolism of cellular contents and resorption of nutrients.

Anyone who has witnessed the yellow, red, and orange autumnal leaf coloration of deciduous trees in some temperate forests can appreciate the phenomenon of senescence. Interestingly, the brilliant colors that may be associated with leaf senescence by trees of northern temperate forests caused by the accumulation of anthocyanins in the palisade parenchyma and/or the unmasking of carotenoids by the degradation of chlorophyll, is widely noted but poorly understood. Why should some species synthesize and accumulate anthocyanins when others do not? Many roles have been hypothesized for the function of anthocyanins including phytoprotection, more efficient chlorophyll breakdown and nitrogen resorption, and increased leaf temperatures among them. Clear evidence supporting each of these hypotheses is scarce, although the nitrogen resorption hypothesis seems best supported among them (Lee et al. 2003). It is also possible that autumnal anthocyanin production in some taxa is simply a peculiar, but beautiful, byproduct of senescence.

Abscission, the final step of senescence, is the process by which an organ separates from the rest of the plant in an organized manner. The zone of abscission develops during plant growth but will not initiate separation of the organ until a signal for abscission is perceived in the zone (Nakano et al. 2013; Ma et al. 2015). Cells in this zone are small and square-shaped, making them morphologically distinct from neighboring cells (Taylor and Whitelaw 2001). In leaves, this abscission zone is located at the base of the petiole and allows the shedding of the entire leaf structure when senescence is complete. Abscission is a four-stage process starting with the development of the abscission zone, followed by the acquired competence of cells in this zone to respond to abscission signals, the activation of abscission, and the development of a protective layer on the newly exposed surface where the organ was attached (Nakano et al. 2013). Genes functioning in hormone signaling, cell wall degradation, and defense are upregulated during abscission. Hydrolytic enzymes, including pectinases and cellulases, degrade the cell walls adhering adjacent cells within the abscission zone, permitting the detachment of the organ distal to the zone (Taylor and Whitelaw 2001).

The importance of phytohormones to abscission has long been recognized. Within the abscission zone, abscission is inhibited by auxin and promoted by ethylene (Nakano et al. 2013). Abscission will not occur when there is a steady stream of auxin transported through the abscission zone from the organ in question; it is only when the stream of auxin is interrupted that cells in the abscission zone become particularly sensitive to ethylene and abscission commences (Ma et al. 2015). Among the developmental and environmental signals that may induce abscission of a leaf are senescence, photoperiod, water stress, wounding or pathogen attack, and chemical pollution (Taylor and Whitelaw 2001). Under these conditions, the supply of ethylene from the leaf, along with a decreased supply of auxin and heightened cell sensitivity to ethylene, promote the process of abscission within the abscission zone. Genes involved in stress responses and ethylene signaling are upregulated, a process that can be triggered experimentally by exposing plants to ethylene or by imposing water stress (Liao et al. 2016). The interactions between auxin and ethylene within this process are complex, as high concentrations of auxin can promote the synthesis of ethylene, and ethylene both inhibits auxin transport and increases cell sensitivity to ethylene itself within the abscission zone (Taylor and Whitelaw 2001). ABA, once believed to be the primary phytohormone promoting abscission, is essential to a variety of stress responses within plants but probably only indirectly related to abscission through its stimulation of ethylene production (Dörffling 2015).

2.5.4 Deciduous and Evergreen Leaves Represent Different Strategies

Leaf longevity, which varies among taxa from several months to years, has long fascinated biologists and ecologists, with questions about why deciduous or evergreen trees might sometimes dominate different habitats and often coexist in a single community. Decades of work on this question point to several factors that might explain it. Evergreen leaves benefit from a longer season of photosynthetic activity, lower leaf construction costs per unit time of longevity, and a durable lamina that resists damage by environmental stressors and herbivores (Givnish 2002). In contrast, deciduous taxa typically have higher rates of photosynthesis relative to leaf mass, as well as low dormant-season costs in terms of both respiration and root activity required to replace water loss. Whereas evergreen leaves have more mechanical tissues and a low leaf specific area that result in internal shading of photosystems and within-leaf competition for CO_2, deciduous leaves are characterized by greater leaf-specific area corresponding to increased capture of light energy and stomatal conductance per unit time (Givnish 2002; Warren and Adams 2004). Deciduous leaves can be seen as opportunistic relative to the more conservative strategy of evergreen leaves; when conditions are favorable, deciduous plants invest heavily in establishing high photosynthetic capacity but with irreversible loss of function as conditions become seasonally unfavorable for such leaves (van Ommen Kloeke et al. 2012).

Leaves with the highest maximum rates of photosynthesis per unit leaf mass are nutrient-rich and short-lived, whereas leaves with conservative nutrient use and multi-year persistence have lower maximum rates of photosynthesis (Givnish 2002; van Ommen Kloeke et al. 2012). Given the tradeoffs in these factors, investigators have long pondered under what combinations of climate and edaphic conditions each strategy might be most beneficial. In environments where nutrient limitation places a substantial burden on plant productivity, a sacrifice in maximum carbon fixation per unit time may be offset by the capacity to retain, over multiple years, functional leaves into which an investment of many carbohydrates and nutrients was made. Evergreens also have an extended seasonal window of photosynthetic activity during any period in which temperature and moisture availability permit carbon fixation, even in seasons frequently characterized by conditions not conducive to photosynthesis (van Ommen Kloeke et al. 2012). In environments where there is greater access to mineral nutrients, deciduous taxa might be expected to dominate. However, the adaptive relationship between leaf habit and nutrient availability remains controversial as deciduous taxa like larch (*Larix* spp.) dominate some nutrient-poor plant communities (Givnish 2002).

Traits associated with leaf longevity may also have a bearing on seasonal drought adaptiveness among woody plants. Deciduous angiosperms, derived from evergreen ancestors, first appeared in the tropics, likely as an adaptation to seasonal drought during diversification into drier habitats bordering tropical rainforests (Axelrod 1966). Where drought is severe, taxa with deciduous leaves are well-suited because leaves are shed at the end of the moist season, leaving the plant with a greatly decreased water loss rate at the onset of the dry season (Givnish 2002; Marchin et al. 2010). Only later did such taxa invade cooler regions of the world, already adapted to a seasonal dormant period. Despite this, a deciduous strategy clearly is not a prerequisite to occupy seasonally cold or dry regions, as deciduous and evergreen plants alike have been incredibly successful at colonizing such habitats around the world (Givnish 2002). In fact, many regions that freeze seasonally are dominated by evergreens, despite the decrease in water availability in frozen soil. Taxonomically, the seasonal timing and extent of leaf abscission varies such that even congeneric species may differ in their evergreen, winter-deciduous, or summer-deciduous nature. Among oaks, for example, winter-deciduous species dominate in cold-temperate forests, whereas evergreen and summer-deciduous forms may be found frequently in ecosystems with a distinct period of summer drought, such as that of a Mediterranean climate (Terradas 1999).

2.6 FLOWER AND FRUIT DEVELOPMENT

2.6.1 Processes of Floral Initiation and Development

Following germination, a plant in the vegetative phase of its life cycle produces roots and shoots, vegetative structures gaining in size and complexity before a transition to floral initiation. This vegetative phase may last several weeks, as in the case of some accessions of *Arabidopsis*, or for years, in the case of many woody perennials (Huijser and Schmid 2011). Reproduction can occur only after the integration of various endogenous and environmental signals at the whole-plant level that prompt a phase transition from juvenile to mature, at which point SAMs are competent to produce flowers. Under promotive conditions, SAMs of plants that have made this transition may develop into inflorescence meristems that subsequently become floral meristems or produce lateral floral meristems or new inflorescence meristems as part of their developmental program (Huijser and Schmid 2011). Five major pathways collectively integrate internal and external factors to decide when to initiate flower development, including a photoperiod pathway, a vernalization pathway, a gibberellin pathway, an age pathway, and an autonomous pathway (Srikanth and Schmid 2011; Teotia and Tang 2015). Levels of sugars within the plant, for which the shoot apex is a strong sink, also may promote the initiation of flowering in conjunction with signaling through these pathways.

The photoperiod pathway in *Arabidopsis* relies on the perception of day length by phytochromes and cryptochrome that ultimately regulate the production of a systemic signal (i.e. a 'florigen') to induce floral development at the SAM (Wilkie et al. 2008; Srikanth and Schmid 2011). The vernalization pathway induces flowering when a chilling requirement, which varies in duration among taxa and among accessions within taxa, is met during the cool season. Through the gibberellin pathway, GA upregulates the expression of genes linked to timing of flowering and floral identity, with significant interactions with both the photoperiod and age pathways (Wilkie et al. 2008; Teotia and Tang 2015). At the center of the age pathway are micro-RNAs miR156 and miR172, which govern the development of reproductive competence of SAMs. The expression of miR156, which promotes juvenility, is abundant in young plants and decreases over time or with other developmental cues like ambient temperature or accumulation of sugars (Teotia and Tang 2015). In contrast, the expression of miR172, which is low in young plants and increases over time, promotes the transition to sexual maturity (Hyun et al. 2017). The autonomous pathway refers to any endogenous signaling that acts alone or in concert with any of the four other pathways to promote flowering (Wilkie et al. 2008; Srikanth and Schmid 2011).

Although there are multiple pathways that time the initiation of flowering, these pathways are not independent (reviewed by Srikanth and Schmid 2011). Extensive crosstalk between the pathways involves a set of common "integrator genes" that activate flower development genes. For example, the FLOWERING LOCUS T (*FT*) gene integrates signals from both the photoperiod and vernalization pathways; the *LEAFY* (*LFY*) gene integrates signals from the photoperiod, GA, and age pathways; and the *SUPPRESSOR OF OVEREXPRESSION OF CO 1* (*SOC1*) gene integrates signals from the photoperiod, vernalization and GA pathways (Srikanth and Schmid 2011). A remaining question is the identity of the florigen, the long-distance signal that integrates information from multiple flowering pathways to induce flowering. In *Arabidopsis*, the protein FT fulfills the role of florigen (Abe et al. 2005), although GA has been implicated as a florigen in some taxa, and it is not clear how many different signaling molecules might function in this role.

2.6.1.1 Mechanisms of Floral Development

The initiation of a floral primordium from the inflorescence meristem requires a local auxin maximum in the L1 layer of the meristem, similar to that needed for the formation of a leaf primordium from a SAM. During the early development of the floral primordium, however, PIN proteins in several cells of the L1 layer orient at the basal end of each cell, redirecting auxin to the center of the primordium, where it accumulates (Vaddepalli et al. 2015). Transcription factors essential to floral initiation and identity (e.g. LFY, APETALA1 (AP1), and their homologues) upregulate the expression of many genes essential to floral development (Liu et al. 2009). A positive feedback between *LFY* and auxin signaling commits the primordium to the floral fate (Vaddepalli et al. 2015). The phytohormone GA promotes the initiation of floral primordia by activating *LFY*, which in turn suppresses GA signaling and activates the expression of *AP1*, a gene regulating meristem identity (Benlloch et al. 2007; Vaddepalli et al. 2015). *AP1* and other genes work synergistically with *LFY* to regulate the development of an organized floral meristem.

Given that most of our knowledge of floral initiation comes from *Arabidopsis*, it remains to be seen to what degree such models can be generalized to other plants. Major similarities are evident, as citrus seedlings transformed to constitutively express either *LFY* or *AP1* genes of *Arabidopsis* may produce fertile flowers in less than one year, dramatically shortening the typical juvenile period of six or more years (Peña et al. 2001). Unlike *Arabidopsis*, woody perennials that have become reproductive must maintain a sufficient number of vegetative meristems for future shoot growth (Tan and Swain 2006).

Among many woody plants, floral initiation takes place in the year prior to anthesis, after terminal bud set occurs. The current hypothesis in apple is that the SAM must produce a minimum number of vegetative primordia (e.g. budscales, leaves, or bracts) before it can commit to the floral identity (Fulford 1966). In trees that bear fruits in alternate years, this process is inhibited by the presence of developing seeds, either through the export of endogenous GAs from the seeds or by competition between seeds and the SAM for hormones that induce floral initiation (Dennis and Neilsen 1999). The rate of primordium formation in the bud is reduced within plants developing fruits, and consequently the minimum bud development necessary for floral initiation does not occur, and the SAM remains vegetative (McLaughlin and Greene 1991). Although many taxa bear fruits in alternate years, others produce flowers and fruits reliably each season.

Flower development in angiosperms proceeds through the formation of concentric whorls of extensively modified leaves (Causier et al. 2010). In a perfect flower, floral organ primordia give rise to the sepals and petals of the perianth, as well as the fertile stamens and carpels of the androecium and gynoecium, respectively. The now-classic ABC model of floral development (Bowman et al. 1991), and a refined ABCE model, describe the role of several classes of floral specification genes that encode MADS-box proteins that are transcription factors that regulate the specification of whorls during flower development. According to the ABC model, A-class genes are expressed in the sepal and petal whorls, B-class genes in the petal and stamen whorls, and C-class genes in the stamen and carpel whorls (Causier et al. 2010). Within each whorl, the unique expression of one of these genes, or a combination of genes, specifies the developmental program for that whorl (Bowman et al. 1991). The model has subsequently been modified to include E-class genes that are expressed in all four whorls and establish the floral context in which the A-, B-, and C-class genes operate. The ABC model has proved invaluable to the determination of genetic mechanisms involved in flower development, despite some challenges associated with it (Causier et al. 2010). In gymnosperms, B-class orthologs are expressed during the development of male and female cones, whereas C-class orthologs are expressed in male cones alone (Bowman et al. 2012). Clearly, the developmental programs for reproductive structures of angiosperms and gymnosperms represent variations on a common developmental theme, despite their differences in reproductive morphology.

2.6.1.2 Sexual Reproduction

For a diploid species, the sporophyte phase of the plant life cycle is diploid (2n) and the gametophyte phase is haploid (1n). Multicellular gametophytes of seed plants are embedded within the reproductive structures of the sporophyte, such as in the ovule and anther of an angiosperm flower or on the scales of a gymnosperm cone. The diploid megaspore mother cell within an ovule undergoes meiosis to produce four haploid megaspores, three of which degenerate and one of which develops into a multi-nucleate megagametophyte called an embryo sac (Liu and Qu 2008). The embryo sac contains eight nuclei, including the egg cell, a central cell with two polar nuclei, three antipodals, and two synergids, the identities of which are determined by a gradient in the concentration of auxin received from the maternal plant (Sundaresan and Alandete-Saez 2010). Within the anther, a microspore mother cell undergoes meiosis to produce four haploid microspores, and each develops by mitosis into a pollen grain containing a tube nucleus and one or more generative nuclei (sperm cells).

Meiosis is a form of cell division responsible for producing genetically variable 1n gametes from a 2n parent, accomplished by one round of chromosomal replication and two rounds of nuclear division (Alberts et al. 2002). During meiosis, each chromosome of the megaspore cell duplicates its genetic material to produce a doubled chromosome with two identical arms (chromatids). During the process of crossing over, homologous chromosomes pair up while in this doubled state and swap portions of DNA between their adjacent chromatids to generate many genetic possibilities for resulting gametes (Alberts et al. 2002). Another source of genetic variation among gametes is the independent assortment of homologous chromosomes to each side of the cell before cytokinesis, such that the assignment of maternal and paternal chromosomes inherited by each daughter cell is random. By these two mechanisms, each haploid daughter cell (gamete) resulting from meiosis inherits a unique combination of maternal and paternal chromosomes and the alleles (distinct gene copies) contained on them from the parent plant (Alberts et al. 2002).

Sexual reproduction in angiosperms requires the transfer of pollen from the anther of a flower to the stigma of a compatible flower that may be on the same plant if the species is sexually self-compatible or on a genetically distinct individual if the species is sexually self-incompatible. When pollen is deposited on a stigma, the tube nucleus of the pollen grain begins to grow into a pollen tube that travels down the style until it reaches the egg cell within the ovule. The generative nuclei from the pollen grain travel down the cytoplasm of the pollen tube, with one nucleus fertilizing the egg to produce a diploid zygote and another fusing with the two polar nuclei of the central cell to produce a triploid endosperm. This fertilization of two cells within the megagametophyte by generative nuclei is termed double fertilization and occurs only in angiosperms. Sexual reproduction in gymnosperms is similar, except that the gametophytes develop on the scales of male and female cones or other structures instead of in flowers, the endosperm remains haploid in the absence of double fertilization, and more than one egg cell may be fertilized by distinct sperm nuclei to initially produce multiple embryos (Baroux et al. 2002).

In both angiosperms and gymnosperms, the successful diploid zygote produced by the union of a generative nucleus (sperm cell) and the egg cell represents the start of a new sporophyte generation that inherits half of its genetic material from each parent. This genetically distinct zygote develops into an embryo contained within the ovule (seed). In gymnosperms, the seed is released directly from the cone scale and not embedded in a fruit. In angiosperms, the seed is enclosed in an ovary that develops into a fruit containing one or more seeds, depending on the number of ovules produced and fertilized.

2.6.1.3 Polyploidy and Apomixis

Although plants are often diploid, with reproduction by seed that is exclusively sexual, two important variations on this theme merit consideration. Many plants are polyploid, meaning their nuclei contain more than two homologous chromosomes. Polyploidy arises when gametes containing more than one set of chromosomes are produced following errors in mitosis or meiosis. When at least one unreduced gamete fuses with another gamete during fertilization, a zygote with more than two homologous chromosomes will result (Comai 2005). Moreover, polyploid plants often reproduce by seed through the asexual pathways of apomixis. Gametophytic apomixis is the development of an embryo from a diploid megagametophyte unreduced by meiosis. Sporophytic apomixis is the adventitious development of an embryo from a somatic cell of the ovule, without the intermediate development of a megagametophyte (Hojsgaard et al. 2014).

Polyploidy and the associated switch to apomictic seed production may confer adaptive benefits. Some of the duplicated genes within polyploid lineages may take on new functionality through mutation and selection, offering polyploids an evolutionary flexibility not present in diploids (Comai 2005; Hojsgaard et al. 2014). Interestingly, polyploids often undergo a gradual diploidization characterized by the extensive loss or modification of redundant genes. In fact, many diploid taxa have arisen from polyploid ancestors, apparently because of the evolutionary potential available to them. Another potential benefit of polyploidy is the increased reproductive opportunity provided by the loss of self-incompatibility and/or the development of apomixis that often come along with polyploidy (Comai 2005). Apomixis is present in a wide variety of taxa belonging to many different families, and has clearly arisen independently among different lineages, since it is found embedded within generally sexual clades (Hojsgaard et al. 2014). Benefits of apomixis might include reproductive assurance in environments characterized by pollen or pollinator limitation, and the clonal perpetuation of well-adapted genotypes without consequences associated with inbreeding among sexually reproducing taxa (Ma et al. 2015). A good example of the link between polyploidy and apomixis can be found in *Amelanchier*, a genus of temperate trees and shrubs in which polyploid species reproduce almost exclusively (ca. 99%) by apomixis, whereas diploid species rarely do so (Burgess et al. 2014).

2.6.2 Fruits and Their Development

The defining characteristic distinguishing angiosperms from gymnosperms is the presence of flowers and fruits in the former and their absence in the latter. The fruit, which comprises the mature ovary and its contents, represents a key novelty in sexual reproduction by angiosperms. Among woody plants, dry fruits may be dehiscent like capsules and legumes that release their seeds at maturity, or indehiscent like nuts, which retain their seeds. Most examples of fleshy fruits are indehiscent, including berries, drupes, various compound fruits, and accessory fruits like pomes, for which much of the overall structure is derived from tissues other than the ovary wall. Gymnosperms do not produce fruits, as the ovules rest naked on the surface of a cone scale or other structure, absent a surrounding ovary. Even *Ginkgo biloba*, a deciduous gymnosperm that produces seeds enclosed in a fleshy "fruit," does not bear true fruits, as the ovule that gives rise to the seed is not contained within the ovary of a flower, and the fleshy covering is the modified testa of the ovule.

2.6.2.1 Fruit Set and Fruit Growth

The first step of fruit development after fertilization is fruit set, in which the fruitlets become determined and are likely to persist to maturation and ripening. This process involves signaling between the developing seeds and the surrounding tissues of the ovary, which will become the pericarp of the fruit. For example, auxin, GA, and cytokinin all have roles in the promotion of fruit development after fertilization. Auxin is synthesized in the ovules and transported to the pericarp, where it induces GA synthesis, promoting fruit growth (McAtee et al. 2013). Interestingly, the application of auxin or GA can stimulate the continued development of fruits in the absence of seeds, which demonstrates the important role of hormone signaling by seeds within the fruit. Increasing cytokinin concentrations promotes the accumulation of auxin and GA in the pericarp and stimulates cell division in the developing fruit (Ding et al. 2013). At the same time, ABA and ethylene decrease within the pericarp. Failure of fruit set may occur when ovules are unfertilized, environmental conditions are stressful, or assimilates available for fruit growth are limited (Marcelis et al. 2004). For example, when ovules are not fertilized, ethylene biosynthesis and signaling within the pericarp is not reduced, leading to abscission of the fruit and the diversion of maternal assimilates to other, fertilized fruits.

The growth of the fruit proceeds through a process of cell division followed by cell expansion that may produce extraordinary growth of the pericarp in many taxa or relatively modest growth in others. Fruit growth is influenced by the proportion of ovules fertilized, with smaller fruits produced when fertilization is less robust (McAtee et al. 2013). In dry, dehiscent fruits, cells within the pericarp become lignified, and one or more dehiscence zones develop to permit the fruit to shatter and release seeds. Cells of fleshy fruits do not lignify, but may expand dramatically as they fill with water during the growth of the fruit (Seymour et al. 2013; Stikić et al. 2015). Many genes are presumably involved in the determination of fruit shape and several such genes have been identified and characterized.

2.6.2.2 Fruit Maturation and Ripening

Fruit maturation must occur before the process of fruit ripening or senescence is initiated. Auxin exported from developing seeds prohibits the ripening of fruit tissues until the seeds are mature. With seed maturation, the supply of auxin to the pericarp diminishes, ABA increases, and ripening proceeds (McAtee et al. 2013; Seymour et al. 2013). In both dry and fleshy fruits, a decrease in auxin seems to be essential for the dehiscence of fruits and/or cellular sensitivity to endogenous ethylene to initiate ripening. In this manner, communication between the seeds and the pericarp yet again regulates developmental growth of fruit tissues. Through the influence of hormones, particularly ABA or ethylene, the process of ripening culminates in dry or fleshy fruits that dehisce or soften, respectively. Additional changes in fleshy fruits may occur with ripening, such as change in color and aroma, and an increase in sweetness caused by the breakdown of stored starches into sugar (Seymour et al. 2013). Fruit ripening can be characterized as climacteric if it is accompanied by an increase in respiration and production of ethylene, or non-climacteric if ripening occurs without the burst in respiration rate and ethylene synthesis (Stikić et al. 2015). Fruits may also display an intermediate type of ripening that indicates the existence of varying degrees of climacteric and nonclimacteric behavior (Paul et al. 2012).

2.6.2.3 Dispersal Strategies

An incredible diversity of mechanisms help to disperse seeds into the landscape, including fruit or seed morphologies that capitalize on gravity, water, wind, animals, and explosive dehiscence (Howe and Smallwood 1982). Following programmed ripening and senescence, fruits of many taxa will abscise from the maternal plant, whereas the fruits of others will remain attached to the maternal plant and only the seeds will disperse by dehiscence. Among gymnosperms, which lack fruits, most seeds have a papery wing that aids in dispersal by wind and gravity once it abscises from a female cone. Some fire-adapted gymnosperms, such as *Pinus banksiana*, produce resinous cones that often remain on the maternal plant and open to release seeds following exposure to high temperatures typical of a forest fire (Beaufait 1960). Likewise, some angiosperms adapted to fire-prone communities produce fruits that release seeds when fire has been present.

A noteworthy aspect of fruit production called masting, or mast seeding, is exhibited by many species but is particularly conspicuous in some woody trees. Large numbers of fruits are synchronously produced in certain years (i.e. mast years) by plants of a particular taxon across vast expanses of habitat, cycling with years in which fruits may be almost entirely absent. Two well-known examples are the *Quercus*-dominated forests of eastern North America and the *Nothofagus*-dominated forests of New Zealand (Kelly et al. 2008). The explanation for this observation is a matter of some controversy, and several hypotheses have been partially supported by data. Depending on the context, it may be simply a response of populations to pulses in resource availability, or a strategy to improve pollination efficiency, or a means by which the proportion of seeds consumed by predators is reduced when the consumptive capacity of predators is overwhelmed (Kelly and Sork 2002). Whatever its causes, seed masting has profound ecological consequences, as it produces dramatic cycles of resource availability and leads to far-reaching trophic consequences that alter community structure in complex ways (Kelly et al. 2008).

2.7 ADAPTATION

2.7.1 Evolution Is a Change in Allele Frequencies

An adaptation is any morphological, anatomical, physiological, or biochemical trait that develops in a population within a given environment by increasing the reproductive success (fitness) of organisms in which it is carried. As a process, adaptation is therefore understood to be the route by which populations of organisms are modified by natural selection to become well-suited to the conditions of their environment. Many examples of plant adaptation can be postulated, including adaptation of plants to grow on serpentine soils low in some mineral nutrients and high in heavy metals; adaptation of plants to the specific photoperiods and low temperatures of tundra; adaptation of plants to the Mediterranean climate characterized by months of summer drought and heat; or adaptation of plants to saltwater swamps in tidal areas along the ocean. Indeed, plants have evolved a striking diversity of mechanisms to respond to environmental stresses and to facilitate their fitness in challenging environments around the world. Essential to understanding adaptation is the concept of biological evolution, which is the change in allele frequencies within populations over time. Rather than comprising static collections of individuals with fixed genetic identities, reproducing populations are constantly in a state of flux with respect to the identities and frequencies of alleles in the population with each new generation. Since some alleles are expected to confer a fitness advantage, environmental pressures "select" for survival of those organisms with particularly adaptive alleles, leading them to become more common or even fixed entirely in subsequent generations. In this way, the novel morphological, anatomical, physiological, or biochemical adaptations that help an organism survive its environment today are the legacy of environmental pressures shaping the evolution of its ancestors.

2.7.2 Adaptation Arises Locally

Adaptation is an inherently local phenomenon. The selective pressures that shape the evolution of a population are locally imposed, and reproduction on the local scale produces populations that become adapted to their immediate environments. It is for this reason that geographically widespread species of trees and shrubs often display variation in growth and development concordant with the environmental conditions of the provenance (geographic origin of germplasm) in which the population has evolved. In fact, the concept of provenance has proven useful in efforts to collect adaptable genotypes for horticulture, agriculture, forestry, and ecological restoration. Contrasting models of adaptive differentiation among provenances have been developed, and it may be the case that different concepts apply in different cases. Ecotypes, more-or-less discrete evolutionary subunits of a species, are locally adapted groups of plants well-suited to specific climatic, edaphic, or biotic environments (Langlet 1971). In contrast to the ecotype, clines describe gradual geographic variation of any traits within a species or species complex (Huxley 1938). Clinal variation may be found within a species along gradients in temperature, photoperiod, edaphic conditions, and so on (Gregor and Watson 1961). In both ecotypic and clinal models of adaptation, selection on the local scale shapes the genotypes and phenotypes of local populations.

Stressful environments may exert strong selection pressures for novel environmental adaptation by prohibiting further population expansion by non-adapted genotypes (Brindle and Vines 2007). For example, populations on ecologically stressful sites like serpentine, alkaline, drought-prone, or flood-prone soils may harbor genotypes more adaptable to such conditions than other populations of the taxon (Snaydon 1970; Rajakaruna 2004). However, it is difficult to predict if, or how rapidly, a population under selection pressure might adapt through the evolution of novel traits (García-Ramos and Kirkpatrick 1997; Vergeer and Kunin 2013). With strong selection pressure, plant populations may undergo rapid evolution, presumably through the selection of alleles that are already present at low frequencies and that incidentally confer stress resistance. The nature and extent of selection pressures, the amount of genetic diversity in such populations, the genetic basis for traits limiting fitness, and the extent of non-adaptive gene flow from core populations to peripheral populations are complex factors that make adaptive responses difficult to predict. Nonetheless, the evaluation of plants from geographically or ecologically distinct populations may prove valuable to plant breeders seeking germplasm with novel adaptations.

2.7.3 Provenance Trials

Identifying and selecting for genetically controlled differences in phenotypes within species can be accomplished by using provenance trials. Propagules from diverse provenances are grown in a common environment to remove the environmental component of phenotypic variation among plants from each provenance. Because the phenotype of an individual plant depends on the interaction between its genotype and the environment in which it is grown, provenance trials (also called "common garden" trials) have the capacity to resolve whether phenotypic differentiation among natural populations is the result of environmental or genetic influences. Without a provenance trial, variation in environmental conditions or phenotypes across the natural range of a taxon is not sufficient evidence to conclude local adaptation, as many genotypes have sufficient phenotypic plasticity to adapt and compete in a variety of environments (Grenier et al. 2016).

When local adaptation is present, a provenance trial illustrates the profound influence of native environments on shaping genetic differentiation for specific growth characteristics and stress tolerances (Gustavsson 2001; Cornelius and Mesén 1997; Smithberg and Weiser 1968; Donselman and Flint 1982; Ginwal et al. 2005). For example, in an evaluation of *Cornus sericea* L., plant canopy shape varied from upright to nearly decumbent, and mean annual stem growth varied more than tenfold among plants originating from provenances that differed markedly in photoperiod and winter temperature (Smithberg and Weiser 1968). Likewise, both drought adaptive traits and cold acclimation in *Cercis canadensis* correlates with environmental variation among provenances differing in drought and cold (Donselman and Flint 1982). Populations of *Betula pubescens* subsp. *czerepanovii* on polluted barrens adjacent to copper-nickel smelters in Russia display both a resistance to heavy metal contamination and a reduced capacity to compete with congeners from unpolluted sites when cultivated in typical soils (Eränen 2008). Research on *Eucalyptus obliqua* has shown that plants originating from calcareous sites may be more adapted to basic soils and experience less nutrient stress under basic conditions than those indigenous to acidic sites (Anderson and Ladiges 1978). Many taxa exhibit such evidence for local adaptation, which may prove useful for horticulture, forestry, and ecological restoration.

2.7.4 Adaptationism

Although adaptation is no doubt a powerful mechanism for the evolution of diversity, some conceptual and practical difficulties with the investigation of adaptation should be considered. In 1979, Gould and Lewontin published a landmark paper in which they cautioned against the practice of "adaptationism," the assumption that the traits of organisms are independently optimized through adaptation by natural selection without consideration for other means by which specific traits might arise. Gould and Lewontin (1979) suggested that processes other than adaptation, such as genetic drift, phylogenetic history, indirect selection through associations of seemingly unrelated traits, and exaptation, should be considered potential alternatives to adaptation when attempting to explain the origins of particular traits (Gould and Lewontin 1979; Pigliucci and Kaplan 2000). For example, genetic drift might be a consequence of small population sizes, with few individuals contributing to the genetic identity of the next generation. In such an environment, allele frequencies may fluctuate dramatically by chance alone, and populations that are genetically and phenotypically distinct may arise even in the absence of natural selection. Phylogenetic history and indirect selection pertain to the possibility that a contemporary trait is merely a product of the history of the lineage (e.g. phylogenetic inertia) or developmental correlations of that trait and others. Finally, exaptation refers to any trait that might have evolved to fit a particular function before being co-opted for some other function due to its usefulness in a particular environment (McLennan 2008). In this way, a trait seemingly tailored to a given environment may actually have arisen in an altogether different selective environment and coincidentally finds utility elsewhere. Even Darwin recognized that adaptation might not be the sole agent of evolutionary change, concluding, "I am convinced that natural selection has been the main but not the exclusive means of modification" (Darwin 1859). In the decades that have followed Gould and Lewontin's critique of adaptationism, evolutionary biologists have worked toward a more comprehensive approach to investigating traits that seem to be adaptations (Rose and Lauder 1996; Pigliucci and Kaplan 2000; Olson 2012).

2.7.5 Challenges in the Study of Adaptation

Several challenges face investigators working on the topic of adaptation. One such challenge is the difficulty in recognizing what traits are truly adaptive, rather than neutral or even harmful byproducts of developmental constraints, phylogenetic history, or genetic drift. For example, it would seem difficult to decide whether the epigeal germination of honeylocust (*Gleditsia*), in which the cotyledons of seedlings are lifted above the ground, is an adaptation or just a byproduct of its embryo morphology. Likewise, there is no obvious means by which to decide whether hypogeal germination of coffeetree (*Gymnocladus*), in which the comparatively large cotyledons remain inside the seed coat resting in the soil, is an adaptation or merely a developmental constraint of seedlings with heavy cotyledons. In each of these legumes, seed size might be adaptively selected through the balance of various evolutionary pressures, but the type of germination might be nothing more than a developmental byproduct.

Biologists also face the challenge of distinguishing between traits that are adaptations (i.e. products of natural selection acting within a particular environment that confer advantages to individuals that possess them) vs. traits that are merely adaptive (i.e. traits that confer a fitness advantage in a particular environment, regardless of their evolutionary origins). For example, it might be tempting to suppose leaf senescence, resorption of leaf nutrients, and abscission of leaves of deciduous plants in the fall is an adaptation to seasonal cold in temperate regions of the world, since most deciduous leaves would not survive the low temperatures of winter. However, if deciduous angiosperms evolved from evergreen ancestors in the warm tropics (Axelrod 1966), deciduousness is merely a suitable but pre-existing strategy present in those deciduous plants that find themselves in seasonally cold environments. The accrual of cold hardiness that may accompany dormancy seems to be a more obvious adaptation to cold. Indeed, given that cold hardiness is present in both deciduous and evergreen taxa, these common mechanisms seem to be the defensible place to draw the line for cold adaptation.

Another challenge in studying adaptation is the identification of specific environments in which putative adaptations arose, and how natural selection actually shaped trait evolution in a particular study system. For example, if an investigator is interested in studying the comparative drought tolerance among congeneric species that occupy dry habitats and mesic habitats, it might turn out to be rather difficult to ascertain whether putative adaptations to drought arose by natural selection in the dry habitat or were pre-existing among the ancestral lineage and simply lost among descendants in the mesic environment. Likewise, one might ask whether it is reasonable to suppose that all plants that are endemic to serpentine soils adapted specifically to the stressful environment imposed by their habitat. It seems likely that some, by historical accident or the presence of mechanisms adapted for some other purpose, have been able to compete on these difficult sites with no need for us to appeal to novel adaptation to serpentine conditions.

An additional consideration with regard to adaptation is that the distribution of plant populations is not necessarily in equilibrium with the distribution of suitable habitat. For example, Florida nutmeg (*Torreya taxifolia*), an evergreen tree restricted to a small portion of the Apalachicola river basin of the southeastern United States, seems to have been unable to re-establish its former northward range in eastern North America following the last glacial maximum (Barlow 2001). Whether the seeds, which seem generally ill-suited to transport by contemporary dispersers, were once dispersed by now-extinct megafauna is a matter of some debate. In any case, Florida nutmeg seems trapped in an unsuitable habitat as a result of past selective pressures for dispersal mechanisms that no longer serve to disperse. Because transplant efforts show that the species is well-suited to more northern habitats in the Appalachian Mountains, and it is a relict in decline in the warm, humid south (Barlow 2001), the species is an extreme example of a taxon out of equilibrium with the distribution of suitable environment. It is worth considering, then, that the contemporary distribution of a taxon may not represent the environment to which it is historically adapted.

2.7.6 Adaptation in this Book

It is easy to appreciate the role of adaptation in plant growth and development, and the difficulties inherent to the study of adaptation. It seems probable that most of the adaptations described in detail within the second half of this book represent genuine adaptations to very specific environmental stressors. However, even if we grant that some of the apparent adaptations discussed in this book are not genuine adaptations evolved in the selective environment to which they seem now to be especially well-suited, the fact remains that these adaptive mechanisms enable the plants that possess them to tolerate remarkable environmental stresses. Through the rest of this book, adaptive traits that confer a survival and reproductive advantage to their hosts will be referred to simply as adaptations.

REFERENCES

Abe, M., Y. Kobayashi, S. Yamamoto, Y. Daimon, A. Yamaguchi, Y. Ikeda, H. Ichinoki, M. Notaguchi, K. Goto, and T. Araki. 2005. FD, a bZIP protein mediating signals from the floral pathway integrator FT at the shoot apex. *Science* 309:1052–1056.

Alberts, B., A. Johnson, J. Lewis, M. Raff, K. Roberts, and P. Walter. 2002. *Molecular Biology of the Cell*, 4th edition. New York: Garland Science.

Almeida, T., G. Pinto, B. Correia, C. Santos, and S. Goncalves. 2013. QsMYB1 expression is modulated in response to heat and drought stresses and during plant recovery in *Quercus suber*. *Plant Physiol. Biochem.* 73:274–281.

Aloni, R., E. Aloni, M. Langhans, and C.I. Ullrich. 2006. Role of cytokinin and auxin in shaping root architecture: Regulating vascular differentiation, lateral root initiation, root apical dominance and root gravitropism. *Ann. Bot.* 97:883–893.

Anderson, C.A. and P.Y. Ladiges. 1978. A comparison of three populations of *Eucalyptus obliqua* L'Hérit. Growing on acid and calcareous soils in southern Victoria. *Aust. J. Bot.* 26:93–109.
Arc, E., J. Sechet, F. Corbineau, L. Rajjou, and A. Marion-Poll. 2013. ABA crosstalk with ethylene and nitric oxide in seed dormancy and germination. *Front. Plant Sci.* 4:63.
Arora, R., L.J. Rowland, and K. Tanino. 2003. Induction and release of bud dormancy in woody perennials: A science comes of age. *HortScience* 38:911–921.
Arora, R. and K. Taulavuori. 2016. Increased risk of freeze damage in woody perennials vis-à-vis climate change: Importance of deacclimation and dormancy response. *Front. Environ. Sci.* 4:44. http://dx.doi.org/10.3389/fenvs.2016.00044.
Arsovski, A.A., A. Galstyan, J.M. Guseman, and J.L. Nemhauser. 2012. Photomorphogenesis. *Arabidopsis Book* 10:e0147. http://dx.doi.org/10.1199/tab.0147.
Axelrod, D.I. 1966. Origin of deciduous and evergreen habits in temperate forests. *Evolution* 20:1–15.
Bar, M. and N. Ori. 2014. Leaf development and morphogenesis. *Development* 141:4219–4230.
Barbier, F.F., J.E. Lunn, and C.A. Beveridge. 2015. Ready, set, go! A sugar hit starts the race to shoot branching. *Curr. Opin. Plant Biol.* 25:39–45.
Barlow, C. 2001. Anachronistic fruits and the ghosts who haunt them. *Arnoldia* 61:14–21.
Baroux, C., C. Spillane, and U. Grossniklaus. 2002. Evolutionary origins of the endosperm in flowering plants. *Genome Biol.* 3:1026.1–1026.5. http://dx.doi.org/10.1186/gb-2002-3-9-reviews1026.
Barra-Jiménez, A. and L. Ragni. 2017. Secondary development in the stem: When *Arabidopsis* and trees are closer than it seems. *Curr. Opin. Plant Biol.* 35:145–151.
Baskin, J.M. and C.C. Baskin. 2004. A classification system for seed dormancy. *Seed Sci. Res.* 14:1–16.
Beaufait, W.R. 1960. Some effects of high temperatures on the cones and seeds of jack pine. *For. Sci.* 6:194–199.
Benlloch, R., A. Berbel, A. Serrano-Mislata, and F. Madueño. 2007. Floral initiation and inflorescence architecture: A comparative view. *Ann. Bot.* 100:659–676.
Boonman, A., E. Prinsen, F. Gilmer, U. Schurr, A.J.M. Peeters, L.A.C.J. Voesenek, and T.L. Pons. 2007. Cytokinin import rate as a signal for photosynthetic acclimation to canopy light gradients. *Plant Physiol.* 143:1841–1852.
Bowman, J.L., D.R. Smyth, and E.M. Meyerowitz. 1991. Genetic interactions among floral homeotic genes of *Arabidopsis. Development* 112:1–20.
Bowman, J.L., D.R. Smyth, and E.M. Meyerowitz. 2012. The ABC model of flower development: Then and now. *Development* 139:4095–4098.
Briggs, W.R. and M.A. Olney. 2001. Photoreceptors in plant photomorphogenesis to date. Five phytochromes, two cryptochromes, one phototropin, and one superchrome. *Plant Physiol.* 125:85–88.
Brindle, J.R. and T.H. Vines. 2007. Limits to evolution at range margins: When and why does adaptation fail? *Trends Ecol. Evolut.* 22:140–147.
Brown, C.L., R.G. McAlpine, and P.P. Kormanik. 1967. Apical dominance and form in woody plants: A reappraisal. *Am. J. Bot.* 54:153–162.
Burgess, M.B., K.R. Cushman, E.T. Doucette, N., Talent, C.T. Frye, and C.S. Campbell. 2014. Effects of apomixis and polyploidy on diversification and geographic distribution in *Amelanchier* (Rosaceae). *Am. J. Bot.* 101:1375–1387.
Cao, D., C.C. Baskin, J.M. Baskin, F. Yang, and Z. Huang. 2014. Dormancy cycling and persistence of seeds in soil of a cold desert haplophyte shrub. *Ann. Bot.* 113:171–179.
Carlsbecker, A. and Y. Helariutta. 2005. Phloem and xylem specification: Pieces of the puzzle emerge. *Curr. Opin. Plant Biol.* 8:512–517.
Catalá, R., J. Medina, and J. Salinas. 2011. Integration of low temperature and light signaling during cold acclimation response in *Arabidopsis. Proc. Natl. Acad. Sci.* 108:16475–16480.
Causier, B., Z. Schwarz-Sommer, and B. Davies. 2010. Floral organ identity: 20 years of ABCs. *Sem. Cell. Dev. Biol.* 21:73–79.
Cline, M.G., N. Bhave, and C.A. Harrington. 2009. The possible roles of nutrient deprivation and auxin repression in apical control. *Trees* 23:489–500.
Cline, M.G. and C.A. Harrington. 2007. Apical dominance and apical control in multiple flushing of temperate woody species. *Can. J. For. Res.* 37:74–83.
Comai, L. 2005. The advantages and disadvantages of being polyploid. *Nat. Rev. Genet.* 6:836–846.
Copete, M.A., J.M. Herranz, P. Ferrandis, and E. Copete. 2015. Annual dormancy cycles in buried seeds of shrub species: Germination ecology of *Sideritis serrata* (Labiatae). *Plant Biol.* 17:798–807.
Cornelius, J.P. and J.F. Mesén. 1997. Provenance and family variation in growth rate, stem straightness, and foliar mineral concentration in *Vochysia guatemalensis. Can. J. For. Res.* 27:1103–1109.

Darwin, C. 1859. *On the Origin of Species by Means of Natural Selection, or, the Preservation of Favoured Races in the Struggle for Life*. Lond.: J. Murray.

Dennis, Jr., F.G. and J.C. Neilsen. 1999. Physiological factors affecting biennial bearing in tree fruit: The role of seeds in apple. *HortTechnology* 9:317–322.

Ding, J., B. Chen, X. Xia, W. Mao, K. Shi, Y. Zhou, and J.Yu. 2013. Cytokinin-induced parthenocarpic fruit development in tomato is partly dependent on enhanced gibberellin and auxin biosynthesis. *PLOS ONE* 8:e70080. http://dx.doi.org/10.1371/journal.pone.0070080.

Dinneny, J.R. 2014. A gateway with a guard: How the endodermis regulates growth through hormone signaling. *Plant Sci.* 214:14–19.

Donselman, H.M. and H.L. Flint. 1982. Genecology of eastern redbud (*Cercis canadensis*). *Ecology* 63:962–971.

Dörffling, K. 2015. The discovery of abscisic acid: A retrospect. *J. Plant Growth Regul.* 34:795–808.

Dun, E.A., B.J. Ferguson, and C.A. Beveridge. 2006. Apical dominance and shoot branching. Divergent opinions or divergent mechanisms? *Plant Physiol.* 142:812–819.

Efroni, I., Y. Eshed, and E. Lifschitz. 2010. Morphogenesis of simple and compound leaves: A critical review. *Plant Cell* 22:1019–1032.

Endo, A., M. Okamoto, and T. Koshiba. 2014. ABA biosynthetic and catabolic pathways. In *Abscisic Acid: Metabolism, Transport and Signaling*, ed. D.-P. Zhang, 21–45. Dordrecht: Springer Science+Business Media. http://dx.doi.org/10.1007/978-94-017-9424-4_2.

Eränen, J.K. 2008. Rapid evolution towards heavy metal resistance by mountain birch around two subarctic copper-nickel smelters. *J. Evol. Biol.* 21:492–501.

February, E.C. and S.I. Higgins. 2016. Rapid leaf deployment strategies in a deciduous savanna. *PLOS ONE* 11:e0157833. http://dx.doi.org/10.1371/journal.pone.0157833.

Finch-Savage, W.E. and G. Leubner-Metzger. 2006. Seed dormancy and the control of germination. *New Phytol.* 171:501–523. http://dx.doi.org/10.1111/j.1469-8137.2006.01787.x.

Finkelstein, R., W. Reeves, T. Ariizumi, and C. Steber. 2008. Molecular aspects of seed dormancy. *Annu. Rev. Plant Biol.* 59:387–415.

Fleming, A.J. 2005. Formation of primordia and phyllotaxy. *Curr. Opin. Plant Biol.* 8:53–58.

Fordham, A.J. 1965. Germination of woody legume seeds with impermeable seed coats. *Arnoldia* 25:1–8.

Fort, C., F. Muller, P. Label, A. Granier, and E. Dreyer. 1998. Stomatal conductance, growth and root signaling in *Betula pendula* seedlings subjected to partial soil drying. *Tree Physiol.* 18:769–776.

Franklin, K.A. and G.C. Whitelam. 2005. Phytochromes and shade-avoidance responses in plants. *Ann. Bot.* 96:169–175.

Franklin, K.A. and G.C. Whitelam. 2007. Light-quality regulation of freezing tolerance in *Arabidopsis thaliana*. *Nat. Genet.* 39:1410–1413.

Fukushima, K. and M. Hasebe. 2013. Adaxial-abaxial polarity: The developmental basis of leaf shape diversity. *Genesis* 52:1–18.

Fulford, R.M. 1966. The morphogenesis of apple buds. III. The inception of flowers. *Ann. Bot.* 30:207–219.

García-Ramos, G. and M. Kirkpatrick. 1997. Genetic models of adaptation and gene flow in peripheral populations. *Evolution* 51:21–28. http://dx.doi.org/10.1002/dvg.22728.

Giehl, R.F.H. and N. von Wirén. 2014. Root nutrient foraging. *Plant Physiol.* 166:509–517.

Ginwal, H.S., S.S. Phartyal, P.S. Rawat, and R.L. Srivastava. 2005. Seed source variation in morphology, germination and seedling growth of *Jatropha curcas* Linn. in central India. *Silvae Genet.* 54:76–80.

Givnish, T.J. 2002. Adaptive significance of evergreen vs. deciduous leaves: Solving the triple paradox. *Silva Fenn.* 36:703–743.

Gould, S.J. and R.C. Lewontin. 1979. The spandrels of San Marco and the Panglossian paradigm: A critique of the adaptationist programme. *Proc. R. Soc. Lond. B* 205:581–598.

Gregor, J.W. and P.J. Watson. 1961. Ecotypic differentiation: Observations and reflections. *Evolution* 15:166–173.

Grenier, S., P. Barre, and I. Litrico. 2016. Phenotypic plasticity and selection: Nonexclusive mechanisms of adaptation. *Scientifica* 2016:1–9. http://dx.doi.org/10.1155/2016/7021701.

Guo, Y. 2013. Towards systems biological understanding of leaf senescence. *Plant Mol. Biol.* 82:519–528.

Gustavsson, B.A. 2001. Genetic variation in horticulturally important traits of fifteen wild lingonberry *Vaccinium vitis-idaea* L. populations. *Euphytica* 120:173–182.

Hacke, U.G., J.S. Sperry, W.T. Pockman, S.D. Davis, and K.A. McCulloh. 2001. Trends in wood density and structure are linked to prevention of xylem implosion by negative pressure. *Oecologia* 126:457–461.

Hartmann, H.T., D.E. Kester, F.T. Davies, and R.L. Geneve. 2011. *Hartmann and Kester's Plant Propagation: Principles and Practices*, 8th edition. Upper Saddle River: Prentice Hall.

Hauvermale, A.L., K.M. Tuttle, Y. Takebayashi, M. Seo, and C.M. Steber. 2015. Loss of *Arabidopsis thaliana* seed dormancy is associated with increased accumulation of the GID1 GA hormone receptors. *Plant Cell Physiol.* 9:1773–1785.

Heide, O.M. and A.K. Prestrud. 2005. Low temperature, but not photoperiod, controls growth cessation and dormancy induction and release in apple and pear. *Tree Physiol.* 25:109–114.

Hidayati, S.N., J.M. Baskin, and C.C. Baskin. 2000. Morphophysiological dormancy in seeds of two North American and one Eurasian species of *Sambucus* (Caprifoliaceae) with underdeveloped spatulate embryos. *Am. J. Bot.* 87:1669–1678.

Hojsgaard, D., S. Klatt, R. Baier, J.G. Carman, and E. Hörandl. 2014. Taxonomy and biogeography of apomixis in angiosperms and associated biodiversity characteristics. *CRC Crit. Rev. Plant Sci.* 33:414–427.

Horvath, D.P., J.V. Anderson, W.S. Chao, and M.E. Foley. 2003. Knowing when to grow: Signals regulating bud dormancy. *Trends Plant Sci.* 8:534–540.

Howe, H.F. and J. Smallwood. 1982. Ecology of seed dispersal. *Ann. Rev. Ecol. Syst.* 13:201–228.

Hu, F., P.P. Mou, J. Weiner, and S. Li. 2014. Contrasts between whole-plant and local nutrient levels determine root growth and death in *Ailanthus altissima* (Simaroubaceae). *Am. J. Bot.* 101:812–819.

Huché-Thélier, L., L. Crespel, J. Le Gourrierec, P. Morel, S. Sakr, and N. Leduc. 2016. Light signaling and plant responses to blue and UV radiations—Perspectives for applications in horticulture. *Environ. Exp. Bot.* 121:22–38.

Huijser, P. and M. Schmid. 2011. The control of developmental phase transitions in plants. *Development* 138:4117–4129.

Huxley, J. 1938. Species formation and geographical isolation. *Proc. Linn. Soc. Lond.* 150:253–264.

Hyun, Y., R. Richter, and G. Coupland. 2017. Competence to flower: Age-controlled sensitivity to environmental cues. *Plant Physiol.* 173:36–46.

Ichihashi, Y. and H. Tsukaya. 2015. Behavior of leaf meristems and their modification. *Front. Plant Sci.* 6:1060.

Ioio, R.D., F.S. Linhares, and S. Sabatini. 2008. Emerging role of cytokinin as a regulator of cellular differentiation. *Curr. Opin. Plant Biol.* 11:23–27.

Janská, A., P. Maršík, S. Zelenková, and J. Ovesná. 2010. Cold stress and acclimation – What is important for metabolic adjustment? *Plant Biol.* 12:395–405.

Javelle, M., V. Vernoud, N. Depège-Fargeix, C. Arnould, D. Oursel, F. Domergue, X. Sarda, and P.M. Rogowsky. 2010. Overexpression of the epidermis-specific homeodomain-leucine zipper IV transcription factor Outer Cell Layerl in maize identifies target genes involved in lipid metabolism and cuticle biosynthesis. *Plant Physiol.* 154:273–286.

Jones, A.R., E.M. Kramer, K. Knox, R. Swarup, M.J. Bennett, C.M. Lazarus, H.M. Ottoline Leyser, and C.S. Grierson. 2008. Auxin transport through non-hair cells sustains root-hair development. *Nat. Cell Biol.* 11:78–84.

Kalberer, S.R., M. Wisniewski, and R. Arora. 2006. Deacclimation and reacclimation of cold-hardy plants: Current understanding and emerging concepts. *Plant Sci.* 171:3–16.

Kelly, D., W.D. Koenig, and A.M. Liebhold. 2008. An intercontinental comparison of the dynamic behavior of mast seeding communities. *Popul. Ecol.* 50:329–342.

Kelly, D. and V.L. Sork. 2002. Mast seeding in perennial plants: Why, how, where? *Annu. Rev. Ecol. Evol. Syst.* 33:427–447.

Kidner, C.A. and M.C.P. Timmermans. 2010. Signaling sides: Adaxial-abaxial patterning in leaves. *Curr. Top. Dev. Biol.* 91:141–168.

Kim, G.-T., S. Yano, T. Kozuka, and H. Tsukaya. 2005. Photomorphogenesis of leaves: Shade-avoidance and differentiation of sun and shade leaves. *Photochem. Photobiol. Sci.* 4:770–774.

Kim, H.J., Y.K. Kim, J.Y. Park, and J. Kim. 2002. Light signalling mediated by phytochrome plays an important role in cold-induced gene expression through the C-repeat/dehydration responsive element (C/DRE) in *Arabidopsis thaliana. Plant Journal* 29:693–704.

Kümpers, B.M.C., S.J. Burgess, I. Reyna-Llorens, R. Smith-Unna, C. Boursnell, and J.M. Hibberd. 2017. Shared characteristics underpinning C4 leaf maturation derived from analysis of multiple C3 and C4 species of *Flaveria. J. Exp. Bot.* 68:177–189. http://dx.doi.org/10.1093/jxb/erw488.

Lang, G.A. 1987. Dormancy: A new universal terminology. *HortScience* 22:817–820.

Langlet, O. 1971. Two hundred years' genecology. *Taxon* 20:653–722.

Lauri, P.E. and J.M. Lespinasse. 1993. The relationship between cultivar fruiting-type and fruiting branch characteristics in apple trees. *Acta Hortic.* 349:259–263.

Lauri, P.E., E. Térouanne, J.M. Lespinasse, J.L. Regnard, and J.J. Kelner. 1995. Genotypic differences in the axillary bud growth and fruiting pattern of apple fruit branches over several years – An approach to regulation of fruit bearing. *Sci. Hort.* 64:265–281.

Lee, D.W., J. O'Keefe, N.M. Holbrook, and T.S. Feild. 2003. Pigment dynamics and autumn leaf senescence in a New England deciduous forest, eastern USA. *Ecol. Res.* 18:677–694.

Lens, F., J.S. Sperry, M.A. Christman, B. Choat, D. Rabaey, and S. Jansen. 2011. Testing hypotheses that link wood anatomy to cavitation resistance and hydraulic conductivity in the genus *Acer. New Phytol.* 190:709–723.

Liao, W., Y. Li, Y. Yang, G. Wang, and M. Peng. 2016. Exposure to various abscission-promoting treatments suggests substantial ERF subfamily transcription factors involvement in the regulation of cassava leaf abscission. *BMC Genomics* 17:538. http://dx.doi.org/10.1186/s12864-016-2845-5.

Libault, M., L. Brechenmacher, J. Cheng, D. Xu, and G. Stacey. 2010. Root hair systems biology. *Trends Plant Sci.* 15:641–650.

Linkohr, B.I., L.C. Williamson, A.H. Fitter, and H.M. Ottoline Leyser. 2002. Nitrate and phosphate availability and distribution have different effects on root system architecture of *Arabidopsis. Plant J.* 29:751–760.

Liscum, E., S.K. Askinosie, D.L. Leuchtman, J. Morrow, K.T. Willenburg, and D.R. Coats. 2014. Phototropism: Growing towards an understanding of plant movement. *Plant Cell* 26:38–55.

Liu, C., Z. Thong, and H. Yu. 2009. Coming into bloom: The specification of floral meristems. *Development* 136:3379–3391.

Liu, J. and L.-J. Qu. 2008. Meiotic and mitotic cell cycle mutants involved in gametophyte development in *Arabidopsis. Mol. Plant* 1:546–574.

Liu, X., H. Zhang, Y. Zhao, Z. Feng, Q. Li, H.-Q. Yang, S. Luan, J. Li, and Z.-H. He. 2013. Auxin controls seed dormancy through stimulation of abscisic acid signaling by inducing ARF-mediated ABI3 activation in *Arabidopsis. Proc. Natl. Acad. Sci. U.S.A.* 110:15485–15490. http://dx.doi.org/10.1073/pnas.1304651110.

López-Bucio, J., A. Cruz-Ramírez, and L. Herrera-Estrella. 2003. The role of nutrient availability in regulating root architecture. *Curr. Opin. Plant Biol.* 6:280–287.

Ma, C., S. Meir, L. Xiao, J. Tong, Q. Liu, M.S. Reid, and C.-Z. Jiang. 2015. A KNOTTED1-LIKE HOMEOBOX protein regulates abscission in tomato by modulating the auxin pathway. *Plant Physiol.* 167:844–853.

Mähönen, A.P., A. Bishopp, M. Higuchi, K.M. Nieminen, K. Kinoshita, K. Törmäkangas, Y. Ikeda, A. Oka, T. Kakimoto, and Y. Helariutta. 2006. Cytokinin signaling and its inhibitor AHP6 regulate cell fate during vascular development. *Science* 311:94–98.

Marcelis, L.F.M., E. Heuvelink, L.R. Baan Hofman-Eijer, J. Den Bakker, and L.B. Xue. 2004. Flower and fruit abortion in sweet pepper in relation to source and sink strength. *J. Exp. Bot.* 55:2261–2268.

Marchin, R., H. Zeng, and W. Hoffmann. 2010. Drought-deciduous behavior reduces nutrient losses from temperate deciduous trees under severe drought. *Oecologia* 163:845–854.

Marhavy, P., M. Vanstraelen, B. De Rybel, D. Zhaojun, M.J. Bennett, T. Beeckman, and E. Benková. 2013. Auxin reflux between the endodermis and pericycle promotes lateral root initiation. *EMBO J.* 32:149–158.

Mason, M.G., J.J. Ross, B.A. Babst, B.N. Wienclaw, and C.A. Beveridge. 2014. Sugar demand, not auxin, is the initial regulator of apical dominance. *Proc. Natl. Acad. Sci.* 111:6092–6097.

McAdam, S.A.M., T.J. Brodribb, and J.J. Ross. 2016. Shoot-derived abscisic acid promotes root growth. *Plant Cell Environ.* 39:652–659.

McAtee, P., S. Karim, R. Schaffer, and K. David. 2013. A dynamic interplay between phytohormones is required for fruit development, maturation, and ripening. *Front. Plant Sci.* 4:1–6.

McLennan, D.A. 2008. The concept of co-option: Why evolution often looks miraculous. *Evol. Educ. Outreach* 1:247–258.

Meyerowitz, E.M. 1997. Genetic control of cell division patterns in developing plants. *Cell* 88:299–308.

McLaughlin, J.M. and D.W. Greene. 1991. Fruit and hormones influence flowering of apple. I. Effect of cultivar. *J. Amer. Soc. Hort. Sci.* 116:446–449.

Miyashima, S. and K. Nakajima. 2011. The root endodermis: A hub of developmental signals and nutrient flow. *Plant Signal. Behav.* 6:1954–1958.

Miyashima, S., J. Sebastian, J.-Y. Lee, and Y. Helariutta. 2013. Stem cell function during plant vascular development. *EBMO J.* 32:178–193.

Möller, B. and D. Weijers. 2009. Auxin control of embryo patterning. *Cold Spring Harb. Perspect. Biol.* 1:a001545. http://dx.doi.org/10.1101/cshperspect.a001545.

Müller, D. and O. Leyser. 2011. Auxin, cytokinin and the control of shoot branching. *Ann. Bot.* 107:1203–1212.

Murray, J.A.H., A. Jones, C. Godin, and J. Traas. 2012. Systems analysis of shoot apical meristem growth and development: Integrating hormonal and mechanical signaling. *Plant Cell* 24:3907–3919.

Nakano, T., M. Fujisawa, Y. Shima, and Y. Ito. 2013. Expression profiling of tomato pre-abscission pedicels provides insights into abscission zone properties including competence to respond to abscission signals. *BMC Plant Biol.* 13:40. http://dx.doi.org/10.1186/1471-2229-13-40.

Nelson, T. and N. Dengler. 1997. Leaf vascular pattern formation. *Plant Cell* 9:1121–1135.

Nicoll, B.C. and D. Ray. 1996. Adaptive growth of tree root systems in response to wind action and site conditions. *Tree Physiol.* 16:891–898.

Nonogaki, H., G.W. Bassel, and J.D. Bewley. 2010. Germination—Still a mystery. *Plant Sci.* 179:574–581.

Olsen, J.E. 2010. Light and temperature sensing and signaling in induction of bud dormancy in woody plants. *Plant Mol. Biol.* 73:37–47.

Olson, M.E. 2012. The developmental renaissance in adaptationism. *Trends Ecol. Evol.* 27:278–287.

Paque, S. and D. Weijers. 2016. Q&A: Auxin: The plant molecule that influences almost anything. *BMC Biol.* 14. http://dx.doi.org/10.1186/s12915-016-0291-0.

Paul, V., R. Pandey, and G.C. Srivastava. 2012. The fading distinctions between classical patterns of ripening in climacteric and non-climacteric fruit and the ubiquity of ethylene—An overview. *J. Food Sci. Technol.* 49:1–21.

Pellegrini, A.F.A., W.R.L. Anderegg, C.E.T. Paine, W.A. Hoffmann, T. Kartzinel, S.S. Rabin, D. Sheil, A.C. Franco, and S.W. Pacala. 2017. Convergence of bark investment according to fire and climate structures ecosystem vulnerability to future change. *Ecol. Lett.* 20:307–316.

Peña, L., M. Martín-Trillo, J. Juárez, J.A. Pina, L. Navarro, and J.M. Martínez-Zapater. 2001. Constitutive expression of *Arabidopsis* LEAFY or APETALA1 genes in citrus reduces their generation time. *Nat. Biotechnol.* 19:263–267.

Penfield, S. 2008. Temperature perception and signal transduction in plants. *New Phytol.* 179:615–628.

Pereira, H. 2011. *Cork: Biology, Production and Uses.* Amsterdam: Elsevier Publications.

Perilli, S., R. Di Mambro, and S. Sabatini. 2012. Growth and development of the root apical meristem. *Curr. Opin. Plant Biol.* 15:17–23.

Phartyal, S.S., T. Kondo, Y. Hoshino, C.C. Baskin, and J.M. Baskin. 2009. Morphophysiological dormancy in seeds of the autumn-germinating shrub *Lonicera caerulea* var. *emphyllocalyx* (Caprifoliaceae). *Plant Species Biol.* 24:20–26.

Pigliucci, M. and J. Kaplan. 2000. The fall and rise of Dr Pangloss: Adaptationism and the *Spandrels* paper 20 years later. *Trends Ecol. Evol.* 15:66–70.

Pitterman, J., J.S. Sperry, J.K. Wheeler, U.G. Hacke, and E.H. Sikkema. 2006. Mechanical reinforcement of tracheids compromise the hydraulic efficiency of conifer xylem. *Plant Cell Environ.* 29:1618–1628.

Plomion, C., G. Leprovost, and A. Stokes. 2001. Wood formation in trees. *Plant Physiol.* 127:1513–1523.

Rajakaruna, N. 2004. The edaphic factor in the origin of plant species. *Intl. Geol. Rev.* 46:471–478.

Read, J. and A. Stokes. 2006. Plant biomechanics in an ecological context. *Am. J. Bot.* 93:1546–1565.

Rose, M.R. and G.V. Lauder. 1996. Post-spandrel adaptationism. In *Adaptation*, ed. M.R. Rose and G.V. Lauder, 1–8. New York: Academic Press.

Rowe, J.H., J.F. Topping, J. Liu, and K. Lindsey. 2016. Abscisic acid regulates root growth under osmotic stress conditions via an interacting hormonal network with cytokinin, ethylene and auxin. *New Phytol.* 211:225–239.

Sanchez, P., L. Nehlin, and T. Greb. 2012. From thin to thick: Major transitions during stem development. *Trends Plant Sci.* 17:113–121.

Sarijeva, G., M. Knapp, and H.K. Lichtenthaler. 2007. Differences in photosynthetic activity, chlorophyll and carotenoid levels, and in chlorophyll fluorescence parameters in green sun and shade leaves of *Ginkgo* and *Fagus*. *J. Plant Physiol.* 164:950–955.

Schiefelbein, J., L. Huang, and X. Zheng. 2014. Regulation of epidermal cell fate in *Arabidospsis* roots: The importance of multiple feedback loops. *Front. Plant Sci.* 5:47. http://dx.doi.org/10.3389%2Ffpls.2014.00047.

Schippers, J.H.M. 2015. Transcriptional networks in leaf senescence. *Curr. Opin. Plant Biol.* 27:77–83.

Seymour, G.B., L. Ostergaard, N.H. Chapman, S. Knapp, and C. Martin. 2013. Fruit development and ripening. *Ann. Rev. Plant Biol.* 64:219–241.

Shimizu-Sato, S. and H. Mori. 2001. Control of outgrowth and dormancy in axillary buds. *Plant Physiol.* 127:1405–1413.

Smithberg, M.H. and C.J. Weiser. 1968. Patterns of variation among climatic races of red- osier dogwood. *Ecology* 49:495–505.

Snaydon, R.W. 1970. Rapid population differentiation in a mosaic environment. I. The response of Anthoxanthum odoratum populations to soils. *Evolution* 24:257–269.

Souter, M. and K. Lindsey. 2000. Polarity and signalling in plant embryogenesis. *J. Exp. Bot.* 51:971–983.

Srikanth, A. and M. Schmid. 2011. Regulation of flowering time: All roads lead to Rome. *Cell. Mol. Life Sci.* 68:2013–2037.

Steemans, P., A.L. Hérissé, J. Melvin, M.A. Miller, F. Paris, J. Verniers, and C.H. Wellman. 2009. Origin and radiation of the earliest vascular land plants. *Science* 324:353.

Stikić, R., Z. Jovanović, B. Vucelic-Radović, M. Marjanović, and S. Savić. 2015. Tomato: A model species for fruit growth and development studies. *Bot. Serb.* 39:95–102.

Su, Y.H., Y.B. Liu, B. Bai, and X.S. Zhang. 2015. Establishment of embryonic shoot-root axis is involved in auxin and cytokinin response during *Arabidopsis* somatic embryogenesis. *Front. Plant Sci.* 5:792. http://dx.doi.org/10.3389/fpls.2014.00792.

Sundaresan, V. and M. Alandete-Saez. 2010. Pattern formation in miniature: The female gametophyte of flowering plants. *Development* 137:179–189.

Tan, F.-C. and S.M. Swain. 2006. Genetics of flower initiation and development in annual and perennial plants. *Physiol. Plant.* 128:8–17.

Taylor, J.E. and C.A. Whitelaw. 2001. Signals in abscission. *New Phytol.* 151:323–339.

Teotia, S. and G. Tang. 2015. To bloom or not to bloom: Role of microRNAs in plant flowering. *Mol. Plant* 8:359–377.

Terashima, I., T. Araya, S.-I. Miyazawa, K. Sone, and S. Yano. 2005. Construction and mainteneance of the optimal photosynthetic systems of the leaf, herbaceous plant and tree: An eco-developmental treatise. *Ann. Bot.* 95:507–519.

Terradas, J. 1999. Holm oak and holm oak forests: An introduction. In *Ecology of Mediterranean evergreen oak forests*, ed. F. Rodà, J. Retana, C.A. Gracia, and J. Bellot, 3–14. Berlin: Springer.

Thomas, T.L. 1993. Gene expression during plant embryogenesis and germination: An overview. *Plant Cell* 5:1401–1410.

Thomashow, M.F. 2010. Molecular basis of plant cold acclimation: Insights gained from studying the CBF cold response pathway. *Plant Physiol.* 154:571–577.

Vaddepalli, P., S. Scholz, and K. Schneitz. 2015. Pattern formation during early floral development. *Curr. Opin. Genet. Dev.* 32:16–23.

Van Ommen Kloeke, A.E.E., J.C. Douma, J.C. Ordoñez, P.B. Reich, and P.M. van Bodegom. 2012. Global quantification of contrasting leaf life span strategies for deciduous and evergreen species in response to environmental conditions. *Glob. Ecol. Biogeogr.* 21:224–235.

Vergeer, P. and W.E. Kunin. 2013. Adaptation at range margins: Common garden trials and the performance of *Arabidopsis lyrata* across its northwestern European range. *New Phytol.* 197:989–1001.

Warren, C.R. and M.A. Adams. 2004. Evergreen trees do not maximize instantaneous photosynthesis. *Trends Plant Sci.* 9:270–274.

Weijers, D. and G. Jürgens. 2005. Auxin and embryo axis formation: The ends in sight? *Curr. Opin. Plant Biol.* 8:32–37.

Weinig, C. 2000. Limits to adaptive plasticity: Temperature and photoperiod influence shade-avoidance responses. *Am. J. Bot.* 87:1660–1668.

Wilkie, J.D., M. Sedgley, and T. Olesen. 2008. Regulation of floral initiation in horticultural trees. *J. Exp. Bot.* 59:3215–3228.

Wilson, B.F. 2000. Apical control of branch growth and angle in woody plants. *Am. J. Bot.* 87:601–607.

Yang, D., D.D. Seaton, J. Krahmer, and K.J. Halliday. 2016. Photoreceptor effects on plant biomass, resource allocation, and metabolic state. *Proc. Natl. Acad. Sci.* 113:7667–7672.

Yao, Q., L.R. Wang, H.H. Zhu, and J.Z. Chen. 2009. Effect of arbuscular mycorrhizal fungal inoculation on root system architecture of trifoliate orange (*Poncirus trifoliata* L. Raf.) seedlings. *Sci. Hortic.* 121:458–461.

3 Photosynthesis and Respiration in Woody Plants

Wenhao Dai and Qi Zhang

CONTENTS

3.1 INTRODUCTION

Plants are gigantic energy producers through the largest synthetic process on earth, photosynthesis. Photosynthesis in plants is a process by which light energy is converted to chemical energy that is stored in plants. Besides providing energy for plant growth and survival, photosynthesis fixes a large amount of carbon that can account for up to 40% the dry mass of a plant. Each year, land plants yield an estimated net primary production (NPP) of ~50 petagrams (10^{15} grams) of carbon that all life, including human beings, on earth depends on for their survival and activities (Field et al., 1998). For example, fossil fuels (coal, oil, and gas) are photosynthetic products made by plants, particularly trees, millions of years ago. Photosynthesis consumes CO_2 and produces O_2 and helps maintain the balance of O_2 and CO_2 in the atmosphere. Photosynthesis provides humans and animals direct food sources (grains, vegetables, fruits), such as carbohydrates, proteins, nutrients, and fibers, and these products can be converted to other food sources such as meat and milk after they are consumed by animals.

Woody plants are a special group in the plant kingdom as pertaining to energy production and accumulation, due to their unique characteristics, such as perennial nature, large size and deep root system, and long life expectancy, as well as being evergreen in some species. These characteristics mean woody species' photosynthesis and respiration can be unique, including non-foliar photosynthesis, winter photosynthesis, and age-related changes in photosynthesis. Photosynthesis and/or respiration occur year-round in woody plants, particularly in temperate, subtropical, and tropical regions; therefore, the environmental conditions in both growing and non-growing seasons play important roles in regulating photosynthesis and respiration, affecting the biomass accumulation and distribution in plants. Key environmental factors that affect photosynthesis and respiration of woody plants are discussed in this chapter and other chapters in this book.

3.2 PHOTOSYNTHESIS

3.2.1 Photosynthesis Site

The leaf is the primary site of photosynthesis, although photosynthesis occurs in all green parts of a plant. The leaf structure appears to be specialized for the complete photosynthetic process. For example, the upper leaf surface is covered by a layer of cuticle cells that prevent water loss. The stomata on the leaf surface allow air exchange between the leaf and the atmosphere while minimizing water loss. Through stomata, plants uptake carbon dioxide (CO_2) and release oxygen (O_2). The leaf stomata are sensitive to changes in environmental conditions including temperature, moisture, light intensity, and air pressure. Plants are able to regulate the photosynthetic process by changing the aperture of stomata. Inside the leaf, the dense palisade cells keep a large number of chloroplasts and a loose arrangement of sponge cells that close to stomata and the vascular system, providing enough intercellular space for importation of air and water and exportation of photosynthetic products. Other green tissues of a plant, such as cotyledon, stem, aerial root, flower, and fruits can be sites of photosynthesis because of the presence of chloroplasts; however, photosynthetic efficiency greatly varies. In woody species, photosynthesis in non-foliar parts, such as bark and fruit, may play an important role in certain periods of plant growth and development, and details of non-foliar photosynthesis are discussed in Section 3.5.1 in this chapter.

Photosynthesis is conducted in chloroplasts where light energy is captured and converted into different forms of chemical energy, such as sugar and other organic molecules that are stored in different plant tissues. Chloroplasts are oval-shaped organelles with a double-membrane coat. Included in the inner membrane, the space that is filled with colorless liquid (matrix) is called stroma within which are stacks of hollow disks called grana. The stroma comprises the most volume of a chloroplast and contains enzymes involved in the process of energy conversion and storage. The stroma contains its own DNA (chloroplast DNA), ribosomes, and RNAs. The genome of chloroplasts has been intensively researched in plants. In higher plants, the grana are dispersed in the stroma. Individual thylakoids in each granum are connected through the lumens (internal space).

Chlorophyll, a green pigment responsible for the green color in plants, is located on the thylakoid membrane. Chlorophyll acts as the photoreceptor in photosynthesis. In plants, there are two forms of chlorophyll, chlorophyll a (*chl a*) and chlorophyll b (*chl b*), which have different absorbance of light wavelengths. The absorption peaks of *chl a* and *chl b* are at 400–500 and 600–700 nanometer (nm), respectively, allowing them to absorb violet-blue and orange-red wavelengths, respectively In plant photosynthesis, *chl a* plays a major role in trapping light energy, and *chl b* makes its contribution to photosynthesis by transferring the trapped energy to *chl a*. Other pigments involved in photosynthesis in plants include carotenes and xanthophyll.

3.2.2 Photosynthesis Stages

A series of reactions are involved in the four stages of an entire photosynthetic process in the plant: light capture, electron transport, ATP generation, and carbon fixation. These reactions can be classified into two groups: light dependent (light) and light independent (dark) reactions.

3.2.2.1 Light Dependent (Light) Reactions

Light dependent (light) reactions occur on the thylakoid membrane. Chlorophylls attached to the proteins on the membrane absorb light energy. The light energy is then used to split water and release oxygen and hydrogen by removing electrons (photolysis). There are two photosystems (centers) in the light reactions: photosystem I (PSI) and photosystem II (PSII). The PSII reaction center is the chlorophyll molecule with an absorption peak at 682 nm (P_{680}), and the PSI center is the chlorophyll dimer with an absorption peak at 700 nm (P_{700}). In PSII, chlorophylls capture light energy to break down water and release electrons. These electrons are transferred through an electron transfer chain from PSII to PSI in the thylakoid membrane, using the energy captured from light. Electrons are eventually accepted by an important metabolic nucleotide, nicotinamide adenine dinucleotide phosphate (NADP), that is reduced to $NADPH_2$. The movement of electrons is coupled with the movement of the proton (H^+) from the stroma to the lumen, generating a pH gradient across the thylakoid membrane. When protons move back to the stroma to balance the pH gradient, a so-called proton-motive force is built-up. This force is used to generate the high-energy adenosine triphosphate (ATP) molecules using adenosine diphosphate (ADP) and inorganic phosphate (Pi). The process of ATP synthesis is called photophosphorylation.

3.2.2.2 Light Independent (Dark) Reactions

Light independent (dark) reactions occur in the stroma, during which carbon is fixed to produce carbohydrates using ATP and NADPH generated from the light reactions. There are three reactions – carboxylation, reduction, and regeneration – involved in carbohydrate synthesis. The three reactions can proceed in either light or dark conditions, as long as sufficient ATP and NADPH are available; therefore, these are termed light independent or dark reactions. During the process of carbon fixation, inorganic CO_2 is bound to ribulose-1,5-bi-phosphate (RuBP) catalyzed by the principal enzyme Rubisco (ribulose-1,5-biphosphate carboxylase/oxygenase) to generate 3-phosphoglyceric acid (3-PGA). The 3-PGA is then reduced to triose-phosphate (G3P) using ATP and NADPH generated from the light reactions. Remaining in the chloroplasts and using ATP, the majority of the G3P is used to regenerate RuBP to capture another CO_2. The process from using RuBP to regenerating RuBP is called the photosynthetic carbon reduction (PCR) cycle or the Calvin–Benson (Calvin) cycle. The rest of G3P generated from carboxylation leaves the chloroplast to produce sucrose and other metabolites in cytosol. In summary, 6 CO_2 are fixed by Rubisco to generate 12 PGA, during which 12 ATP and 12 NADPH are used to reduce 12 PGA to 12 G3P. Two of the G3P are used to synthesize one molecule of sucrose and the other 10 G3P are used to regenerate 6 RuBP that remain in the Calvin cycle for further CO_2 fixation.

The principal enzyme in CO_2 fixation is Rubisco (ribulose-1,5-biphosphate carboxylase/oxygenase) and is the most abundant enzyme on earth. Rubisco controls carboxylation efficiency, and its activity is regulated by pH and Mg^+ in the stroma and atmospheric CO_2. Rubisco is more active at pH of 8 and high Mg^+. Light influences Rubisco activity through its regulation of pH and Mg^+. Under the dark condition, the stroma's pH is low (~5), while in light, the stroma's pH is high (~8) and the level of Mg^+ in stroma also increases. Rubisco is an inefficient enzyme and needs to be activated by a CO_2 molecule binding to an allosteric site on Rubisco; therefore, the higher the CO_2 level, the more active the Rubisco. Rubisco is not substrate-specific and can catalyze fixation of both CO_2 and O_2. In the presence of O_2, light stimulates production of CO_2 by oxygenating RuBP to one molecule of PGA that is available for further metabolism in the PCR cycle and another 2-C molecule of phosphoglycolate that is toxic to plants. The phosphoglycolate is then metabolized in a series of reactions in the peroxisome and the mitochondrion, resulting in the release of a molecule of CO_2 and recovery of the remaining carbon by the PCR cycle. This process of RuBP oxygenation is called photorespiration (PR), as it involves the reoxidation of products previously assimilated in photosynthesis. The ratio of CO_2 to O_2 (CO_2/O_2) and temperature play roles in regulating photorespiration. A high CO_2/O_2 increases carboxylation, while a low ratio of CO_2/O_2 stimulates oxygenation. High temperature favors photorespiration as the solubility of O_2 is less reduced by high temperature than that of CO_2.

3.2.2.3 C4 (Hatch–Slack) and Crassulacean Acid Metabolism (CAM) Pathway

Plants that perform photosynthesis solely through the PCR or Calvin cycle are called C3 plants because the first stable carbon product in carboxylation reaction is a three-carbon acid, PGA. Another group of plants are called C4 plants because the first detectable carbon products are four-carbon compounds such as aspartic, malic, and oxaloacetic acids. This type of photosynthesis system is known as C4 or the Hatch–Slack pathway. C4 plants have a special leaf structure to separate the site of O_2 production from the Rubisco enzyme. This special structure is called Kranz anatomy which is where a sheath of thick-walled cells (i.e. a bundle sheath) containing large chloroplasts surrounds the vascular bundle. These large chloroplasts in the bundle cells contain a large volume of stroma, but few stacking thylakoids, indicating a poorly developed PSII system (less O_2 generated). The bundle cells are connected with the adjacent thin-walled mesophyll cells that contain very few chloroplasts and large intercellular space. The C4 plants initiate their carboxylation in the mesophyll cells where the bicarbonate ion HCO_3^- is bonded to phosphoenolpyruvate (PEP) catalyzed by phosphoenolpyruvate carboxylase (PEPcase) to produce C4 organic acids (OAA) (Figure 3.1A). Located in the cytosol, PEPcase is a light-activated enzyme that catalyzes a rapid fixation of carbon to PEP at a very low concentration of HCO_3^-. The organic acids are then diffused to the bundle sheath cell. In the bundle sheath cell, the organic acid is decarboxylated by reducing $NADP^+$ to NADPH and releasing CO_2 and a C3 organic acid (pyruvate or alanine). The released CO_2 enters into the Calvin cycle taking place in the bundle sheath chloroplast. The C3 organic acid diffuses back to the mesophyll cell and is converted to pyruvate to regenerate PEP. Because the bundle sheath cells are very close to the vascular bundle, removal of the Calvin cycle products is quick.

The CO_2 assimilation rate can be 2–3 times higher in C4 plants than in C3 plants; however, two more ATP are needed to regenerate PEP for each CO_2 being assimilated in C4 plants. The Calvin cycle does not occur in the mesophyll cell in C4 plants due to a lack of Rubisco, and CO_2 molecules are transported in the form of organic acid (malate or aspartate) to the bundle sheath cell. The CO_2 level in the bundle sheath cell can be up to ten-fold higher than that in C3 plants, and any CO_2 escaping from the Calvin cycle may be recaptured in the mesophyll cell. The dense chloroplasts and high concentration of CO_2 in the bundle sheath cell allow a very high efficiency of carbon fixation. The concentration of CO_2 in the bundle sheath cell is so high that any oxygenation reactions of Rubisco are inhibited.

Photorespiration has been considered as a costly and inefficient process regarding energy input (5 ATP and 3 NADPH) and carbon acquisition (a net loss of carbon) compared to CO_2 reduction in the Calvin cycle (3 ATP and 2 NADPH, a net gain of carbon) (Hopkins and Hüner, 2009). Although the leaf photosynthesis rate of C3 plants is usually greater that C4 plants, C4 plants show little advantage at whole planting level than those that C3 plants (Snaydon, 1991). It may be because (1) photorespiration protects C3 plants from photooxidative damage, which may occur under conditions of high light when there is limited CO_2 supply, by permitting continued operation of the electron transport system (Hopkins and Hüner, 2009); (2) photorespiration produces amino acid serine, which has a fundamental role in plant metabolism and signaling (Gregory, 1989); and (3) CO_2 produced in photorespiration may be refixed (Gregory, 1989; Pallardy, 2008, and references therein); or simply because of the week light intensity in dense plant stands (Snaydon, 1991).

Many succulents and a few other plants perform another type of carbon fixation, Crassulacean acid metabolism (CAM), to deal with very hot and dry conditions. These plants close their stomata when the temperature is very high in the daytime to reduce water loss and open the stomata at night when the temperature is low. At night, CO_2 is fixed in cytosol by PEPcase to produce malate. Malate

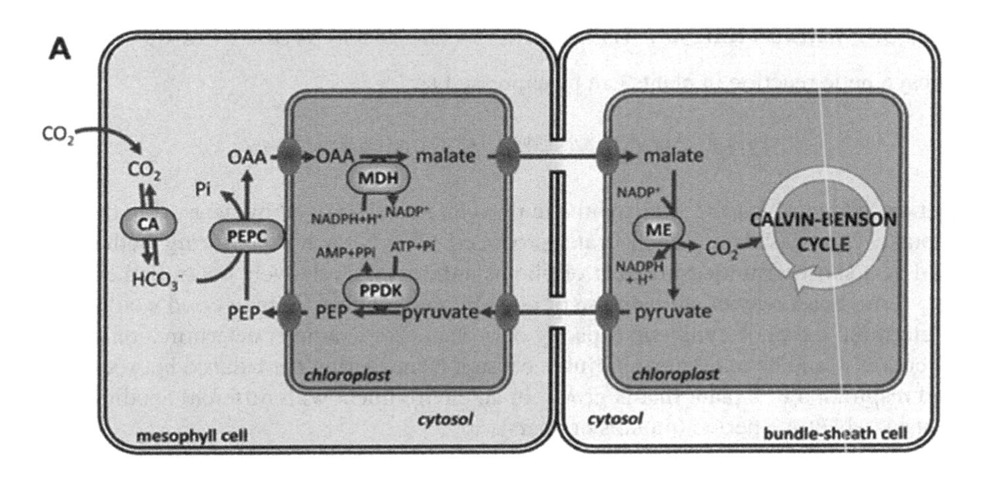

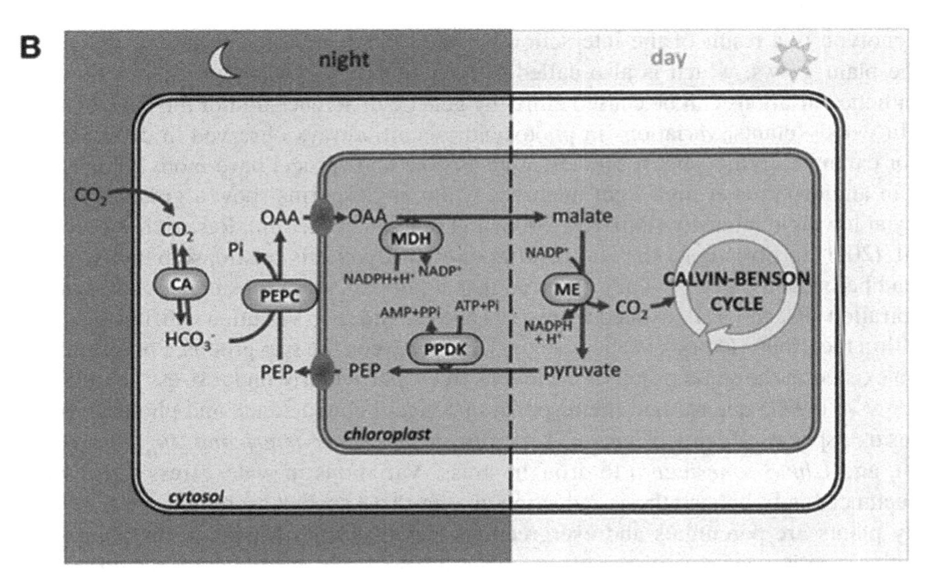

FIGURE 3.1 (A) C4 (Hatch–Slack) pathway. CO_2 is fixed in mesophyll cells by PEPC to form oxaloacetate (OAA). OAA is reduced to malate in mesophyll cells and then export to bundle sheath cells. Malate is decarboxylated to yield CO_2 that is fixed by Rubisco to enter the Calvin cycle. (B) (CAM) Pathway. At night, CO_2 is fixed PEPC to produce OAA in cytosol and then reduced to malic acid that is transported to the vacuole. In daylight, malic acid is released from the vacuole and decarboxylated to release CO_2 that is fixed by Rubisco to enter the Calvin cycle. CA-carbonic anhydrase; PEPC-phosphoenolpyruvate carboxylase; MDH-NADP malate dehydrogenase; ME-malic enzyme; PPDK-pyruvate phosphate dikinase; OAA-oxaloacetate; PEP-phosphoenolpyruvate (From Michelet et al., 2013. *Front. Plant Sci.* 4:470. doi: 10.3389/fpls.2013.00470).

is transported to the vacuole and becomes malic acid. In daylight, ATP and NADPH are generated in the light-dependent reactions. Malic acid is released from the vacuole, then decarboxylated, yielding CO_2 to be fixed by Rubisco to enter the Calvin cycle (Figure 3.1B). In both C4 and CAM plants, PEP and HCO_3^- are catalyzed by PEPcase to produce C4 organic acids that are subsequently decarboxylated to provide CO_2 for the Calvin cycle. These two processes are spatially separated in the C4 plants (mesophyll and bundle sheath cells) but occur in the same type of cells (mesophyll) in CAM plants. The CAM process has been only detected in one genus of trees (*Clusia* spp.) (Pallardy, 2008, and references therein).

3.2.3 Photosynthesis Rate and Its Influencing Factors in Woody Plants

The photosynthetic reaction in plants can be expressed as:

$$12H_2O + 6CO_2 \xrightarrow{\text{Sunlight}} 6O_2 + C_6H_{12}O_6 + 6H_2O$$

The equation indicates that the photosynthetic rate can be determined by measuring the quantities of CO_2 uptake, O_2 released, or carbohydrates produced. Clearly, such a measuring method may only be applied as a short-term measurement of photosynthesis in a relatively controlled environment. For a long-term measurement, an increase of plant biomass within a time period would be a better way to determine the photosynthetic capacity of a plant. For example, determination of a net dry weight increase would be more meaningful because it is measuring the balance between photosynthesis and respiration of a plant that is grown in an environment with different conditions/stresses over a considerable time period (months or years).

3.2.3.1 Variations in Photosynthesis Rate

Plant phenotype is a result of the interaction between the genetic make-up and the environment where the plant grows, which is also called G (genetic) × E (environment) interaction; therefore, photosynthetic variations can be caused either by genetic difference and/or the environmental conditions. In woody plants, variations in photosynthesis are always observed in different species or types. For example, gymnosperm species, such as pine and spruce, have more efficient photosynthesis than angiosperms at high light intensity, while angiosperms show a greater photosynthetic efficiency at low light intensity (Pallardy, 2008, and references therein). Research was conducted by Chu et al. (2011) to investigate the leaf respiration/photosynthesis relationship and variation in 39 woody and herbaceous species. They reported that significant variations in net photosynthesis (P_n), dark respiration (R), and R/P_n existed across 39 species, and such variations (differences) were even larger within the groups (woody vs. herbaceous) than between the two groups. Photosynthetic variations also existed in the same group of deciduous trees, particularly under stress conditions. Davies and Kozlowski (1977) determined the responses (stomatal conductance and photosynthesis) of six deciduous tree species (*Acer saccharum*, *Acer rubrum*, *Fraxinus americana*, *Juglans nigra*, *Cornus amomum*, and *Ulmus americana*) to drought stress. Variations in water stress sensitivity, stomatal conductance, and photosynthesis reduction among these species were observed. A large group of woody plants are perennials and evergreen species that keep leaves on the plant year-round. Photosynthetic activities in evergreen plants occur during dormant and winter time; therefore, evergreen plants have a longer time period of photosynthetic activity than deciduous plants, which helps to compensate for an overall low photosynthesis in evergreen species during the growing season averaged over a year (Pallardy, 2008, and references therein).

3.2.3.2 Diurnal and Seasonal Variations

The photosynthesis rate changes throughout the day. Under regular environmental conditions (temperature, light intensity, and air humidity), most woody species follow the daily change of the double-peaked daily pattern in which the first and second peaks of the daily photosynthesis rates occur generally around noon and late afternoon., The causes of such a pattern or diurnal variations include the daily changes of light intensity and temperature and other factors, such as humidity, wind, rainfall, cloud, and fog all of which play their roles. These variables may regulate the operation of stomatal aperture, water loss, and photosynthetic capacity of mesophyll cells (Berry and Björkman, 1980; Bernacchi et al., 2002; Yamori et al., 2014). For example, water loss by extreme high temperature and high wind speed may cause stomatal closure and thus a midday reduction in photosynthesis (Berry and Björkman, 1980; Yamori et al., 2014).

Seasonal variation in photosynthesis is one of the specific characteristics of woody species due to their perennial nature. There are significant differences in photosynthesis between deciduous and evergreen plants (Pallardy, 2008 and references therein). Variations in the same functional group (deciduous vs. evergreen) are mainly due to seasonal changes of temperature, light intensity, photoperiod, and water availability. For deciduous species, fast spring re-foliation and fall defoliation play important roles in the rapid increase in photosynthesis in spring and sudden decrease in fall. The peak of photosynthesis often occurs in the summer and early fall in deciduous species due to high temperature, high light intensity, and long light time. In summary, the annual photosynthetic capacity of a deciduous plant largely depends on leaf development in the early growing season and the retention of leaves in late fall. The later green foliage is held, the higher the photosynthetic rate of the plant. For evergreen woody species, the photosynthesis rate gradually increases in spring along with the increase of air and soil temperature. Photosynthetic capacity reaches its maximum in the summer and then gradually decreases in fall when the temperature gets lower. Winter photosynthesis is often seen in evergreen species. The rate of winter photosynthesis in evergreens mainly depends on the temperature, particularly night temperature. Often time for most evergreen species, if the night temperature is below freezing point, photosynthesis becomes negligible (Pallardy, 2008 and references therein); therefore, the annual photosynthetic capacity of an evergreen plant is largely dependent on low temperature and its duration during the winter. Seasonal variation in photosynthesis in woody plants is also affected by environmental conditions. Diurnal and seasonal variations in photosynthetic performance related to water/drought condition in two evergreen Mediterranean shrubs (*Erica multiflora* and *Globularia alypum*) was studied by Llorens et al. (2003). Differences in maintaining water potential year-round between the two species was observed. Seasonal fluctuations in water potential and leaf gas exchange rates were found in *G. alypum*, with the lowest values recorded in summer, while *E. multiflora* did not experience such significant fluctuations between seasons. That overall water potential in *G. alypum* was lower than in *E. multiflora* indicates that they use different strategies in response to drought stress in which *G. alypum* tolerated lower water potentials and *E. multiflora* appeared to be a drought-avoiding species. Zhang et al. (2007) researched the response of six woody species with different leaf phonology, evergreen, winter deciduous (WD), and drought deciduous (DD), to the drought condition. Two strategies for water use in response to drought were also used by these species. Evergreen and WD species employed drought avoidance strategy by closing their stomata to maintain a constant relative water content (RWC). In contrast, DD species used an avoidance strategy to first respond to drought by decreasing stomatal conductance and RWC to maintain a high photosynthetic capacity, and then drought tolerance strategy of shoot dieback when the drought was prolonged. Tucci et al. (2010) conducted a study of diurnal and seasonal variations in photosynthesis of peach palms (*Bactris gasipaes*), a tropical species under subtropical conditions that faced large seasonal variations in temperature and rainfall. Results showed that the highest photosynthesis rate (P_n, net CO_2 assimilation) was observed in mid-summer followed by fall, spring, and winter. The same pattern of diurnal variations in P_n was observed in all seasons, i.e. the peak value occurred at 8:00 to 9:00, then gradually declined from 11:00 to late afternoon. The pattern of diurnal change of stomatal conductance also followed the same pattern as P_n, indicating that the diurnal and seasonal variations in photosynthesis of peach palms might be stomatal control regulated by the environmental conditions.

3.2.3.3 Environmental Factors Affecting Photosynthesis Rate

The photosynthesis rate is generally affected by the plant parts including the leaf (or other photosynthetically functional parts) structure and age, growth vigor/condition, and also environmental factors, such as light/radiation, CO_2 concentration, temperature regimes, air humidity/fog, and other physiological factors.

Irradiance on photosynthesis – Plants capture energy from light for carbon assimilation; therefore, the light condition has a significant impact on photosynthesis. The amount of light available to the plant for photosynthesis can be expressed using photosynthetic active radiation (PAR). PAR light is the light with a wavelength of 400 to 700 nanometers (nm). The wavelength of light generally determines light energy: the shorter the wavelength, the greater the light energy. While PAR provides the range of light for plant photosynthesis, the amount of PAR a lighting system produces in a certain time period can be measured using photosynthetic photon flux (PPF). The unit of PPF is micromoles produced per second ($\mu mol \cdot s^{-1}$). The amount of PAR that actually arrives at the plant or a plant received is called PPFD (photosynthetic photon flux density). PPFD is a direct expression of the irradiance level and often be described as micromoles produced per square meter per second ($\mu mol \cdot m^{-2} \cdot s^{-1}$). On a sunny summer day, the PPFD is about 2000 $\mu mol \cdot m^{-2} \cdot s^{-1}$, while in a plant tissue culture room, the PPFD is around 30–100 $\mu mol \cdot m^{-2} \cdot s^{-1}$.

Light response curve – The rate of photosynthesis generally increases along with an increase of the irradiance level within a certain range; however, the increase of light intensity does not always increase the photosynthetic rate. When light intensity increases to a point/level, the photosynthesis rate does not increase (called light saturation point). When the irradiance level is below a certain point, there is no net photosynthetic product accumulated (called light compensation point). It is known that the cause of light saturation is mainly due to limited CO_2 supply at high light intensities, and the cause of light compensation is due to an increase of carbohydrate consumption by respiration, particularly at low light intensities. Different plants or even different leaves in the same plant show variations in response to light intensity. Such variations indicate the difference in photosynthetic efficiency of different plants/leaves that can be measured using the slope of the linear phase of the curve. For example, sun plants have a higher light compensation point than shade plants, indicating that sun plants need more strong light to reach their peak photosynthesis, and also sun plants assimilate more CO_2 at saturating light intensities. In general, the greater the curve slope is, the higher photosynthetic efficiency will be. While plants show species-specific characteristics in response to light intensities, individual plants have the ability to adjust to changes in light conditions.

Photodamage and its protection – Excess irradiance causes plant damage (photodamage) caused by the excess energy captured by photosystems (Aro et al., 1993). Zavafer et al. (2015) reported that the photodamage to PSII might be initially caused by excess light energy absorbed by the oxygen evolving complex (OEC), while the excess energy captured by photosynthetic pigments may inhibit the repair of damaged PSII (Takahashi and Murata, 2005). Excess light energy that causes damage is often through destruction of the photosynthetic apparatus, such as membranes and chlorophyll. Plants have the ability to develop certain mechanisms to mitigate the damage caused by excess irradiance or other stresses. It has been well documented that there is an accumulation of reactive oxygen species (ROS) at the cellular level, when plants are under various stresses including high light intensity. Mechanisms exist in plants to avoid damage by excess ROS. The xanthophyll cycle is one protective mechanism that plants have developed to protect against oxidative stress generated by high-light intensity. Six types of xanthophyll cycle are engaged in antioxidant defense in plant cells (Latowski et al., 2004, 2011). Some products are effective quenchers of ROS and others facilitate dissipation of captured light energy in light-harvesting complexes (LHC) of the PSII system. Xanthophylls act as accessory light-harvesting pigments that can help the de-excitation of chlorophyll, protecting for cellular damage from the toxic effects of light (Horton et al., 1996; Niyogi et al., 1997).

Low light adaptation – Plants have the ability to adapt to different environmental conditions including light intensity and photoperiod. Depending on their response to light intensity, plants are usually classified into two groups: shade-tolerant (shade) or shade-intolerant (sun) plants. Shade-tolerant plants are able to grow well under relatively low light intensity, while shade-intolerant plants grow poorly and even die under low light intensity conditions. Often time plant performance/growth is determined by the plant's ability to conduct photosynthesis under low light intensity to compensate for respiration and sustain growth. Such differences in light requirements are also seen in different leaves of the same plant. Evidence of such adaptation can be seen in the change of morphology and anatomical structure of the leaf in different plants and different locations of leaves on the same plant. For example, shade leaves are often thinner and larger than sun leaves. Morphologically, shade leaves have fewer palisade cells, more sponge mesophyll cells, and more chlorophyll. Thin and large leaf surfaces of shade leaves help light penetration and reduce respiration because there are fewer deep cells. The special structure of shade leaves help enhance photosynthetic efficiency by harvesting more sunlight at a low light level. Sun leaves, on the other hand, are usually thicker and have more developed palisade cells and less space in sponge mesophyll cells.

Temperature – Temperature affects plant growth and development through the enzyme-involved biological processes including the photosynthetic process. Plants have adapted to a relatively wide range of temperatures. Some plant species conduct photosynthetic activities at a temperature close to freezing, and others can conduct photosynthesis above 35 to 40°C (Berry and Björkman, 1980). However, photosynthesis can be inhibited under low or high temperatures due to the damage of the photosynthetic system including membranes of cells/organelles and enzymes. Significant differences in photosynthetic responses to temperatures among plant species can be traced back to the environment of their origins. Plants acquired a series of adaptations to temperature (and other environmental conditions) during a long period of evolution. These adaptations help plants maintain the functional integrity of the photosynthetic apparatus when they face environmental stresses. As mentioned, the temperature impact on photosynthesis is largely executed by regulating the activities of a series of photosynthetic enzymes. Temperature extremes have other impacts on photosynthesis. An extremely low temperature, particularly a temperature below freezing point, causes cell/tissue damage through the formation of ice, which may kill the plants that are not hardy or have not acclimatized for the winter (Berry and Björkman, 1980; Allen and Ort, 2001). Non-freezing low temperatures (chilling) will usually not cause death of the plant, but can inhibit synthesis in the photosynthetic apparatus. For example, the low temperature of 10°C slowed down the accumulation of chlorophylls in developing chloroplasts (Allen and Ort, 2001). Low temperature and light intensity may interactively contribute photoinhibition, which is defined as the light-dependent decrease in the photosynthetic rate. Photoinhibition may occur when irradiance is in excess of that required for the evolution of O_2 or the assimilation of CO_2 (Murata et al., 2007 and references therein). A classic explanation for photoinhibition is that the low temperature lowered the light saturation point, therefore increasing the sensitivity of PSII to light intensity, indicating that low temperature enhanced photoinhibition of PSII under strong light (Berry and Björkman, 1980). However, Murata et al. (2007) determined that low temperature actually suppresses the synthesis of proteins that are used for the repair of PSII rather than causing damage to PSII.

High temperatures (heat) above a certain threshold level can cause damage to plant growth and development. Such threshold levels vary among species and different developmental stages (Wahid et al., 2007 and references therein). During plant growth, high temperature affects a series of biological processes in the plant including photosynthesis and respiration. High temperature extremes can cause death of plant tissues and the whole plant (Wahid et al., 2007). Photosynthetic activities/rate can be affected through inhibiting synthesis of chlorophyll, reducing photosynthetic pigments and proteins, and degrading photosynthetic apparatus by heat-induced ROS (Camejo et al., 2006). Under the heat condition, CO_2 assimilation is decreased by reduced stomatal conductance and inactivated enzymes such as Rubisco (Crafts-Brandner and Salvucci, 2000), while the rate of respiration is increased; therefore, net photosynthesis decreases (Nakamoto and Hiyama, 1999). The temperature response of the plant is also species-specific. C4 plants in general have a higher optimal temperature (30 to 45°C) than C3 plants (20 to 25°C), which is partially due to the higher temperature requirement of PEPcase (30 to 35°C) as compared to Rubisco (20 to 25°C) (Hopkins and Hüner, 2009). Sage and Kubien (2007) reviewed the differential temperature responses of C3 and C4 photosynthesis in a normal range of temperatures. The capacities of a few photosynthetic enzymes such as Rubisco, RuBP regeneration, and P_i regeneration are factors that limit photosynthetic efficiency. The response of photosynthetic enzymes to temperature is related to the concentration of CO_2. For example, in C3 plants, under low CO_2 conditions, the capacity of Rubisco is the limiting factor in photosynthesis, while at high CO_2, the capacity of P_i regeneration, electron transport, or Rubisco activase might be limiting factors. Rubisco capacity appears to be the factor limiting net CO_2 assimilation in chilling-tolerant C4 plants at temperatures $<$20°C; however, it is uncertain that Rubisco capacity is still the limiting factor when temperatures are elevated. Sage and

Kubien's review (2007) pointed out that a better understanding of plant response to temperature stress will put us in a better position to predict and mitigate the effects of global climatic change on photosynthesis in higher plants.

3.3 RESPIRATION

Plants need energy and carbon skeletons to sustain their growth and development. As discussed in the previous part of this chapter, plants capture and convert light energy into chemical energy through photosynthesis. The captured energy is stored in the plant as complex compounds, such as sugar, starch, protein, etc. The process plants use to release energy is called respiration.

3.3.1 Aerobic Respiration

Aerobic respiration is expressed as:

$$C_6H_{12}O_6 + 6O_2 + 6H_2O \rightarrow 6CO_2 + 12H_2O + \text{Energy}\ (36\ \text{ATP})$$

In aerobic respiration, the organic molecule glucose is oxidized to release energy (ATP) in the presence of O_2. The O_2 itself is reduced to form H_2O, and the carbon atoms of glucose are released as CO_2. The energy released from respiration is available to the plant in the form of adenosine triphosphate (ATP) that is involved in almost all biological processes in living plant cells.

There are four major stages in the aerobic respiration process. The first stage is glycolysis in which one 6C-glucose is split into two 3C-pyruvates. Glycolysis occurs in the cytosol and can be seen in all living cells. In plants, glucose is derived from sucrose (12C) and other carbohydrates, such as starch. A series of enzymes are involved in the process of glycolysis (Figure 3.2). In glycolysis, ATP is used in two steps at the early stage, while ATP and NADPH are synthesized as the energy

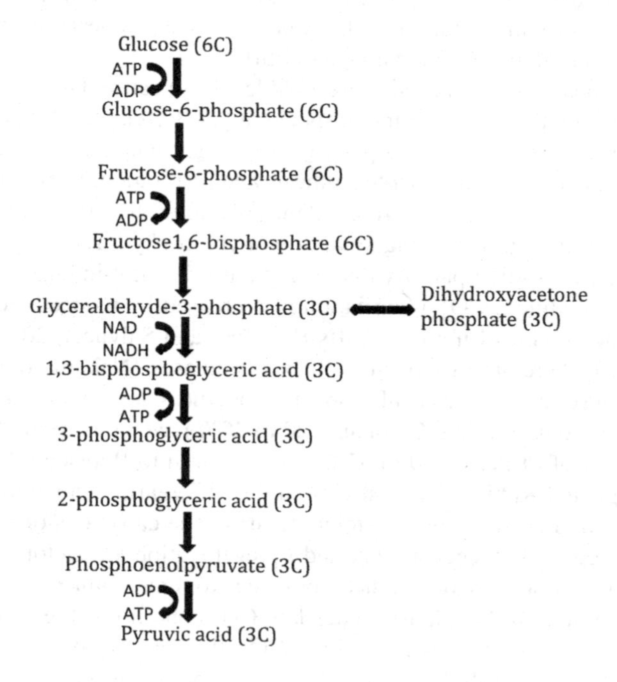

FIGURE 3.2 Steps of glycolysis.

being released from the breakdown of the chemical bond. In summary, two molecules of ATP are used for the initial 'jump start' of the aerobic respiration process and four molecules of ATP and two molecules of NADPH are produced at the later steps; therefore, two molecules of net ATP and two molecules of NADPH can be released from one molecule of glucose in the process of glycolysis.

Pyruvates generated from glycolysis diffuse from the cytosol to the mitochondrion. In mitochondrion, with the presence of O_2, each of two pyruvates is oxidized into 2C-acetyl coenzyme A (acetyl CoA) during which one NAD^+ is reduced to one NADH and one CO_2 is released.

The tricarboxylic acid cycle (TAC), also called Krebs cycle or citric acid cycle, occurs in the inner mitochondrial matrix. The TAC is an aerobic process. The first step of the TAC cycle is the combination of acetyl CoA (2C) and oxaloacetic acid (OAA) (4C) catalyzed by citrate synthase. The first product of the TAC is a 6C citrate. The second step is to change citrate to iso-citrate followed by a series of chemical reactions that result in the release of CO_2, NADH, and $FADH_2$. The last step of TAC is to regenerate OAA that is ready to accept another acetyl CoA to continue the TAC. In one step where succinyl-CoA is converted to succinic acid, one molecule of ATP is produced directly from ADP and inorganic phosphate (i.e. substrate-level phosphorylation). In summary, each TAC uses one acetyl CoA, three NADH and one $FADH_2$, and one ATP were produced and two CO_2 are released. The tricarboxylic acid cycle (TAC) can be described as:

$$\text{Acetyl CoA} + 3H_2O + 3NAD^+ + FAD^+ + ADP + Pi$$

$$\rightarrow 2CO_2 + 3NADH + 3H^+ + FADH_2 + ATP$$

The last stage of aerobic respiration is to release the energy that is stored in NADH and $FADH_2$ generated from glycolysis and the TAC cycle. Such energy release is achieved through an oxidation process of NADH and $FADH_2$ (also known as oxidative phosphorylation). NADH and $FADH_2$ are also known as the hydrogen carriers because they carry all the hydrogen from the original glucose. The hydrogen carriers move to the inner membrane of the mitochondrion. The inner membrane contains a series of electron carriers and enzymes that help the passage of electrons from one carrier to another. The oxidation process is coupled with the release of photons and electrons from the reduced NADH and $FADH_2$. A photon is pumped across the inner membrane to the inner membrane space using the energy from the reduction of NADH and $FADH_2$, and the electrons are passed along electron carriers, and this electron passing system is called the electron transport chain (ETC). The electrons finally reach the negatively charged oxygen molecule, the final acceptor of electrons at the end of the chain, to form water ($2H^+ + 2e^- + \frac{1}{2} O_2 \rightarrow H_2O$). During the process of electron transfer along the chain, hydrogen ions (H^+) are continuously pumped into the inner membrane space and accumulated which results in a gradient difference between the two sides of the inner membrane matrix. This process is similar to the production of protons in the intrathylakoid space in photosynthesis. ATP generation in aerobic respiration has the same principles as photosynthesis: (1) a proton gradient difference across the membrane; (2) the membrane is impermeable to protons; and (3) protons return to the matrix through an integral membrane protein complex, ATP synthase and ATP are produced (Hopkins and Hüner, 2009). Hydrogen ions move back to the inner membrane matrix due to the H^+ gradient across the inner membrane. The movement of H^+ from the inner membrane space to the inner membrane matrix is coupled with the release of energy that is used for the production of ATP from ADP and Pi catalyzed by ATP synthase. The number of ATP produced depends on the electron donor and the condition. Under an ideal condition (aerobic), one molecule of NADH produces three molecules of ATP, and one molecule of $FADH_2$ produces two molecules of ATP.

3.3.2 Anaerobic Respiration (Fermentation)

In the anaerobic condition, the pyruvate produced by glycolysis remains in cytosol and is fermented into alcohol (in yeast or plant) or lactate (in animal) and CO_2. Although no ATP is produced, fermentation plays a major role in recycling NADH to NAD^+, which is used in the oxidation of glyceraldehyde-3-P in glycolysis so that the small amount of ATP generated in glycolysis may help the plant survive under anaerobic conditions.

3.3.3 Energy Usage

Plant growth and development can be attributed to cell division, enlargement, and differentiation. Any activities involved in these biological processes in living cells need energy. Plants make energy via photosynthesis and store it in plant tissues in the forms of sugar, starch, and other carbohydrates. When plants need energy for their activities, it is released from those complex compounds and stored in ATP through phosphorylation that arises through the formation of ATP. There are two ways to achieve phosphorylation, substrate-level phosphorylation and oxidative phosphorylation. The difference between the two types of phosphorylation is the process and mechanism of ATP formation. In substrate level phosphorylation that occurs in glycolysis (in cytosol) and the tricarboxylic acid cycle (TAC) (in mitochondria matrix), a phosphate group (Pi) is physically added to ADP to form ATP. The Pi is directly removed from a substrate and transferred into ADP and NAD and FAD. In oxidative phosphorylation occurring along the ETC on the inner membrane of mitochondria, ATP is synthesized indirectly using the energy released from the ETC when the high energy molecules NADH and $FADH_2$ are oxidized.

3.3.4 Respiration Regulation at a Whole Plant Level

Metabolic activities involved in any stage of plant growth and development are supported by the energy released from respiration. Plants have the capability to adjust their own use of energy. When plants need more energy for growth or development, such as seed germination, fast vegetative growth, and flower and fruit production, respiration speeds up; when plant growth slows down, respiration rate decreases. Plant respiration is also regulated by the substrate level. Plants are able to regulate respiration rate in response to change in substrate. When substrate is limited, respiration slows down, when substrate level is high, respiration increases. Under certain conditions, the substrate level can rapidly increase, which stimulates a much higher respiration rate that may result in energy waste. To mitigate wasting energy resulting from unnecessarily high respiration or to regulate the energy balance in response to a changing environment, some plant species as well as algae, fungi, and yeasts have developed special mechanisms for regulating respiration. A typical mechanism in plants is cyanide-resistant respiration (CRR), an alternative pathway that transfers electrons from the ubiquinone pool to oxygen, bypassing the cytochrome chain via an alternative oxidase (AOX) on the inner membrane of mitochondria. AOX is not sensitive to the conventional inhibitors of cytochrome respiration, such as cyanide (CN^-). Under stress conditions, such as heat, cold, and other oxidative stresses, activities of the cyanide-resistant respiration increase to correct the imbalance of respiratory carbon metabolism and electron transport caused by changes in the supply or demand of carbon, reducing power and ATP (Moore and Siedow, 1991; Vanlerberghe and McIntosh, 1997). Cyanine-resistance respiration consumes oxygen and produces water and heat without synthesis of ATP. Heat production from CRR benefits seed germination and plant flowering in cold conditions. CRR can also enhance fruit ripening and disease resistance. Overall, cyanine-resistance respiration help plants maintain normal activities by releasing excess energy and regulating glycolysis, the TCA cycle, and oxidation of substrates, particularly under stress conditions.

3.3.5 Respiration Measurement and Rate

The respiratory quotient (RQ) is used to measure respiration efficiency and expressed as (Pallardy, 2008):

$$RQ = CO_{2\,(released)}/O_{2\,(used)}$$

The ratio of CO_2 production to O_2 consumption reflects the oxidation level of the substrate including carbohydrates, proteins, and fatty acids. The characteristics of the substrate dramatically affect the RQ. The examples were provided by Pallardy (2008) to calculate RQ. When a molecule of glucose is oxidized, six molecules of O_2 will be used and six CO_2 molecules will be released; therefore, the RQ is 1:

$$C_6H_{12}O_6 + 6O_2 \rightarrow 6CO_2 + 6H_2O;$$

$$RQ = 6CO_2 / 6O_2 = 1.0$$

When other substrates, such as proteins and fatty acids are oxidized, more O_2 will be needed; the RQ will be less than 1. For example, to oxidize tripalmitin, more O_2 is needed, therefore, the RQ is about 0.7:

$$C_{51}H_{98}O_6 + 72.5O_2 \rightarrow 51CO_2 + 79H_2O;$$

$$RQ = 51CO_2 / 72.5O_2 = 0.7$$

In certain instances, the RQ will be greater than 1 because less O_2 is used to oxidize the substrate. For example, three molecules of O_2 are needed to oxidize one molecule of malic acid into four molecules of CO_2, so the RQ is 1.33:

$$C_4H_6O_5 + 3O_2 \rightarrow 4CO_2 + 3H_2O;$$

$$RQ = 4CO_2 / 3O_2 = 1.33$$

In an anaerobic condition, such as fermentation, the RQ will be much higher than in the aerobic condition because no O_2 is available and only CO_2 is produced.

The respiration rate can also be expressed as the concentration/volume of O_2 (Q_{O2}) used or CO_2 (Q_{CO2}) released from a certain weight of fresh or dry tissue/protoplast within a time period, such as μmol/g/h (CO_2), μmol/mg/h (CO_2), or μl/g/h (O_2).

3.4 EFFECTS OF ENVIRONMENTAL CONDITIONS ON RESPIRATION

Regulation and control can occur at any stage and at any biological level of respiration. Under regular conditions, the yield of ATP is mainly regulated by mitochondrial oxidative phosphorylation; however, under environmental stress conditions, the regulation and control of respiration may be specific to certain respiration stages and stresses. For example, under flooding conditions (oxygen deficiency or hypoxia), regulation at the glycolysis stage becomes essential for ATP production (Fernie et al., 2004; van Dongen et al., 2011; António et al., 2016). The effects of environmental conditions on respiration and photosynthesis have been popular topics for numerous research projects, and the literature on these topics is considerable. While the effects of individual stresses on plant respiration and photosynthesis and their regulation are discussed in chapters in this book, this chapter only briefly summarizes the overall effects of the major environmental factors at whole-plant level. Readers are therefore referred to a few reviews on these topics (Atkin et al., 2005; Smith and Dukes, 2013).

3.4.1 Temperature

A series of enzymes are involved in respiration; therefore temperature is the major factor affecting the rate of respiration. In general, within a certain range of temperature, the higher the temperature, the greater the respiration rate in plants. For most plant species, the temperatures of 25 to 35°C are most suitable for respiration. Temperatures above 50°C can inhibit the activity of many enzymes, while at temperatures below 10°C, respiration becomes very weak.

3.4.2 Oxygen and Carbon Dioxide

Generally, plant respiration increases with an increase of O_2 concentration in the environment. Concentration of O_2 determines respiration direction, anaerobic or aerobic respiration. If O_2 concentration is below 10%, aerobic respiration sharply decreases and anaerobic respiration will start; if O_2 concentration is 10 to 20%, respiration is always aerobic. If O_2 concentration is too high (>20%), respiration rate may decrease. Concentration of CO_2 is also an important factor influencing respiration. When CO_2 is above 5%, respiration rate starts decreasing; therefore, accumulation of CO_2 in the root area caused by flooding or compacted soil significantly inhibits plant respiration.

3.4.3 Water

Water is solvent of enzymes that have to be activated in the presence of water; therefore, the water content in plant tissues/cells is a limiting factor for respiration. If the water in plant tissues or cells is too low, such as in mature seeds and old tissues, respiration in these tissues will be very slow. However, in certain stages of plant growth and development, such as fast-growing tissues and organs, water stress may enhance respiration because more energy is needed or may increase the activity of the alternative path of respiration.

3.5 UNIQUE FEATURES IN WOODY PLANTS

3.5.1 Non-Foliar Photosynthesis

In higher plants, photosynthesis takes place predominantly in green leaves because the leaf structure is perfectly specialized for the photosynthetic process; however, any other chlorophyll-containing parts are potentially able to conduct photosynthesis (Aschan and Pfanz, 2003 and references therein). In woody plants, some non-foliar structures including stem, root, and fruit contain a certain amount of chlorophyll and are capable of conducting photosynthesis that plays a significant role in carbon assimilation in these plants (Pfanz et al., 2002). A clear difference in the CO_2 source for photosynthesis between the leaf and non-foliar tissue has been observed. Certain cells in cortex, ray cells, and pith of a stem mainly used CO_2 that is released from respiration. This process of how the CO_2 is used is called carbon re-fixation, while the CO_2 used for photosynthesis in the leaf comes mainly from the atmosphere.

Saveyn et al. (2010) reported that photosynthesis in the trunk in all three tested tree species contributed to trunk growth and development. Their result also showed that stem photosynthesis not only contributed to the trunk's direct carbon gain, but also contributed to the development of new leaves after manual defoliation, indicating that no-foliar photosynthesis is one of the mechanisms that plants employ to survive stresses, particularly defoliation-causing stresses. The research on aspen (*Populus*) bark photosynthesis indicated that bark photosynthesis provided about 10 to 15% carbon during the mid-summer vegetative period and a greater contribution from bark photosynthesis might be seen when leaf photosynthesis was limited under certain conditions of environmental stress (Kharouk et al., 1995; Pfanz et al., 2002; Aschan and Pfanz, 2003).

The primary functions of the root system are to anchor the plant and uptake water and nutrients from the soil. However, some plant species, particularly tropical species including some orchid species, lack leaves and even stems; therefore, their carbon assimilation largely depends on aerial roots (Aschan and Pfanz, 2003 and references therein). These aerial roots contain chlorophyll capable of conducting photosynthesis. In some cases, root photosynthesis is the only carbon source for growth and development. These species are called epiphytes, including ferns, orchids, bromeliads, and aroids. Root tissues of some epiphytic species, such as orchids, contain a large amount of chlorophyll that produces sufficient photosynthetic products from the root to support the needs of life. Similar to stem photosynthesis, root photosynthesis is also a recycling process of internally respired CO_2 due to a lack of gas-exchange structures on the root surface. Some specialized plant species, mainly in the orchid group, have no developed leaf and/or stem structure, and they can only perform photosynthesis in the green root (Aschan and Pfanz, 2003).

Fruit is another major organ that provides an additional contribution of carbon assimilation to the plant (Blanke and Lenz, 1989; Aschan and Pfanz, 2003 and references therein). Fruit photosynthesis is particularly important for many tree fruit species because it is directly involved in the development of fruit yield and quality. In fruit, chlorophyll is contained in the surface layers (epidermis and hypodermis) and inner perivascular tissues (tissues surrounding the vascular bundles). The fruit chlorophyll content varies in various fruit species and the ratio of *chl a* to *chl b* is also different in different fruit tissues (Aschan and Pfanz, 2003). Photosynthetic activity in the fruit is affected by the quantity of chlorophyll, the amount of light the fruit chlorophyll received, and CO_2 intensity at the photosynthetic site. During fruit development, young fruits always contain more chlorophyll than ripe fruits, and light penetration allows for greater accumulation of pigment and compound in the fruit skin. Most importantly, the frequency of stomata on the fruit surface is much higher in young fruit than in mature fruit. The outer layer of the fruit contains stomata, although the number is much lower (1 to 10% less) than that in the abaxial surface of the respective leaf. Because the number of stomata on a leaf or a fruit for a particular plant species does not change, the stomatal frequency on the fruit decreases as the fruit expands. This can explain why the efficiency of carbon assimilation in young fruit is much greater than that in mature fruit (Blanke et al., 1987; Blanke and Lenz, 1989). More recently, Xu et al. (2016) reported that fruit photosynthesis significantly contributed to the fruit development of hickory (*Carya cathayensis*) and pecan (*Carya illinoensis*), including enhancement of fruit growth and seed development. A significant difference in fruit photosynthesis between the two species was observed. Pavel and Dejong (1993) reported that fruit photosynthesis of a late maturing peach (*Prunus persica*) 'Cal Red' contributed about 3 to 9% and 8 to 15% of carbohydrates of weekly requirements at the early and midseason fruit development stages, respectively. Mature fruits contributed 3 to 5% of the total fruit carbohydrate requirements. Hieke et al. (2002) reported that in lychee trees (*Litchi chinensis*), leaf photosynthesis mainly in recently expanded dark green leaves contributed >95% of the carbohydrates and fruit photosynthesis provided about 3% for fruit growth during the fruit development period. Hiratsuka et al. (2012) conducted research on the effect of fruit bagging on CO_2 assimilation by Satsuma mandarin (*Citrus unshiu*) fruit. They found that gross photosynthesis in the fruit peaked at 99 days after full bloom, which was comparable to leaf photosynthesis. The net photosynthetic rate was undetectable in the bagged fruit due to the decline in chlorophyll content. They also found that less sugar and organic acid were accumulated in the bagged fruit, and further analysis found that bagging inhibited carbon assimilation through photosynthesis and PEPcase, causing low sugar content in the mature fruit.

3.5.2 Age-Related Changes in Photosynthesis

Flowering plants can be classified into annual, biannual, and perennial based on their life cycle. A typical life cycle of a flowering plant is from seed to seed, including the stages of seed germination, vegetative growth, reproductive growth, and seed production. The period from seed germination to flowering is called the juvenile/asexual stage, and the rest of the life cycle is called the mature/sexual stage. For some plant species, it only takes a few weeks or months to complete the life cycle, while others may take years and decades to finish their life cycle. For example, annual species complete their life cycle within a year or growing season, while biannual species need two years or growing seasons to complete the life cycle, and often time the first year for vegetative and the second year for reproductive growth.

Perennials, a large group of plant species, complete their life cycle in more than three years. The majority of woody species are perennial although some biannual plants have the woody structure. Trees and shrubs dominate the perennial group. In some perennial species, a portion of a plant will die at the end of each growing season. For example, the leaf and/or the entire above-ground part die during the winter, and new growth starts from the crown or root system below the ground in the following spring. With other woody plants, such as evergreen species, the leaf remains on the plant for a few years. No matter if it is a deciduous or evergreen species, woody perennials have two contrasting stages of development. The stage before the plant reaches its reproductive capacity is called the vegetative/juvenile stage. Once the plant is able to fully produce flowers/fruits/seeds, it is considered to have reached the reproductive/sexual stage or what is usually called 'mature stage'. Besides acquiring reproductive capability, there are also other significant changes in morphology and physiology between the two stages, resulting in significant changes in the structure and function of the plant. Such changes will be continuously taking place during the whole life of the plant.

The concept of aging or age-related changes are various depending on the subject, even in the same species group. For example, the leaf of a deciduous species completes its growing and aging process in one growing season, while the leaf of an evergreen plant may take more than one year to finish the transition process of 'young' to 'old'; therefore, changes in leaf morphology (size, shape, or chlorophyll content) and physiology (photosynthesis and respiration) vary between these plants. At a whole plant level, differences in aging and age-related changes between different plant types clearly exist. For most of the woody species, the current year's growth (new growth) starts from the last or last several years' growth (old growth); therefore, the age of the plant can significantly affect the new growth and may influence biomass accumulation and reproductive capacity (yield and quality) through photosynthesis and respiration.

Age-related changes in photosynthesis have been studied for a long time. Singh and Lal (1935) found that the photosynthesis rate of flax, sugar cane, and wheat increased during leaf maturation and then decreased when the leaf was aging. Research by Freeland (1952) investigated the effect of leaf age on photosynthetic efficiency in six conifer species. The results showed that maximum photosynthesis occurred at leaf maturation time in the first growing season, and the rate of photosynthesis decreased with the increasing age of the leaf. The study also found that 3-year-old conifer leaves still retained the photosynthetic capacity. Research has been conducted to investigate the aging effect on photosynthesis in woody species at the whole plant level (Bond, 2000, and references therein). The review summarized that photosynthesis and stomatal conductance of woody plants were generally reduced with increasing age with a few exceptions, i.e. photosynthesis increased with increased plant age. These exceptions can be explained by the nature of woody plants, particularly some tree species because the size (tree height, root depth) increases with the increase in age. The taller the tree, the more sun light it can receive, and the deeper the root, the more water/nutrients it can uptake, particularly at times of environmental stresses including drought and shading (DeSoyza et al., 1996).

Aging is a developmental process during which changes in plant metabolism take place as the plant ages. These changes are often associated with a specific developmental period, such as maturity. Potential mechanisms of age-related changes in photosynthesis include

1. the differential expression of genes in different development states, particularly between juvenile and mature plants – Fox example, some genes associated with chlorophyll binding, anthocyanin production, and proline-rich proteins expressed differently between juvenile and adult larch (*Larix*) and English ivy (*Hedera helix*) plants, and changes in synthesis of plant hormones after the plant reached the mature state indicated that certain metabolism pathways alternate (Greenwood, 1995); however, such changes of hormones in other species have not been evidenced;
2. the increase of plant size is always associated with age increase – Plant size increase results in an increased hydraulic resistance because of the increased distance and the number of obstacles between roots and shoot apices and the increase in gravity, which eventually limit water supply for transpiration and could limit stomatal conductance and photosynthesis in big and old trees (Hubbard et al., 1999; Bond, 2000);
3. the aging effect on photosynthesis involved in the increased oxidative stress in chloroplasts – Munné-Bosch and Alegre (2002) compared a few bio-physiological parameters between

1- and 3-year-old (young) and 7-year-old (mature) *Cistus clusii* plants. The result showed that net photosynthesis was reduced in 7-year-old plants, while no difference was observed between 1- and 3-year-old plants. The age-related reduction in photosynthesis was associated with higher stomatal closure and reduction in the relative efficiency of PSII photochemistry. They also found that enhanced lipid oxidation in chloroplasts increased the level of malondialdehyde, MDA, and decreased chlorophylls and chloroplastic antioxidant defense levels in 7-year-old plants. In summary, oxidation stress in chloroplasts or other organelles slowly accumulates as plants age.

3.5.3 Winter Photosynthesis in Evergreen Species

Plants generally show no or very limited photosynthetic activities during the winter because of the annual leaf senescence in deciduous species and low temperature that is not suitable for photosynthesis. However, winter photosynthesis is often seen in evergreen plants, a group of plant species on which the foliage remains on the plant throughout the year (Wyka and Oleksyn, 2014, and references therein). Individual leaves of evergreen plants also go through the aging process, and the old leaves are always replaced with new growth; therefore, a completely green foliage can be seen year around. The majority of evergreen plants, particularly in the angiosperm group, belong to tropical species that favor warm and temperate climates. As discussed in previous paragraphs, photosynthesis can be conducted in non-foliar parts of a plant, such as bark; therefore, photosynthetic activities may be observed in deciduous plants during the winter, but are limited to warm regions. Winter photosynthesis mainly refers to evergreen species. The leaves of evergreen species often develop special structures and physiological adaptions to survive more than one season, particularly during low winter temperatures. Winter photosynthesis can compensate for energy loss by respiration but also provide energy for overwintering. The stored energy made during the winter will be used for bud breaking and fast growth in spring (Wyka and Oleksyn, 2014).

REFERENCES

Allen, D. J., and D. R. Ort. 2001. Impacts of chilling temperatures on photosynthesis in warm-climate plants. *Trends Plant Sci.* 6:36–42.

António, C., C. Päpke, M. Rocha, H. Diab, A. M. Limami, T. Obata, A. R. Fernie, and J. T. van Dongen. 2016. Regulation of primary metabolism in response to low oxygen availability as revealed by carbon and nitrogen isotope redistribution. *Plant Physiol.* 170:43–56.

Aro, E. M., I. Virgin, and B. Andersson. 1993. Photoinhibition of photosystem II. Inactivation, protein damage and turnover. *Biochim. Biophys. Acta* 1143:113–134.

Aschan, G., and H. Pfanz. 2003. Non-foliar photosynthesis—A strategy of additional carbon acquisition. *Flora* 198:81–97.

Atkin, O. K., D. Bruhn, and M. G. Tjoelker. 2005. Response of plant respiration to changes in temperature: Mechanisms and consequences of variations in Q_{10} values and acclimation. In *Plant Respiration. Advances in Photosynthesis and Respiration*, ed. H. Lambers and M. Ribas-Carbo, 95–135. Springer, Dordrecht, The Netherlands.

Bernacchi, C. J., A. R. Portis, H. Nakano, S. von Caemmerer, and S. P. Long. 2002. Temperature response of mesophyll conductance. Implications for the determination of RuBisCO enzyme kinetics and for limitations to photosynthesis in vivo. *Plant Physiol.* 130:1992–1998.

Berry, J., and O. Björkman. 1980. Photosynthetic response and adaptation to temperature in higher plants. *Annu. Rev. Plant Physiol.* 31:491–543.

Blanke, M. M., D. P. Hucklesby, and B. A. Notton. 1987. Distribution and physiological significance of photosynthetic phosphoenolpyruvate carboxylase in developing apple fruit. *J. Plant Physiol.* 129:319–325.

Blanke, M. M., and F. Lenz. 1989. Fruit photosynthesis. *Plant Cell Environ.* 12:31–46.

Bond, B. J. 2000. Age-related changes in photosynthesis of woody plants. *Trend Plant Sci.* 5:349–353.

Camejo, D., A. Jiménez, J. J. Alarcón, W. Torres, J. M. Gómez, and F. Sevilla. 2006. Changes in photosynthetic parameters and antioxidant activities following heat-shock treatment in tomato plants. *Funct. Plant Biol.* 33:177–187.

Chu, Z., Y. Lu, J. Chang, M. Wang, H. Jiang, J. He, C. Peng, and Y. Ge. 2011. Leaf respiration/photosynthesis relationship and variation: An investigation of 39 woody and herbaceous species in east subtropical China. *Trees* 25:301–310. doi: 10.1007/s00468-010-0506-x.

Crafts-Brandner, S. J., and M. E. Salvucci. 2000. RuBisCO activase constrains the photosynthetic potential of leaves at high temperature and CO_2. *Proc. Natl. Acad. Sci. U.S.A.* 97:13430–13435.

Davies, W. J., and T. T. Kozlowski. 1977. Variations among woody plants in stomatal conductance and photosynthesis during and after drought. *Plant Soil* 46:435–444.

DeSoyza, A. G., A. C. Franc, R. A. Virjinia, J. E. Reynolds, and W. G. Whitford. 1996. Effects of plant size on photosynthesis and water relations in the desert shrub *Prosopis glandulosa* (Fabaceae). *Am. J. Bot.* 83:99–105.

Fernie, A. R., F. Carrari, and L. J. Sweetlove. 2004. Respiratory metabolism: Glycolysis, the TCA cycle and mitochondrial electron transport. *Curr. Opin. Plant Biol.* 7:254–261.

Field, C. B., M. J. Behrenfeld, J. T. Randerson, and P. Falkowski. 1998. Primary production of the biosphere: Integrating terrestrial and oceanic components. *Science* 281:237–240. doi: 10.1126/science.281.5374.237.

Freeland, R. O. 1952. Effect of age of leaves upon the rate of Photosynthesis in some conifers. *Plant Phyiol.* 27:685–690.

Greenwood, M. S. 1995. Juvenility and maturation in conifers: Current concepts. *Tree Physiol.* 15:433–438.

Gregory, R. P. F. 1989. *Biochemistry of Photosynthesis*. 3rd ed. Wiley, New York, NY.

Hieke, S., C. M. Menzel, and P. Lüdders. 2002. Effects of leaf, shoot and fruit development on photosynthesis of lychee trees (*Litchi chinensis*). *Tree Physiol.* 22:955–961.

Hiratsuka, S., Y. Yokoyama, H. Nishimura, T. Miyazaki, and K. Nada. 2012. Fruit photosynthesis and phosphoenolpyruvate carboxylase activity as affected by lightproof fruit bagging in satsuma mandarin. *J. Am. Soc. Hort. Sci.* 137:215–220.

Hopkins, W. G., and N. P.A. Hüner. 2009. *Introduction to Plant Physiology*. 4th ed. John Wiley & Sons, Inc., Hoboken, NJ.

Horton, P., A. V. Ruban, and R. G. Walters. 1996. Regulation of light harvesting in green plants. *Plant Physiol.* 47:415–420.

Hubbard, R. M., B. J. Bond, and M. G. Ryan. 1999. Evidence that hydraulic conductance limits photosynthesis in old *Pinus ponderosa* trees. *Tree Physiol.* 19:165–172.

Kharouk, V. I., E. M. Middleton, S. L. Spencer, B. N. Rock, and D. L. Williams. 1995. Aspen bark photosynthesis and its significance to remote sensing and carbon budget estimate in the boreal ecosystem. *Water Air Soil Pollut.* 82:483–497.

Latowski, D., J. Grzyb, and K. Strzalka. 2004. The xanthophyll cycle – molecular mechanism and physiological significance. *Acta Physiol. Plant.* 26:197–212. doi: 10.1007/s11738-004-0009-8.

Latowski, D., P. Kuczyńska, and K. Strzałka. 2011. Xanthophyll cycle – A mechanism protecting plants against oxidative stress. *Redox Rep.* 16:78–90. doi: 10.1179/174329211X13020951739938.

Llorens, L., J. Peñuelas, and I. Filella. 2003. Diurnal and seasonal variations in the photosynthetic performance and water relations of two co-occurring Mediterranean shrubs, *Erica multiflora* and *Globularia alypum*. *Physiol. Plant.* 118:84–95.

Lodish, H., A. Berk, C. A. Kaiser, M. Krieger, A. Bretscher, H. Ploegh, A. Amon, and M. P. Scott. 2008. *Molecular Cell Biology*. 6th ed. W.H. Freeman and Company, New York, NY.

Michelet, L., M. Zaffagnini, S. Morisse, F. Sparla, M. E. Pérez-Pérez, F. Francia, A. Danon, C. H. Marchand, S. Fermani, P. Trost, and S. D. Lemaire. 2013. Redox regulation of the Calvin–Benson cycle: Something old, something new. *Front. Plant Sci.* 4:470. doi: 10.3389/fpls.2013.00470.

Moore, A. L., and J. N. Siedow. 1991. The regulation and nature of the cyanide-resistant alternative oxidase of plant mitochondria. *Biochim. Biophys. Acta* 1059:121–140.

Munné-Bosch, S., and L. Alegre. 2002. Plant aging increases oxidative stress in chloroplasts. *Planta* 214:608–615.

Murata, N., S. Takahashi, Y. Nishiyama, and S. I. Allakhverdiev. 2007. Photoinhibition of photosystem II under environmental stress. *Biochim. Biophys. Acta* 1767:414–421.

Nakamoto, H., and T. Hiyama. 1999. Heat-shock proteins and temperature stress. In *Handbook of Plant and Crop Stress*, ed. M. Pessarakli, 399–416. Marcel Dekker, New York.

Niyogi, K. K., O. Bjorkman, and A. R. Grossman. 1997. The roles of specific xanthophylls in photoprotection. *Proc. Natl. Acad. Sci. U.S.A.* 94:14162–14167.

Pallardy, S. G. 2008. *Physiology of Woody Plants*. 3rd Ed. Elsevier Inc., San Diego, CA.

Pavel, E. W., and T. M. Dejong. 1993. Estimating the photosynthetic contribution of developing peach (*Prunus persica*) fruits to their growth and maintenance carbohydrate requirements. *Physiol. Plant.* https://doi.org/10.1111/j.1399-3054.1993.tb05507.x.

Pfanz, H., G. Aschan, R. Langenfeld-Heyser, C. Wittmann, and M. Loose. 2002. Ecology and ecophysiology of tree stems: Corticular and wood photosynthesis. *Naturwissenschaften* 89:147–162. doi: 10.1007/s00114-002-0309-z.

Sage, R. F., and D. S. Kubien. 2007. The temperature response of C_3 and C_4 photosynthesis. *Plant Cell Environ.* 30:1086–1106.

Saveyn, A., K. Steppe, N. Ubierna, and T. E. Dawson. 2010. Woody tissue photosynthesis and its contribution to trunk growth and bud development in young plants. *Plant Cell Environ.* 33:1949–1958.

Singh, B. N., and K. N. Lal. 1935. Investigations of the effect of age on assimilation of leaves. *Ann. Bot.* 49:291–307.

Smith, N. G., and J. S. Dukes. 2013. Plant respiration and photosynthesis in global-scale models: Incorporating acclimation to temperature and CO_2. *Glob. Change Biol.* 19:45–63. doi: 10.1111/j.1365-2486.2012.02797.x.

Snaydon, R. W. 1991. The productivity of C3 and C4 plants-A reassessment. *Funct. Ecil.* 5:321–330.

Takahashi, S., and N. Murata. 2005. Interruption of the Calvin cycle inhibits the repair of photosystem II from photodamage. *Biochim. Biophys. Acta* 1708:352–361.

Tucci, M. L. S., N. M. Erismann, E. C. Machado, and R. V. Ribeiro. 2010. Diurnal and seasonal variation in photosynthesis of peach palms grown under subtropical conditions. *Photosynthetica* 48:421–429.

van Dongen, J. T., K. J. Gupta, S. J. Ramírez-Aguilar, W. L. Araújo, A. Nunes-Nesi, and A. R. Fernie. 2011. Regulation of respiration in plants: A role for alternative metabolic pathways. *J. Plant Physiol.* 168:1434–1443.

Vanlerberghe, G. C., and L. McIntosh. 1997. Alternative oxidase: From gene to function. *Annu. Rev. Plant Physiol. Plant Mol. Biol.* 48:703–734.

Wahid, A., S. Gelani, M. Ashraf, and M. R. Foolad. 2007. Heat tolerance in plants: An overview. *Environ. Exp. Bot.* 61:199–223.

Wyka, T. P., and J. Oleksyn. 2014. Photosynthetic ecophysiology of evergreen leaves in the woody angiosperms – A review. *Dendrobiology* 72:3–27.

Xu, Q., J. Wu, Y. Cao, X. Yang, Z. Wang, J. Huang, G. Xia, Q. Zhang, and Y. Hu. 2016. Photosynthetic characteristics of leaves and fruits of Hickory (*Carya cathayensis* Sarg.) and Pecan (*Carya illinoensis* K.Koch) during fruit development stages. *Trees* 30:1523–1534. doi: 10.1007/s00468-016-1386-5.

Yamori, W., K. Hikosaka, and D. A. Way. 2014. Temperature response of photosynthesis in C_3, C_4, and CAM plants: Temperature acclimation and temperature adaptation. *Photosynth. Res.* 119:101–117. https://doi.org/10.1007/s11120-013-9874-6.

Zavafer, A., M. H. Cheah, W. Hillier, W. S. Chow, and S. Takahashi. 2015. Photodamage to the oxygen evolving complex of photosystem II by visible light. *Sci. Rep.* 5:16363. doi: 10.1038/srep16363.

Zhang, J. L., J. J. Zhu, and K. F. Cao. 2007. Seasonal variation in photosynthesis in six woody species with different leaf phenology in a valley savanna in southwestern China. *Trees* 21:631–643. doi: 10.1007/s00468-007-0156-9.

4 Plant Growth Regulation

Yoo Gyeong Park, Abinaya Manivannan, Prabhakaran Soundararajan, and Byoung Ryong Jeong

CONTENTS

4.1 INTRODUCTION

Plant hormones and environmental factors play central roles in the regulation of growth, development, nutrient allocation, and source/sink transitions. Plant hormones play important functions in regulating developmental processes and signaling networks involved in plant responses to a wide range of biotic and abiotic stresses. The response and adaptation of plants are determined by complex interactions with metabolites, plant growth regulators (PGRs), secondary messengers, and downstream defense and/or homeostasis genes. Plants produce a wide variety of hormones that include abscisic acid (ABA), auxins, brassinosteroids, cytokinins, ethylene, gibberellins, jasmonic acid (JA), methyl jasmonate (MeJA), and salicylic acid (SA). Apart from the PGRs, vital environmental cues, such as light and temperature, also entrain several physiological processes in plants. Hence, in this chapter, the various roles of environmental factors and plant hormones in the regulation of plant growth and development are discussed.

4.2 PLANT HORMONES AND PLANT GROWTH REGULATORS (PGRS)

Plant hormones are small molecules that regulate plant growth and development, as well as responses to changing environmental conditions (Colebrook et al., 2014). Plant development is regulated and coordinated through the action of several classes of small molecules (plant hormones) that may act either close to or remote from their sites of synthesis to mediate genetically programmed developmental changes or responses to environmental stimuli (Davies, 2010).

Post-embryonic plant growth and development are sustained by the meristem, a source of undifferentiated cells that gives rise to the adult plant structures (Moubayidin et al., 2009). Two hormones, cytokinins and auxins, are well-known to act antagonistically in controlling meristem activities (Moubayidin et al., 2009). Auxins, cytokinins, and auxin-cytokinin interactions are usually considered important for regulating growth and organized development in plant tissues and organ cultures, as these two hormones are generally required (Evans et al., 1981).

In recent years, plant hormone biology research has been successful not only in elucidating the function of recently discovered hormones, such as brassinosteroids and JA and the novel hormone strigolactone (Gomez-Roldan et al., 2008), but also making major advances in understanding hormone perception and signal transduction, such as the discovery of a receptor for auxin (Kepinski and Leyser, 2005), gibberellin (GA) (Ueguchi-Tanaka et al., 2005) and ABA (Pandey et al., 2009).

Plant hormones play important roles in diverse growth and developmental processes, as well as various biotic and abiotic stress responses in plants (Bari and Jones, 2009). The signal molecules SA and JA or MeJA are endogenous plant growth substances that regulate plant growth and development, and responses to environmental stresses (Yao and Tian, 2005). JA and its MeJA are cyclopentanone compounds, a group of naturally occurring PGRs (Sembdner and Parthier, 1993). In this chapter, recent advances in the field of plant growth regulation of woody plants is reviewed for a better understanding of hormones and environmental factors in the modulation of plant growth, development, and regulation.

4.3 ROLES OF PGRS IN PLANT GROWTH AND REGULATION

4.3.1 Abscisic Acid (ABA)

Abscisic acid (ABA) was discovered in the 1950s to be a phytohormone affecting leaf abscision and bud dormancy (Skriver and Mundy, 1990). It is involved in the regulation of many aspects of plant growth and development including seed germination, embryo maturation, leaf senescence, stomatal aperture, and adaptation to environmental stresses (Wasilewska et al., 2008). ABA plays an important role in transducing the signal from root to shoot when plants are experiencing stressful conditions around the roots, giving rise to a water-saving anti-transpirant activity such as stomatal closure and reduced leaf/canopy expansion (Davies et al., 2002). ABA functions in controlling transpiration, dehydration tolerance, and other water status-associated processes, such as seed maturation and dormancy, and winter bud dormancy (Takezawa et al., 2011). ABA is also involved in processes of negative control of growth and development including inhibition of lateral root growth and inflorescence formation (Milborrow, 1974). The ABA treatment of apple fruit at lower concentrations promoted ethylene production in mature apple fruit (*Malus pumila* var. *domestica*), but higher concentrations inhibited it (Thimann et al., 1989). Involvement of ABA in growth cessation is unclear, and there is evidence for a role for ABA in differentiation of leaf and bud scale primordia in bud development in the deciduous shrub, Saskatoon berry (*Amelanchier alnifolia*) (Baldwin et al., 2000; Rohde et al., 2002).

4.3.2 Auxins

The phytohormone auxin has been implicated in diverse developmental processes throughout the life cycle of plants, including apical dominance, tropic responses, vascular development, organ patterning, flower development, and fruit development (Ghanashyam and Jain, 2009). A major site for auxin synthesis is the shoot apex and young leaf, from which it is transported down the plant in the polar transport stream (Ljung et al., 2001). Auxin in high concentrations applied to the decapitated or girdled stem of a woody plant can replace apical control and inhibit branch diameter growth, upward bending, and production of newly formed leaves in free growth species (Wilson and Archer, 1983; Timell, 1986). Auxin moving in this way can inhibit shoot branching because either blocking polar auxin transport or removing the shoot apex promotes shoot branching, with the branch promoting effect of removing the apex negated by the addition of auxin to the decapitated stump (Cline, 1991). Auxin indole-3-acetic acid (IAA) has long been recognized as an important hormone regulating a wide array of responses in plant growth and development, including cell division, elongation, and differentiation of vascular tissues as well as initiation of buds and lateral roots, and control of lateral branching (Thimann, 1969). Outgrowth of lateral buds closest to the cutting point was found by Wilson and Kelty (1994) in black oak (*Quercus velutina*). Furthermore, their finding that the removal of all large buds in black oak enhanced the outgrowth of epicormic buds was also supported. Harmer and Baker (1995) determined that decapitation of terminal buds in *Quercus petraea* was most effective in promoting branching when carried out during the second flush. Auxin also mediates plant responses to light and gravity (Moore and Evans, 1986).

4.3.3 Brassinosteroids

Brassinosteroids are a unique class of plant hormones that are structurally similar to animal steroid hormones and are involved in the regulation of growth, development, and various physiological responses in plants (Bajguz, 2007). Brassinosteroids have pleotropic effects and can induce a broad spectrum of cellular responses including stem elongation, pollen tube growth, leaf bending and epinasty, root inhibition, induction of ethylene biosynthesis, proton pump activation, xylem differentiation, and regulation of gene expression (Arteca and Arteca, 2001). Besides its positive effect on branch elongation, the 5F-HCTS- (5 alpha monofluro derivative-homocastasterone) driven enhancement of branch elongation is potentially useful to improve micropropagation techniques for woody species, in which branch elongation is typically a constraint for efficient micropropagation (Pereira-Netto et al., 2006). Bao et al. (2004) demonstrated a synergistical promotion of lateral root formation by brassinosteroids and auxin. The exogenous application of brassinosteroids increased plant tolerance to stresses of low and high temperatures, drought, and high moisture (Ali et al., 2008). Brassinosteroids increase biomass, chlorophyll contents, photosynthesis, and the potential of antioxidants and detoxification to improve plant stress tolerance (Divi and Krishna, 2009; Hasan et al., 2011).

4.3.4 Cytokinins

Cytokinins have the ability to induce plant cell division (Miller et al., 1955) and were discovered in the 1950s. In plants, cytokinins regulate numerous biological processes, including responses to environmental stresses via a complex network of cytokinin signaling (Ha et al., 2012) and other developmental processes including leaf senescence, apical dominance, chloroplast development, anthocyanin production, and regulation of cell division (Debi et al., 2005; Park and Jeong, 2012). Abdalla et al. (1985) reported that application of kinetin to *Adonis autumnalis* L. (Pheasant's-eye) increased plant height, stem diameter, and fruit weight, and caused early flowering. Clarke et al. (1994) proposed that cytokinin may inhibit the catalytic effects of ethylene on senescence at the level of its perception, as exogenous cytokinin had a stronger effect on the delay of senescence than did treatment with silver ions. Cytokinin, together with auxin, plays an essential role in plant morphogenesis, having a profound influence on the formation of roots and shoots and their relative growth (Debi et al., 2005; Park and Jeong, 2012). Cytokinin is known to promote the outgrowth of dormant axillary buds, thus antagonizing the activity of auxin (Werner and Schmülling, 2009). In *Picea abies*, a fixed-growth tree, there was a positive correlation between cytokinin (zeatin riboside) content and lateral bud size during the critical period of bud formation that determines shoot length (Chen et al., 1996).

4.3.5 Ethylene

Ethylene can accelerate leaf senescence and fruit ripening (Abeles et al., 1992). The gaseous phytohormone ethylene is a signal molecule that is active during both plant development and senescence. Ethylene is synthesized as a response to several biotic and abiotic stresses (Tingey et al., 1976). When up-regulated under stressful conditions such as heat and drought, ethylene can inhibit root growth and development, and it can also reduce shoot/leaf expansion (Pierik et al., 2006). Ethylene may activate processes related to plant defense such as the production of phytoalexins (Fan et al., 2000), pathogenesis-related proteins (Rodrigo et al., 1993), the induction of the phenylpropanoid pathway (Chappell et al., 1984), and cell wall alterations (Bell, 1981). Dandekar et al. (2004) reported differential regulation of ethylene with respect to fruit quality components in apple. During banana fruit ripening ethylene production triggers a developmental cascade that is accompanied by a huge conversion of starch to sugars, an associated burst of respiratory activity, and an increase in protein synthesis (Bapat et al., 2010). Liu et al. (1999) have analyzed the expression of ACC synthase gene in association with ethylene biosynthesis and ripening in banana. Ethylene accelerated leaf senescence under low O_3, but under acute O_3 elevation ethylene signaling seemed to be required for protection from necrotic cell death (Vahala et al., 2003).

4.3.6 Gibberellins (GA)

GA was originally identified as a substance secreted from the fungus *Gibberella fujikuroi* that causes 'bakanae' (or foolish seedling) disease in rice (Kurosawa, 1926). The plant hormone GA has long been known to modulate development throughout the plant life cycle (Fleet and Sun, 2005). They can stimulate growth of most organs through enhanced cell division and cell elongation, and also promote developmental phase transitions, including those between seed dormancy and germination, juvenile and adult growth phases, and vegetative and reproductive development (Colebrook et al., 2014). Gibberellic acid (GA_3) has also been used to stimulate flower development in callas (Cohen, 1981). Shifts in the relative concentrations of plant hormones also occur at maturation, and gibberellins can promote flowering or cause mature tissues to revert to juvenile conditions (Greenwood, 1995).

4.3.7 Jasmonic Acid (JA)

Jasmonic acid (JA) and its derivatives are, collectively referred to as JAs, are linolenic, acid-derived, cyclopentanone-based compounds and are widely distributed in plants (Wasternack and Kombrink, 2009). They were first characterized as potent regulators mediating defense responses, for example, resistance to insect attack and bacterial and fungal infection (Turner et al., 2002; Wasternack, 2007). The effect of JAs was also observed in several aspects of plant growth and development, including promotion of flower and fruit development, induction of tuberization and tendril coiling, inhibition of seed and pollen germination, and restriction of root growth (Wasternack, 2007). The JA mediated increase in endogenous ABA concentration is still under study in woody plant species like apple (Kondo et al., 2001).

4.3.8 MeJA

MeJA, collectively referred to as jasmonates, are important cellular regulators involved in diverse developmental processes, such as seed germination, root growth, fertility, fruit ripening, and senescence (Creelman and Rao, 2002). MeJA have been reported to modulate chlorophyll degradation and anthocyanin formation (Creelman and Mullet, 1997), aroma development (Olias et al., 1992), and ethylene production (Khan and Singh, 2007; Lalel et al., 2003). MeJA effectively alleviate the adverse effects of NaCl, although the extent of mitigating NaCl effect was not significant in growth attributes of soybean (Yoon et al., 2009). Earlier reports suggested that salt stress reduced the biomass of tomato (Kaya et al., 2001), pea (Ahmad and Jhon, 2005), and rice (Yeo et al., 1999), although shoot dry weight was more sensitive to salinity than root dry weight (Essa, 2002). MeJA induced extensive biochemical and anatomical changes in Norway spruce, *Picea abies* (L.) Karst., similar to those caused by pathogens or artificial wounding (Franceschi et al., 2002; Martin et al., 2002, 2003).

4.3.9 Salicylic Acid (SA)

Salicylic acid (SA) is a phenolic compound widely distributed in plants. Its concentration widely differs among species (up to 100-fold differences) (Raskin et al., 1990). SA plays a central role as a signaling molecule involved in local defense reaction and systemic resistance induction (Durner et al., 1997). SA also plays a role in plant response to abiotic stresses such as drought (Chini et al., 2004), chilling (Kang and Saltveit, 2002), heavy metal toxicity (Freeman et al., 2005), heat (Larkindale et al., 2005), and osmotic stress (Borsani et al., 2001). In this sense, SA appears to be, just like in mammals, an 'effective therapeutic agent' for plants (Rivas-San Vicente and Plasencia, 2011). Previous studies have also shown that SA could ameliorate the damaging effects of heavy metals on membranes in rice (Mishra and Choudhuri, 1999), reduce low temperature stress in maize and banana (Janda et al., 1999; Kang et al., 2003), and improve thermotolerance of tall fescue seedlings (He et al., 2002). Pre-harvest application of SA tends to suppress postharvest anthracnose disease severity caused by *Colletotrichum gloeosporioides* in mango fruit (Zainuri et al., 2001) and the application of acibenzolar-S-methyl, a functional analogue of SA, on melon plant prior to flowering could effectively inhibit infections of several postharvest fungal diseases in melon fruit (Huang et al., 2000).

4.4 ENVIRONMENTAL REGULATION OF PLANT GROWTH AND DEVELOPMENT

In general, plants inherit the ability to respond to an extensive variety of external or environmental signals. Among the environmental factors, light and temperature are the major cues that impact plant growth and development.

4.4.1 Light-Mediated Entrainment of Plant Growth and Development

Light is vital for plant life because the light environment determines growth, morphology, and developmental changes in plants. The regulation of plant growth and development in correspondence to light is interlinked with endogenous hormone levels and molecular regulations. In general, the light signal acts as a crucial cue for several developmental phenomena in plants, including seed germination, photo-morphogenesis, and flowering response.

4.4.1.1 Impact of Light Intensity on Plant Growth and Development

Plants are entrained by light intensity, spectrum, direction, and photoperiod. Light intensity is the degree of brightness or the amount of light sensed by plants (Stuefer and Huber, 1998). Light intensity tends to change with the time of the day, season, geographic location, and weather. Plants are able to respond to light intensity changes at a very low fluence (~100 pmol·m^{-2}). For instance, light intensity induces transcription of photosynthetic genes, low fluence (1 mmol·m^{-2}) response needed for seed germination, and high irradiance (more than 10 mmol·m^{-2}) needed for inhibition of hypocotyl elongation (Fankhauser and Chory, 1997). Though plants inherit the ability to sense light signals, the optimum light intensity differs from species to species. Light intensity greatly influences the photosynthetic process in plants. For instance, an increase of light intensity enhances the rate of photosynthesis (Anderson et al., 1995; Walters and Horton, 1994; Weston et al., 2000), while a decrease in light intensity reduces the photosynthesis process, resulting in the reduction of plant growth and development (Bohning and Burnside, 1956; Dennison and Alberte, 1982). Likewise, excessive light intensity can cause scorching of plant tissues (Ma et al., 2010; Varade et al., 1970), reduction in crop yield (Fahlgren et al., 2015; Loomis and Williams, 1963), decrease in chlorophyll content (Danesi et al., 2004; Lichtenthaler et al., 1981; Sato et al., 2015), and increase in plant temperature (Berry and Bjorkman, 1980; Chabot and Chabot, 1977).

4.4.1.2 Impact of Photoperiod on Plant Growth and Development

The term 'photoperiod' denotes the amount of time a plant receives light. Previously, the concept of photoperiod was recognized by the length of light periods that instigate the flowering response. However, later researches revealed that floral development was critically regulated by the length of uninterrupted dark periods. Thus, photoperiod acts as a vital factor that controls flowering in many plants (Heide, 1977; Ito and Saito, 1962; Park et al., 2015; Piringer and Cathey, 1960; Schaffer et al., 1998; Somers et al., 1998). Broadly, plants are classified into three major groups depending on their requirement for light duration for flowering, including short day, long day, and day neutral plants. In general, short day plants (or long night plants), such as chrysanthemum and poinsettia, need a short day light time for flowering (Duca, 2015; Higuchi et al., 2012; Moe et al., 1992; Somers et al., 1998). On the contrary, long day plants (or short night plants), such as beet, radish, lettuce, spinach, and potato, flower when the day length exceeds the critical day length (Lang et al., 1977; Suge and Rappaport, 1968; van Dijk and Hautekeete, 2007; Vince et al., 1964; Vreugdenhil and Struik, 1989; Zeevaart, 1971). Day neutral plants induce flowering regardless of day length. A well-known example of day neutral plants are miniature roses (Cerny et al., 2003; Hwang et al., 2005). Taken together, the duration of light influences the regulation of flowering response in plants. Similarly, light direction or phototropism is also involved in the entrainment of plant growth and morphogenesis via the regulation of endogenous phytohormones (Whippo and Hangarter, 2006). Moreover, the duration of light influences the synthesis of phytochemicals with vital functions in plant growth and development.

4.4.1.3 Impact of Spectral Quality on Plant Growth and Development

Light quality denotes light color or wavelength perceived by the plant. Broadly, plants respond to a wide range of spectral quality ranging from ultra violet (UV) B to far-red (FR), in which red and blue light are two primary wave lengths influencing plant growth and development (Fankhauser and Chory, 1997; Manivannan et al., 2015; Park et al., 2012). Plants absorb a comparatively smaller amount of green light, and it is considered the least effective for growth and development in plants (Mirkovic and Scholes, 2015). In most plants, vegetative growth is primarily influenced by blue light (Fankhauser and Chory, 1997; Manivannan et al., 2015). Similarly, red light also affects stem length, leaf morphology, and flowering in combination with blue light (Ani et al., 2015; Chen et al., 2015; Park et al., 2015).

4.4.1.4 Photoreceptors – Masters Behind Light Perception

Plants perceive numerous light signals from the environment through a set of chromoproteins with photo-sensory reception called photoreceptors (Galvão and Fankhauser, 2015). According to Smith (2000), photoreceptors are distributed throughout the plant surface, and the structural component of photoreceptors consists of an apoprotein attached to different types of chromophores. The apoprotein-chromophore complex determines the nature of the spectrum to be absorbed by the plant (Fankhauser and Chory, 1997; Galvão and Fankhauser, 2015; Smith, 2000). Several photoreceptors corresponding to perception of red/far-red to blue/UV light have been discovered. The red/far red spectrum is perceived by the phytochromes (phyA-E). In terrestrial plants, phytochromes play roles in seed germination, de-etiolation process, stomatal development, flowering response, senescence, phototropism, gravitrophism, and shade avoidance (Chen et al., 2015; Kami et al., 2010; Leivar and Monte, 2014; Parks et al., 1996; Quail et al., 1995; Robson and Smith, 1996; Sakuraba et al., 2014; Smith, 1995). In addition, phytochromes act as a major component that entrains the circadian rhythm of plants by offering temporal signals to maintain the developmental stage to be initiated at an appropriate point (Smith, 2000). Phytochromes are also involved in regulating secondary metabolites and pigment synthesis in plants (Engelsma and Meijer, 1965; Fankhauser and Chory, 1997). A phytochrome consists of an amino terminal photo-sensory region covalently bonded to a phytochromobilin tetrapyrrol chromophore and a carboxyl terminal (Fankhauser and Chory, 1997). Phytochromes are initially produced as an inactive compound (Pr) that is activated by red light absorption at 660 nm and then converted to the active form (Pfr) with the major absorption peak at 730 nm (Butler et al., 1959). Two types of phytochromes, type I (*phyA*) and type II (*phyB* and *phyC*), have been discovered in higher plants. However, very little information is available regarding *phyD* and *phyE* (Clack et al., 1994; Quail et al., 1995).

Blue light has also been seen to stimulate several physiological responses in plants, such as retardation of hypocotyl growth, flowering control, phototropic curvature, regulation of circardian rhythm, cotyledon expansion, movement of stomata, and stimulation of transcription factors (Fankhauser and Chory, 1997). Blue light is absorbed by the photoreceptors called cryptochromes that were first discovered in *Arabidopsis* (Ahmad and Cashmore, 1993; Koornneef et al., 1980). On the contrary to phytochromes, only two cryptochromes have been identified as cry1 and cry2 with partially overlapping roles in *Arabidopsis* (Ahmad and Cashmore, 1993; Fankhauser and Chory, 1997; Lin et al., 1995). The cry1 plays an important function in de-etiolation, flowering response, seed dormancy, and senescence (Chaves et al., 2011; Kami et al., 2010). Cryptochromes function in varying ranges of blue light spectrum (320 to 400 nm). Cryptochrome is initially activated by blue light; however, prior to illumination, cryptochromes exist as homodimers with an amino terminal consisting of an oxidized FAD chromophore (Chaves et al., 2011; Conrad et al., 2015). Upon being activated by blue light, the electron transfer chain is initiated, and the reduction of the chromophore to the signaling state is triggered. This cascade reaction causes structural modulations in the amino and carboxyl terminals of cryptochromes that result in the interaction between the signaling intermediate and the photoreceptor (Conrad et al., 2014; Engelhard et al., 2014). Other than cryptochromes, phototropins (a type of photoreceptor) also play a vital role in the perception of blue light in plants. Overall, a complex range of biological molecules and networks are involved in light perception from environment to the conception of light signals in the photosynthetic process and photomorphogenesis in woody plants. Better understanding of the molecular regulation of photoreceptors during the growth and development of plants could unravel the complex mechanisms involved in light mediated growth entrainment.

4.4.2 Temperature-Mediated Entrainment of Plant Growth and Development

In nature, plants are subjected to numerous changes in temperature even in a single day. Such modulations in temperature could have a considerable influence on plant physiology. For instance, a chilling temperature reduces the activity of enzymes, membrane rigidity, perturbation of protein complexes, and generation of reactive oxygen species (ROS), hinders photosynthesis, and induces electrolyte leakage across membranes (Ruelland and Zachowski, 2010). Similarly, an increase in temperature also leads to several lethal consequences to plants. Temperature, particularly in the extreme, may not only result in deleterious effects to plants, it also plays a vital role in regulating physiological functions, such as movement of corolla (van Doorn and van Meeteren, 2003), synchronization of internal biological clock (Thines and Harmon, 2010), flowering (bud vernalization) (Kim et al., 2009), and seed germination (stratification) (Finch-Savage and Leubner-Metzger, 2006).

Much research has focused on temperature stress tolerance, but and very little information exists on how temperature mediates plant development, particularly in the field of molecular regulation and sensing mechanisms in plants. It has been reported that variations in temperature can be perceived by the plant, which influences a few molecular changes (Ruelland and Zachowski, 2010).

Temperature variations induced modulation in the flow of membrane fluidity in *Arabidopsis* (Seo et al., 2010). The up shift or down shift in temperature may result in the unfolding of protein molecules (Pastore et al., 2007). Moreover, heat-induced conformational changes of proteins were observed in several plants, such as *Arabidopsis* (Lee et al., 2009), alfalfa (*Medicago saliva*) (Sangwan et al., 2002), and barley (*Hordeum vulgare*) (Xue, 2003). According to Hardham and Gunning (1978), a decrease in temperature stimulated de-polymerization of microfilaments and tubules in plants. Evidently, the disassembly of microtubules occurred in tobacco cells after cold treatment (0°C) followed by the disappearance of radial actin filaments (Pokorna et al., 2004). Likewise, temperature affects the activity of enzymes involved in some primary metabolic processes including photosynthesis (Ruelland and Zachowski, 2010).

4.4.2.1 Temperature-Mediated Flowering Response

Temperature along with photoperiod signals also play a key role in plant transition from a vegetative to reproductive stage of plants. Several signaling networks regulate flowering, particularly through the expression of the Flowering Locus (*FT*) gene (Michaels, 2009). A synergistic response to temperature and phytochromes on the FT-dependent flowering control were studied in *Arabidopsis* (Michaels, 2009). Among phytochromes (phy), phyB possessed the prominent function of shade detection, and a phyB mutant displayed temperature-dependent early flowering (Halliday et al., 2003). Other experiments revealed an association between phytochromes and temperature; however, molecular regulation of temperature sensing is still under investigation (McClung and Davis, 2010). Apart from the phytochromes, microRNAs also acted as potential regulators in the signaling pathways of the temperature-mediated transition in flowering (Lee et al., 2010). Modulations in ambient temperatures may trigger a change in the level of miRNAs such as miR172. Such a change can stimulate expression of the *FT* gene, resulting in early flowering (Lee et al., 2010).

4.4.2.2 Temperature-Mediated Auxin Control

Plant size is primarily determined by hormone-based cell division and elongation. Auxin plays a potential role in both cell division and elongation (Mockaitis and Estelle, 2008). Interestingly, the ability of auxin to influence cell division or elongation is highly dependent on temperature. The temperature-mediated promotion of auxin response requires *TRANSPORT INHIBITOR RESPONSE2* (*TIR2*) that encodes an enzyme required for auxin production (Yamada et al., 2009). The absence of *TIR2* hinders growth at elevated temperatures (Gray et al., 1998). Although the relationship between temperature and hormonal regulation of plant growth was evidenced, the molecular rationale is still under study.

Taken together, plant growth and development are extensively influenced by environmental cues including light and temperature. These two factors entrain vital physiological processes and molecular regulations in plants. Thus, better understanding of the relationship between plants and the environment may help identify novel traits with economic importance in plant breeding.

4.5 ROLE OF PLANT GROWTH REGULATORS DURING STRESS ALLEVIATION OF PLANTS

Plants possess various biochemical and physiological mechanisms to overcome stress. The response and adaptation of plants is determined by a series of complex interactions with metabolites, PGRs, secondary messengers, and the downstream defense and/or homeostasis genes (Olivoto et al., 2017; Verslues, 2017; Sewelam et al., 2016). Perennials evolved with more, different mechanisms than annuals for their survival under various stress conditions. For example, the water status of a tree is controlled by its stomatal conductance and osmoregulation; water homeostasis is sensed by PGRs; and plant growth regulator signals are either in conflict or complementary to each other, especially under stress conditions. A higher level of ABA decreased the level of free auxins due to inducing formation of conjugated auxins (Popko et al., 2010). In other cases, SA activated the accumulation of ABA under stress conditions (Miura and Tada, 2014). JA and ethylene-insensitive mutants led to the accumulation of ABA in plants (Cvetkovska et al., 2005). Singh and Gautam (2013) reported that ABA is involved in the regulation of cytokinins, auxins, ethylene, and brassinosteriods. The ROS acts as the secondary messenger of PGR signaling (Coupe et al., 2006). Presumably, PGRs are responsible for the downstream movement of ROS signals. Under stress conditions, ROS accumulation leads to oxidative stress. Meanwhile, PGRs can sense the temporal- and spatial-localization of ROS and activate the defense mechanism. In the next section, the main roles of SA JA and MeJA in response to various abiotic and biotic stresses in woody species are discussed.

4.5.1 Salicylic Acid in Stress Alleviation

Salicylic acid (SA) (is a phenolic compound naturally present in plants. It is involved in signal transduction processes in plants (Sticher et al., 1997). SA plays an important role in alleviating various stresses including salinity, heat, and cold. (Dat et al., 1998; Singh and Usha, 2003; Szalai and Janda, 2009; Horváth et al., 2007). SA alleviates a stress condition by ion exclusion, compartmentalization, osmotic adjustments, reduction of lipid peroxidation, and stress-induced synthesis of protein kinases (SIPK) (Sekmen et al., 2007). Usually, salt movement and accumulation in the plant are controlled by the transpirational stream. Transpiration flux is controlled by stomatal function. Krasavina (2007) mentioned that SA mediated the stomatal closure which mitigates the ion flux in the shoot. Other reports also showed that SA regulates H^+-ATPases (Sze et al., 1999), pyrophosphatses (PPases), Ca^{2+} ATPases, plasma membrane Na^+/H^+ antiporters, and salt overly sensitive (SOS) pathway under a stress condition (Mahajan et al., 2008; Qiu et al., 2002). Singh and Gautam (2013) explained the detailed role of SA-induced protein kinases (SAIPK), a protein belonging to the MAPK family activated upon osmotic stress that resulted in the up-regulation of abscisic aldehyde oxidase synthesis in *Solanum lycopersicum*. Several studies, including one by Zhu (2003), reported that pretreatment with SA overcomes the loss of intercellular water and rapid accumulation of organic solutes. Plants pretreated with SA showed enhanced antioxidant mechanisms, decreased Na accumulation, increased K^+ ratio over Na^+ ratio, and enhanced contents of proline and glycine betaine for stress alleviation (Badawi et al., 2004; Chen and Dickman, 2005; Raza et al., 2007). An increase in proline and glycine betaine stabilize the cell membrane. Treatment with SA keeps ROS under the signaling level to amplify stress mechanisms. Garretón et al. (2002) reported the involvement of SA in transcription of the defense response genes. In particular, transcripts of aldose reductase, glutathione S-transferase, glycosyl transferase, peroxiredoxin, peroxidase, thioredoxin, glutaredoxin, and pathogen resistance genes were upregulated by SA (Roxas et al., 1997; Vanderauwera et al., 2005; Wang et al., 2005). Cell signaling/signal transduction acts as a mediator of osmotic adjustment. Osmoticum changes the functions of plasma membrane, cytosol, and cellular components. Efflux of Na^+ across the plasma membrane and vacuolar compartmentalization is modulated by SOS1-Na^+/H^+ exchanger and NHX1 antiporter, respectively (Qiu et al., 2002). Decrease in IAA and cytokinin due to SA application increased cell division rate in the root apical meristem region (Aloni et al., 2006).

Application of SA in pear (*Pyrus bretschnideri*) enhanced its tolerance to *Penicillium epansum*, causing fruit decay during the post-harvest storage (Cao et al., 2006). Enhanced resistance to *Colletotrichum gloeosporioides* in mango and melon fruit by applying SA and acibenzolar-S-methyl (an analogue of SA) was reported (Huang et al., 2000; Joyce et al., 2001). An increase in resistance triggered by SA application before fruit harvest lasted to the post-harvest period (Cao et al., 2006). Research found that application of SA that decreased disease incidence and lesion diameter on pear fruit was associated with rapid accumulation of ROS. Transient localization of H_2O_2 acts as a signaling molecule under stress conditions. High levels of H_2O_2 stimulate the activity of peroxidases and the oxidation of phenolic compounds. Therefore, a prompt signaling mechanism of H_2O_2 protects cell walls against pathogen intrusion (Cao et al., 2006). Other enzymes associated with peroxidases to protect/strengthen cell walls are phenylalanine ammonia lyase, chitinase, β-1,3-glucanase, and glutathione reductase. Phenylalanine ammonia lyase is a precursor to phytoalexins and phenolic compounds involved in defense mechanisms (Nicholson and Hammerschmidt, 1992). Chitinase and β-1,3-glucanase is a key enzyme that acts directly against pathogens (Patil et al., 2000). Glutathione reductase protects cell organelles during high accumulation of ROS (Mittler et al., 2004). As a consequence of high H_2O_2 levels, activities of CAT and APX were temporarily decreased during the initial period of stress (Cao et al., 2006).

Vacuum infiltration of SA to apricot (*Prunus armeniaca*) fruit enhanced postharvest quality (Dragovic-Uzelac et al., 2005). The SA treatment increased the local and systemic resistance against pathogens and inhibited the formation of ethylene in peach (*Prunus persica*) fruit (Chan et al., 2007). Prevention of senescence by SA was related to the enhancement of hydrophilic total antioxidant activity (H-TAA) and lyphophilic TAA (L-TAA) in sweet cherry (*Prunus avium*) (Valero et al., 2011). In apricot, SA increased H-TAA more than L-TAA (Wang et al., 2015). Application of SA increased PAL, flavonoids, and H_2O_2 and decreased O_2^-. Meanwhile, a high SOD (superoxide dismutase) in SA treatment was correlated with an increase of H_2O_2. On the other hand, decreasing activities of H_2O_2-detoxifying enzymes such as CAT and APX in the SA treatment could stimulate the signaling process. Accumulation of H_2O_2 was tightly controlled via POD (peroxidase) activity. ROS metabolism delays the rate of respiration as well as ethylene synthesis (Wang et al., 2015). The addition of SA suppressed cell wall degradation enzymes (Sayyari et al., 2011); therefore, titratable acidity, soluble solid contents, and firmness of apricot fruit were enhanced by SA treatment (Wang et al., 2015).

Foliar spray of SA improved nutrient uptake of grape (*Vitis vinifera* L.) under salt stress, which was directly related to the absorption of essential elements (Amiri et al., 2015). Pre-treatment of grape with SA minimized lipid peroxidation and relative electrolyte leakage under heat or cold stress. Activity of antioxidant enzymes, such as APX, GR, and MDHA, and redox ratio of ascorbate-glutathione, were enhanced due to SA (Wang and Li, 2006). Further, Wang and Li (2006) found that the cytosolic Ca^{2+} level was higher in the SA treated mesophyll cells under heat or cold stress. In addition, SA protected chloroplast structure during stress by pumping back the accumulated Ca^{2+} into vacuole or intercellular space. According to Ding et al. (2007), SA attributed to the reduction of AsA and glutathione accompanied by H_2O_2 accumulation for the alleviation of chilling stress in the fruit of *Mangifera indica* L. 'Zill'.

Oxidative stress created by high mercury (Hg^{2+}) was reduced by pre-treatment with SA in alfalfa (*Medicago sativa*). A decrease of LPO and ROS accumulation in alfalfa by SA was associated with higher activities of NADH oxidase, APX, and POD, and an increase of AsA, GSH, and proline, along with decreased SOD activity during the Hg^{2+} stress (Zhou et al., 2009).

Molecular insight of SA in grapevine leaves during heat stress revealed that pretreatment of SA prevented the decline of net photosynthesis by maintaining the Rubisco activation state and heat shock protein 21 (HSP21) (Wang et al., 2010). Pre-harvest application of SA to navel orange (*Citrus sinensis* L. 'Cara') increased carotenoids, AsA, glutathione, phenolics, flavonoids, and other antioxidant compounds in pulp and peel during storage (Huang et al., 2008b). In the post-harvest period, fruits of 'Cara' immersed in SA solution showed reduced lipid peroxidation (LPO) and accelerated H_2O_2. Associated with high H_2O_2, antioxidant enzymes, such as CAT, SOD, GR and DHAR, increased with treatment of SA (Huang et al., 2008a). Therefore, SA has a vital role in enhanced resistance and improved fruit of fruit trees and other woody plants.

4.5.2 Jasmonic Acid in Stress Alleviation

Jasmonic acids are oxylipins that are derived from the oxygenation of polyunsaturated fatty acid. They are involved in a wide range of biological processes such as plant fertility, floral development, senescence induction, growth inhibition, and fruit ripening (Creelman and Mullet, 1995). During the developmental process or environmental stimuli, synthesis and accumulation of JA are transient (Stintzi et al., 2001). Signaling transduction of JA is finely tuned by complex mechanisms triggered by various transcription factors and downstream genes. Browse (2005) reported that JA is involved in the fertility process at three levels (filament elongation, anther dehiscence, and pollen viability). Later Song et al. (2011) showed JA controlling male fertility and regulating stamen elongation and anther development via myeloblastosis (MYB) transcription factors such as MYB24 and MYB21. Responses of plants to biotic stresses depends upon the cross-talk between SA, JA, and ethylene. JA mediates local and systemic defense against biotropic pathogens, whereas JA and ET (ethylene) are involved in resistance against necrotrophs (Glazebrook, 2005). Involvement of JA during insect attack is regulated by MYC2 mediated via vegetative storage protein 2 (*VSP2*). Induction of MYC2 mediated by ABA redirects the JA pathway. The *aba2* mutant failed to activate the MYC2 gene (Takagi et al., 2016). Most of the priming reactions involved with ABA and JA depend upon the signaling of SA. In poplar leaves, generation of JA and α-linoleic acid were observed after the treatment of (Z)-3-hexenal primes. Similar to SA-mediated priming, involvement of ROS and mitogen-activated protein kinases (MAPKs) were noticed with JA-dependent signals. During insect invasion, activation of MAPKs accumulates JA content. For the initiation of SAR (systemic acquired resistance), rapid transient accumulation of JA is necessary, and it is involved in the long distance information transformation. Spoel et al. (2007) reported that JA and SA act in a tandem fashion. Transient accumulation of JA was followed by SA-mediated defense responses. In addition, ethylene acts as a good supporter for JA–SA cross-talk (Jia et al., 2013). Resistance to *Pseudomonas syringae* by a negative regulation of JA through interaction with a MYC2 protein was promoted by TIME FOR COFFEE (TIC) protein (Shin et al., 2012). During the *P. syringae* infection, SA-independent callose accumulation was suppressed by phytotoxin COR (coronatine) which is structurally similar to JA-Ile. The DELLAs proteins, such as GA insensitive 1–3 (GA1-3), repressor of ga1-3 (RGA1), RGA2 and phytochrome interacting factors (PIFs), are also involved in the crosstalk between GA and JA (Kazan and Manners, 2012; Navarro et al., 2008). Adie et al. (2007) mentioned that for the synthesis of JA and its defense function, ABA is mandatory. Callose deposition during the pathogen intervention was highly regulated by JA and ABA. Concentration of JA increased to the highest level in citrus root immediately after salt stress. The expression of key enzymes involved in the JA biosynthetic pathway, such as *Ctlox3*, *Ctaos*, and *Ctaoc*, was induced in JA-treated plants (Arbona et al., 2010). Similarly in *Vitis vinifera*, JAZ/TIFY family-related genes, such as *JAZ1/TIFY10a* and *JAZ3/TIFY6b*, were expressed more during salt stress (Ismail et al., 2012). JA also plays an important role in the induction of chilling stress.

The JA-mediated signaling process was widely reported in citrus. Among PGRs, ABA played a central role in the signaling process. Accumulation of ABA was triggered by JA. Under drought stress in citrus roots, rapid accumulation of JA induced progressive segregation of ABA. Immediate to the drought stress, ABA induced stomatal closure and organ drop, and modulated other physiological responses. During salt stress, ABA synthesis is correlated with decrease of Na^+ in photosynthetically active organs. Responses of plants are intricated with the complex network of plant hormones. In the biosynthetic pathway of JA-ABA, up-regulation of genes, such as *lipoxygenase 3* (*LOX3*), *allene oxide synthase* (*AOS*), and *9-cis-epoxycarotenoid dioxygenase* (*NCED1*), was observed immediately during the onset of drought stress (de Ollas et al., 2013). The salicylhydroxamic acid (SHAM), an inhibitor of JA, reduced the biosynthesis of JA and directly reduced the ABA content to 70% of the control. Though NFZ (norflurazon), an ABA inhibitor, decreased the ABA level, the content of JA was not significantly affected in citrus. This denotes the transient burst of JA could help in the progressive accumulation of ABA. JA signals the ABA receptor pyrabactin resistance-like (PYL) 4-protein to regulate metabolic reprogramming of plants. Interaction of JA and ABA was modulated by the TOPLESS co-repressor proteins (Lackman et al., 2011). For the cross-talk between JA and ABA, MYC2 plays a central role. JA signaling to the ABA-induced metabolic changes is partly identified to occur in the transcriptional level. Increased JA improved the endogenous level of ABA. Signals of JA transiently improved the pharmacological and rescuing mechanism against stress in plants (de Ollas et al., 2013). The NPR1 (Pathogenesis Resistance 1) mutant plays an important role in interactions between SA and JA (Spoel et al., 2007).

Downstream of NPR1, WRKY TFs induce SA and simultaneously prevent the expression of JA (Mao et al., 2007). The receptor, MPK4, mediates the antagonistic mechanism between SA and JA (Brodersen et al., 2006). During the defense process, SA induces NPR1 in order to activate GRX480 and form the complex with TGA factors. Meanwhile, the activation of GRX480 suppresses the other JA-responsive mechanisms (Meyer et al., 2008). The JA-inducible protein epithiospecifying senescence regulator (ESR) antagonistically regulates the cross-talk between JA and SA (Miao and Zentgraf, 2007). Ismail et al. (2012) reported that JA acted as the signaling element to improve salt tolerance in grapevine by modulating the expression of Na^+/H^+ Exchanger (*NHX1*), *resveratrol synthase* (*RS*), and *stilbene synthase* (*StsY*). Transcription factors of the NACs family, such as NAM (no apical meristem), ATAF (*Arabidopsis thaliana* activation factor), and CUC (cup-shaped cotyledon) are involved in the interaction with ABA and JA (Mauch-Mani and Flors, 2009). More insight on the detailed mechanisms of involvement of JA with other plant hormones would help understanding of the sense and signaling process of woody plants.

4.5.3 Methyl Jasmonate in Stress Alleviation

Similar to JA, application of a functionally similar molecule, MeJA, showed effects, on herbivores pathogens. MeJA increased the anatomical and chemical defenses in pine trees. The effect of MeJA showed increased resin in the xylem and needles rather than the bark (Heijari et al., 2005). Reduction in growth and physiology observed in the MeJA-treated plants indicates a shift in carbon allocation. Synthesis of terpene resin in specific tissues could increase MeJA to protect against *Hylobius abietis*, a large pine weevil attracted by the α-pinene and β-pinene (Zagatti et al., 1997). Conifers synthesizing chemical defense terpenoids are attractive as well as toxic. Application of MeJA produces traumatic resin ducts in xylem and parenchyma cells by increased activities of terpene synthases. Either mechanical wounds or insect attack increased resin synthesis and induced the complex changes in the stem anatomy. Occurrence of *H. abietis* was reduced to 62% in the MeJA-treated pine seedlings over the control plants. Involvement of monoterpene and α-pinene is high in the xylem tissues and needles. However, net photosynthesis decreased in the MeJA treatment. Interestingly, formation of resin ducts was found in the continuous band in the newly developed xylem tissues. Resin duct numbers were almost 5.8-fold higher than in the control plants. Meanwhile, annual diameter growth and tracheid cell lumen area were smaller in the MeJA-treated pine trees. Monoterpene and diterpene induced the defense against *H. aboetis*. Terpenoid resins are vital for plant and herbivore interactions. Changes in tissue quality could also play an important role in depressed growth of insect populations. Needle browning due to high accumulation of terpenes could reduce infestation by insects or pathogens. Increased terpene biosynthesis in response to MeJA could be involved by induction of secondary metabolites-related genes. MeJA could also lead to a reduction in the cell expansion of the vascular cambium. Therefore, for cell length, cell wall chemistry, and cell wall structure, MeJA plays a vital role along with resistance against herbivorous insects (Heijari et al., 2005). During chilling stress, electrolytic leakage and water loss ameliorated by the application of MeJA helped increase hydraulic conductivity (Lee et al., 1997).

Long term studies on MeJA application in pine trees suggested that MeJA influences the monoterpenes, sesquiterpenes and tricyclic resic acids, and helped eradicate European pine sawfly and common pine fly (Heijari et al., 2008). Gould et al. (2009) suggested that MeJA increased the soluble guaiacol peroxidase (POX) activity, polyphenol oxidase (PPO), α-pinene and β-pinene during wounding at low than high concentrations. Comparing the undamaged *H. abietis* plants with the damaged plants, increase in monoterpenes and sesquiterpenes was observed in Scots pine (*Pinus sylvestris*) (Heijari et al., 2011). Higher resistance in Monterey pine (*Pinus radiate*) due to the application of MeJA was reported against *Diplodia pinea* (Gould et al., 2008). In Jack pine (*Pinus banksiana*), MeJA treatment increased monoterpenes in both the juvenile and mature trees against the fungal infection (Erbilgin and Colgan, 2012). Hence, MeJA influences growth, physiology, and resistance to insects and diseases.

REFERENCES

Abdalla, N. M., El Gengaihi, S. E., Solomos, T., and Al Badawy, A. A. (1985). The effects of kinetin and Alar-85 applications on the growth and flowering of *Adonis autumnalis* L. In *Proceedings Annual Meeting Plant Growth Regulator Society of America.*

Abeles, F. B., Morgan, P. W., and Saltveit, M. E. (1992). *Ethylene in Plant Biology*, 2nd Ed. Academic Press, San Diego.

Adie, B. A., Pérez-Pérez, J., Pérez-Pérez, M. M., Godoy, M., Sánchez-Serrano, J. J., Schmelz, E. A., and Solano, R. (2007). ABA is an essential signal for plant resistance to pathogens affecting JA biosynthesis and the activation of defenses in Arabidopsis. *The Plant Cell, 19*(5), 1665–1681.

Ahmad, M., and Cashmore, A. R. (1993). HY4 gene of *Arabidopsis thaliana* encodes a protein with characteristics of a blue-light photoreceptor. *Nature, 366*, 162–166.

Ahmad, P., and Jhon, R. (2005). Effect of salt stress on growth and biochemical parameters of *Pisum sativum* L. *Archives of Agronomy and Soil Science, 51*(6), 665–672.

Ali, B., Hasan, S. A., Hayat, S., Hayat, Q., Yadav, S., Fariduddin, Q., and Ahmad, A. (2008). A role for brassinosteroids in the amelioration of aluminium stress through antioxidant system in mung bean (*Vigna radiata* L. Wilczek). *Environmental and Experimental Botany, 62*(2), 153–159.

Aloni, R., Aloni, E., Langhans, M., and Ullrich, C. I. (2006). Role of cytokinin and auxin in shaping root architecture: Regulating vascular differentiation, lateral root initiation, root apical dominance and root gravitropism. *Annals of Botany, 97*(5), 883–893.

Amiri, J., Eshghi, S., Tafazoli, E., Kholdebarin, B., and Abbaspour, N. (2015). Ameliorative effects of salicylic acid on mineral concentrations in roots and leaves of two grapevine (*Vitis vinifera* L.) cultivars under salt stress. *VITIS-Journal of Grapevine Research, 53*(4), 181.

Anderson, J. M., Chow, W. S., and Park, Y. I. (1995). The grand design of photosynthesis: Acclimation of the photosynthetic apparatus to environmental cues. *Photosynthesis Research, 46*(1–2), 129–139.

Ani, N. N., Harun, A. N., Samsuri, S. F. M., and Ahmad, R. (2015). Effect of red and blue lights on photomorphogenesis in *Brassica chinensis*. In: *The Malaysia-Japan Model on Technology Partnership* (pp. 49–58). ed. K.A. Hamid Osamu, O. Ona, A.M. Bostamam, and A.P.A. Ling. Springer, Japan.

Arbona, V., Argamasilla, R., and Gómez-Cadenas, A. (2010). Common and divergent physiological, hormonal and metabolic responses of Arabidopsis thaliana and *Thellungiella halophila* to water and salt stress. *Journal of Plant Physiology, 167*(16), 1342–1350.

Arteca, J. M., and Arteca, R. N. (2001). Brassinosteroid-induced exaggerated growth in hydroponically grown Arabidopsis plants. *Physiologia Plantarum, 112*(1), 104–112.

Badawi, G. H., Kawano, N., Yamauchi, Y., Shimada, E., Sasaki, R., Kubo, A., and Tanaka, K. (2004). Overexpression of ascorbate peroxidase in tobacco chloroplasts enhances the tolerance to salt stress and water deficit. *Physiologia Plantarum, 121*(2), 231–238.

Bajguz, A. (2007). Metabolism of brassinosteroids in plants. *Plant Physiology and Biochemistry, 45*(2), 95–107.

Baldwin, B., Bandara, M. S., and Tanino, K. (2000). Bud scale maturation in Saskatoon berry (*Amelanchier alnifolia* Nutt). Plantlets following in vitro hormonal treatments. *Acta Horticulturae, 520*, 203–208.

Bao, F., Shen, J., Brady, S. R., Muday, G. K., Asami, T., and Yang, Z. (2004). Brassinosteroids interact with auxin to promote lateral root development in Arabidopsis. *Plant Physiology, 134*(4), 1624–1631.

Bapat, V. A., Trivedi, P. K., Ghosh, A., Sane, V. A., Ganapathi, T. R., and Nath, P. (2010). Ripening of fleshy fruit: Molecular insight and the role of ethylene. *Biotechnology Advances, 28*(1), 94–107.

Bari, R., and Jones, J. D. (2009). Role of plant hormones in plant defense responses. *Plant Molecular Biology, 69*(4), 473–488.

Bell, A. A. (1981). Biochemical mechanisms of disease resistance. *Annual Review of Plant Physiology, 32*(1), 21–81.

Berry, J., and Bjorkman, O. (1980). Photosynthetic response and adaptation to temperature in higher plants. *Annual Reviews in Plant Physiology, 31*(1), 491–543.

Bohning, R. H., and Burnside, C. A. (1956). The effect of light intensity on rate of apparent photosynthesis in leaves of sun and shade plants. *American Journal of Botany*, 43(8), 557–561.

Borsani, O., Valpuesta, V., and Botella, M. A. (2001). Evidence for a role of salicylic acid in the oxidative damage generated by NaCl and osmotic stress in Arabidopsis seedlings. *Plant Physiology, 126*(3), 1024–1030.

Brodersen, P., Petersen, M., Nielsen, H. B., Zhu, S., Newman, M. A., Shokat, K. M., Rietz, S., Parker, J., and Mundy, J. (2006). Arabidopsis MAP kinase 4 regulates salicylic acid- and jasmonic acid/ethylene-dependent responses via EDS1 and PAD4. *The Plant Journal, 47*(4), 532–546.

Browse, J. (2005). Jasmonate: An oxylipin signal with many roles in plants. *Vitamins and Hormones, 72*, 431–456.

Butler, W. L., Norris, K. H., Siegelman, H. W., and Hendricks, S. B. (1959). Detection, assay, and preliminary purification of the pigment controlling photoresponsive development of plants. *Proceedings of the National Academy of Sciences of the United States of America, 45*, 703–708.

Cao, J., Zeng, K., and Jiang, W. (2006). Enhancement of postharvest disease resistance in Ya Li pear (*Pyrus bretschneideri*) fruit by salicylic acid sprays on the trees during fruit growth. *European Journal of Plant Pathology, 114*(4), 363–370.

Cerny, T. A., Faust, J. E., Layne, D. R., and Rajapakse, N. C. (2003). Influence of photoselective films and growing season on stem growth and flowering of six plant species. *Journal of American Society for Horticultural Science, 128*(4), 486–491.

Chabot, B. F., and Chabot, J. F. (1977). Effects of light and temperature on leaf anatomy and photosynthesis in *Fragaria vesca. Oecologia, 26*(4), 363–377.

Chan, Z., Qin, G., Xu, X., Li, B., and Tian, S. (2007). Proteome approach to characterize proteins induced by antagonist yeast and salicylic acid in peach fruit. *Journal of Proteome Research, 6*(5), 1677–1688.

Chappell, J., Hahlbrock, K., and Boller, T. (1984). Rapid induction of ethylene biosynthesis in cultured parsley cells by fungal elicitor and its relationship to the induction of phenylalanine ammonia-lyase. *Planta, 161*(5), 475–480.

Chaves, I., Pokorny, R., Byrdin, M., Hoang, N., Ritz, T., Brettel, K., and Ahmad, M. (2011). The cryptochromes: Blue light photoreceptors in plants and animals. *Annual Reviews in Plant Biology, 62*, 335–364.

Chen, C., and Dickman, M. B. (2005). From The Cover: Proline suppresses apoptosis in the fungal pathogen *Colletotrichum trifolii. Proceedings of the National Academy of Sciences of the United States of America, 102*(9), 3459–3464.

Chen, D., Wan, Z., Zhou, Y., Xiang, W., Zhong, J., Ding, M., Yu, H., and Ji, Z. (2015). Tuning into blue and red: Europium single-doped nano-glass-ceramics for potential application in photosynthesis. *Journal of Materials Chemistry C, 13*: 3141–3149.

Chen, H. J., Bollmark, M., and Eliasson, L. (1996). Evidence that cytokinin controls bud size and branch form in Norway spruce. *Physiologia Plantarum, 98*(3), 612–618.

Chini, A., Grant, J. J., Seki, M., Shinozaki, K., and Loake, G. J. (2004). Drought tolerance established by enhanced expression of the CC–NBS–LRR gene, ADR1, requires salicylic acid, EDS1 and ABI1. *The Plant Journal, 38*(5), 810–822.

Clack, T., Mathews, S., and Sharrock, R. A. (1994). The phytochrome apoprotein family in Arabidopsis is encoded by five genes: The sequences and expression of PHYD and PHYE. *Plant Molecular Biology, 25*(3), 413–427.

Clarke, S. F., Jameson, P. E., and Downs, C. (1994). The influence of 6-benzylaminopurine on post-harvest senescence of floral tissues of broccoli (*Brassica oleracea* var italica). *Plant Growth Regulation, 14*(1), 21–27.

Cline, M. G. (1991). Apical dominance. *The Botanical Review, 57*(4), 318–358.

Cohen, D. (1981). Micropropagation of Zantedeschia hybrids. *Combined Proceedings-International Plant Propagators' Society, 31*, 312–316.

Colebrook, E. H., Thomas, S. G., Phillips, A. L., and Hedden, P. (2014). The role of gibberellin signaling in plant responses to abiotic stress. *Journal of Experimental Biology, 217*(1), 67–75.

Conrad, K. S., Manahan, C. C., and Crane, B. R. (2014). Photochemistry of flavoprotein light sensors. *Nature Chemical Biology, 10*(10), 801–809.

Coupe, S. A., Palmer, B. G., Lake, J. A., Overy, S. A., Oxborough, K., Woodward, F. I., Gray, J. E., and Quick, W. P. (2006). Systemic signalling of environmental cues in Arabidopsis leaves. *Journal of Experimental Botany, 57*(2), 329–341.

Creelman, R. A., and Mullet, J. E. (1995). Jasmonic acid distribution and action in plants: Regulation during development and response to biotic and abiotic stress. *Proceedings of the National Academy of Sciences, 92*(10), 4114–4119.

Creelman, R. A., and Mullet, J. E. (1997). Biosynthesis and action of jasmonates in plants. *Annual Review of Plant Biology, 48*(1), 355–381.

Creelman, R. A., and Rao, M. V. (2002). The oxylipin pathway in Arabidopsis. In: *The Arabidopsis Book*, C. R. Somervile, E. M. Meyerowitz, eds. American Society of Plant Biologists, USA.

Cvetkovska, M., Rampitsch, C., Bykova, N., and Xing, T. (2005). Genomic analysis of MAP kinase cascades in Arabidopsis defense responses. *Plant Molecular Biology Reporter, 23*(4), 331–343.

Dandekar, A. M., Teo, G., Defilippi, B. G., Uratsu, S. L., Passey, A. J., Kader, A. A., John, R., Colgan, R. J., and James, D. J. (2004). Effect of down-regulation of ethylene biosynthesis on fruit flavor complex in apple fruit. *Transgenic Research, 13*(4), 373–384.

Danesi, E. D. G., Rangel-Yagui, C. O., Carvalho, J. C. M., and Sato, S. (2004). Effect of reducing the light intensity on the growth and production of chlorophyll by *Spirulina platensis. Biomass and Bioenrgy*, *26*(4), 329–335.

Dat, J. F., Lopez-Delgado, H., Foyer, C. H., and Scott, I. M. (1998). Parallel changes in H2O2 and catalase during thermotolerance induced by salicylic acid or heat acclimation in mustard seedlings. *Plant Physiology*, *116*(4), 1351–1357.

Davies, P. J. (2010). The plant hormones: Their nature, occurrence, and functions. In *Plant Hormones* (pp. 1–15). Springer, Netherlands.

Davies, W. J., Wilkinson, S., and Loveys, B. (2002). Stomatal control by chemical signalling and the exploitation of this mechanism to increase water use efficiency in agriculture. *New Phytologist*, *153*(3), 449–460.

de Ollas, C., Hernando, B., Arbona, V., and Gómez-Cadenas, A. (2013). Jasmonic acid transient accumulation is needed for abscisic acid increase in citrus roots under drought stress conditions. *Physiologia Plantarum*, *147*(3), 296–306.

Debi, B. R., Taketa, S., and Ichii, M. (2005). Cytokinin inhibits lateral root initiation but stimulates lateral root elongation in rice (*Oryza sativa*). *Journal of Plant Physiology*, *162*(5), 507–515.

Dennison, W. C., and Alberte, R. S. (1982). Photosynthetic responses of *Zostera marina* L. (eelgrass) to *in situ* manipulations of light intensity. *Oecologia*, *55*(2), 137–144.

Ding, Z. S., Tian, S. P., Zheng, X. L., Zhou, Z. W., and Xu, Y. (2007). Responses of reactive oxygen metabolism and quality in mango fruit to exogenous oxalic acid or salicylic acid under chilling temperature stress. *Physiologia Plantarum*, *130*(1), 112–121.

Divi, U. K., and Krishna, P. (2009). Brassinosteroid: A biotechnological target for enhancing crop yield and stress tolerance. *New Biotechnology*, *26*(3), 131–136.

Dragovic-Uzelac, V., Pospišil, J., Levaj, B., and Delonga, K. (2005). The study of phenolic profiles of raw apricots and apples and their purees by HPLC for the evaluation of apricot nectars and jams authenticity. *Food Chemistry*, *91*(2), 373–383.

Duca, M. (2015). Plant biorhythms. In: *Plant Physiology* (pp. 231–246). Springer International Publishing.

Durner, J., Shah, J., and Klessig, D. F. (1997). Salicylic acid and disease resistance in plants. *Trends in Plant Science*, *2*(7), 266–274.

Engelhard, C., Wang, X., Robles, D., Moldt, J., Essen, L. O., Batschauer, A., Bittl, R., and Ahmad, M. (2014). Cellular metabolites enhance the light sensitivity of Arabidopsis cryptochrome through alternate electron transfer pathways. *The Plant Cell*, *26*(11), 4519–4531.

Engelsma, G., and Meijer, G. (1965). The influence of light of different spectral regions on the synthesis of phenolic compounds in gherkin seedlings in relation to photomorphogenesis II. Indoleacetic acid oxidase activity and growth. *Acta Botanica Neerlandica*, *14*(1), 73–92.

Erbilgin, N., and Colgan, L. J. (2012). Differential effects of plant ontogeny and damage type on phloem and foliage monoterpenes in jack pine (*Pinus banksiana*). *Tree Physiology*, *32*(8), 946–957.

Essa, T. A. (2002). Effect of salinity stress on growth and nutrient composition of three soybean (*Glycine max* L. Merrill) cultivars. *Journal of Agronomy and Crop Science*, *188*(2), 86–93.

Evans, D. A., Sharp, W. R., and Flick, C. E. (1981). Growth and behavior of cell cultures: Embryogenesis and organogenesis. In: *Plant Tissue Culture: Methods and Application in Agriculture*. ed. Trevor A. Thorpe, Academic Press, New York.

Fahlgren, N., Gehan, M. A., and Baxter, I. (2015). Lights, camera, action: High-throughput plant phenotyping is ready for a close-up. *Current Opinion in Plant Biology*, *24*, 93–99.

Fan, X., Mattheis, J. P., and Roberts, R. G. (2000). Biosynthesis of phytoalexin in carrot root requires ethylene action. *Physiologia Plantarum*, *110*(4), 450–454.

Fankhauser, C., and Chory, J. (1997). Light control of plant development. *Annual Reviewa in Cellular Developmental Biology*, *13*(1), 203–229.

Finch-Savage, W. E., and Leubner-Metzger, G. (2006). Seed dormancy and the control of germination. *New Phytologist*, *171*(3), 501–523.

Fleet, C. M., and Sun, T. P. (2005). A DELLAcate balance: The role of gibberellin in plant morphogenesis. *Current Opinion in Plant Biology*, *8*(1), 77–85.

Franceschi, V. R., Krekling, T., and Christiansen, E. (2002). Application of methyl jasmonate on *Picea abies* (Pinaceae) stems induces defense-related responses in phloem and xylem. *American Journal of Botany*, *89*(4), 602–610.

Freeman, J. L., Garcia, D., Kim, D., Hopf, A., and Salt, D. E. (2005). Constitutively elevated salicylic acid signals glutathione-mediated nickel tolerance in Thlaspi nickel hyperaccumulators. *Plant Physiology*, *137*(3), 1082–1091.

Galvão, V. C., and Fankhauser, C. (2015). Sensing the light environment in plants: Photoreceptors and early signaling steps. *Current Opinion in Neurobiology*, *34*, 46–53.

Garretón, V., Carpinelli, J., Jordana, X., and Holuigue, L. (2002). The as-1 promoter element is an oxidative stress-responsive element and salicylic acid activates it via oxidative species. *Plant Physiology*, *130*(3), 1516–1526.

Ghanashyam, C., and Jain, M. (2009). Role of auxin-responsive genes in biotic stress responses. *Plant Signaling and Behavior*, *4*(9), 846–848.

Glazebrook, J. (2005). Contrasting mechanisms of defense against biotrophic and necrotrophic pathogens. *Annual Review in Phytopathology*, *43*, 205–227.

Gomez-Roldan, V., Fermas, S., Brewer, P. B., Puech-Pagès, V., Dun, E. A., Pillot, J. P., Letisse, F., Matusova, R., Danoun, S., Portais, J. C., Bouwmeester, H., Bécard, G., Beveridge, C. A., Rameau, C., and Rochange, S. F. (2008). Strigolactone inhibition of shoot branching. *Nature*, *455*(7210), 189–194.

Gould, N., Reglinski, T., Northcott, G. L., Spiers, M., and Taylor, J. T. (2009). Physiological and biochemical responses in *Pinus radiata* seedlings associated with methyl jasmonate-induced resistance to *Diplodia pinea*. *Physiological and Molecular Plant Pathology*, *74*(2), 121–128.

Gould, N., Reglinski, T., Spiers, M., and Taylor, J. T. (2008). Physiological trade-offs associated with methyl jasmonate-induced resistance in *Pinus radiata*. *Canadian Journal of Forest Research*, *38*(4), 677–684.

Gray, W. M., Ostin, A., Sandberg, G., Romano, C. P., and Estelle, M. (1998). High temperature promotes auxin-mediated hypocotyl elongation in Arabidopsis. *Proceedings of the National Academy of Sciences of the United States of America*, *95*, 7197–7202.

Greenwood, M. S. (1995). Juvenility and maturation in conifers: Current concepts. *Tree Physiology*, *15*(7–8), 433–438.

Ha, S., Vankova, R., Yamaguchi-Shinozaki, K., Shinozaki, K., and Tran, L. S. P. (2012). Cytokinins: Metabolism and function in plant adaptation to environmental stresses. *Trends in Plant Science*, *17*(3), 172–179.

Halliday, K. J., Salter, M. G., Thingnaes, E., and Whitelam, G. C. (2003). Phytochrome control of flowering is temperature sensitive and correlates with expression of the floral integrator FT. *The Plant Journal*, *33*, 875–885.

Hardham, A. R., and Gunning, B. E. S. (1978). Structure of cortical microtubule arrays in plants. *The Journal of Cell Biology*, *77*, 14–34.

Harmer, R., and Baker, C. (1995). An evaluation of decapitation as a method for selecting clonal *Quercus petraea* (Matt) Liebl with different branching intensities. *Annales des Sciences Forestières*, *52*(2), 89–102 (EDP Sciences).

Hasan, S. A., Hayat, S., and Ahmad, A. (2011). Brassinosteroids protect photosynthetic machinery against the cadmium induced oxidative stress in two tomato cultivars. *Chemosphere*, *84*(10), 1446–1451.

He, Y. L., Liu, Y. L., Chen, Q., and Bian, A. H. (2002). Thermotolerance related to anti oxidation induced by salicylic acid and heat hardening in tall fescue seedlings. *Journal of Plant Physiology and Molecular Biology*, *28*(2), 89–95.

Heide, O. M. (1977). Photoperiod and temperature interactions in growth and flowering of strawberry. *Physiologia Plantaraum*, *40*(1), 21–26.

Heijari, J., Blande, J. D., and Holopainen, J. K. (2011). Feeding of large pine weevil on *Scots pine* stem triggers localised bark and systemic shoot emission of volatile organic compounds. *Environmental and Experimental Botany*, *71*(3), 390–398.

Heijari, J., Nerg, A. M., Kainulainen, P., Viiri, H., Vuorinen, M., and Holopainen, J. K. (2005). Application of methyl jasmonate reduces growth but increases chemical defence and resistance against *Hylobius abietis* in Scots pine seedlings. *Entomologia Experimentalis et Applicata*, *115*(1), 117–124.

Heijari, J., Nerg, A. M., Kainulainen, P., Vuorinen, M., and Holopainen, J. K. (2008). Long-term effects of exogenous methyl jasmonate application on Scots pine (*Pinus sylvestris*) needle chemical defence and diprionid sawfly performance. *Entomologia Experimentalis et Applicata*, *128*(1), 162–171.

Higuchi, Y., Sumitomo, K., Oda, A., Shimizu, H., and Hisamatsu, T. (2012). Day light quality affects the night-break response in the short-day plant chrysanthemum, suggesting differential phytochrome-mediated regulation of flowering. *Journal of Plant Physiology*, *169*(18), 1789–1796.

Horváth, E., Pál, M., Szalai, G., Páldi, E., and Janda, T. (2007). Exogenous 4-hydroxybenzoic acid and salicylic acid modulate the effect of short-term drought and freezing stress on wheat plants. *Biologia Plantarum*, *51*(3), 480–487.

Huang, R. H., Liu, J. H., Lu, Y. M., and Xia, R. X. (2008a). Effect of salicylic acid on the antioxidant system in the pulp of 'Cara cara'navel orange (*Citrus sinensis* L. Osbeck) at different storage temperatures. *Postharvest Biology and Technology*, *47*(2), 168–175.

Huang, R., Xia, R., Lu, Y., Hu, L., and Xu, Y. (2008b). Effect of preharvest salicylic acid spray treatment on post-harvest antioxidant in the pulp and peel of 'Cara' navel orange (*Citrus sinenisis* L. Osbeck). *Journal of the Science of Food and Agriculture*, *88*(2), 229–236.

Huang, Y., Deverall, B. J., Tang, W. H., Wang, W., and Wu, F. W. (2000). Foliar application of acibenzolar-S-methyl and protection of postharvest Rock melons and Hami melons from disease. *European Journal of Plant Pathology*, *106*(7), 651–656.

Hwang, S. J., Moon, H. H., and Jeong, B. R. (2005). Effects of difference between day and night temperatures on growth and flowering of miniature roses. *Horticulture Environmental and Biotechnology*, *46*(5), 321–328.

Ismail, A., Riemann, M., and Nick, P. (2012). The jasmonate pathway mediates salt tolerance in grapevines. *Journal of Experimental Botany*, *63*(5), 2127–2139.

Ito, H., and Saito, T. (1962). Studies on the flower formation in the strawberry plants. I. Effects of temperature and photoperiod on the flower formation. *Tohoku Journal of Agricultural Research*, *13*, 191–203.

Janda, T., Szalai, G., Tari, I., and Páldi, E. (1999). Hydroponic treatment with salicylic acid decreases the effects of chilling injury in maize (*Zea mays* L.) plants. *Planta*, *208*(2), 175–180.

Jia, C., Zhang, L., Liu, L., Wang, J., Li, C., and Wang, Q. (2013). Multiple phytohormone signalling pathways modulate susceptibility of tomato plants to *Alternaria alternata* f. sp. *lycopersici*. *Journal of Experimental Botany*, *64*(2), 637–650.

Joyce, D. C., Wearing, H., Coates, L., and Terry, L. (2001). Effects of phosphonate and salicylic acid treatments on anthracnose disease development and ripening of 'Kensington Pride' mango fruit. *Animal Production Science*, *41*(6), 805–813.

Kami, C., Lorrain, S., Hornitschek, P., and Fankhauser, C. (2010). Light-regulated plant growth and development. *Current Topics in Developmental Biology*, *91*, 29–66.

Kang, G. Z., Wang, C. H., Sun, G. C., and Wang, Z. X. (2003). Salicylic acid changes activities of H_2O_2-metabolizing enzymes and increases the chilling tolerance of banana seedlings. *Environmental and Experimental Botany*, *50*(1), 9–15.

Kang, H. M., and Saltveit, M. E. (2002). Chilling tolerance of maize, cucumber and rice seedling leaves and roots are differentially affected by salicylic acid. *Physiologia Plantarum*, *115*(4), 571–576.

Kaya, C., Kirnak, H., and Higgs, D. (2001). Enhancement of growth and normal growth parameters by foliar application of potassium and phosphorus in tomato cultivars grown at high (NaCl) salinity. *Journal of Plant Nutrition*, *24*(2), 357–367.

Kazan, K., and Manners, J. M. (2012). JAZ repressors and the orchestration of phytohormone crosstalk. *Trends in Plant Science*, *17*(1), 22–31.

Kepinski, S., and Leyser, O. (2005). The Arabidopsis F-box protein TIR1 is an auxin receptor. *Nature*, *435*(7041), 446–451.

Khan, A. S., and Singh, Z. (2007). Methyl jasmonate promotes fruit ripening and improves fruit quality in Japanese plum. *The Journal of Horticultural Science and Biotechnology*, *82*(5), 695–706.

Kim, D. H., Doyle, M. R., Sung, S., and Amasino, R. M. (2009). Vernalization: Winter and the timing of flowering in plants. *Annual Reviews in Cellular Development*, *25*(1), 277–299.

Kondo, S., Tsukada, N., Niimi, Y., and Seto, H. (2001). Interactions between jasmonates and abscisic acid in apple fruit, and stimulative effect of jasmonates on anthocyanin accumulation. *Journal of the Japanese Society for Horticultural Science*, *70*(5), 546–552.

Koornneef, M., Rolff, E., and Spruit, C. J. P. (1980). Genetic control of light-inhibited hypocotyl elongation in *Arabidopsis thaliana* (L.) Heynh. *Zeitschrift für Pflanzenphysiologie*, *100*, 147–160.

Krasavina, M. S. (2007). Effect of salicylic acid on solute transport in plants. In *Salicylic Acid: A Plant Hormone* (pp. 25–68). Springer Netherlands.

Kurosawa, E. (1926). Experimental studies on the nature of the substance secreted by the "bakanae" fungus. *Transactions, Natural History Society of Formosa*, *16*, 213–227.

Lackman, P., González-Guzmán, M., Tilleman, S., Carqueijeiro, I., Pérez, A. C., Moses, T., Thevelein, J. M., Kanno, Y., Hakkinen, S. T., Van Montagu, M. C. E., Thevelein, J. M., Maaheimo, H., Oksman-Caldentey, K.-M., Rodriguez, P. L., Rischer, H., and Goossens, A. (2011). Jasmonate signaling involves the abscisic acid receptor PYL4 to regulate metabolic reprogramming in Arabidopsis and tobacco. *Proceedings of the National Academy of Sciences*, *108*(14), 5891–5896.

Lalel, H. J. D., Singh, Z., and Tan, S. C. (2003). The role of methyl jasmonate in mango ripening and biosynthesis of aroma volatile compounds. *The Journal of Horticultural Science and Biotechnology*, *78*(4), 470–484.

Lang, A., Chailakhyan, M. K., and Frolova, I. A. (1977). Promotion and inhibition of flower formation in a dayneutral plant in grafts with a short-day plant and a long-day plant. *Proceedings of the National Academy of Sciences, 74*(6), 2412–2416.

Larkindale, J., Hall, J. D., Knight, M. R., and Vierling, E. (2005). Heat stress phenotypes of Arabidopsis mutants implicate multiple signaling pathways in the acquisition of thermotolerance. *Plant Physiology, 138*(2), 882–897.

Lee, H., Yoo, S. J., Lee, J. H., Kim, W., Yoo, S. K., Fitzgerald, H., Carrington, J. C., and Ahn, J. H. (2010). Genetic framework for flowering-time regulation by ambient temperature-responsive miRNAs in Arabidopsis. *Nucleic Acids Research, 38*, 3081–3093.

Lee, J. R., Lee, S. S., Jang, H. H., Lee, Y. M., Park, J. H., Park, S. C., Moon, J. C., Park, S. K., Kim, S. Y., Lee, S. Y., Chae, H. B., Jung, Y. J., Kim, W. Y., Shin, M. R., Cheong, G. W., Kim, M. G., Kang, K. R., Lee, K. O., Yun, D. J., and Lee, S. Y. (2009). Heat-shock dependent oligomeric status alters the function of a plant-specific thioredoxin-like protein, AtTDX. *Proceedings of the National Academy of Sciences, 106*, 5978–5983.

Lee, T. M., Lur, H. S., and Chu, C. (1997). Role of abscisic acid in chilling tolerance of rice (*Oryza sativa* L.) seedlings: II. Modulation of free polyamine levels. *Plant Science, 126*(1), 1–10.

Leivar, P., and Monte, E. (2014). PIFs: Systems integrators in plant development. *The Plant Cell, 26*(1), 56–78.

Lichtenthaler, H. K., Buschmann, C., Döll, M., Fietz, H. J., Bach, T., Kozel, U., Meier, D., and Rahmsdorf, U. (1981). Photosynthetic activity, chloroplast ultrastructure, and leaf characteristics of high-light and low-light plants and of sun and shade leaves. *Photosynthetic Research, 2*, 115–141.

Lin, C., Robertson, D. E., Ahmad, M., Raibekas, A. A., Jorns, M. S., Dutton, P. L., and Cashmore, A. R. (1995). Association of Flavin adenine dinucleotide with the Arabidopsis blue light receptor CRY1. *Science, 5226*, 968–970.

Liu, X., Shiomi, S., Nakatsuka, A., Kubo, Y., Nakamura, R., and Inaba, A. (1999). Characterization of ethylene biosynthesis associated with ripening in banana fruit. *Plant Physiology, 121*(4), 1257–1266.

Ljung, K., Bhalerao, R. P., and Sandberg, G. (2001). Sites and homeostatic control of auxin biosynthesis in Arabidopsis during vegetative growth. *The Plant Journal, 28*(4), 465–474.

Loomis, R. S., and Williams, W. A. (1963). Maximum crop productivity: An extimate. *Crop Scicence, 3*(1), 67–72.

Ma, Z., Li, S., Zhang, M., Jiang, S., and Xiao, Y. (2010). Light intensity affects growth, photosynthetic capability, and total flavonoid accumulation of Anoectochilus plants. *HortScience, 45*, 863–867.

Mahajan, S., Pandey, G. K., and Tuteja, N. (2008). Calcium-and salt-stress signaling in plants: Shedding light on SOS pathway. *Archives of Biochemistry and Biophysics, 471*(2), 146–158.

Manivannan, A., Soundararajan, P., Halimah, N., Ko, C. H., and Jeong, B. R. (2015). Blue LED light enhances growth, phytochemical contents, and antioxidant enzyme activities of *Rehmannia glutinosa* cultured in vitro. *Horticulture, Environment, and Biotechnology, 56*(1), 105–113.

Mao, P., Duan, M., Wei, C., and Li, Y. (2007). WRKY62 transcription factor acts downstream of cytosolic NPR1 and negatively regulates jasmonate-responsive gene expression. *Plant and Cell Physiology, 48*(6), 833–842.

Martin, D., Gershenzon, J., and Bohlmann, J. (2003). Induction of volatile terpene biosynthesis and diurnal emission by methyl jasmonate in foliage of Norway spruce. *Plant Physiology, 132*(3), 1586–1599.

Martin, D., Tholl, D., Gershenzon, J., and Bohlmann, J. (2002). Methyl jasmonate induces traumatic resin ducts, terpenoid resin biosynthesis, and terpenoid accumulation in developing xylem of Norway spruce stems. *Plant Physiology, 129*(3), 1003–1018.

Mauch-Mani, B., and Flors, V. (2009). The ATAF1 transcription factor: At the convergence point of ABA-dependent plant defense against biotic and abiotic stresses. *Cell Research, 19*(12), 1322–1323.

McClung, C. R., and Davis, S. J. (2010). Ambient thermometers in plants: from physiological outputs towards mechanisms of thermal sensing. *Current Biology, 20*(24), R1086–R1092.

Meyer, Y., Siala, W., Bashandy, T., Riondet, C., Vignols, F., and Reichheld, J. P. (2008). Glutaredoxins and thioredoxins in plants. *Biochimica et Biophysica Acta (BBA)-Molecular Cell Research, 1783*(4), 589–600.

Miao, Y., and Zentgraf, U. (2007). The antagonist function of Arabidopsis WRKY53 and ESR/ESP in leaf senescence is modulated by the jasmonic and salicylic acid equilibrium. *The Plant Cell, 19*(3), 819–830.

Michaels, S. D. (2009). Flowering time regulation produces much fruit. *Current Opinion in Plant Biology, 12*, 75–80.

Milborrow, B. V. (1974). The chemistry and physiology of abscisic acid. *Annual Review of Plant Physiology, 25*, 259–307.

Miller, C. O., Skoog, F., Von Saltza, M. H., and Strong, F. M. (1955). Kinetin, a cell division factor from deoxyribonucleic acid1. *Journal of the American Chemical Society, 77*(5), 1392–1392.

Mirkovic, T., and Scholes, G. D. (2015). Photosynthetic light harvesting. In: *Photobiology* (pp. 231–241). Springer New York.

Mishra, A., and Choudhuri, M. A. (1999). Effects of salicylic acid on heavy metal-induced membrane deterioration mediated by lipoxygenase in rice. *Biologia Plantarum*, *42*(3), 409–415.

Mittler, R., Vanderauwera, S., Gollery, M., and Van Breusegem, F. (2004). Reactive oxygen gene network of plants. *Trends in Plant Science*, *9*(10), 490–498.

Miura, K., and Tada, Y. (2014). Regulation of water, salinity, and cold stress responses by salicylic acid. *Frontiers in Plant Science*, *5*, 4.

Mockaitis, K., and Estelle, M. (2008). Auxin receptors and plant development: A new signaling paradigm. *Annual Reviews in Cellular Developmental Biology*, *24*, 55–80.

Moe, R., Fjeld, T., and Mortensen, L. M. (1992). Stem elongation and keeping quality in poinsettia (*Euphorbia pulcherrima* Willd.) as affected by temperature and supplementary lighting. *Scientia Horticulturae*, *50*(1), 127–136.

Moore, R., and Evans, M. L. (1986). How roots perceive and respond to gravity. *American Journal of Botany*, *73*(4), 574–587.

Moubayidin, L., Di Mambro, R., and Sabatini, S. (2009). Cytokinin–auxin crosstalk. *Trends in Plant Science*, *14*(10), 557–562.

Navarro, L., Bari, R., Achard, P., Lisón, P., Nemri, A., Harberd, N. P., and Jones, J. D. (2008). DELLAs control plant immune responses by modulating the balance of jasmonic acid and salicylic acid signaling. *Current Biology*, *18*(9), 650–655.

Nicholson, R. L., and Hammerschmidt, R. (1992). Phenolic compounds and their role in disease resistance. *Annual Review of Phytopathology*, *30*(1), 369–389.

Olias, J. M., Sanz, L. C., Rios, J. J., and Perez, A. G. (1992). Inhibitory effect of methyl jasmonate on the volatile ester-forming enzyme system in Golden Delicious apples. *Journal of Agricultural and Food Chemistry*, *40*(2), 266–270.

Olivoto, T., Nardino, M., Carvalho, I. R., Follmann, D. N., Szareski, V. I. J., Ferrari, M., de Pelegrin, A. J., and de Souza, V. Q. O. (2017). Plant secondary metabolites and its dynamical systems of induction in response to environmental factors: A review. *African Journal of Agricultural Research*, *12*(2), 71–84.

Pandey, S., Nelson, D. C., and Assmann, S. M. (2009). Two novel GPCR-type G proteins are abscisic acid receptors in Arabidopsis. *Cell*, *136*(1), 136–148.

Park, Y. G., and Jeong, B. R. (2012). Growth and early yield of stenting-propagated domestic roses are not affected by cytokinins applied at transplanting. *Flower Research Journal*, *20*(2), 55–63.

Park, Y. G., Muneer, S., and Jeong, B. R. (2015). Morphogenesis, flowering, and gene expression of *Dendranthema grandiflorum* in response to shift in light quality of night interruption. *International Journal of Molecular Sciences*, *16*, 16497–16513.

Park, Y. G., Park, J. E., Hwang, S. J., and Jeong, B. R. (2012). Light source and CO_2 concentration affect growth and anthocyanin content of lettuce under controlled environment. *Horticulture, Environment, and Biotechnology*, *53*, 460–466.

Parks, B. M., Quail, P. H., and Hangarter, R. P. (1996). Phytochrome A regulates red-light induction of phototropic enhancement in Arabidopsis. *Plant Physiology*, *110*(1), 155–162.

Pastore, A., Martin, S. R., Politou, A., Kondapalli, K. C., Stemmler, T., and Temussi, P. A. (2007). Unbiased cold denaturation: Low-and high-temperature unfolding of yeast frataxin under physiological conditions. *Journal of American Chemical Society*, *129*, 5374–5375.

Patil, R. S., Ghormade, V., and Deshpande, M. V. (2000). Chitinolytic enzymes: An exploration. *Enzyme and Microbial Technology*, *26*(7), 473–483.

Pereira-Netto, A. B., Cruz-Silva, C. T. A., Schaefer, S., Ramírez, J. A., and Galagovsky, L. R. (2006). Brassinosteroid-stimulated branch elongation in the marubakaido apple rootstock. *Trees*, *20*(3), 286–291.

Pierik, R., Tholen, D., Poorter, H., Visser, E. J., and Voesenek, L. A. (2006). The Janus face of ethylene: Growth inhibition and stimulation. *Trends in Plant Science*, *11*(4), 176–183.

Piringer, A. A., and Cathey, H. M. (1960). Effect of photoperiod, kind of supplemental light and temperature on the growth and flowering of petunia plants. *Proceedings of American Society of Horticultural Science*, *76*, 649–660.

Pokorna, J., Schwarzerova, K., Zelenkova, S., Petrasek, J., Janotova, I., Capkova, V., and Opatrny, Z. (2004). Sites of actin filament initiation and reorganization in cold treated tobacco cells. *Plant, Cell and Environment*, *27*, 641–653.

Popko, J., Hänsch, R., Mendel, R. R., Polle, A., and Teichmann, T. (2010). The role of abscisic acid and auxin in the response of poplar to abiotic stress. *Plant Biology*, *12*(2), 242–258.

Qiu, Q. S., Guo, Y., Dietrich, M. A., Schumaker, K. S., and Zhu, J. K. (2002). Regulation of SOS1, a plasma membrane Na^+/H^+ exchanger in *Arabidopsis thaliana*, by SOS2 and SOS3. *Proceedings of the National Academy of Sciences*, *99*(12), 8436–8441.

Quail, P. H., Boylan, M. T., Parks, B. M., Short, T. W., Xu, Y., and Wagner, D. (1995). Phytochromes: Photosensory perception and signal transduction. *Science*, *268*(5211), 675–680.

Raskin, I., Skubatz, H., Tang, W., and Meeuse, B. J. D. (1990). Salicylic acid levels in thermogenic and non-thermogenic plants. *Annals of Botany*, *66*(4), 376–373.

Raza, S. H., Athar, H. R., Ashraf, M., and Hameed, A. (2007). Glycinebetaine-induced modulation of antioxidant enzymes activities and ion accumulation in two wheat cultivars differing in salt tolerance. *Environmental and Experimental Botany*, *60*(3), 368–376.

Robson, P. R., and Smith, H. (1996). Genetic and transgenic evidence that phytochromes A and B act to modulate the gravitropic orientation of *Arabidopsis thaliana* hypocotyls. *Plant Physiology*, *110*, 211–216.

Rodrigo, I., Vera, P., Tornero, P., Hernández-Yago, J., and Conejero, V. (1993). cDNA cloning of viroid-induced tomato pathogenesis-related protein P23 (characterization as a vacuolar antifungal factor). *Plant Physiology*, *102*(3), 939–945.

Rohde, A., Prinsen, E., De Rycke, R., Engler, G., van Montagu, M., and Boerjan, W. (2002). PtABI3 impinges on the growth and differentiation of embryonic leaves during bud set in Poplar. *The Plant Cell*, *14*(8), 1885–1901.

Roxas, V. P., Smith, R. K., Allen, E. R., and Allen, R. D. (1997). Overexpression of glutathione S-transferase/glutathione peroxidase enhances the growth of transgenic tobacco seedlings during stress. *Nature Biotechnology*, *15*(10), 988–991.

Ruelland, E., and Zachowski, A. (2010). How plants sense temperature. *Environmental and Experimental Botany*, *69*, 225–232.

Sakuraba, Y., Jeong, J., Kang, M. Y., Kim, J., Paek, N. C., and Choi, G. (2014). Phytochrome-interacting transcription factors PIF4 and PIF5 induce leaf senescence in Arabidopsis. *Nature Communications*, *14*(5), 4636.

Sangwan, V., Orvar, B. L., Beyerly, J., Hirt, H., and Dhindsa, R. S. (2002). Opposite changes in membrane fluidity mimic cold and heat stress activation of distinct plant MAP kinase pathways. *The Plant Journal*, *31*, 629–638.

Sato, R., Ito, H., and Tanaka, A. (2015). Chlorophyll b degradation by chlorophyll b reductase under high-light conditions. *Photosynthetic Research*, *146*, 249–259.

Sayyari, M., Castillo, S., Valero, D., Díaz-Mula, H. M., and Serrano, M. (2011). Acetyl salicylic acid alleviates chilling injury and maintains nutritive and bioactive compounds and antioxidant activity during post-harvest storage of pomegranates. *Postharvest Biology and Technology*, *60*(2), 136–142.

Schaffer, R., Ramsay, N., Samach, A., Corden, S., Putterill, J., Carré, I. A., and Coupland, G. (1998). The late elongated hypocotyl mutation of Arabidopsis disrupts circadian rhythms and the photoperiodic control of flowering. *Cell*, *93*, 1219–1229.

Sekmen, A. H., Türkan, I., and Takio, S. (2007). Differential responses of antioxidative enzymes and lipid peroxidation to salt stress in salt-tolerant *Plantago maritima* and salt-sensitive *Plantago media*. *Physiologia Plantarum*, *131*(3), 399–411.

Sembdner, G. A. P. B., and Parthier, B. (1993). The biochemistry and the physiological and molecular actions of jasmonates. *Annual Review of Plant Biology*, *44*(1), 569–589.

Seo, P. J., Kim, M. J., Song, J. S., Kim, Y. S., Kim, H. J., and Park, C. M. (2010). Proteolytic processing of an Arabidopsis membrane-bound NAC transcription factor is triggered by cold-induced changes in membrane fluidity. *Biochemistry Journal*, *427*, 359–367.

Sewelam, N., Kazan, K., and Schenk, P. M. (2016). Global plant stress signaling: Reactive oxygen species at the cross-road. *Frontiers in Plant Science*, *7*, 187.

Shin, J., Heidrich, K., Sanchez-Villarreal, A., Parker, J. E., and Davis, S. J. (2012). TIME FOR COFFEE represses accumulation of the MYC2 transcription factor to provide time-of-day regulation of jasmonate signaling in Arabidopsis. *The Plant Cell*, *24*(6), 2470–2482.

Singh, B., and Usha, K. (2003). Salicylic acid induced physiological and biochemical changes in wheat seedlings under water stress. *Plant Growth Regulation*, *39*(2), 137–141.

Singh, P. K., and Gautam, S. (2013). Role of salicylic acid on physiological and biochemical mechanism of salinity stress tolerance in plants. *Acta Physiologiae Plantarum*, *35*(8), 2345–2353.

Skriver, K., and Mundy, J. (1990). Gene expression in response to abscisic acid and osmotic stress. *The Plant Cell*, *2*(6), 503.

Smith, H. (1995). Physiological and ecological function within the phytochrome family. *Annual Reviews in Plant Biology*, *46*(1), 289–315.

Smith, H. (2000). Phytochromes and light signal perception by plants—An emerging synthesis. *Nature, 407*, 585–591.

Somers, D. E., Devlin, P. F., and Kay, S. A. (1998). Phytochromes and cryptochromes in the entrainment of the Arabidopsis circadian clock. *Science, 282*(5393), 1488–1490.

Song, S., Qi, T., Huang, H., Ren, Q., Wu, D., Chang, C., Xie, D., Liu, Y., Peng, J., and Xie, D. (2011). The jasmonate-ZIM domain proteins interact with the R2R3-MYB transcription factors MYB21 and MYB24 to affect jasmonate-regulated stamen development in Arabidopsis. *The Plant Cell, 23*(3), 1000–1013.

Spoel, S. H., Johnson, J. S., and Dong, X. (2007). Regulation of tradeoffs between plant defenses against pathogens with different lifestyles. *Proceedings of the National Academy of Sciences, 104*(47), 18842–18847.

Sticher, L., Mauch-Mani, B., and Métraux, A. J. (1997). Systemic acquired resistance. *Annual Review of Phytopathology, 35*(1), 235–270.

Stintzi, A., Weber, H., Reymond, P., Farmer, E. E., and Farmer, E. E. (2001). Plant defense in the absence of jasmonic acid: The role of cyclopentenones. *Proceedings of the National Academy of Sciences, 98*(22), 12837–12842.

Stuefer, J. F., and Huber, H. (1998). Differential effects of light quantity and spectral light quality on growth, morphology and development of two stoloniferous Potentilla species. *Oecologia, 117*, 1–8.

Suge, H., and Rappaport, L. (1968). Role of gibberellins in stem elongation and flowering in radish. *Plant Physiology, 43*, 1208–1214.

Szalai, G., and Janda, T. (2009). Effect of salt stress on the salicylic acid synthesis in young maize (*Zea mays* L.) plants. *Journal of Agronomy and Crop Science, 195*(3), 165–171.

Sze, H., Li, X., and Palmgren, M. G. (1999). Energization of plant cell membranes by H^+-pumping ATPases: Regulation and biosynthesis. *The Plant Cell, 11*(4), 677–689.

Takagi, H., Ishiga, Y., Watanabe, S., Konishi, T., Egusa, M., Akiyoshi, N., Shimada, H., Mori, I. C., Hirayama, T., Kaminaka, H., Shimada, H., and Sakamoto, A. (2016). Allantoin, a stress-related purine metabolite, can activate jasmonate signaling in a MYC2-regulated and abscisic acid-dependent manner. *Journal of Experimental Botany, 67*(8), 2519–2532.

Takezawa, D., Komatsu, K., and Sakata, Y. (2011). ABA in bryophytes: How a universal growth regulator in life became a plant hormone? *Journal of Plant Research, 124*(4), 437–453.

Thimann, K. V. (1969). *The auxins. The Physiology of Plant Growth and Development*, M. B. Wilkins, ed., p. 345. McGraw-Hill, London.

Thimann, K. V., Zhi-Yi, and Thimann, K. V. (1989). The roles of carbon dioxide and abscisic acid in the production of ethylene. *Physiologia Plantarum, 75*(1), 13–19.

Thines, B., and Harmon, F. G. (2010). Ambient temperature response establishes ELF3 as a required component of the core Arabidopsis circadian clock. *Proceedings of the National Academy of Sciences, 107*, 3257–3262.

Timell, T. E. (1986). *Compression Wood in Gymnosperms* (pp. 1189–1193). Springer Verlag, New York, USA.

Tingey, D. T., Standley, C., and Field, R. W. (1976). Stress ethylene evolution: A measure of ozone effects on plants. *Atmospheric Environment, 10*(11), 969–974.

Turner, J. G., Ellis, C., and Devoto, A. (2002). The jasmonate signal pathway. *The Plant Cell, 14*(suppl 1), S153–S164.

Ueguchi-Tanaka, M., Ashikari, M., Nakajima, M., Itoh, H., Katoh, E., Kobayashi, M., Chow, T. Y., Hsing, Y. I., Kitano, H., Yamaguchi, I., and Matsuoka, M. (2005). GIBBERELLIN-INSENSITIVE DWARF1 encodes a soluble receptor for gibberellin. *Nature, 437*(7059), 693–698.

Vahala, J., Keinänen, M., Schützendübel, A., Polle, A., and Kangasjärvi, J. (2003). Differential effects of elevated ozone on two hybrid aspen genotypes predisposed to chronic ozone fumigation. Role of ethylene and salicylic acid. *Plant Physiology, 132*(1), 196–205.

Valero, D., Díaz-Mula, H. M., Zapata, P. J., Castillo, S., Guillén, F., Martínez-Romero, D., and Serrano, M. (2011). Postharvest treatments with salicylic acid, acetylsalicylic acid or oxalic acid delayed ripening and enhanced bioactive compounds and antioxidant capacity in sweet cherry. *Journal of Agricultural and Food Chemistry, 59*(10), 5483–5489.

van Dijk, H., and Hautekeete, N. (2007). Long day plants and the response to global warming: Rapid evolutionary change in day length sensitivity is possible in wild beet. *Journal of Evolutionary Biology, 20*, 349–357.

van Doorn, W. G., and van Meeteren, U. (2003). Flower opening and closure: A review. *Journal of Experimental Botany, 54*, 1801–1812.

Vanderauwera, S., Zimmermann, P., Rombauts, S., Vandenabeele, S., Langebartels, C., Gruissem, W., and Van Breusegem, F. (2005). Genome-wide analysis of hydrogen peroxide-regulated gene expression in

Arabidopsis reveals a high light-induced transcriptional cluster involved in anthocyanin biosynthesis. *Plant Physiology*, *139*(2), 806–821.

Varade, S. B., Stolzy, L. H., and Letey, J. (1970). Influence of temperature, light intensity, and aeration on growth and root porosity of wheat, *Triticum aestivum*. *Agronomy Journal*, *62*, 505–507.

Verslues, P. E. (2017). Time to grow: Factors that control plant growth during mild to moderate drought stress. *Plant, Cell & Environment*, *40*(2), 177–179.

Vicente, M. R.-S., and Plasencia, J. (2011). Salicylic acid beyond defence: Its role in plant growth and development. *Journal of Experimental Botany*, *62*(10), 3321–3338.

Vince, D., Blake, J., and Spencer, R. (1964). Some effects of wave-length of the supplementary light on the photoperiodic behaviour of the long-day plants, carnation and lettuce. *Physiologia Plantarum*, *17*(1), 119–125.

Vreugdenhil, D., and Struik, P. C. (1989). An integrated view of the hormonal regulation of tuber formation in potato (*Solanum tuberosum*). *Physiologia Plantarum*, *75*, 525–531.

Walters, R. G., and Horton, P. (1994). Acclimation of Arabidopsis thaliana to the light environment: Changes in composition of the photosynthetic apparatus. *Planta*, *195*, 248–256.

Wang, D., Weaver, N. D., Kesarwani, M., and Dong, X. (2005). Induction of protein secretory pathway is required for systemic acquired resistance. *Science*, *308*(5724), 1036–1040.

Wang, L. J., Fan, L., Loescher, W., Duan, W., Liu, G. J., Cheng, J. S., Luo, H. B., and Li, S. H. (2010). Salicylic acid alleviates decreases in photosynthesis under heat stress and accelerates recovery in grapevine leaves. *BMC Plant Biology*, *10*(1), 1.

Wang, L. J., and Li, S. H. (2006). Salicylic acid-induced heat or cold tolerance in relation to Ca^{2+} homeostasis and antioxidant systems in young grape plants. *Plant Science*, *170*(4), 685–694.

Wang, Z., Ma, L., Zhang, X., Xu, L., Cao, J., and Jiang, W. (2015). The effect of exogenous salicylic acid on antioxidant activity, bioactive compounds and antioxidant system in apricot fruit. *Scientia Horticulturae*, *181*, 113–120.

Wasilewska, A., Vlad, F., Sirichandra, C., Redko, Y., Jammes, F., Valon, C., Frey, N. F., and Leung, J. (2008). An update on abscisic acid signaling in plants and more.... *Molecular Plant*, *1*(2), 198–217.

Wasternack, C. (2007). Jasmonates: An update on biosynthesis, signal transduction and action in plant stress response, growth and development. *Annals of Botany*, *100*(4), 681–697.

Wasternack, C., and Kombrink, E. (2009). Jasmonates: Structural requirements for lipid-derived signals active in plant stress responses and development. *ACS Chemical Biology*, *5*(1), 63–77.

Werner, T., and Schmülling, T. (2009). Cytokinin action in plant development. *Current Opinion in Plant Biology*, *12*(5), 527–538.

Weston, E., Thorogood, K., Vinti, G., and López-Juez, E. (2000). Light quantity controls leaf-cell and chloroplast development in *Arabidopsis thaliana* wild type and blue-light-perception mutants. *Planta*, *211*, 807–815.

Whippo, C. W., and Hangarter, R. P. (2006). Phototropism: Bending towards enlightenment. *Plant Cell*, *18*, 1110–1119.

Wilson, B. F., and Archer, R. R. (1983). Apical control of branch movements and tension wood in black cherry and white ash trees. *Canadian Journal of Forest Research*, *13*(4), 594–600.

Wilson, B. F., and Kelty, M. J. (1994). Shoot growth from the bud bank in black oak. *Canadian Journal of Forest Research*, *24*(1), 149–154.

Xue, G. P. (2003). The DNA-binding activity of an AP2 transcriptional activator HvCBF2 involved in regulation of low-temperature responsive genes in barley is modulated by temperature. *The Plant Journal*, *33*, 373–383.

Yamada, M., Greenham, K., Prigge, M. J., Jensen, P. J., and Estelle, M. (2009). The TRANSPORT INHIBITOR RESPONSE2 gene is required for auxin synthesis and diverse aspects of plant development. *Plant Physiology*, *151*, 168–179.

Yao, H., and Tian, S. (2005). Effects of pre-and post-harvest application of salicylic acid or methyl jasmonate on inducing disease resistance of sweet cherry fruit in storage. *Postharvest Biology and Technology*, *35*(3), 253–262.

Yeo, A. R., Flowers, S. A., Rao, G., Welfare, K., Senanayake, N., and Flowers, T. J. (1999). Silicon reduces sodium uptake in rice (*Oryza sativa* L.) in saline conditions and this is accounted for by a reduction in the transpirational bypass flow. *Plant, Cell and Environment*, *22*(5), 559–565.

Yoon, J. Y., Hamayun, M., Lee, S. K., and Lee, I. J. (2009). Methyl jasmonate alleviated salinity stress in soybean. *Journal of Crop Science and Biotechnology*, *12*(2), 63–68.

Zagatti, P., Lemperiere, G., and Malosse, C. (1997). Monoterpenes emitted by the large pine weevil, *Hylobius abietis* (L.) feeding on Scots pine, *Pinus sylvestris* L. *Physiological Entomology*, *22*(4), 394–400.

Zainuri, D. C. J., Wearing, A. H., Coates, L., Terry, L., and Terry, L. (2001). Effects of phosphonate and salicylic acid treatments on anthracnose disease development and ripening of 'Kensington Pride' mango fruit. *Australian Journal of Experimental Agriculture*, *41*(6), 805–813.
Zeevaart, J. A. (1971). Effects of photoperiod on growth rate and endogenous gibberellins in the long-day rosette plant spinach. *Plant Physiology*, *47*, 821–827.
Zhou, Z. S., Guo, K., Elbaz, A. A., and Yang, Z. M. (2009). Salicylic acid alleviates mercury toxicity by preventing oxidative stress in roots of *Medicago sativa*. *Environmental and Experimental Botany*, *65*(1), 27–34.
Zhu, J. K. (2003). Regulation of ion homeostasis under salt stress. *Current Opinion in Plant Biology*, *6*(5), 441–445.

5 Plant Response to Drought Stress

Rakefet David-Schwartz, Hanan Stein, Eran Raveh, David Granot, Nir Carmi, and Tamir Klein

CONTENTS

5.1 INTRODUCTION

Life on Earth evolved under water, and water is the most basic requirement for living organisms. When we look for signs of life on other planets, we first search for water. Over three billion years of marine life passed before the first living organisms were able to survive out of water; the first terrestrial plants appeared just 360 million years ago. All living organisms need to maintain a stable internal environment in order to survive (homeostasis). To that end, the major evolutionary breakthroughs that allowed plant life to survive on dry land were the development of structures for water uptake, water transport and the maintenance of water relations between the plant and the terrestrial environment, specifically, roots, xylem conduits and an outer seal (i.e. cuticle, much like our own skin). As these early terrestrial plants improved their contacts with both soil and air environments, they were no longer adapted to life underwater. All plants need water for their daily functioning and desiccation is an ever-present risk. In this chapter, we present the response of woody plants to this risk. We start by describing principles of water transport and its importance, followed by a discussion of the effect of water stress on tree performance and the main physiological strategies that trees use to cope with this stress. We then review the molecular mechanisms that trees use to cope with water stress, with illustrations of how those mechanisms can be manipulated to induce drought tolerance. We also mention the use of water stress as an agricultural tool, and conclude with suggestions and predictions for future directions in research on water stress in woody species.

5.1.1 Water Relations of Woody Plants in a Nutshell: Water Transport in Plants

Plants need to ensure that adequate amounts of water reach most, if not all, of their cells. However, on land, water is not found everywhere. A river or a lake may not be far off, but plants are sessile organisms and so have had to develop roots to extract water from soil. Water is then distributed to all cells through a large network of thin conduits called the xylem. But, what drives the water out of the soil and into the different plant compartments? This driving force is particularly important for woody plants whose leaves may be located over 100 m above the ground. This question haunted scientists for hundreds of years, and to some extent, still does (Angeles et al. 2004; Zimmermann et al. 2004). The first hypotheses were intimately related to the aforementioned compartments, namely that (1) roots pump water out of soil; and that (2) if the plant's "pipes" are sufficiently narrow, capillary forces push the water upward. While both of these processes are important for transpiration, they can explain only a small portion of the water's ascent. Water is unique in its strong cohesion and adhesion forces, but the so-called pump is not down in the roots, but rather up in the air surrounding the foliage. This pulling force is generated by evaporation on the surface of leaves and is transmitted downward through water columns under tension (Dixon 1896). Water movement in plants is explained by the cohesion-tension theory and can be described as *transpirational lift*.

5.1.2 Role of Water in Plant Growth and Development

Before we delve into the secrets of plant water transport, let's remind ourselves of the multiple roles that water plays in plant growth and development and the multiple water sources that woody plants exploit. In the previous section, we highlighted the need for any living cell, and specifically plant cells, to stay wet. Water provides the medium for all biochemical reactions in a living cell; it is nature's ultimate solvent. Plant cells are unique in that they are linked to one another by microscopic channels called plasmodesmata, this arrangement allows water to be actively absorbed by cells and to move between neighboring cells (the symplastic pathway). Water also passively moves through intercellular spaces and the permeable rigid cell walls, and this passive transport of water (the apoplastic pathway) is faster than the symplastic pathway. The transcellular pathway is another route in which water may pass through aquaporins (AQPs), which are transmembrane water channels that facilitate the transport of water across membranes and determine the permeability of membranes, cells and tissues to water (Maurel et al. 2008; Heinen et al. 2009). Cell walls provide rigidity to any plant tissue, but without water to hold the tissue erect, wilting occurs. In summary, plants need water (1) as a medium for biochemistry, (2) to transport material between different cells and plant parts and (3) to maintain turgor at the cell level and physical stability at the tissue level. But, remarkably, these three functions together account for less than 1% of plant water use. The vast majority of the water used by plants (4) returns to the atmosphere through transpiration, a consequence of stomata opening for photosynthesis. When water evaporates from leaf surfaces, it plays another role (5) as it cools down the foliage, helping the plant release excess heat.

Water sources for plants are as varied as water functions. Although most water is extracted from the soil by roots, these plastic plant organs can tap into any reservoir, including groundwater, runoff, drainage and irrigation water. In some woody species, water is also taken up by non-root tissues such as leaves (Limm et al. 2009) and bark (Katz et al. 1989). Finally, many woody species can store water in their stems and other tissues. Over a short to medium timescale (less than a few years), this internal water storage can be regarded as an additional water source (Scoffoni et al. 2012; Klein et al. 2016).

5.1.3 Water Potential and Stomatal Behavior

Of all plants, tall trees astonish us the most. If we think about the tension at which water is held inside their xylem conduits and account for the hydrostatic pressure, we know that tension must increase as water moves up the tree. This relates to the concept of water potential: Water will always move from higher free energy to lower free energy. Therefore, water transport in plants follows a water potential (Ψ_p) gradient. Measuring Ψ_p inside the xylem is very difficult, although some interesting attempts have been made (Zimmermann et al. 2004). However, if a whole leaf or shoot is held under pressure, the pressure required to drive water out from its cut petiole should be equal in magnitude and opposite in direction to the tension (i.e. Ψ_p) in the leaf (Scholander et al. 1965). Using a pressure chamber and a canopy crane, one can follow the Ψ_p of leaves and terminal shoots of tall forest trees (Figure 5.1). In any tree, the Ψ_p becomes more negative with height. Still, Ψ_p values vary among tree species due to differences in xylem design and changes in stomatal function.

Yet another implication of transpirational lift is that water transport in plants can be described in terms of two different Darcy's Law analogies. In the gas phase, we define:

$$T = g_s \times \text{VPD} \tag{5.1}$$

where T, transpiration (g m^{-2} s^{-1}), is the product of g_s, stomatal conductance (g m^{-2} s^{-1}), and VPD, vapor pressure deficit (kPa kPa^{-1}).

In the liquid phase, we define:

$$\text{SF} = K \times \left(\Psi_s - \Psi_l - 0.01h\right) \tag{5.2}$$

where SF, sap flow (g m^{-2} s^{-1}), is the product of K (g m^{-2} s^{-1} MPa^{-1}), xylem conductivity, and the difference between Ψ_s, soil water potential (MPa), and Ψ_l, leaf water potential (MPa), minus the gravitational component that is a function of h, tree height (m).

Trees can adjust their water use and level of hydraulic risk through the coordination of hydraulic and stomatal regulation (Sperry 2000; Brodribb and McAdam 2011; Choat et al. 2012). While xylem properties (e.g. conduit diameter and pit membrane structure) provide long-term control, stomata dynamically adjust leaf transpiration as they open and close (Tyree and Sperry 1989; Cochard et al. 1996; Woodruff et al. 2008; Meinzer et al. 2009; Brodribb et al. 2016). Upon desiccation, stomata may respond to a multitude of physical and chemical signals, including leaf hydration (turgor loss) and the levels of the phytohormone abscisic acid (ABA) (Brodribb and McAdam 2013; Speirs et al. 2013) in xylem and leaves. Ψ_l also plays an important role in controlling guard cell movement (Cochard et al. 1996; Salleo et al. 2000; Brodribb et al. 2003). In fact, leaf g_s of water vapor and Ψ_l interact via a feedback mechanism, through which stomatal sensitivity to Ψ_l (i.e. reduction in g_s as Ψ_l decreases) results in an increase in Ψ_l (Oren et al. 1999; Buckley 2005).

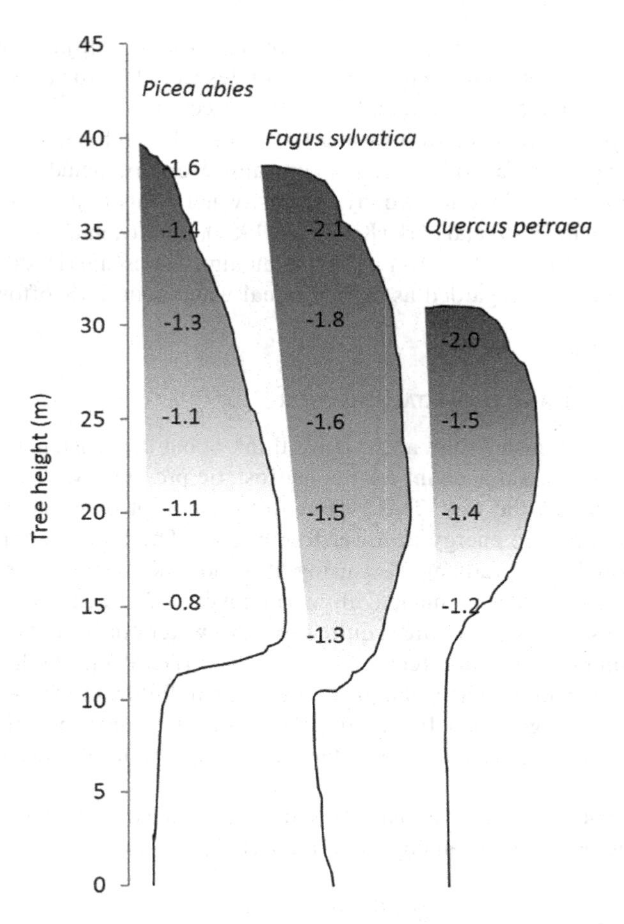

FIGURE 5.1 Leaf water potential (MPa) height profile of three key European tree species at the Swiss Canopy Crane site near Basel, Switzerland. All measurements were performed on sunlit leaves over 1 h on 19 September 2013 (air temperature = 17°C, photosynthetically active radiation = 150–300 μmol m^{-2} s^{-1}, vapor pressure deficit = 2.2 kPa).

5.1.4 Definitions of Drought

Water transport in plants can be viewed in terms of mass transfer from mildly negative Ψ_s (e.g. −0.1 MPa) to very negative Ψ in the air (e.g. Ψ_a = −100 MPa). In the soil–plant–atmosphere continuum, plants can tolerate Ψ of −2 MPa or up to −10 MPa among conifer trees of the Cupressaceae (Brodribb et al. 2014). This limitation, set by xylem design, applies to liquid water and sets an explicit threshold on Ψ_s. In the gas phase, Ψ_a can theoretically decrease below −300 MPa (when relative humidity < 10%) without directly disturbing plant water transport. On the contrary, drier air increases the atmospheric pull (VPD) and thus the rate of transport (Eq. 5.1). The water flux will increase up to the limit of the maximum xylem conductivity. Beyond that point, increased flux will cause the leaf Ψ to decrease (Eq. 5.2), again putting the xylem at risk. Since changes in Ψ_s and Ψ_a act on the plant Ψ in different ways, it is rather confusing to describe both decreases in soil moisture and VPD using the single term 'drought'. It is necessary to distinguish between soil drought and atmospheric drought, as is commonly done in controlled experiments (e.g. in a greenhouse). However, few field studies have decoupled these types of drought and their effects on tree function, despite the fact that such scenarios are not uncommon. For example, in the eastern Mediterranean region, short and intense heat waves occur year-round, with VPD rising over 4 kPa under a wide range of soil moisture conditions. These events strongly reduce gas exchange fluxes in a pine forest, with the response weakening as soil moisture decreases (Tatarinov et al. 2016).

Another source of confusion around the definition of drought (also referred to as water deficit or desiccation) is a lack of global metrics. Environmental drivers of Ψ_l are Ψ_s and Ψ_a, which are primarily influenced by precipitation, evapotranspiration and VPD. These parameters vary not only over time, but also among biomes. Annual trends of daily VPD were measured at eight flux-tower sites representing various biomes across the Americas. VPD greater than 4 kPa was recorded in only two biomes (desert grassland and tallgrass prairie); whereas at the boreal, temperate, and tropical forest sites, VPD was never above 2 kPa. To some extent, the effect of high VPD on tree function should vary among biomes, as tree species in biomes that are less prone to atmospheric drought are less adapted to high VPD. Yet, Ψ_l is regulated by stomatal closure and when VPD increases, g_s tends to converge across species (Oren et al. 1999). That would suggest that the effect of high VPD should be rather similar among trees in both dry and humid biomes. Tree vulnerability to drought is mostly determined by the VPD reaching (or remaining below) a critical threshold. While VPD maxima may change from one year to the next or as the climate changes, it is assumed that not all forest biomes are similarly threatened by atmospheric drought.

In the soil, water availability is often studied locally without reference to a standard scale, so that the term 'soil drought' can have different meanings in different studies (Vicca et al. 2012). This is because the availability of water for transpiration depends on the local soil and plant hydraulic properties (i.e. soil water retention curves and the Ψ_s threshold for root uptake) (Sinclair et al. 2005). Soil water retention changes with soil type, which is mostly independent of the forest biome type and, therefore, would be expected to account for additional variation among biomes in terms of their susceptibility to soil drought. Knowledge of soil water-retention characteristics and the Ψ_s threshold for root uptake for a specific soil structure and tree species allows for the calculation of the actual amount of transpirable soil water content (tSWC). It has recently been demonstrated that changes in tSWC can predict episodic tree mortality in a semi-arid pine forest to a fairly high degree of precision (Klein et al. 2014b). Quantification of tSWC (and its decay) offers a general framework for measuring the extent of soil drought, which allows comparisons of the hydraulic vulnerability of different forest ecosystems.

5.2 EFFECT OF DROUGHT STRESS ON TREE PERFORMANCE

Our review of the effects of drought on trees starts with a discussion of phenotypic responses and then delves into the complex physiological and molecular mechanisms that underlie those effects. The somewhat anthropocentric term *tree performance*, which is commonly used in forestry, usually refers to two basic parameters: tree growth (including the rate of growth and the overall health) and tree survival. As simple as these two parameters may seem, they each include multiple nuances and complex interactions. Trees change with age, year and season and respond to many different factors. Here, we explore some of the major drought responses that are directly linked to performance.

5.2.1 Plastic Morphological Responses

Being perennial, trees cannot escape drought conditions as annual plants do (by completing their life cycle during the wet season). The need to avoid or tolerate prolonged and temporary water deficits has led trees to develop special responses to drought stress. In order to minimize water loss under prolonged drought stress, trees change the shapes of their organs. These plastic responses, some of which are described below, involve leaves, stem and roots.

Under severe drought, trees may shed their leaves or, alternatively, increase their leaf dry mass per unit area (Niinemets 2001). Mild drought results in leaf shrinkage, which has been found to be associated with a decline in leaf hydraulic conductance. Interestingly, this acclimation process occurs earlier in species from mesic sites than in species from xeric sites (Scoffoni et al. 2011, 2013).

In conifers such as a Mediterranean pine species (Klein et al. 2014c), needle length can serve as an indicator of drought stress. Among leaf parameters, leaf thickness and density are highly responsive to environmental cues. A survey of 597 woody species suggested that leaf thickness is positively correlated with assimilation capacity; whereas leaf density is negatively correlated with assimilation capacity (Niinemets 1999). Under drought conditions, trees maintain their leaf thickness better than herbaceous species (Kennedy and Booth 1958). In a study involving 558 broadleaf and 39 needle-leaf shrubs and trees from 182 geographical sites, Niinemets (2001) found that, among the broadleaf species, leaf density was negatively and strongly correlated with mean precipitation during the three driest months of the year. In contrast, among the needle-leaf species, leaf density was correlated with total annual precipitation, probably reflecting the evergreen habit of these species and their greater drought resistance, as compared to broadleaf species. When comparing habitats, leaf-thickness of woody species from xeric sites was maintained better than that of species from mesic sites, mainly due to the greater negative osmotic potential and higher cell wall elasticity (CWE) of those species (Regier et al. 2009; Scoffoni et al. 2011). Similarly, low CWE was correlated with greater shrinkage of leaf thickness among citrus trees under limited water availability (Syvertsen and Levy 1982). The response of leaf thickness to drought is so robust that leaf thickness can be a suitable indicator for irrigation management (Syvertsen and Levy 1982; Seelig et al. 2012).

Underground, roots also have plastic response to water constraints (Brunner et al. 2015). In order to increase the uptake of soil water and reduce water losses due to transpiration during periods of prolonged drought stress, root-to-shoot ratio is increased (Hsiao and Xu 2000). In a meta-analysis, Poorter et al. (2012) showed that this increase in the root-to-shoot ratio correlates with drought severity. In general, forest trees from xeric sites have higher root-to-shoot ratios than trees from mesic sites (Mokany et al. 2006; Markesteijn and Poorter 2009). However, variations in this trait have been reported among populations of poplar (*Populus kangdingensis* and *P. cathayana*) and oak (*Quercus robur*, *Q. petraea* and *Q. pubescens*) (Yin et al. 2004; Arend et al. 2011).

Perhaps the most studied plastic response in forest trees exposed to a water deficit is cambial activity as reflected in tree rings (Chave et al. 2009; Baas 2013). Tree rings, which are rhythmic variations in wood density, are composed of xylem elements from the cambial tissue and are formed during the rainy season of each year in regions with rhythmic changes in climate. Xylem elements are fixed once they are formed, and thus they reflect tree growth conditions and can be used to evaluate relative water availability with each passing year (Fonti et al. 2010; Dorman et al. 2015). The cambium has intrinsic potential for year-round activity; however, such activity is strongly affected by water availability and temperature. When temperatures are too low, no cambial activity takes place; however, once temperatures rise and the soil is saturated with water, maximal cambial activity is achieved, and wide xylem conduits are generated to form early wood tissue. As soon as water becomes less available, narrower xylem conduits are formed to create the late wood. In other words, trees develop large cells for conducting water when water is readily available and, once water is less readily available, xylem production decreases until it ceases. The duration and intensity of water availability determine the relative amounts of different tissues, conduit dimensions and the density and the thickness of cell walls (Chave et al. 2009; Fonti et al. 2010). In addition to annual cambial activity, there are also intra-annual fluctuations in cambial density (Fritts 1976; Zalloni et al. 2016). In those cases, a layer of wide cells is formed within a late wood zone, or a layer of narrow cells is formed within the early wood zone. These fluctuations affect wood quality and occur when climate conditions are unstable. These fluctuations have become more frequent due to climate change, particularly in regions that have distinct wet and dry seasons (Zalloni et al. 2016).

5.2.2 Drought-Induced Mortality

Despite the many responses and strategies that trees use to withstand water stress, drought-induced mortality is inevitable at times. The largest forest losses in the past decade, aside from those due to deforestation activities, are probably those in central Russia, following wildfires in that region in 2010, and losses in northwestern North America, due to the mountain pine beetle. In both of these cases, tree mortality was not the direct result of drought, but the 2010 wildfires were certainly related to an exceptional heat wave (de Groot et al. 2013), and the dynamics of the beetle infestation are well-correlated with a warming trend (Bentz et al. 2010). Identifying the underlying causes of tree mortality events involves the use of field measurements to decipher the eco-physiological pathways leading to mortality. Three main tree mortality mechanisms have been proposed: hydraulic failure, carbon starvation as result of prolonged stomatal closure and increased attacks by biotic agents promoted by reduced plant defense abilities (McDowell et al. 2008; Martínez-Vilalta 2014). The definition and scope of carbon starvation is still being discussed, specifically, whether trees might lose all of their carbon reserves (McDowell 2011) or lose the ability to allocate carbon to certain tissues (Sala et al. 2010). Other discussions have addressed whether xylem cavitation and its reversal are uncommon (Cochard and Delzon 2013) or rather routine in certain species under prolonged drought conditions (Klein et al. 2014c).

In order to partition the specific contribution of each of the three mechanisms in the field, interactions must be carefully studied. Efforts to untangle potential mortality mechanisms must consider two different facets of carbon and water in plants: (1) they are both necessary for plant life at the whole-tree scale (i.e. as fluxes) and (2) they are also both necessary for specific tissues that fulfill specific functions. Drought can act on shoot hydration levels, arresting meristematic growth through the loss of turgor, and can affect whole-tree hydraulic conductivity via xylem cavitation. Similarly, carbon economy is vulnerable in both whole-tree transport, through phloem-flow disorders, and specific storage pools, such as root starch (Figure 5.2a). Biotic stressors can act on any of the tree compartments in parallel with the abiotic stressors of reduced moisture or humidity and elevated temperatures.

Tree mortality usually involves functional damage to one or more of the water and carbon fluxes and tissue levels, further exacerbated by biotic attacks (Figure 5.2a). The dynamics of water and carbon fluxes depend on the duration of the drought, tree drought responses and biotic stressors. For example, when stomata close, the subsequent reductions in sap flow and carbon transport can be stopped through partial defoliation (Figure 5.2b), which reduces the total transpiring area. The water status of the remaining leaves might recover, while carbon reserves in specific compartments

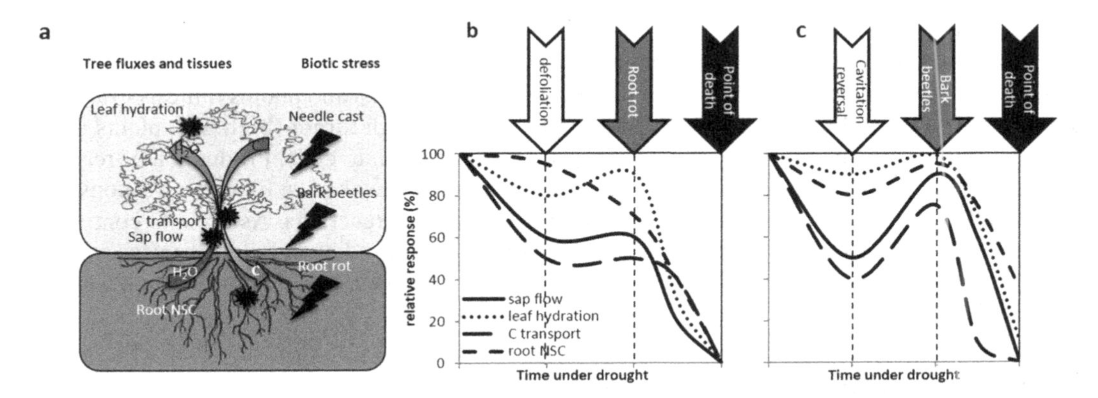

FIGURE 5.2 A schematic representation of major elements of mortality mechanisms in a drought-stressed tree (a) and two arbitrary mortality scenarios (b, c) showing the effects of tree drought responses (black arrows) and biotic stressors (gray arrows) on water and carbon fluxes (sap flow and carbon transport, respectively) and tissue levels [leaf hydration and root nonstructural carbohydrates (NSC), respectively]. Tree mortality usually involves damage to one or more of these four sensitive elements, further exacerbated by biotic attacks. The dynamics of the water and carbon fluxes depend on the balance between the duration of the drought, tree drought responses and biotic stressors. (Reprinted from Klein (2015) with permission from Oxford University Press.)

(e.g. the roots) continue to decrease, possibly due to the transport of that stored carbon to sink tissues (e.g. root growth for soil water exploration), as shown in Aguadé et al. (2015). Fungal root rot infection imposes an immediate limitation on water uptake and has a delayed effect on the root non-structural carbohydrate (NSC) pool and the flow of substances through the phloem. If drought persists, all four components might reach a zero level at mortality (Figure 5.2b) (Poyatos et al. 2013). Depending on the tree species, a different scenario might include hydraulic recovery by cavitation reversal (Brodersen and McElrone 2013), which might lead to the recovery of leaf hydration and root NSC levels (Figure 5.2c). Biotic stresses, such as a bark beetle attack, might severely affect the movement of substances through the phloem, including carbon transport. If the cavitation-reversal mechanism is assisted by sugar secretion, then the loss of a carbohydrate supply might arrest that process, leading to a loss of hydraulic conductivity. Limited carbon re-mobilization also means that trees might die with residual amounts of NSC (Figure 5.2c), as has sometimes been observed (Sala et al. 2010; Klein et al. 2014c).

The simplified, speculative mortality scenarios described here demonstrate some of the complexities eco-physiologists face as they try to untangle the causes of tree mortality. First, the carbon and water economies of trees are closely linked: the xylem and phloem are tightly linked and, at the leaf level, stomata balance carbon gain and water loss. Therefore, it is hard to separate out discrete drought effects on specific fluxes and tissues. Second, tree drought responses interact with these economies and hence their effects are often difficult to measure. Third, pathogens may exert systemic effects on carbon and water economies even when their immediate effects are restricted to a specific compartment, such as the root system. In addition, warming and drying may affect biotic stressors independently. For example, the activity of fungal pathogens might decline with desiccation; whereas bark beetles might benefit from warmer temperatures (Bentz et al. 2010). For the sake of clarity, the scenarios in Figure 5.2 present a sequential order of events, but in reality, physiological drought responses and biotic attacks can occur simultaneously. In addition to these uncertainties, efforts to assess the cause of mortality may come too late, necessitating a post-mortem investigation.

5.3 PLASTIC PHYSIOLOGICAL RESPONSES

When plants sense water stress, they defend themselves by activating physiological mechanisms at multiple levels, depending on the severity and duration of the drought. Physiological responses to water deficit begin with reductions in cell growth due to turgor loss and decreases in g_s that are inevitably accompanied by lower rates of CO_2 uptake and photosynthesis (Taiz and Zeiger 2010; Perez-Martin et al. 2014). In order to maintain high turgor (high Ψp), plants make active osmotic adjustments to facilitate the movement of water into cells. Production of reactive oxygen species (ROS) following reductions in photosynthesis evokes an oxidative response. In addition, hormonal changes seem to be involved in all of these reactions. As drought continues, reduced cell expansion and changes in the allocation of carbon become dominant processes, with morphological and anatomical consequences. Some of these physiological responses are described below.

5.3.1 Maintaining Turgor through Osmotic Adjustment

Maintaining turgor potential during drought stress ensures the homeostasis necessary for growth and survival. Turgor potential depends on osmotic potential and the elasticity of cells and tissues. Osmotic potential is affected by the accumulation of solutes in the process of osmoregulation, which is also referred to as osmotic adjustment (OA). Following exposure to water stress, plants accumulate solutes (osmolytes/compatible solutes) in order to regulate the osmotic potential within their cells until the intracellular osmotic potential is equal to the potential of the medium outside the cells. Water potential at the turgor loss point (π_{tlp}) is equal to the osmotic potential and is considered to be a reliable proxy for drought resistance (Bartlett et al. 2014). The more negative the π_{tlp}, the more drought-tolerant the individual is presumed to be. More negative π_{tlp} values have been shown to correlate with species growing at sites with less available water (Bartlett et al. 2012b; Maréchaux et al. 2015). Since osmotic potential at full turgor has been shown to strongly correlate with osmotic potential at the turgor loss point, in many species, it is possible to estimate drought tolerance via osmometer measurements (Bartlett et al. 2012a, b; Maréchaux et al. 2015).

Plant species vary in the osmolytes they use for OA. Free amino acids, various ions, soluble sugars and free polyamines are some of the substances used as osmolytes while the free amino acid proline is the most common osmolyte (Verbruggen and Hermans 2008; Krasensky and Jonak 2012). These molecules are considered compatible with the internal cell environment since they are not toxic even when they accumulate to relatively high concentrations, chiefly in the cytoplasm, but also in the vacuole. Inorganic ions, mainly Na^+, K^+, Ca^{2+}, and Cl^-, which can also act as osmotic solutes in plants, may accumulate to relatively high levels in the vacuole, but not in the cytoplasm, where high levels of such ions are toxic. Some solutes protect plants from stress by detoxifying ROS to protect membrane integrity and by stabilizing proteins under low-water-content conditions (Turner and Jones 1980; Chen and Jiang 2010).

Osmotic adjustment (OA) is generally accepted as a mechanism for plant adaptation to drought. Blum (2016) described a positive and significant association between OA and yield under drought conditions across 26 published studies and concluded that OA is an important mechanism for sustaining crop yields under drought stress.

In trees, proline accumulation has been shown to be an adaptive trait that is dependent upon genetic factors and plant age. Poplar hybrids adapted to dry conditions displayed higher proline concentrations in their older leaves, but not in their younger leaves or roots (Xiao et al. 2008; Cocozza et al. 2010). Similar observations have been reported in other, but not all poplar hybrids, suggesting that other osmolytes or other adjustment mechanisms, such as differential allocation of resources or better retention of water by other means, are involved (Barchet et al. 2014). In two different experiments in which poplar populations from wet and dry regions were subjected to prolonged drought stress, the accumulation of soluble sugars and free proline was reported. Populations from dry regions exhibited earlier OA, suggesting stronger tolerance to drought stress as compared with populations from a wet climate (Cramer et al. 2007; Xiao et al. 2008). Under drought stress, increased proline levels have also been observed in other tree species such as *Fagus sylvatica, Ziziphus mauritiana, Laurus nobilis, Prunus* spp. and *Ceratonia silique* (Diamantoglou and Rhizopoulou 1992; Clifford et al. 1998; Vincent et al. 2007; De Diego et al. 2013b; Jiménez et al. 2013).

Coniferous species also accumulate proline in response to drought, as shown for several pine species including *Pseudotsuga menziesii, Pinus halepensis, Pinus radiata* and *Pinus pinaster* (Diamantoglou and Rhizopoulou 1992; Schraml and Rennenberg 2000; Corcuera et al. 2012; De Diego et al. 2013b; Du et al. 2015). However, *Pinus pinaster* and *P. menziesii* genotypes from wet areas accumulated higher levels of proline in response to drought than genotypes from drier sites (Schraml and Rennenberg 2000; Corcuera et al. 2012; Du et al. 2015). Species from drier sites may utilize alternative physiological mechanisms or, alternatively, may have evolved structural adaptations to water deficit, as has been shown for the xylem elements of *Pinus halepensis* (David-Schwartz et al. 2016).

Other osmolytes, such as soluble sugars, accumulate in parallel to ABA in response to drought stress, as was reported for *Laurus azorica* trees (Buchanan and Balmer 2005). Similarly, sucrose, glucose, fructose and myoinositol were reported to accumulate following drought in some poplar hybrids (Tschaplinski and Tuskan 1994; Sibout and Guerrier 1998).

Recent studies involving gas chromatography/mass-spectrometry have revealed and confirmed the existence of genotype-specific metabolic profiles in response to drought stress in poplar (Barchet et al. 2014; Hamanishi et al. 2015). Some hybrids demonstrate high levels of solute accumulation, implying the activation of an osmotic response, while others use a reactive-oxygen-scavenging approach to resist drought (Barchet et al. 2014). Variation is also seen in the overall metabolomes of six genotypes of *Pinus balsamifera*, with particular differences in the accumulation of sugars such as glucose and galactinol (Hamanishi et al. 2015). Use of metabolomic analysis to various tree species from different environments should further contribute to our understanding of osmotic regulation in woody species.

5.3.2 Stomatal Closure in Response to Various Stimuli

The primary role of stomata, which are composed of two guard cells, is to prevent dehydration and control the movement of gases between the internal leaf air spaces and the external atmosphere. Water vapor within leaves evaporates to the atmosphere through open stomata at a faster pace than CO_2 is taken up and incorporated through photosynthesis, with the result that 98% of the water taken up by the roots is usually lost to the atmosphere via stomata. In light of this reality, the guard cells must operate to ensure an appropriate balance between the uptake of CO_2 for photosynthesis and water loss. Accordingly, guard cells are sensitive to Ψ_l, air humidity (VPD), light intensity and CO_2 concentration. The stomata close when leaf water potential is low, humidity is low (high VPD), light intensity is low and the concentration of CO_2 is high. The sensitivity of guard cells to these stimuli varies among plant species. Stomatal responses to changes in Ψ_l are more intense than their responses to any other stimuli, and stomatal responses to changes in air humidity are more intense than responses to changes in photosynthetic factors (Aasamaa and Sober 2011).

It is not yet clear how guard cells sense changes in Ψ_l, VPD, light intensity or CO_2 concentrations. Stomata close in order to preserve Ψ_l and maintain plant turgor, yet under drought conditions, loss of leaf turgor may actually cause stomata to close, due to low water potential outside the guard cells, pulling water out of the guard cells. Indeed, in some tree species, Ψ_l declines with the onset of stomatal closure, so that stomatal closure may occur immediately in response to changes in Ψ_l (Loewenstein and Pallardy 1998).

Low light intensity almost inevitably decreases photosynthesis, increasing the concentration of CO_2 in the stomatal cavity. Accordingly, the same stomatal sensing mechanism may be involved in the responses to low light intensity and high concentrations of CO_2, although that mechanism is still unclear. Roles for carbonic anhydrase (CA) and AQP in sensing high levels of CO_2 and stomatal closure were recently suggested in the context of a study of olive trees (Perez-Martin et al. 2014). CAs facilitate the movement of CO_2 into the guard cells by catalyzing the conversion of CO_2 into carbonic acid. As such, CAs are thought to sense CO_2 concentrations and affect stomatal closure (Hu et al. 2010) and may also activate AQPs that facilitate the movement of water out of the guard cells.

With regard to the sensing of VPD, it has been suggested that guard cells sense the increased transpiration rate resulting from reduced humidity levels. Accordingly, changes in transpiration rates are among the immediate stimuli that may indicate air-drying conditions (Mott and Parkhurst 1991). In many plant species, there is a daily midday drop in transpiration, which may be the result of increased VPD caused by air warming over the course of the day. In contrast to low air humidity, which has an immediate physical effect, soil drying decreases the rate of transpiration via stimulated ABA production in the roots (Loewenstein and Pallardy 1998). This ABA is then carried by the transpiration stream toward the guard cells and closes the stomata (Pospíšilová 2003). This process is a slower than the stomatal closure stimulated by accelerated transpiration due to increased VPD, as it requires the production of ABA and the transport of ABA from the roots to the leaves. Variation in stomatal sensitivity to ABA levels has been reported for *Populus trichocarpa* ecotypes and *P. koreana* × *trichocarpa* 'Peace', which possess stomata that are insensitive to exogenous ABA applications (Braatne et al. 1992; Ridolfi et al. 1996). These trees might also respond better to reductions in hydraulic conductance. Indeed, a study with olive (*Olea europaea* L.) trees found that the reduction in stomatal g_s during imposed water stress was more closely related to the loss of hydraulic functioning in the most distal organs of the plant (i.e. roots and leaves) than to increased ABA levels at the leaf, stem and root levels (Torres-Ruiz et al. 2014).

In most plant species (except for plants such as cacti, which employ crassulacean acid metabolism), stomata open at dawn in response to blue light, allowing the leaves to start absorbing CO_2 for red light-activated photosynthesis. However, if light intensity is low because of clouds, shade or high latitude during the winter, the rate of photosynthesis will be lower and stomata will close to avoid unnecessary water loss. This closure is probably stimulated by increased CO_2 concentrations within the stomatal cavity that are sensed by the guard cells (Ward et al. 2013; Xu et al. 2016). Coordination of photosynthesis with the transpiration rate might be even more important under drought conditions and should happen not only at low light-intensity levels, but also under high light intensity, when sugar production exceeds the plant's ability to transport and use that sugar. That may happen through the feedback inhibition of photosynthetic gene expression by excess sugar, which is not an immediate effect as it may require transcriptional down-regulation of genes related to photosynthesis (Rolland et al. 2006; McCormick et al. 2008). Yet, an excess of sugars might also close stomata rapidly via sugar-sensing within guard cells, as has recently been shown for citrus trees (Lugassi et al. 2015). For many years, sugars were thought to be osmolytes that open stomata. However, it was recently shown that mesophyll-produced sugars carried toward the stomata by the transpiration stream are sensed by the guard cells and stimulate stomatal closure (Kelly et al. 2013). Interestingly, stomatal closure in response to sugar level is dependent on ABA produced within guard cells and so exploits the same mechanism as the drought-induced ABA-mediated stomatal closure effect. That happens mainly when light intensity levels and the rate of photosynthesis are high, as has been shown in citrus leaves, and allows for sugar production to be coordinated with transpiration (Lugassi et al. 2015). This effect might be exaggerated under water-stress conditions since the export of sugar to sink tissues, a process that is dependent on water availability and phloem pressure flow, might be halted under such conditions, causing increased sugar concentrations in the leaves that might accelerate stomatal closure.

5.3.3 Oxidative Response

One of the primary and cardinal changes in response to drought stress is stomatal closure and the consequent reduction in photosynthesis due to limited CO_2 absorption (Chaves et al. 2002). Subsequently, photorespiration becomes dominant and ROS accumulate (Novaes et al. 2010; Noctor et al. 2014). 70% of ROS that are generated under drought stress are produced through photorespiration (Noctor et al. 2002). ROS can damage many cell components and eventually lead to cell death. However, enhanced cellular ROS levels also act as an alarm signal that triggers the plant's defense pathways and responses, enabling it to adapt to the changing environment (Choudhury et al. 2016; Noctor et al. 2014). One of the signaling roles of ROS is to trigger the scavenging system that reduces damage caused by oxidative stress. These scavenging systems include antioxidant enzymes such as superoxide dismutase (SOD) that converts superoxide to hydrogen peroxide and enzymes that convert hydrogen peroxides to water and oxygen, including ascorbate peroxidase (APX), glutathione peroxidase (GPX), guaiacol peroxidase (POD) and catalase (CAT). These reactions are accompanied by antioxidant compounds like ascorbate, tocopherol, thioredoxin and reduced glutathione (Miller et al. 2010; Kapoor et al. 2015). The balance between ROS generation and ROS scavenging ultimately determines the oxidative load that can lead to cell damage if the generation of ROS exceeds the scavenging of those species (Novaes et al. 2010).

Trees also have a system of ROS modulation that is similar to that of annual plants. High SOD activity confers trees with the ability to survive drought conditions in relatively dry climates. Following reductions in net photosynthesis, the antioxidative enzymes SOD and APX are up-regulated in olive leaves and roots, and their expression continues to increase as drought intensity increases. A greater increase in APX activity was recorded in shoots as compared to roots, implying that APX is active mainly in the chloroplast, probably protecting the photosynthetic machinery (Sofo et al. 2005). Similar observations of SOD and APX enhancement were reported for wilting poplar cuttings (Morabito and Guerrier 2000). In a comparison of drought-adapted and non-adapted poplar genotypes, the drought-adapted genotypes had more SOD activity in their roots; whereas there was more APX activity among the non-adaptive genotypes (Regier et al. 2009). High levels of SOD activity were also associated with drought resistance in a poplar population from a dry climate (Xiao et al. 2008). Similarly, one-year-old English walnut (*Juglans regia*) trees subjected to drought treatment had increased antioxidant-enzyme activity, with APX acting as the dominant enzyme in those plants (Maurel et al. 2015).

5.3.4 Use of Carbon Reserves

Plant growth and maintenance depend on sufficient carbon assimilation. An internal balance between carbon supply and consumption is achieved through the use of reserves of non-structural carbohydrates (NSC), especially under stress-driven carbon-deficit conditions (Klein and Hoch 2015; Hartmann and Trumbore 2016). As opposed to structural carbon, which is stored as cellulose and lignin, NSCs include soluble sugars, starch, lipids or fructans, all of which accumulate in the living tissues of stems, branches and roots (Sala et al. 2012). Carbon reserve levels are dynamic over a tree's life cycle. Seedlings invest most of the carbon that they assimilate in growth and development, while at maturity, trees increase NSC pools (Niinemets 2010). As a result, seedlings are expected to be more vulnerable to prolonged drought stress as they rely strongly on immediate photosynthesis; whereas mature and large trees rely on their NSC pools (Niinemets 2010). Decreased water potential following drought stress causes turgor loss and consequent growth inhibition that minimizes growth-related carbon demand. As described in Section 5.3.2 and thereafter, one of the first responses to drought stress is stomatal closure, which leads to a decrease in the supply of carbon. However, growth is inhibited before there is any decrease in photosynthesis, and thus NSC pools may, counter-intuitively, increase in size during early stages of stress. As drought progresses, continued demand for NSCs for respiration and osmotic adjustment may reduce NSC pools to levels that lead to mortality (Hartmann and Trumbore 2016). Although numerous studies have been conducted to decipher carbon dynamics under drought stress, our understanding of this topic is still limited. One reason for this is the fact that NSC is not easily measured *in vivo* at the whole-tree level, and so concentrations are measured in dry samples. These snapshot measurements are later interpolated to construct models of seasonal dynamics (Klein and Hoch 2015).

Depending on the tree's age and size, a negative carbon balance can develop within days to weeks of the initiation of drought stress (McDowell et al. 2008; Klein et al. 2011). Such observations led to the important hypothesis of drought-induced carbon starvation (McDowell et al. 2008; McDowell 2011) specifically, that exhaustion of carbon reserves can be fatal to trees. In young Norway spruce (*Picea abies*) trees subjected to drought, a negative net carbon balance was observed between Weeks 5 and 7 of the stress (Hartmann et al. 2015). However, newly assimilated carbon continued to increase NSC pools in the aboveground tissues of these trees and this NSC was transported into the root system even during later phases of drought (Hartmann et al. 2015). Perhaps allocating carbon to NSC pools in tissues during drought allows the plant to maintain carbon pools and support its belowground organs. The maintenance of the NSC storage pool size under drought stress was also noted in adult *P. halepensis* trees in a semi-arid forest, which had dramatic decreases in growth rates (Klein et al. 2014a). In *Pinus sylvestris*, larger carbon reserves before severe drought, as well as a high carbon-assimilation rate under stressful conditions did not result in aboveground growth, but were nonetheless critical for tree survival (Garcia-Forner et al. 2016). Dramatically low levels of NSCs in branches and stems were evident in the dead trees after this severe drought, as compared to surviving trees (Garcia-Forner et al. 2016). Thus, critical levels of NSCs in aboveground tissues are essential for tree survival under severe drought conditions. Other factors, such as stomatal response and the resistance of xylem to embolism, also contribute to tree survival as the surviving trees fully recovered after irrigation (Garcia-Forner et al. 2016). High NSC levels have been shown to decrease the effect of drought stress, likely by allowing a higher osmolality to be maintained (O'Brien et al. 2014). More research is required to evaluate the role of NSCs in osmotic adjustment in response to drought stress (O'Brien et al. 2014; Delzon 2015). Carbon metabolism is linked to whole-plant hydraulic functions, and both carbon and turgor should be prioritized in research into drought-resistance strategies.

5.3.5 Hormonal Changes

Plant hormones play crucial roles in plant development and physiology. Phytohormones include ABA, auxin, cytokinin, ethylene, gibberellin, jasmonate, salicylic acid, brassinosteroids and strigolactone, and it is likely that more phytohormones will be discovered (Vanstraelen and Benková 2012). Recent studies of phytohormone status in response to drought stress illustrate possible interactions among hormones (O'Leary et al. 2011). Hormone-related studies of stomatal closure upon drought in model herbaceous species showed that ABA and jasmonate positively regulate stomatal closure, while auxin and cytokinins have negative effects (Daszkowska-Golec and Szarejko 2013). A comprehensive study of *P. pinaster* that compared the hormonal responses at two contrasting environmental sites revealed increases in ABA and in the auxin precursor, indole acetic acid (IAA), in needles in response to water stress, but found that the levels of salicylic acid were unaffected by the examined environmental conditions (Corcuera et al. 2012). The findings of a study involving hybrid poplar (*P. tremula* × *P. alba*) suggest that gibberellin catabolism and signaling may be important for drought response (Zawaski and Busov 2014). To date, gibberellins have not been studied extensively in other woody species in the context of drought stress.

Jasmonate is essential for plant responses to injury and biotic stress (Howe and Jander 2008; Koo and Howe 2009), but its role in plant responses to drought stress is not fully understood. A negative effect of drought conditions on jasmonate levels was reported in *P. pinaster* populations, although an opposite effect was observed among some other populations, suggesting no dominant pattern for this hormone (Pedranzani et al. 2007; Corcuera et al. 2012). In contrast to jasmonate, ethylene is induced by drought stress in roots and shoots and can move from root to shoot to inhibit the shoot growth rate (Schachtman and Goodger 2008). Ethylene affects the stomatal apertures of *Arabidopsis thaliana* plants under drought stress (Desikan et al. 2006; Daszkowska-Golec and Szarejko 2013). In trees, ethylene in needles of *Pinus banksiana* (jack pine) seedlings exposed to drought was correlated with a membrane injury index and decreased in response to the application of ABA (Rajasekaran and Blake 1999). This ABA-induced inhibition of ethylene has been suggested as an explanation for the high ABA levels observed in roots, which diminish the growth-inhibition effect of ethylene (Mahajan and Tuteja 2005). The ethylene precursor ACC (1-aminocyclopropane-1-carboxylic acid), which has been shown to move from root to shoot in response to drought, accumulated in roots of drought-treated mandarin seedlings in an ABA-dependent manner and was translocated to the shoot to induce leaf abscission only after re-watering (Gomez-Cadenas et al. 1996).

Changes in IAA levels have been reported in most plants subjected to drought stress. Changes in IAA are often due to crosstalk with other hormones and regulate stomatal closure (Pospíšilová 2003), ethylene synthesis (Hansen and Grossmann 2000) and generation of ROS (Tognetti et al. 2012). IAA can affect the biosynthesis of ethylene and that ethylene can affect the biosynthesis of IAA (Swarup et al. 2002; Tsuchisaka and Theologis 2004; Santner and Estelle 2009).

Knowledge regarding the involvement of auxin in trees' responses to drought is limited. In *Pinus radiata*, auxin is the phytohormone that is most affected by drought, with increased IAA levels in needles correlating with reductions in relative water content (RWC), reductions in maximum quantum yield of PSII photochemistry (Fv/Fm) and increased electrolyte leakage, which indicates membrane damage (De Diego et al. 2012). IAA is also known to stimulate the growth of new lateral roots under drought stress. An immunolocalization study done in *P. radiata* suggested that both ABA and IAA accumulate in the root cortex under drought conditions. However, in that study, ABA levels in the root cortex decreased as ABA was translocated to the shoot, while IAA levels increased in a manner that might affect the formation of lateral roots (De Diego et al. 2013a).

As opposed to ABA, cytokinins (CK) act to promote cell division (Pospíšilová 2003) and CK levels are reduced upon drought stress. On the other hand, CKs also delay leaf senescence (Argueso et al. 2009). Decreases in CK levels were noted in apple (*Malus domestica*) trees after two drought and re-watering cycles, while CK levels increased following the third cycle (Zhu et al. 2004). Increasing the endogenous levels of CK through reverse genetics involving inducible promoters has led to drought tolerance in herbaceous species (Rivero et al. 2007; Ma 2008). One effort to manipulate CK levels in trees included the expression of genes from CK-producing bacteria and resulted in unregulated CK production that severely affected plant development; however, drought resistance was not examined in that study (Kües 2007). The successful modulation of CK levels via a drought-inducible promoter would increase our understanding of the role of CK in drought response in trees.

ABA is by far the most studied phytohormone with respect to drought response. ABA can be a long-distance signal transmitter, but while acting locally, it induces stomatal closure (Cutler et al. 2010; van Dongen et al. 2011; Krasensky and Jonak 2012; McAdam et al. 2016). While ABA confers drought tolerance in the short term, continuous accumulation of ABA under prolonged drought stress can have negative effects on plant growth and development (Sreenivasulu et al. 2012).

ABA-derived stomatal closure has been extensively investigated in woody species. Differences in ABA accumulation and stomatal closure were demonstrated in two contrasting parental clones of *Populus nigra* (Cocozza et al. 2010). Drought treatment resulted in ABA accumulation in the less drought-resistant parental clone, which reacted by closing its stomata. Under the same conditions, the more drought-resistant clone accumulated less ABA, but did so earlier than the sensitive clone, with no change in stomatal behavior (Cocozza et al. 2010). There were increased ABA concentrations in the xylem of Scots pine (*P. sylvestris*) as predawn Ψ_l decreased during both the wet and dry seasons (Lundin et al. 2007). Similarly, increased ABA concentrations were observed into the dry season in evergreen and semi-deciduous tree species in the wet-dry tropics of Australia; whereas no changes in ABA levels were observed among deciduous tree species from the same area (Tcherkez et al. 2012). A comparison of European beech (*Fagus sylvatica* L.) trees from a mesic site with those from a xeric site in a common garden revealed that drought induced a stronger ABA response in genotypes from the mesic site (Carsjens et al. 2014). Differences in foliar ABA levels were also found among different eucalyptus (*Eucalyptus globulus* and *E. viminalis*) provenances in response to mild water stress. However, those differences were not related to rainfall patterns (McKiernan et al. 2016).

Girdling (ring-barking) of *P. radiata* caused ABA to accumulate in leaves, which resulted in decreased g_s, but there were no changes in tree water potential. These alterations were not correlated with the transient increases in NSC, suggesting that foliage-derived ABA accumulation alone was responsible for the observed decrease in g_s (Mitchell et al. 2016).

A comparison of angiosperm tree species, as well as herbaceous species suggested that stomatal closure occurs in response to a change in the VPD through the rapid synthesis of ABA, which is triggered by a decrease in leaf turgor (McAdam and Brodribb 2016). Interestingly, trees from a mesic site were more sensitive to external pressure (i.e. decreased turgor) and accumulated ABA more rapidly than trees from a xeric site (McAdam and Brodribb 2016).

Coniferous species all close their stomata upon the accumulation of ABA, but each species uses one of two distinct methods to maintain stomatal closure during sustained drought stress. Pinaceae and Araucariaceae depend on high ABA levels; whereas Cupressaceae depend on leaf desiccation for prolonged stomatal closure (Brodribb and McAdam 2013; Brodribb et al. 2014). Cupressaceae species resist embolism better than other coniferous species due to their xylem structure and thus not only keep their stomata open under more stressful conditions, but, following re-watering, also recover more quickly after prolonged stomatal closure (Delzon et al. 2010; Brodribb et al. 2014). It has been suggested that improved xylem construction and consequently better embolism resistance in conifers may be an adaptive trait in dry habitats (David-Schwartz et al. 2016). In general, it seems that genotypes that are less adapted to a dry climate demonstrate stronger and more rapid ABA responses than drought-adapted trees, probably due to other pre-activated resistance mechanisms in the drought-adapted trees.

5.3.6 Root Adaptation and Mycorrhizae

Mycorrhizae are the result of an ancient symbiosis between fungi and plant roots that dates back to the early Devonian period when plants first colonized land. There are two types of mycorrhizal symbiosis, ectomycorrhizae and endomycorrhizae (Parniske 2008; Lehto and Zwiazek 2011). Both types are abundant and are associated with most plant species.

Arbuscular mycorrhizal fungi (AMF) are endomycorrhizal fungi that penetrate the cortical cells of the roots of a vascular plant, colonize them and establish a symbiotic association (Parniske 2008). This symbiosis helps the plant take up water (Parniske 2008; Wu et al. 2009). In horticultural trees, research into the role of mycorrhizae under drought stress began more than 30 years ago, when Levy and Krikun (1980) first reported the effects of AMF on the g_s, photosynthesis and proline content of rough lemon (*Citrus jambhiri* Lush) seedlings during the development of a soil water deficit and recovery from that stress. The presence of AMF also increases leaf transpiration, root hydraulic conductivity, root length, biomass production and mineral uptake (Levy et al. 1983; Ortas et al. 2002). However, these effects vary between different species of AMF (Ortas et al. 2002), although the mechanisms involved in this tolerance are still not fully understood. Possible pathways include the enhancement of water uptake via extra-radical hyphae (Peterson et al. 2004), increased mineral uptake (Wu and Zou 2009) via improved root-system architecture (e.g. increases in root length, number of lateral roots, root diameter, root branching density and root mortality (Espeleta and Eissenstat 1998; Boukcim et al. 2001; Paszkowski and Boller 2002; Wu et al. 2012)), effects on the production of root polyamines (Wu et al. 2012) and/or increased osmotic adjustment capability (accumulation of organic and inorganic low-molecular-weight solutes (Kubikova et al. 2001; Wu and Xia 2006; Wu et al. 2007; Abbaspour et al. 2012). The effects of mycorrhizae on the drought tolerance of plants might also involve the ability of AMF to enhance a tree's antioxidant defense system under drought stress (Wu et al. 2006). Since drought stress can happen under many guises (e.g. different magnitude, different season of the year, different physiological stages of the tree, etc.), the relative contribution of AMF to increases in tree drought tolerance is not clear and could vary.

Unlike AMF, ectomycorrhizal fungi do not penetrate individual cells within the root (Lehto and Zwiazek 2011). Their hyphae cover the root tip and extend into the root, where they surround the plant cells within the root cortex. Similar to AMF, this symbiosis also includes the exchange of water, minerals and carbon between the host plant and the fungi. Since the cytoplasmic connections between the ectomycorrhizal fungus and the host plant root cells are indirect, the transport of water, carbons and minerals is not very efficient. Yet, since these fungi usually cover the root tip, their significant surface area increases the efficiency of their absorption of water and minerals. As a result, in some cases, the effect of ectomycorrhizal fungi on the host plant's water and mineral status can be greater than that of AMF (Osonubi et al. 1991). Ectomycorrhizae can contribute to drought tolerance by enhancing the expression of AQP genes in poplar trees (Marjanovic et al. 2005), and through antioxidant production in the roots of the host plant (Alvarez et al. 2009). The contribution of ectomycorrhizae to the host tree's drought tolerance varies based on environmental conditions and the specific interaction between the tree and the fungi, as well as the type of ectomycorrhizae.

5.3.7 Whole-Plant Strategies for Drought Tolerance

In preceding sections, we discussed the cellular mechanisms of the responses of woody plants to water stress. Often, these different response mechanisms operate in parallel, rather than separately. Therefore, at the whole-plant level, we find syndromes or trait ensembles. Together, these combinations of responses form drought-resistance strategies. However, the need to resist drought stress is not absolute. Many organisms, including seasonal plants and geophytes, are spatially or temporally absent from water-stress situations (the 'escape' strategy). Woody plants, on the contrary, do not have this luxury: with long-lasting canopies and root systems, they must endure or perish. The case of re-sprouting is one partial exception to this rule (Zeppel et al. 2015) but lies outside the scope of this chapter. Traditionally, resistance to any type of stress takes the form of tolerance or avoidance, depending on the plant's ability to withstand the stress intrinsically or to delay its effects to protect its most sensitive organs (Delzon 2015). To these two resistance categories, we add the ability of plants to recover from loss of function related to water stress. In Section 5.1, we learned that, besides the direct desiccation of plant organs, the most crucial sensitivity of whole plants to water stress lies in the water transport system (i.e. the xylem). As Ψ_p decreases, the likelihood of xylem embolism (cavitation) increases, causing the risk of hydraulic failure that may lead to plant mortality. The percent loss of hydraulic conductivity (PLC) at decreasing Ψ_p levels is measured in order to construct vulnerability curves. Such curves have been constructed for hundreds of woody plant species (Choat et al. 2012). Xylem design, specifically the anatomical structure of conduits and the pits connecting them to one another, is responsible for large variations in drought tolerance among tree species.

Avoiding drought stress is one way to survive under dry conditions. Tree drought-stress-avoidance strategies range from the leaf scale to the whole-tree level, and from immediate responses to seasonal changes. In many tree species, stomata close in the presence of very low Ψ_p levels, in order to protect the xylem against the development of embolisms (Cruziat et al. 2002; Brodribb and Holbrook 2004; Yang et al. 2012). The Ψ_p of stomatal closure is usually higher than that of 50% PLC (P50), thus creating a species-specific hydraulic safety margin (Choat et al. 2012). In parallel, the rate of photosynthesis decreases, though residual and often sufficient carbon gain is maintained without risking excessively low Ψ_p (Larcher et al. 1981; Klein et al. 2013b; Kuster et al. 2013; Wolf et al. 2013). Gas exchange can also be maintained by homeostatic adjustments including the ratio of leaf area to sapwood area (McDowell et al. 2002, 2006; Mencuccini 2003). Alternatively, carbon supply can be facilitated by the movement of carbon between storage organs and sink tissues (Guehl et al. 1993; Canham et al. 1999; Körner 2003; Sala et al. 2010). To further avoid drought stress, trees may shift their growth toward wetter periods (Grünzweig et al. 2003; Rotenberg and Yakir 2010; Klein et al. 2013a) and/or adjust their root architecture, mycorrhizal colonization, root distribution in the soil profile (Bréda et al. 2006) and xylem vessel anatomy (Eilmann et al. 2009, 2011). Trees might also shed leaves during dry periods, according to the sensitivity of their leaves to drought (Méndez-Alonzo et al. 2012). In the latter strategy, the canopy size is adjusted in response to a water shortage.

The capacity of xylem refilling to reverse embolism provides another route for survival under drought that can overcome the narrow hydraulic safety margin. Non-woody angiosperm species are similar to angiosperm tree species in their P50 values, which are more negative than the midday water potential (Lens et al. 2016). Embolism formation and repair apparently are not routine and mainly occur under water-deficit conditions. On the other hand, accumulated evidence from experiments and field measurements points to the ability of trees to survive low Ψ_p and high levels of hydraulic conductivity loss and xylem embolism (Cochard 2006; Taneda and Sperry 2008; Klein et al. 2011). This high survival rate suggests either no routine embolism or, alternatively, the existence of efficient recovery mechanisms, which may be more common than previously assumed (Brodersen and McElrone 2013). Empiric evidence for the reversibility of conductivity loss and embolism on a diurnal time-scale has been available for over a decade (Tyree 1999). Recently, sub-diurnal cycles of embolism and refilling have been visualized in the lab for grapevine (*Vitis vinifera*) (Brodersen et al. 2010; Zufferey et al. 2011), bamboo (*Sinarundinaria nitida*) (Yang et al. 2012) and poplar (Secchi and Zwieniecki 2011). Routine embolism repair was observed in the field for a desert woody shrub (Schenk and Espino 2011) and pine trees (Klein et al. 2016). However, it is still unclear whether de novo refilling is a common mechanism representing the norm – or rather the exception (Cochard and Delzon 2013).

A recent study using X-ray micro-CT observations provided evidence that hydraulic vulnerability segmentation in grapevine, rather than embolism repair, contributes to stem protection in that species (Charrier et al. 2016). In other species, hydraulic conductance can also be recovered through the development of new xylem vessels (Cochard et al. 2001; Améglio et al. 2002; Eilmann et al.

2010). These mechanisms enable activity and survival at or beyond the hydraulic safety margin, giving it a less important role. Nevertheless, there are constraints to recovery from embolism. First, xylem refilling is limited by the amount of water that can be taken up by roots. Second, repeated cavitation and its repair may result in the deterioration of the plant's hydraulic system over time (Hacke et al. 2001; Anderegg et al. 2013). Third, in some tree species, the evidence for refilling must be verified with the most current methodology (Wheeler et al. 2013). Fourth, in various tree species, no recovery has been demonstrated (Brodersen and McElrone 2013). Interspecific variation is yet another indication that similarity in forest vulnerability to drought cannot be expected.

The relationship between g_s and Ψ_l is key to understanding how plants function under changing climatic conditions. The variability among tree species has given rise to selection toward two contrasting water-management strategies: isohydric and anisohydric. Tardieu and Simonneau (1998) distinguished between isohydric species, in which stomatal regulation maintains a fairly consistent minimum Ψ_l from day to day, and anisohydric species, in which Ψ_l markedly decreases with changes in evaporative demand. The different behavior of isohydric and anisohydric plants is driven by the sensitivity of their respective guard cells to Ψ_l (Sade et al. 2012). As isohydric plants are highly sensitive to changes in Ψ_l, they close their stomata in response to mild drought stress and maintain a constant water potential. In contrast, anisohydric plants are not as sensitive to Ψ_l and maintain higher g_s and CO_2 assimilation rates under low Ψ_l, which allows them to be more productive under stress conditions (Sade et al. 2012). The effect of stomatal regulation (isohydric/anisohydric) on the ability of trees to survive adverse conditions has been addressed in numerous studies (Bonal and Guehl 2001; West et al. 2007; McDowell et al. 2008; Klein et al. 2011; Meinzer et al. 2014). Although many species are characterized as either isohydric or anisohydric, both strategies can exist in one species (Schultz 2003; Attia et al. 2015). However, in spite of the many studies of the isohydric/anisohydric behavior of trees, there is still no mathematic definition for this important plant trait.

Classification into one of these two categories is usually based on the relationship between g_s and Ψ_l (Fisher et al. 2006; McDowell et al. 2008). Notably, differences among species in the range of Ψ_l values are mainly affected by the minimum Ψ_l values, since maximum Ψ_l is rather similar and always between 0 and −1 MPa. Therefore, potential definitions of a water-management strategy (i.e. isohydry/anisohydry) can take the form of one of these alternatives: (D1) the minimum Ψ_l at which g_s is permitted, (D2) the slope of the change in g_s as function of Ψ_l; or (D3) the extent of variation in Ψ_l over the course of the day. However, none of these definitions is without its weaknesses. The first option requires a very unified, sensitive measurement of g_s, to account for the asymptotic nature of the sensitivity of g_s to Ψ_l [$g_s(\Psi_l)$] approaching stomatal closure. Then, one needs to define the Ψ_l value that serves as the dividing line between isohydric and anisohydric behavior, which is potentially any arbitrary value between −2 MPa and −6 MPa, with little physiological meaning. The second option is complicated by the sigmoidal nature of the $g_s(\Psi_l)$ curve, which has a non-singular slope. Intrinsic in the third option is a need to address the question of variation and delineate (arbitrarily again) a threshold of variation, to distinguish between the two types of behavior.

g_s of water vapor integrates three different effects: (1) stomatal density over the leaf surface (i.e. the number of stomata per unit leaf area), (2) the typical stoma pore size and (3) aperture size. Tradeoffs between these different effects have shaped plant life throughout evolution (Franks and Beerling 2009). Notably, while the first two factors might vary only along long-term time scales ranging from weeks to a plant's full-life cycle (and are usually species-specific), stomatal aperture is attuned dynamically on the time-scale of minutes. Therefore, changes in g_s as function of Ψ_l largely reflect stomatal aperture and stomatal-closure dynamics. Differences in stomatal density and typical pore size among tree species affect the maximum g_s when stomata are fully open. At the other end of the curve, the minimum g_s indicates the species-specific epidermal or cuticular conductance, which is usually a minor residual of g_s.

Ψ_l is largely determined by the water potential of the surrounding air (Ψ_a, more commonly measured as VPD) and that of the rhizosphere (Ψ_r, often estimated from pre-dawn Ψ_l, Ψ_{pd};

(Bonal and Guehl 2001). To experimentally construct a $g_s(\Psi_l)$ curve, g_s must be measured under a large range of Ψ_l values. Such a range is obtained by exposing the plant to changes in the water potentials of the air, the rhizosphere or both. This can be achieved experimentally by a drying procedure or by measuring Ψ_l and g_s under decreasing Ψ_a (increasing VPD) or decreasing Ψ_{pd}. Stomata are highly sensitive to Ψ_l, but also respond to changes in other environmental factors including CO_2 concentration, VPD, air temperature and light. Therefore, to relate changes in g_s to changes in Ψ_l, all other factors must be kept constant. Clearly, an experimental approach is advantageous here, yet limited in terms of the technical aspects of controlling the environment of mature trees, as well as the transferability of collected results back to the field. Repeated, coupled measurements of Ψ_l and g_s taken on sunny days at similar air temperatures and light levels can produce a robust $g_s(\Psi_l)$ curve. In addition, $g_s(\Psi_l)$ is not a state function and curve hysteresis has been shown (Blackman et al. 2009; Brodribb and McAdam 2011). While significant hysteresis was shown in *P. radiata*, in other species, this hysteresis is minor compared to the differences between species-specific curves (Blackman et al. 2009) and sometimes is even negligible (e.g. for *Calitris rhomboidea*) (Brodribb and McAdam 2013).

5.4 MOLECULAR RESPONSES TO WATER STRESS

Molecular responses possess multiple layers of regulation, starting at the DNA level, moving to the RNA level and ending up at the protein-function level, with several sublevels in between. Here, we refer to the protein and RNA concentrations as these are the most frequently measured genetic elements in most plant species.

The molecular response to drought stress has been investigated in a variety of plant species, including model herbaceous and tree species and, as expected, it has been found to be complex and multigenic (Shinozaki and Yamaguchi-Shinozaki 2007; Hamanishi and Campbell 2011). While most studies have reported a high degree of conservation between herbaceous and tree species, many transcripts are distinctively expressed in trees. The basis of this distinction is almost certainly due to the special characteristics of woody species such as longevity, secondary growth and defoliation. Thus, in addition to stress severity, molecular responses to drought depend on tree age, tissue type and phenological stage (Watkinson et al. 2003; Niinemets 2010; Ding and Nilsson 2016).

Under mild drought conditions, the main molecular response to drought is the activation of pathways responsible for stress signals and mechanisms, including hormones, ROS and osmolytes. At the same time, molecular mechanisms that involve growth and photosynthesis are usually suppressed. As the intensity of the drought increases, additional molecular mechanisms are activated, in order to prevent tissue and cellular damage and to better adapt the tree to future constraints.

Currently, hundreds of genes have been identified in relation to drought response, while specific drought-related functions have been identified for only a portion of them. The challenge is to identify drought-related genes that can be used in breeding programs and biotechnological approaches aimed at improving drought resistance, especially among forest tree species, which are completely dependent on environmental conditions. Early studies on the molecular responses of trees to drought stress were often restricted to a few or several genes/proteins at a time due to technological limitations (Chang et al. 1996; Sakr et al. 2003). Later, more advanced techniques such as cDNA libraries and suppression subtractive hybridization (SSH) enabled the investigation of tens of genes simultaneously (Dubos and Plomion 2003; Dubos et al. 2003; Sathyan et al. 2005). These partial representations of the complex molecular response to drought were improved through the use of microarray techniques, in those species for which genomic information was available (Rensink and Buell 2005). For example, a microarray of 6,340 ESTs was used to study the molecular response of *Populus euphratica* Oliv. and revealed 70 genes in leaves and 40 genes in roots that were differentially expressed following drought (Bogeat-Triboulot et al. 2007). More advanced microarrays are now available for various species, *Pinus pinaster*, *P. taeda*, *Quercus* ssp. and poplar, for which there are more than 56,000 representatives of full transcripts (Cohen et al. 2010).

Microarrays of conifers, however, are less representative, as they possess transcripts representing mostly wood-forming tissue and thus provide only a limited picture of the distinctive drought responses of conifers (Heath et al. 2002; Watkinson et al. 2003; Perdiguero et al. 2013). Lorenz et al. (2006) attempted to analyze drought-specific transcripts in loblolly pine (*P. taeda*) and succeeded in building an EST library of more than 6,000 unique transcripts from drought-treated root tissue and identifying several genes that had been characterized previously. The same research group later published an improved microarray of 26,496 cDNAs that they used to analyze gene expression in response to drought in loblolly pine roots (Lorenz et al. 2011).

This situation of partial genomic representation has begun to change thanks to the use of more accurate and comprehensive next-generation sequencing (NGS) techniques. NGS allows the investigation of the transcriptome (all expressed genes in a given tissue at a given time) of any species in the absence of any prior genetic information (Goodwin et al. 2016). Expression profiles have begun to provide information on gene networks and regulators, which is useful for the dissection of temporal and spatial responses to drought at the molecular level. Furthermore, new advanced methods for the study of quantitative changes in proteins in a given tissue, like isobaric tags for relative and absolute quantitation (iTRAQ), now allowing the identification and quantitation of thousands of proteins (Vélez-Bermúdez et al. 2016). Use of integrated advanced approaches at the protein and transcript levels will further our understanding of complex drought responses.

5.4.1 Proteome Response to Drought Stress

Early studies of protein profiles mainly utilized the two-dimensional polyacrylamide gel electrophoresis (2D-PAGE) approach. Differences in protein expression in response to drought stress have been reported in needles of maritime pine (*Pinus pinaster*). In those pine needles, proteins involved in mechanisms such as photosynthesis (Rubisco activase) and cell growth and elongation (actin) were down-regulated in the presence of drought stress, while proteins involved in ROS scavenging (GPX and SOD), lignin synthesis (caffeoyl CoA-O-methyltransferase) and membrane stability (low-molecular-weight HSP) were up-regulated (Costa et al. 1998).

The protein profile of the leaf of drought-stressed holm oak (*Quercus ilex*) was also examined. That examination revealed 14 proteins that were differentially expressed between well-watered and drought-stressed two-year-old saplings. The six identified up-regulated proteins (out of eight) included the RubisCO small subunit, globulin, MW glutenin, fructose-bisphosphate aldolase, β-amylase and a stress-related protein. The two identified down-regulated proteins (out of six) were RubisCO and RubisCO activase (Jorge et al. 2006). Proteomic profiling of the leaf of *Populus tremula* × *P. alba* genotypes revealed up-regulation of two cell membrane-related (plastid-lipid-associated) proteins and two chloroplastic ATPase-related proteins, as well as down-regulation of two malate dehydrogenase transcripts, indicating reduced TCA cycle activity (Durand et al. 2011). Proteomic profiling of the cambium of the same individuals revealed up-regulation of two sucrose synthases, a putative elongation factor EF-2, two glycine-rich proteins and a serine hydroxymethyltransferase, suggesting active growth in cambium despite the stress (Durand et al. 2011). In addition, down-regulated proteins in the cambium included GDSL motif proteins, two glutathione-S-transferases, one class 1 ascorbate peroxidase, a 28-kDa bark storage protein and a vacuolar ATPase that might be linked to cell turgor (Durand et al. 2011).

A recent study that included an iTRAQ-based proteomic analysis identified 4,078 proteins in the leaf of apple trees, 594 of which were differentially expressed upon drought treatment (Zhou et al. 2015). The results of that study allowed the evaluation of the involvement of photosynthesis-related pathways in the maintenance of water-use efficiency under drought stress (Zhou et al. 2015).

5.4.2 Transcriptome Response to Drought Stress

Protein analyses are more complicated to perform than RNA expression analyses and so, even though the actual activity of proteins is not fully reflected at the RNA level, researchers may take advantage of the fact that RNA expression analyses provide a reasonable indication of drought-induced molecular mechanisms. One of the first genes to be activated upon exposure to drought is the osmosensor histidine kinase (*At*HK1) that was identified in Arabidopsis (Urao et al. 1999). *At*HK1 is a membrane protein that senses the osmotic change in the cell upon turgor loss and elicits a signal transduction cascade that involves mitogen-activated protein kinase (MAPK) to activate drought-related genes (Tran et al. 2007). Early expression of HK1 following drought treatment has been detected in cacao (*Theobroma cacao*) trees and eucalyptus (Liu et al. 2001; Bae et al. 2009). An HK1 homolog was also identified in poplar following osmotic stress (Chefdor et al. 2006). As HK1 seems to be a key factor in drought-stress signaling, it is a potential candidate for use in biotechnology programs to improve drought tolerance in trees. Currently available and future transcriptome databases from other tree species are an untapped source for the identification of HK1 homologs.

Flavonoids are known for their ability to protect plants from oxidative stress and genes related to flavonoids are expressed during mild, but not severe, drought stress (Heath et al. 2002; Watkinson et al. 2003). For example, levels of flavonoid-related enzymes such as chalcone isomerase and naringenin-2-oxo dioxygenase were positively correlated with photosynthetic acclimation in loblolly pine (Watkinson et al. 2003).

Genes related to ABA synthesis, signaling and degradation are good candidates for use as agents to confer improved drought tolerance. This has been shown in Arabidopsis for genes such as the ABA biosynthesis enzymes 9-cisepoxycarotenoid dioxygenase (NCED) and for the ABA 8'-hydroxylase that degrades ABA (Iuchi et al. 2001; Umezawa et al. 2006). In citrus trees, NCED genes are expressed in a tissue-dependent manner. While NCED3 is expressed in leaves and other photosynthetic tissues, NCED5 is mainly expressed in ripening fruits (Agustí et al. 2007). Knowledge regarding the spatial and temporal expression patterns of any candidate gene is crucial for any biotechnological approach. Care should be taken in selecting the specific gene, as well as its specific promoter, since enhancement of ABA biosynthesis under a global promoter such as the 35S, can lead to over-accumulation of ABA, which will result in growth inhibition. Instead, inducible promoters, such as those that are induced by drought, should be considered.

Another gene involved in ABA signaling is *PP2C* (protein phosphatase 2C), as noted in *Fagus sylvatica*. An examination of ABA and oxidative-stress-related genes under drought conditions revealed that *PP2C* expression varies in response to drought in five *F. sylvatica* populations (Carsjens et al. 2014). Populations from mesic sites exhibited more pronounced ABA-related gene expression than populations from xeric sites (Carsjens et al. 2014). A transcriptome analysis of the *Pinus taeda* root revealed differential regulation of major regulators and gene clusters in response to drought. This included three up-regulated regulatory genes including cardiolipin synthase/phosphatidyltransferase involved in phospholipid metabolism, zeatin o-glucosyltransferase involved in cytokinin availability and the ABA-related gene *NCED* (Lorenz et al. 2011). Similarly, a gene-network analysis of the root transcriptome of cork oak (*Quercus suber*) identified induction of ABA-responsive genes upon the activation of an ABA-signaling pathway involving PP2C-SnRK2-ABF components (Magalhães et al. 2015).

Transcript analysis based on the cDNA library of roots of *P. halepensis* trees subjected to drought stress revealed eight genes that could be identified, including a myeloblastosis (MYB) transcription factor, a late embryogenesis abundant (LEA) protein and a sucrose synthase (Sathyan et al. 2005). More comprehensive transcriptome analysis of roots from two willow genotypes gave rise to a core set of 28 genes that are up-regulated upon drought stress (Pucholt et al. 2015). These genes included *MYBs*, *WRKYs* and *bZIP* (basic leucine zipper) transcription factors known to be part of the ABA response (Abe et al. 2003). Other identified genes code for heat-stress transcription factors, dehydrin and chlorophyll a/b binding proteins that likely help protect the plant in the presence of very intense light (Pucholt et al. 2015). Transcriptomic analyses of various species and comparative analyses of the identified genes across different species are needed to identify the transcription factors that regulate drought-related mechanisms in woody species.

As drought stress progresses, preserving water becomes a critical issue for the tree. One response to prolonged drought stress is increased cuticle formation. In *P. pinaster* trees under drought conditions, increases in cuticular wax content were accompanied by changes in cuticle-related gene expression (Le Provost et al. 2013).

Other late responsive genes code for AQPs that are channel proteins functioning in the transport of water, CO_2 and small neutral solutes through all types of membranes (Heath et al. 2002). The different groups of AQPs include plasma membrane-intrinsic proteins (PIPs) and tonoplast-intrinsic proteins (TIPs) (Kaldenhoff et al. 2008; Maurel et al. 2015), which are the natural candidates for use in molecular studies of plant responses to drought stress (Kaldenhoff et al. 2007; Moshelion et al. 2015). In many plant species, various AQPs are expressed and function differently depending on the tissue in which they are expressed, as well as their sub-cellular location (Alexandersson et al. 2005, 2010; Heckwolf et al. 2011; Maurel et al. 2015). A study of olive trees revealed down-regulation of one *TIP* and two *PIP* genes in response to a severe drought treatment (–3 MPa) in leaves, twigs and roots, probably to prevent water loss (Secchi et al. 2007). A later study in drought-treated olive trees demonstrated transient up-regulation of *PIP1-2*, as well as gradual down-regulation of *PIP2-1* that was linearly correlated with mesophyll conductance of CO_2 (Perez-Martin et al. 2014). In *Populus trichocarpa* leaves, 12 *TIP* and *PIP* genes were down-regulated in response to a drought treatment. However, only some of these *TIP* genes were strongly up-regulated following re-watering (Laur and Hacke 2014). In a hybrid poplar, three *PIP*1 genes were up-regulated in correlation with increases in root water flow during recovery from drought (Laur and Hacke 2013). Another study that analyzed AQP expression in the root of two contrasting poplar genotypes demonstrated that *PIP1-2* and *TIP1-3* were up-regulated in the less drought-adapted genotype, probably to facilitate the movement of water into the vascular tissue, and were down-regulated in the more drought-adapted genotype, probably to keep water in the roots (Cocozza et al. 2010).

AQPs located in the xylem domain are interesting with regard to xylem function under drought stress and recovery. Xylem parenchyma cells of poplar trees include TIP proteins, and the level of those TIP proteins is positively correlated with leaf hydraulic conductance (Laur and Hacke 2014). The expression of *TIP* genes has also been reported to be positively correlated with leaf hydraulic conductance in grapevine (McDowell et al. 2008). Similarly, xylem parenchyma cells in walnut trees express *PIP2* genes during embolism refilling (Sakr et al. 2003). Studies using genetic transformation have shown that *PIP2* overexpression enhances water transport, but does not improve drought tolerance in grapevine or eucalyptus (Tsuchihira et al. 2010; Perrone et al. 2012). The latter studies, however, did not investigate embolism refilling in the examined transgenic plants. Secchi and Zwieniecki (2014) used a transgenic approach to demonstrate the importance of AQPs for embolisms. They revealed that down-regulation of *PIP1* genes in a poplar hybrid increases vulnerability to drought-induced embolisms and also reduces the tree's ability to recover (Secchi and Zwieniecki 2014). In addition, in poplar trees, transcription of some AQPs is regulated through the circadian clock; whereas the transcription of other AQP genes is not (Lopez et al. 2013). This regulation might play a role in diurnal changes in water conductance.

In general, genes related to embolism recovery are of great interest for efforts to increase trees' tolerance of drought stress. Efforts to identify such genes have been made using a poplar microarray representing over 56,000 genes that was used to analyze xylem parenchyma tissue surrounded by air-filled vessels (Secchi et al. 2011). Increased expression of genes encoding AQPs and ion transporters as well as genes related to carbohydrate metabolism was evident, while genes related to oxidative response were down-regulated. These results support the earlier hypothesis that parenchyma cells in poplar might sense an embolism from neighboring cells via osmoticum build-up. However, these results also led researchers to an additional, new hypothesis that response to an embolism might be triggered by reductions in hypoxic stress and reduced ROS formation (Secchi et al. 2011). In spite of the progress that has been made in understanding the involvement of AQPs in responses to drought in herbaceous species, we are far from a fully integrated view (Sade et al. 2010). Even less is known about the role of the AQPs in drought-stressed woody species and thus more research efforts toward reverse genetics of specific AQPs in various woody species are required.

Transcriptome-profiling analyses of xeric-adapted woody species are a good source of candidate drought-resistance genes. For example, transcriptome profiling of the evergreen broadleaf *Ammopiptanthus mongolicus* under drought stress revealed a list of candidate genes in roots and leaves (Zhou et al. 2012; Pang et al. 2015). Following the induction of drought on *A. mongolicus* seedlings with 20% PEG-6000 for 72 h, root transcriptome revealed up-regulation of hormone-related genes [ethylene and auxin receptors, as well as the auxin response component IAA9 (AUX/indole-3-acetic acid)], genes regulating the light-mediated root growth response upon drought (blue light photoreceptor NPH3 and an interacting protein of NPH1), ion channel and transporter genes, as well as several anti-oxidant genes (Zhou et al. 2012). In the other study, in which mild drought was induced for four weeks in soil, there was little overlap between the leaf transcriptome and the root transcriptome and differentially expressed genes included 25 transcription factors, 26 genes related to signal transduction, asparagine synthesis-related genes and genes that code for fatty acid desaturase-6 and chaperonin 20 (Pang et al. 2015). Furthermore, genes coding for 1-pyrroline-5-carboxylate synthetase, plasma membrane ATPase and cell wall-associated hydrolases were highly up-regulated. In addition, as has been shown for other species, in these trees, AQPs were differentially expressed, probably due to their cellular locations. Specifically, three AQPs were up-regulated in response to drought stress and two were down-regulated (Pang et al. 2015). As findings accumulate from more transcription analyses of different woody species that use different drought-resistance strategies, performed at different developmental stages and involving different tissues and different levels of drought intensity, it will become possible to identify key genes for each category.

5.5 GENETIC ENGINEERING FOR DROUGHT TOLERANCE IN TREES

A thorough understanding of the molecular mechanisms involved in the drought tolerance of trees is fundamental for genetically manipulating their responses to drought (Harfouche et al. 2014). Efforts to improve our knowledge in this area should include the examination of woody species that are considered to be highly resistant to drought such as *Tamarix aphylla* (Dawalibi et al. 2015) or *Zygophyllum xanthoxylum*, a perennial woody succulent xerophyte that is extremely drought-tolerant (Ma et al. 2012), as well as *A. mongolicus*.

5.5.1 Improving the Oxidative Response

Efforts to improve oxidative defenses of plants during drought have included the manipulation of oxidative-related genes. A cytosolic APX transgene (cyt*apx*) that was introduced into plum (*Prunus domestica*) enhanced its tolerance to water stress by retaining photosynthetic activity and water-use efficiency, compared to the wild type (Diaz-Vivancos et al. 2016). In another effort to improve oxidative defenses, Chu et al. (2016) reported that the expression of the C_2H_2-type zinc finger gene *ZxZF*, under the control of the drought-inducible promoter *rd29A* improved drought tolerance in transgenic poplar. Under stress conditions, the transgenic trees with *ZxZF* induced by PEG6000 exhibited improved photosynthetic function and elevated chlorophyll content, as well as up-regulation of both SOD activity and POD activity (Chu et al. 2016).

5.5.2 Improvement of Osmotic Status

Improving the osmotic status of the plant is another way of increasing drought tolerance (Ahmad et al. 2013). Glycine betaine (GB) is a solute that effectively stabilizes the structure and function of macromolecules under stress conditions (Giri 2011). Transgenic plants overproducing GB had greater tolerance of different abiotic stresses (Ahmad et al. 2013). Choline oxidase (codA), a GB-synthesizing enzyme, has been widely used for GB production in transgenic plants such as rice (*Oryza sativa*), potato (*Solanum tuberosum*) and alfalfa (*Medicago sativa*) (Ahmad et al. 2008; Kathuria et al. 2009; Li et al. 2014). Poplar trees expressing the *codA* gene under the control of the oxidative stress-inducible promoter *SWPA2* showed improved drought resistance (Ke et al. 2016). In that work, transgenic poplar trees accumulated higher levels of GB than control plants and exhibited increased tolerance to drought and salt stress, which was associated with increased efficiency of its photosystem II activity (Ke et al. 2016).

Osmotin, which belongs to the fifth group of pathogenesis-related (PR) proteins, is associated with osmotic adaptation in plant cells and aids the accumulation of the osmolyte proline during drought stress. Das et al. (2011) showed that expression of tobacco osmotin in transgenic mulberry plants under the control of the stress-inducible *rd29A* promoter led to improved drought tolerance. When water was withheld for 20 days, leaf necrosis and senescence were substantially greater in the non-transgenic plants than in the transgenic plants. Transgenic plants with the osmotin gene showed better cell membrane stability and accumulated more proline than wild-type plants at different durations of the stress condition (Das et al. 2011). In another attempt to increase proline content, Molinari et al. (2004) generated transgenic citrus rootstock Carrizo citrange (*Citrus sinensis* Osb. × *Poncirus trifoliata* L. Raf.) expressing 1-pyrroline-5-carboxylate synthetase (*P5CS*), a key enzyme required for proline synthesis and its feedback-inhibited by proline. When an insensitive-to-feedback-inhibition version of *P5CS* was expressed, there was osmotic adjustment due to high proline levels and a markedly higher photosynthetic rate than that observed in the control plants when water was withheld for 15 days (Molinari et al. 2004).

5.5.3 Manipulation of Drought-Responsive Genes

Improved response to drought through the manipulation of transcription factors has been reported in various plant species. Movahedi et al. (2015) demonstrated that overexpression of the chickpea *CarNAC3* gene enhanced tolerance of salinity and drought in transgenic poplars. Some NAC transcription factors function in plant development, while a stress-inducible NAC gene family has been shown to be involved in abiotic stress tolerance. In that study, *CarNAC3* was transformed into hybrid poplar plants (*Populus deltoides* × *P. euramericana* cv. Nanlin895). The transgenic plants maintained their normal rooting and stem growth rates under drought stress, while the rooting and stem growth of wild-type plants were suppressed. Furthermore, increases in proline and photosynthetic pigment levels and in antioxidant enzyme activity were observed in the transgenic lines (Movahedi et al. 2015).

Dehydration-responsive element-binding (DREB) proteins are a subfamily of AP2/ERF transcription factors that are stress-induced and play a central role in the regulation of gene expression during drought. Li et al. (2011) overexpressed the *DREB2* gene from *Festuca arundinacea* under a constitutive promoter in paper mulberry (*Broussonetia papyrifera* L. Vent), an economically important tree. Drought resistance was evident in the transgenic seedlings, which exhibited higher leaf water and chlorophyll contents, accumulated more free proline and soluble sugars and had less membrane damage as compared to the wild type.

Manipulation of the *DREB* gene also affected apple tree response to drought, using the (*Malus sieversii*) homolog, *MsDREB6.2* (Liao et al. 2016). Overexpression of *MsDREB6.2* reduced CK levels by increasing the expression of *MdCKX4a* (a key CK catabolism gene). In addition, decreases in the shoot:root ratio and decreases in stomatal aperture and density were observed in transgenic apple plants. As a result, increased root hydraulic conductance, as well as increased expression of the *AQP* genes led to enhanced drought tolerance (Liao et al. 2016).

Manipulating signal transduction pathways is another way to confer drought resistance. The MAPK cascades play essential roles in plant growth, development and stress responses. They include MAPK kinases (MAPKKs) that mediate various stress responses in plants. Wang et al. (2014) showed that overexpression of *PtMKK4* improved drought tolerance in poplar. In that work, *PtMKK4* was overexpressed in hybrid poplar (*P. alba* × *P. glandulosa*) plants that showed lower levels of hydrogen peroxide and higher antioxidant enzyme activity than the control plants. Another potential protein kinase is the GSK3/shaggy-like kinase-homolog, which is involved in various signal transduction pathways. Transgenic poplar overexpressing the *Arabidopsis thaliana GSK1* exhibited higher photosynthesis rates, transpiration rates and g_s than wild-type plants following drought stress (Han et al. 2013).

Late embryogenesis abundant (LEA) proteins play central roles in stress-associated responses. Many LEA genes have been shown to be associated with tolerance of water deficiency and salt, osmotic and freezing stress. Overexpression of a LEA gene from *Tamarix androssowii* (*TaLEA*) improved salt and drought tolerance in transgenic Xiaohei poplar (*Populus simonii* × *P. nigra*) (Gao et al. 2013). In addition, levels of malondialdehyde, which represent the extent of lipid peroxidation, were significantly lower in these transgenic plants as compared to non-transgenic plants, indicating that the *TaLEA* gene may enhance salt and drought tolerance by protecting cell membranes from damage (Gao et al. 2013). Another LEA gene that confers drought tolerance is *Hva1* from barley which was introduced into mulberry trees under the stress-inducible *rd29A* promoter (Checker et al. 2012). Transgenic lines displayed tolerance to drought and salt stress as indicated by better membrane protection (reduction of membrane injury during the stress), photosynthetic yield and increased proline accumulation (Checker et al. 2012).

Dehydrins are LEA proteins that interact with cellular macromolecules to protect those macromolecules from dehydration associated with drought, heat or cold stress. Wang et al. (2011) showed that overexpression of a SK2-type dehydrin gene from *P. euphratica* in a *P. tremula* × *P. alba* hybrid increased drought tolerance by allowing the trees to maintain a higher water capacity and a higher relative water content under drought stress. Manipulation of stress-signaling molecules such as fatty acids and their derivatives also induces drought resistance. Overexpression of a fatty acyl-acyl carrier protein thioesterase (*CpFATB*) gene in poplar improved its drought resistance, as reflected by less growth retardation than was observed in the controls and higher SOD, POD and CAT activity, as well as higher total chlorophyll content (Zhang et al. 2013).

These reports on manipulating drought resistance in trees suggest that multiple-drought stress tolerance properties in important tree species may be simultaneously improved to generate substantial agronomical and ecological drought tolerance. Nevertheless, even though several hundreds of articles have been published regarding genetic manipulation of plants toward drought resistance, the vast majority of the efforts made as a result have had a very limited impact on plant performance in the field (Blum 2014). Results obtained in the laboratory with young plants are quite promising, but intensive, large-scale field trials are necessary to validate these traits.

5.6 WATER STRESS AS AN AGRICULTURAL TOOL

Intentional water stress is used to induce flowering in trees in agricultural systems, as well as to save water. Once the plant is exposed to water stress, chemical signals are produced in the root and sent to the shoot via the xylem sap. The identity of these signals is a subject of debate (e.g. ABA, pH, CKs, a precursor of ethylene, malate and/or other substances) due to differing responses between species (Schachtman and Goodger 2008). In all cases, the result is clear: increased agronomic water-use efficiency (namely the amount of yield per unit of water). Yet, water stress that induces a high crop load or saves water during the current year is not an option if those benefits come at the price of reducing the future health of the tree. Therefore, whenever we apply water stress to trees, the applied stress level should be carefully calibrated and controlled, and its long-term effects should always be evaluated. In any case of applied water stress in an agronomic crop, the effect of the imposed stress will be evaluated from an economic point of view.

5.6.1 Induction of Flowering

All plants are characterized by yearly cycles. An annual cycle of perennial plants includes vegetative, reproductive and dormant stages. In general, all of these stages are affected by environmental conditions. Water availability strongly affects plant production (both vegetative and reproductive). Under moderate water stress, overall biomass production and fruit production are slightly inhibited. However, severe water stress leads to the induction of flowering (the reproductive stage). Once water stress is relieved, a flush of flowers appears, even if it is not the natural time for flowering (Raveh 2008). From an evolutionary point of view, this ensures the production of seeds, which are important for the survival of the species.

The ability of water stress to produce the signal for flowering is also used in commercial orchards. Lemon (*Citrus limon*) trees usually bloom during the spring, and their fruits are harvested the following winter (Raveh 2008). Summer yields due to natural out-of-season flowering are usually rare and such yields are usually small. However, the market value of the out-of-season fruits is ten times more than that of the winter crop. In order to increase summer yields, growers use the Verdelli treatment (Barbera et al. 1985; Barbera and Carimi 1988), in which they withhold water in summer (July to August) until the trees begin to wilt. When irrigation is then resumed, trees respond with a massive autumn flowering flush, followed by the production of out-of-season summer fruit. Since this procedure weakens the tree, it can be applied only once every two years.

Water stress can also be used to induce flowering when temperatures are not sufficiently cold, as a replacement for vernalization (Raveh and Nobel 1999). This agro-technique is used to induce the flowering of mango (*Magnifera indica*) trees in the low-latitude tropics of Northern Australia where temperatures are not sufficiently low (Chacko 1986; Lu and Chacko 2000). However, this technique does not work in all cases or for all trees (Chaikiattiyos et al. 1994). In avocado (*Persea americana*) and litchi (*Litchi chinensis*) trees grown in subtropical Australia (Lat. 28°S; average monthly precipitation of 54 ± 2 mm per month), reducing pre-dawn Ψ_l from about −0.5 MPa to about −2.7 MPa did not induce flower production. Unlike lemon trees, low temperature is essential for flowering in these trees (temperatures below 25°C for avocado and below 20°C for litchi) and cannot be replaced by water stress.

5.6.2 Water Conservation

Over the last 50 years, a world-wide water crisis has developed. Drought areas have doubled, irrigated area has doubled and water withdrawals have tripled (FAO 2015). As a result, farmers are trying to save water by exposing their crops to deficit irrigation (DI) during different periods of the year in a way that will allow them to maintain their incomes while saving water. DI treatments are usually applied during periods in which they will not have any negative effects on yield, usually after fruit set (the most sensitive stage in fruit development) or after fruit is picked (when the risk of income loss is negligible). Yet, before applying a DI treatment, the level of allowed stress should be calibrated for the specific variety and growth conditions, as well as different fruit developmental stages and different crop loads (Spreer et al. 2009; Chenafi et al. 2013).

Additionally, DI requires that growers continuously monitor the water potential of their trees with a pressure chamber (usually about once a week), which is time-consuming. As a result, DI is not very common (Fereres and Soriano 2007), and when it is used, it is usually by large-scale growers. Recently, other tools have been developed for monitoring tree water status (e.g. sap flow measurement, infrared images, leaf patch clamp pressure, time-domain reflectometry and others) (Raveh 2012), yet these tools have never been calibrated in a way that would allow them to replace the accurate measurements of a pressure chamber.

Another approach for saving water involves a method referred to as partial root drying (PRD) (dos Santos et al. 2003; Raveh 2008; Schachtman and Goodger 2008). In this method, part of the root system is exposed to water stress conditions and so supplies a signal for flowering (via the production of high levels of ABA), while the rest of the root system is kept under well-watered conditions. It is quite clear that while a tree develops roots in the wet zone, over time, it will cease its root activity in the dry zone. Therefore, about every two to three weeks, farmers switch the wet and the dry zones by opening the irrigation system on the dry side of the root while closing it in the previously wet zone. In general, PRD procedures supply each plant with 50% of its regular amount of irrigation water. However, the results observed for PRD treatment are similar to those observed for regular DI treatments in which the water supply was cut to the same extent (dos Santos et al. 2003; Raveh 2008; Schachtman and Goodger 2008) and its advantage over regular DI has not been proven.

5.7 CONCLUSIONS AND FUTURE DIRECTIONS

Today, the need to understand tree responses to water stress goes beyond scientific curiosity. As many regions in the world undergo desertification and continued drying, drought-induced mortality is affecting an ever-increasing number of forests and woodlands (McDowell et al. 2008; Klein 2015). Drought-resistance mechanisms include resistance to embolism that is governed by structural elements, use of osmolytes and carbon reserves to control water status and growth, regulation of stomatal aperture, altered hormonal status and other mechanisms. These mechanisms are the result of molecular activity that determines adaptive traits, as well as plastic responses. Typically, woody species (cultivated and forest trees) have great variability in their various species-specific drought-avoidance mechanisms, and this variability is reflected in the outcomes of their drought-resistance strategies. Trees vary in their growth habits, growth requirements, structure (especially at the tissue level), phenology, plastic traits and genetics. There are, however, common features that are unique to woody as opposed to herbaceous species, such as longevity, secondary growth and defoliation, and these features allow us to group species and to identify shared drought-resistance mechanisms. There are several parameters that can serve as indicators of water-stress resistance. However, those parameters can be identified only in large-scale experiments or among adult trees. The best scenario would be to identify resistance-related factors that can be evaluated under optimal conditions at the seedling stage.

Over the last two decades, advances have been made in phenotyping techniques and physiological measurements, as well as molecular methods. These improvements indicate that, in the near future, the costs and throughput rates for -omics techniques will no longer be a bottleneck for research efforts. Considering that research in woody species is time-consuming and involves enormous labor efforts, it is crucial to initiate interdisciplinary research toward deciphering drought resistance. Advanced phenotyping techniques and physiological measurements should be accompanied by studies in the -omics domain such as genetics, transcriptomics, proteomics and metabolomics. New insights gained using this approach should facilitate the identification of specific genetic elements to be used in selection, breeding or genetic engineering toward woody species with improved tolerance of drought stress.

ACKNOWLEDGEMENTS

We thank our colleagues Joshua D. Klein and Gilor Kelly for their valuable comments and suggestions.

REFERENCES

Aasamaa, K., and A. Sober. 2011. Responses of stomatal conductance to simultaneous changes in two environmental factors. *Tree Physiology* 31:855–64.

Abbaspour, H., S. Saeidi-Sar, H. Afshari and M. A. Abdel-Wahhab. 2012. Tolerance of mycorrhiza-infected pistachio (*Pistacia vera* L.) seedling to drought stress under glasshouse conditions. *Journal of Plant Physiology* 169:704–9.

Abe, H., T. Urao, T. Ito, M. Seki, K. Shinozaki and K. Yamaguchi-Shinozaki. 2003. Arabidopsis AtMYC2 (bHLH) and AtMYB2 (MYB) function as transcriptional activators in abscisic acid signaling. *Plant Cell* 15: 63–78.

Aguadé, D., R. Poyatos, M. Gómez, J. Oliva and J. Martínez-Vilalta. 2015. The role of defoliation and root rot pathogen infection in driving the mode of drought-related physiological decline in Scots pine (*Pinus sylvestris* L.). *Tree Physiology* 35:229–42.

Agustí, J., M. Zapater, D. J. Iglesias, M. Cercós, F. R. Tadeo and M. Talón. 2007. Differential expression of putative 9-cis-epoxycarotenoid dioxygenases and abscisic acid accumulation in water stressed vegetative and reproductive tissues of citrus. *Plant Science* 172:85–94.

Ahmad, R., M.D. Kim, K.-H. Back et al. 2008. Stress-induced expression of choline oxidase in potato plant chloroplasts confers enhanced tolerance to oxidative, salt, and drought stresses. *Plant Cell Reports* 27:687–98.

Ahmad, R., C. J. Lim and S.-Y. Kwon. 2013. Glycine betaine: A versatile compound with great potential for gene pyramiding to improve crop plant performance against environmental stresses. *Plant Biotechnology Reports* 7:49–57.

Alexandersson, E., J. A. H. Danielson, J. Råde et al. 2010. Transcriptional regulation of aquaporins in accessions of Arabidopsis in response to drought stress. *The Plant Journal* 61:650–60.

Alexandersson, E., L. Fraysse, S. Sjovall-Larsen et al. 2005. Whole gene family expression and drought stress regulation of aquaporins. *Plant Molecular Biology* 59:469–84.

Alvarez, M., D. Huygens, C. Fernandez et al. 2009. Effect of ectomycorrhizal colonization and drought on reactive oxygen species metabolism of *Nothofagus dombeyi* roots. *Tree Physiology* 29:1047–57.

Améglio, T., C. Bodet, A. Lacointe and H. Cochard. 2002. Winter embolism, mechanisms of xylem hydraulic conductivity recovery and springtime growth patterns in walnut and peach trees. *Tree Physiology* 22:1211–20.

Anderegg, W. R. L., L. Plavcová, L. D. L. Anderegg, U. G. Hacke, J. A. Berry and C. B. Field. 2013. Drought's legacy: Multiyear hydraulic deterioration underlies widespread aspen forest die-off and portends increased future risk. *Global Change Biology* 19:1188–96.

Angeles, G., B. Bond, J. S. Boyer et al. 2004. The cohesion-tension theory. *New Phytologist* 163:451–2.

Arend, M., T. Kuster, M. S. Günthardt-Goerg and M. Dobbertin. 2011. Provenance-specific growth responses to drought and air warming in three European oak species (*Quercus robur*, *Q. petraea* and *Q. pubescens*). *Tree Physiology* 31:287–97.

Argueso, C.T., F. J. Ferreira and J. J. Kieber. 2009. Environmental perception avenues: The interaction of cytokinin and environmental response pathways. *Plant, Cell & Environment* 32, 1147–60.

Attia, Z., J.-C. Domec, R. Oren, D. A. Way and M. Moshelion. 2015. Growth and physiological responses of isohydric and anisohydric poplars to drought. *Journal of Experimental Botany* 66:4373–81.

Baas, P. 2013. *Wood Structure in Plant Biology and Ecology*. Leiden: Brill.

Bae, H., R. C. Sicher, M. S. Kim et al. 2009. The beneficial endophyte *Trichoderma hamatum* isolate DIS 219b promotes growth and delays the onset of the drought response in *Theobroma cacao*. *Journal of Experimental Botany* 60:3279–95.

Barbera, G., and F. Carimi. 1988. Effects of different level of water stress on yield and quality of lemon trees. In *Proceedings of the Sixth International Citrus Congress*, ed. F. Goren and K. Mendel, 717–22. Tel Aviv: Balaban.

Barbera, G., G. Fatta del Bosco and B. Lo Cascio. 1985. Effects of water stress on lemon summer bloom: The Forzatura technique in the Sicilian citrus industry. In *Proceedings of the Xth Meeting of the International Society for Horticultural Science* (ISHS), 391–8. Leuven: ISHS.

Barchet, G. L. H., R. Dauwe, R. D. Guy et al. 2014. Investigating the drought-stress response of hybrid poplar genotypes by metabolite profiling. *Tree Physiology* 34:1203–19.

Bartlett, M. K., C. Scoffoni, R. Ardy et al. 2012a. Rapid determination of comparative drought tolerance traits: Using an osmometer to predict turgor loss point. *Methods in Ecology and Evolution* 3:880–8.

Bartlett, M. K., C. Scoffoni and L. Sack. 2012b. The determinants of leaf turgor loss point and prediction of drought tolerance of species and biomes: A global meta-analysis. *Ecology Letters* 15:393–405.

Bartlett, M. K., Y. Zhang, N. Kreidler et al. 2014. Global analysis of plasticity in turgor loss point, a key drought-tolerance trait. *Ecology Letters* 17:1580–90.

Bentz, B. J., J. Régnière, C. J. Fettig et al. 2010. Climate change and bark beetles of the western United States and Canada: Direct and indirect effects. *BioScience* 60:602–13.

Blackman, C. J., T. J. Brodribb and G. J. Jordan. 2009. Leaf hydraulics and drought stress: Response, recovery and survivorship in four woody temperate plant species. *Plant, Cell & Environment* 32:1584–95.

Blum, A. 2014. Genomics for drought resistance - getting down to earth. *Functional Plant Biology* 41:1191–8.

Blum, A. 2016. Osmotic adjustment is a prime drought stress adaptive engine in support of plant production. *Plant, Cell & Environment* 40:4–10.

Bogeat-Triboulot, M. B., M. Brosche, J. Renaut, L. Jouve, D. Thiec and P. Fayyaz. 2007. Gradual soil water depletion results in reversible changes of gene expression, protein profiles, ecophysiology, and growth performance in *Populus euphratica*, a poplar growing in arid regions. *Plant Physiology* 143: 876–92.

Bonal, D., and J. M. Guehl. 2001. Contrasting patterns of leaf water potential and gas exchange responses to drought in seedlings of tropical rainforest species. *Functional Ecology* 15:490–6.

Boukcim, H., L. Pages, C. Plassard and D. Mousain. 2001. Root system architecture and receptivity to mycorrhizal infection in seedlings of *Cedrus atlantica* as affected by nitrogen source and concentration. *Tree Physiology* 21:109–15.

Braatne, J. H., T. M. Hinckley and R. F. Stettler. 1992. Influence of soil water on the physiological and morphological components of plant water balance in *Populus trichocarpa*, *Populus deltoides* and their F1 hybrids. *Tree Physiology* 11:325–39.

Bréda, N., R. Huc, A. Granier and E. Dreyer. 2006. Temperate forest trees and stands under severe drought: A review of ecophysiological responses, adaptation processes and long-term consequences. *Annals of Forest Science* 63:625–44.

Brodersen, C., and A. McElrone. 2013. Maintenance of xylem network transport capacity: A review of embolism repair in vascular plants. *Frontiers in Plant Science* 4:108.

Brodersen, C. R., A. J. McElrone, B. Choat, M. A. Matthews and K. A. Shackel. 2010. The dynamics of embolism repair in xylem: In vivo visualizations using high-resolution computed tomography. *Plant Physiology* 154:1088–95.

Brodribb, T. J., and M. Holbrook. 2004. Stomatal protection against hydraulic failure: A comparison of coexisting ferns and angiosperms. *New Phytologist* 162:663–70.

Brodribb, T. J., N. M. Holbrook, E. J. Edwards and M. V. Gutierrez. 2003. Relations between stomatal closure, leaf turgor and xylem vulnerability in eight tropical dry forest trees. *Plant, Cell and Environment* 26:443–50.

Brodribb, T. J., and S. A. McAdam. 2011. Passive origins of stomatal control in vascular plants. *Science* 331:582–5.

Brodribb, T. J., and S. A. McAdam. 2013. Abscisic acid mediates a divergence in the drought response of two conifers. *Plant Physiology* 162:1370–7.

Brodribb, T. J., S. A. M. McAdam and M. R. Carins Murphy. 2016. Xylem and stomata, coordinated through time and space. *Plant, Cell & Environment*. doi: 10.1111/pce.12817.

Brodribb, T. J., S. A. M. McAdam, G. J. Jordan and S. C. V. Martins. 2014. Conifer species adapt to low-rainfall climates by following one of two divergent pathways. *Proceedings of the National Academy of Sciences of the United States of America* 111:14489–93.

Brunner, I., C. Herzog, M. A. Dawes, M. Arend and C. Sperisen. 2015. How tree roots respond to drought. *Frontiers in Plant Science* 6:547.

Buchanan, B. B., and Y. Balmer. 2005. Redox regulation: A broadening horizon. *Annual Review of Plant Biology* 56:187–220.

Buckley, T. N. 2005. The control of stomata by water balance. *New Phytologist* 168:275–92.

Canham, D. C., K. R. Kobe, F. E. Latty and L. R. Chazdon. 1999. Interspecific and intraspecific variation in tree seedling survival: Effects of allocation to roots versus carbohydrate reserves. *Oecologia* 121:1–11.

Carsjens, C., Q. Nguyen Ngoc, J. Guzy et al. 2014. Intra-specific variations in expression of stress-related genes in beech progenies are stronger than drought-induced responses. *Tree Physiology* 34:1348–61.

Chacko, E. 1986. Physiology of vegetative and reproductive growth in mango (*Mangifera indica* L.) trees. In *Proceedings of the First Australian Mango Research Workshop*, Commonwealth Scientific and Industrial Research Organization (CSIRO), Melbourne, 54–70.

Chaikiattiyos, S., C. M. Menzel and T. S. Rasmussen. 1994. Floral induction in tropical fruit trees: Effects of temperature and water supply. *Journal of Horticultural Science* 69:397–415.

Chang, S., J. D. Puryear, M. A. D. L. Dias, E. A. Funkhouser, R. J. Newton and J. Cairney. 1996. Gene expression under water deficit in loblolly pine (*Pinus taeda*): Isolation and characterization of cDNA clones. *Physiologia Plantarum* 97:139–48.

Charrier, G., J. M. Torres-Ruiz, E. Badel et al. 2016. Evidence for hydraulic vulnerability segmentation and lack of xylem refilling under tension. *Plant Physiology* 172:1657–68.

Chave, J., D. Coomes, S. Jansen, S. L. Lewis, N. G. Swenson and A. E. Zanne. 2009. Towards a worldwide wood economics spectrum. *Ecology Letters* 12:351–66.

Chaves, M. M., J. S. Pereira, J. Maroco et al. 2002. How plants cope with water stress in the field? Photosynthesis and growth. *Annals of Botany* 89:907–16.
Checker, V. G., A. K. Chhibbar and P. Khurana. 2012. Stress-inducible expression of barley *Hva1* gene in transgenic mulberry displays enhanced tolerance against drought, salinity and cold stress. *Transgenic Research* 21:939–57.
Chefdor, F., H. Bénédetti, C. Depierreux, F. Delmotte, D. Morabito and S. Carpin. 2006. Osmotic stress sensing in *Populus*: Components identification of a phosphorelay system. *FEBS Letters* 580:77–81.
Chen, H., and J. G. Jiang. 2010. Osmotic adjustment and plant adaptation to environmental changes related to drought and salinity. *Environmental Reviews* 18:309–19.
Chenafi, A., P. Monney, M. Ceymann, E. Arrigoni and A. Boudoukha. 2013. Influence of regulated deficit irrigation for apple trees cv. Gala on yield, fruit quality and water use. *Revue Suisse de Viticulture, Arboriculture, Horticulture* 45:92–101.
Choat, B., S. Jansen, T. J. Brodribb et al. 2012. Global convergence in the vulnerability of forests to drought. *Nature* 491:752–6.
Choudhury, F. K., R. M. Rivero, E. Blumwald and R. Mittler. 2016. Reactive oxygen species, abiotic stress and stress combination. *The Plant Journal.* doi:10.1111/tpj.13299.
Chu, Y. G., W. X. Zhang, B. Wu, Q. J. Huang, B. Y. Zhang and X. H. Su. 2016. Overexpression of the novel *Zygophyllum xanthoxylum* C2H2-type zinc finger gene ZxZF improves drought tolerance in transgenic Arabidopsis and poplar. *Biologia* 71:769–76.
Clifford, S. C., S. K. Arndt, J. E. Corlett et al. 1998. The role of solute accumulation, osmotic adjustment and changes in cell wall elasticity in drought tolerance in *Ziziphus mauritiana* (Lamk.). *Journal of Experimental Botany* 49:967–77.
Cochard, H. 2006. Cavitation in trees. *Comptes Rendus Physique* 7:1018–26.
Cochard, H., N. Bréda and A. Granier. 1996. Whole-tree hydraulic conductance and water loss regulation in *Quercus* during drought: Evidence for stomatal control of embolism? *Annals of Forest Science* 53:197–206.
Cochard, H., and S. Delzon. 2013. Hydraulic failure and repair are not routine in trees. *Annals of Forest Science* 70:659–61.
Cochard, H., D. Lemoine, T. Améglio and A. Granier. 2001. Mechanisms of xylem recovery from winter embolism in *Fagus sylvatica*. *Tree Physiology* 21:27–33.
Cocozza, C., P. Cherubini, N. Regier, M. Saurer, B. Frey and R. Tognetti. 2010. Early effects of water deficit on two parental clones of *Populus nigra* grown under different environmental conditions. *Functional Plant Biology* 37:244–54.
Cohen, D., M. B. Bogeat-Triboulot, E. Tisserant, S. Balzergue, M. L. Martin-Magniette and G. Lelandais. 2010. Comparative transcriptomics of drought responses in *Populus*: A meta-analysis of genome-wide expression profiling in mature leaves and root apices across two genotypes. *BMC Genomics* 11:630.
Corcuera, L., E. Gil-Pelegrin and E. Notivol. 2012. Aridity promotes differences in proline and phytohormone levels in *Pinus pinaster* populations from contrasting environments. *Trees* 26:799–808.
Costa, P., N. Bahrman, J.-M. Frigerio, A. Kremer and C. Plomion. 1998. Water-deficit-responsive proteins in maritime pine. *Plant Molecular Biology* 38:587–96.
Cramer, G. R., A. Ergul, J. Grimplet et al. 2007. Water and salinity stress in grapevines: Early and late changes in transcript and metabolite profiles. *Functional & Integrative Genomics* 7:111–34.
Cruziat P., H. Cochard, T. Ameglio. 2002. Hydraulic architecture of trees: Main concepts and results. *Annals of Forest Science* 59: 723–52.
Cutler, S. R., P. L. Rodriguez, R. R. Finkelstein and S. R. Abrams. 2010. Abscisic acid: Emergence of a core signaling network. *Annual Review of Plant Biology* 61:651–79.
Das, M., H. Chauhan, A. Chhibbar, Q. M. R. Haq and P. Khurana. 2011. High-efficiency transformation and selective tolerance against biotic and abiotic stress in mulberry, *Morus indica* cv. K2, by constitutive and inducible expression of tobacco osmotin. *Transgenic Research* 20:231–46.
Daszkowska-Golec, A., and I. Szarejko. 2013. Open or close the gate – Stomata action under the control of phytohormones in drought stress conditions. *Frontiers in Plant Science* 4:138.
David-Schwartz, R., I. Paudel, M. Mizrachi et al. 2016. Indirect evidence for genetic differentiation in vulnerability to embolism in *Pinus halepensis*. *Frontiers in Plant Science* 7:768.
Dawalibi, V., M. Monteverdi, S. Moscatello, A. Battistelli and R. Valentini. 2015. Effect of salt and drought on growth, physiological and biochemical responses of two *Tamarix* species. *Forest Biogeosciences and Forestry* 8:772–9.
De Diego, N., F. Pérez-Alfocea, E. Cantero, M. Lacuesta and P. Moncaleán. 2012. Physiological response to drought in radiata pine: Phytohormone implication at leaf level. *Tree Physiology* 32:435–49.

De Diego, N., J. L. Rodríguez, I. C. Dodd, F. Pérez-Alfocea, P. Moncaleán and M. Lacuesta. 2013a. Immunolocalization of IAA and ABA in roots and needles of radiata pine (*Pinus radiata*) during drought and rewatering. *Tree Physiology* 33:537–49.

De Diego, N., M. C. Sampedro, R. J. Barrio, I. Saiz-Fernández, P. Moncaleán and M. Lacuesta. 2013b. Solute accumulation and elastic modulus changes in six radiata pine breeds exposed to drought. *Tree Physiology* 33:69–80.

de Groot, W. J., A. S. Cantin, M. D. Flannigan, A. J. Soja, L. M. Gowman and A. Newbery. 2013. A comparison of Canadian and Russian boreal forest fire regimes. *Forest Ecology and Management* 294:23–34.

Delzon, S. 2015. New insight into leaf drought tolerance. *Functional Ecology* 29:1247–9.

Delzon, S., C. Douthe, A. Sala and H. Cochard. 2010. Mechanism of water-stress induced cavitation in conifers: Bordered pit structure and function support the hypothesis of seal capillary-seeding. *Plant, Cell & Environment* 33:2101–11.

Desikan, R., K. Last, R. Harrett-Williams et al. 2006. Ethylene-induced stomatal closure in Arabidopsis occurs via AtrbohF-mediated hydrogen peroxide synthesis. *The Plant Journal* 47:907–16.

Diamantoglou, S., and S. Rhizopoulou. 1992. Free proline accumulation in sapwood, bark and leaves of three evergreen sclerophylls and a comparison with an evergreen conifer. *Journal of Plant Physiology* 140:361–5.

Diaz-Vivancos, P., L. Faize, E. Nicolas et al. 2016. Transformation of plum plants with a cytosolic ascorbate peroxidase transgene leads to enhanced water stress tolerance. *Annals of Botany* 117:1121–31.

Ding, J., and O. Nilsson. 2016. Molecular regulation of phenology in trees—because the seasons they are a-changin'. *Current Opinion in Plant Biology* 29:73–9.

Dixon, H. H. 1896. Transpiration into a saturated atmosphere. *Proceedings of the Royal Irish Academy (1889–1901)* 4:627–35.

Dorman, M., T. Svoray, A. Perevolotsky, Y. Moshe and D. Sarris. 2015. What determines tree mortality in dry environments? A multi-perspective approach. *Ecological Applications* 25:1054–71.

dos Santos, T. P., C. M. Lopes, M. L. Rodrigues et al. 2003. Partial rootzone drying: Effects on growth and fruit quality of field-grown grapevines (*Vitis vinifera*). *Functional Plant Biology* 30:663–71.

Du, B., K. Jansen, A. Kleiber et al. 2015. A coastal and an interior Douglas fir provenance exhibit different metabolic strategies to deal with drought stress. *Tree Physiology* 36:148–63.

Dubos, C., G. Le Provost, D. Pot et al. 2003. Identification and characterization of water-stress-responsive genes in hydroponically grown maritime pine (*Pinus pinaster*) seedlings. *Tree Physiology* 23:169–79.

Dubos, C., and C. Plomion. 2003. Identification of water-deficit responsive genes in maritime pine (*Pinus pinaster* Ait.) roots. *Plant Molecular Biology* 51:249–62.

Durand, T. C., K. Sergeant, J. Renaut et al. 2011. Poplar under drought: Comparison of leaf and cambial proteomic responses. *Journal of Proteomics* 74:1396–410.

Eilmann, B., N. Buchmann, R. Siegwolf, M. Saurer, P. Cherubini and A. Rigling. 2010. Fast response of Scots pine to improved water availability reflected in tree-ring width and $\delta^{13}C$. *Plant, Cell & Environment* 33:1351–60.

Eilmann, B., R. Zweifel, N. Buchmann, P. Fonti and A. Rigling. 2009. Drought-induced adaptation of the xylem in Scots pine and pubescent oak. *Tree Physiology* 29:1011–20.

Eilmann, B., R. Zweifel, N. Buchmann, E. Graf Pannatier and A. Rigling. 2011. Drought alters timing, quantity, and quality of wood formation in Scots pine. *Journal of Experimental Botany* 62:2763–71.

Espeleta, J. F., and D. M. Eissenstat. 1998. Responses of citrus fine roots to localized soil drying: A comparison of seedlings with adult fruiting trees. *Tree Physiology* 18:113–9.

FAO. 2015. Food and Agriculture Organization of the United Nations - Water - Irrigation. http://www.fao.org/nr/water/topics_irrigation.html.

Fereres, E., and M. A. Soriano. 2007. Deficit irrigation for reducing agricultural water use. *Journal of Experimental Botany* 58:147–59.

Fisher, R. A., M. Williams, R. L. Do Vale, A. L. Da Costa and P. Meir. 2006. Evidence from Amazonian forests is consistent with isohydric control of leaf water potential. *Plant, Cell & Environment* 29:151–65.

Fonti, P., G. von Arx, I. García-González et al. 2010. Studying global change through investigation of the plastic responses of xylem anatomy in tree rings. *New Phytologist* 185:42–53.

Franks P. J., and D. J. Beerling. 2009. Maximum leaf conductance driven by CO_2 effects on stomatal size and density over geologic time. *Proceedings of the National Academy of Sciences of the United States of America* 106:10343–7.

Fritts, H. 1976. *Tree Rings and Climate*. London: Academic Press.

Gao, W. D., S. Bai, Q. M. Li et al. 2013. Overexpression of *TaLEA* gene from *Tamarix androssowii* improves salt and drought tolerance in transgenic poplar (*Populus simonii* x *P. nigra*). *PLOS ONE* 8:67462.

Garcia-Forner, N., A. Sala, C. Biel, R. Savé, J. Martínez-Vilalta and D. Whitehead. 2016. Individual traits as determinants of time to death under extreme drought in *Pinus sylvestris* L. *Tree Physiology* 36:1196–209.
Giri, J. 2011. Glycinebetaine and abiotic stress tolerance in plants. *Plant Signaling & Behavior* 6:1746–51.
Gomez-Cadenas, A., F. R. Tadeo, M. Talon and E. Primo-Millo. 1996. Leaf abscission induced by ethylene in water-stressed intact seedlings of Cleopatra mandarin requires previous abscisic acid accumulation in roots. *Plant Physiology* 112:401–8.
Goodwin, S., J. D. McPherson and W. R. McCombie. 2016. Coming of age: Ten years of next-generation sequencing technologies. *Nature Reviews Genetics* 17:333–51.
Grünzweig, J. M., T. Lin, E. Rotenberg, A. Schwartz and D. Yakir. 2003. Carbon sequestration in arid-land forest. *Global Change Biology* 9:791–9.
Guehl, J. M., A. Clement, P. Kaushal and G. Aussenac. 1993. Planting stress, water status and non-structural carbohydrate concentrations in Corsican pine seedlings. *Tree Physiology* 12:173–83.
Hacke, U. G., V. Stiller, J. S. Sperry, J. Pittermann and K. A. McCulloh. 2001. Cavitation fatigue. Embolism and refilling cycles can weaken the cavitation resistance of xylem. *Plant Physiology* 125:779–86.
Hamanishi, E. T., G. L. Barchet, R. Dauwe, S. D. Mansfield and M. M. Campbell. 2015. Poplar trees reconfigure the transcriptome and metabolome in response to drought in a genotype- and time-of-day-dependent manner. *BMC Genomics* 16:1–16.
Hamanishi, E. T., and M. M. Campbell. 2011. Genome-wide responses to drought in forest trees. *Forestry* 84:273–83.
Han, M. S., E. W. Noh and S. H. Han. 2013. Enhanced drought and salt tolerance by expression of *AtGSK1* gene in poplar. *Plant Biotechnology Reports* 7:39–47.
Hansen, H., and K. Grossmann. 2000. Auxin-induced ethylene triggers abscisic acid biosynthesis and growth inhibition. *Plant Physiology* 124:1437–48.
Harfouche, A., R. Meilan and A. Altman. 2014. Molecular and physiological responses to abiotic stress in forest trees and their relevance to tree improvement. *Tree Physiology* 34:1181–98.
Hartmann, H., N. G. McDowell and S. Trumbore. 2015. Allocation to carbon storage pools in Norway spruce saplings under drought and low CO_2. *Tree Physiology* 35:243–52.
Hartmann, H., and S. Trumbore. 2016. Understanding the roles of nonstructural carbohydrates in forest trees - from what we can measure to what we want to know. *New Phytologist* 211:386–403.
Heath, L. S., N. Ramakrishnan, R. R. Sederoff et al. 2002. Studying the functional genomics of stress responses in loblolly pine with the Expresso microarray experiment management system. *Comparative and Functional Genomics* 3:226–43.
Heckwolf, M., D. Pater, D. T. Hanson and R. Kaldenhoff. 2011. The *Arabidopsis thaliana* aquaporin AtPIP1;2 is a physiologically relevant CO_2 transport facilitator. *The Plant Journal* 67:795–804.
Heinen, R. B., Q. Ye and F. Chaumont. 2009. Role of aquaporins in leaf physiology. *Journal of Experimental Botany* 60:2971–85.
Howe, G. A., and G. Jander. 2008. Plant immunity to insect herbivores. *Annual Review of Plant Biology* 59:41–66.
Hsiao, T. C., and L. K. Xu. 2000. Sensitivity of growth of roots versus leaves to water stress: Biophysical analysis and relation to water transport. *Journal of Experimental Botany* 51:1595–616.
Hu, H., A. Boisson-Dernier, M. Israelsson-Nordstrom et al. 2010. Carbonic anhydrases are upstream regulators of CO_2-controlled stomatal movements in guard cells. *Nature Cell Biology* 12(suppl.):87–93; sup pp 1–18.
Iuchi, S., M. Kobayashi, T. Taji et al. 2001. Regulation of drought tolerance by gene manipulation of 9-*cis*-epoxycarotenoid dioxygenase, a key enzyme in abscisic acid biosynthesis in Arabidopsis. *The Plant Journal* 27:325–33.
Jiménez, S., J. Dridi, D. Gutiérrez et al. 2013. Physiological, biochemical and molecular responses in four *Prunus* rootstocks submitted to drought stress. *Tree Physiology* 33:1061–75.
Jorge, I., R. M. Navarro, C. Lenz, D. Ariza and J. Jorrín. 2006. Variation in the holm oak leaf proteome at different plant developmental stages, between provenances and in response to drought stress. *Proteomics* 6:S207–14.
Kaldenhoff, R., A. Bertl, B. Otto, M. Moshelion and N. Uehlein. 2007. Characterization of plant aquaporins. *Methods in Enzymology* 428:505–31.
Kaldenhoff, R., M. Ribas-Carbo, J. F. Sans, C. Lovisolo, M. Heckwolf and N. Uehlein. 2008. Aquaporins and plant water balance. *Plant, Cell & Environment* 31:658–66.
Kapoor, D., R. Sharma, N. Handa et al. 2015. Redox homeostasis in plants under abiotic stress: Role of electron carriers, energy metabolism mediators and proteinaceous thiols. *Frontiers in Environmental Science* 3:13.

Kathuria, H., J. Giri, K. N. Nataraja, N. Murata, M. Udayakumar and A. K. Tyagi. 2009. Glycinebetaine-induced water-stress tolerance in *codA*-expressing transgenic *indica* rice is associated with up-regulation of several stress responsive genes. *Plant Biotechnology Journal* 7:512–26.

Katz, C., R. Oren, E. D. Schulze and J. A. Milburn. 1989. Uptake of water and solutes through twigs of *Picea abies* (L.) Karst. *Trees* 3:33–7.

Ke, Q. B., Z. Wang, C. Y. Ji et al. 2016. Transgenic poplar expressing *codA* exhibits enhanced growth and abiotic stress tolerance. *Plant Physiology and Biochemistry* 100:75–84.

Kelly, G., M. Moshelion, R. David-Schwartz et al. 2013. Hexokinase mediates stomatal closure. *The Plant Journal* 75:977–88.

Kennedy, J. S., and C. O. Booth. 1958. Water relations of leaves from woody and herbaceous plants. *Nature* 181:1271–2.

Klein, T. 2015. Drought-induced tree mortality: From discrete observations to comprehensive research. *Tree Physiology* 35:225–8.

Klein, T., S. Cohen, I. Paudel, Y. Preisler, E. Rotenberg and D. Yakir. 2016. Diurnal dynamics of water transport, storage and hydraulic conductivity in pine trees under seasonal drought. *iForest-Biogeosciences and Forestry* 9:710–9.

Klein, T., S. Cohen and D. Yakir. 2011. Hydraulic adjustments underlying drought resistance of *Pinus halepensis*. *Tree Physiology* 31:637–48.

Klein, T., G. Di Matteo, E. Rotenberg, S. Cohen and D. Yakir. 2013a. Differential ecophysiological response of a major Mediterranean pine species across a climatic gradient. *Tree Physiology* 33:26–36.

Klein, T., and G. Hoch. 2015. Tree carbon allocation dynamics determined using a carbon mass balance approach. *New Phytologist* 205:147–59.

Klein, T., G. Hoch, D. Yakir and C. Korner. 2014a. Drought stress, growth and nonstructural carbohydrate dynamics of pine trees in a semi-arid forest. *Tree Physiology* 34:981–92.

Klein, T., E. Rotenberg, E. Cohen-Hilaleh et al. 2014b. Quantifying transpirable soil water and its relations to tree water use dynamics in a water-limited pine forest. *Ecohydrology* 7:409–19.

Klein, T., I. Shpringer, B. Fikler, G. Elbaz, S. Cohen and D. Yakir. 2013b. Relationships between stomatal regulation, water-use, and water-use efficiency of two coexisting key Mediterranean tree species. *Forest Ecology and Management* 302:34–42.

Klein, T., D. Yakir, N. Buchmann and J. M. Grünzweig. 2014c. Towards an advanced assessment of the hydrological vulnerability of forests to climate change-induced drought. *New Phytologist* 201:712–6.

Koo, A. J. K., and G. A. Howe. 2009. The wound hormone jasmonate. *Phytochemistry* 70:1571–80.

Körner, C. 2003. Carbon limitation in trees. *Journal of Ecology* 91:4–17.

Krasensky, J., and C. Jonak. 2012. Drought, salt, and temperature stress-induced metabolic rearrangements and regulatory networks. *Journal of Experimental Botany* 63:1593–608.

Kubikova, E., L. M. Jennifer, H. O. Bonnie, D. M. Michael and M. R. Augé. 2001. Mycorrhizal impact on osmotic adjustment in *Ocimum basilicum* during a lethal drying episode. *Journal of Plant Physiology* 158:1227–30.

Kües, U. 2007. *Wood Production, Wood Technology, and Biotechnological Impacts*. Göttingen: Universitätsverlag Göttingen.

Kuster, T. M., M. Arend, P. Bleuler, M. S. Günthardt-Goerg and R. Schulin. 2013. Water regime and growth of young oak stands subjected to air-warming and drought on two different forest soils in a model ecosystem experiment. *Plant Biology* 15:138–47.

Larcher, W., J. A. P. V. de Moraes and H. Bauer. 1981. Adaptive responses of leaf water potential, CO_2-gas exchange and water use efficiency of *Olea europaea* during drying and rewatering. In *Components of productivity of Mediterranean-climate regions Basic and applied aspects: Proceedings of the International Symposium on Photosynthesis, Primary Production and Biomass Utilization in Mediterranean-Type Ecosystems* held in Kassandra, Greece, September 13–15, 1980, ed. N. S. Margaris and H. A. Mooney, 77–84. Dordrecht: Springer Netherlands.

Laur, J., and U. G. Hacke. 2013. Transpirational demand affects aquaporin expression in poplar roots. *Journal of Experimental Botany* 64:2283–93.

Laur, J., and U. G. Hacke. 2014. The role of water channel proteins in facilitating recovery of leaf hydraulic conductance from water stress in *Populus trichocarpa*. *PLOS ONE* 9:e111751.

Lehto, T., and J. J. Zwiazek. 2011. Ectomycorrhizas and water relations of trees: A review. *Mycorrhiza* 21:71–90.

Lens, F., C. Picon-Cochard, C. E. L. Delmas et al. 2016. Herbaceous angiosperms are not more vulnerable to drought-induced embolism than angiosperm trees. *Plant Physiology* 172:661–7.

Le Provost, G., F. Domergue, C. Lalanne et al. 2013. Soil water stress affects both cuticular wax content and cuticle-related gene expression in young saplings of maritime pine (*Pinus pinaster* Ait). *BMC Plant Biology* 13:95–106.

Levy, Y., and J. Krikun. 1980. Effect of vesicular-arbuscular mycorrhiza on *Citrus jambhiri* water relations. *New Phytologist* 85:25–31.

Levy, Y., J. P. Syvertsen and S. Nemec. 1983. Effect of drought stress and vesicular-arbuscular mycorrhiza on citrus transpiration and hydraulic conductivity of roots. *New Phytologist* 93:61–6.

Li, H., Z. Wang, Q. Ke et al. 2014. Overexpression of *codA* gene confers enhanced tolerance to abiotic stresses in alfalfa. *Plant Physiology and Biochemistry* 85:31–40.

Li, M. R., Y. Li, H. Q. Li and G. J. Wu. 2011. Ectopic expression of *FaDREB2* enhances osmotic tolerance in paper mulberry. *Journal of Integrative Plant Biology* 53:951–60.

Liao, X., X. Guo, Q. Wang et al. 2016. Overexpression of *MsDREB6.2* results in cytokinin-deficient developmental phenotypes and enhances drought tolerance in transgenic apple plants. *The Plant Journal.* doi:10.1111/tpj.13401.

Limm, E. B., K. A. Simonin, A. G. Bothman, and T. E. Dawson. 2009. Foliar water uptake: A common water acquisition strategy for plants of the redwood forest. *Oecologia* 161:449–59.

Liu, W., D. J. Fairbairn, R. J. Reid and D. P. Schachtman. 2001. Characterization of two *HKT1* homologues from *Eucalyptus camaldulensis* that display intrinsic osmosensing capability. *Plant Physiology* 127:283–94.

Loewenstein, N. J., and S. G. Pallardy. 1998. Drought tolerance, xylem sap abscisic acid and stomatal conductance during soil drying: A comparison of canopy trees of three temperate deciduous angiosperms. *Tree Physiology* 18:431–9.

Lopez, D., J.-S. Venisse, B. Fumanal et al. 2013. Aquaporins and leaf hydraulics: Poplar sheds new light. *Plant and Cell Physiology* 54:1963–75.

Lorenz, W. W., R. Alba, Y.-S. Yu, J. M. Bordeaux, M. Simões and J. F. Dean. 2011. Microarray analysis and scale-free gene networks identify candidate regulators in drought-stressed roots of loblolly pine (*P. taeda* L.). *BMC Genomics* 12:264.

Lorenz, W. W., F. Sun, C. Liang et al. 2006. Water stress-responsive genes in loblolly pine (*Pinus taeda*) roots identified by analyses of expressed sequence tag libraries. *Tree Physiology* 26:1–16.

Lu, P., and E. K. Chacko. 2000. Effect of water stress on mango flowering in low latitude tropics of northern Australia. *Acta Horticulturae 509*:283–90.

Lugassi, N., G. Kelly, L. Fidel et al. 2015. Expression of Arabidopsis hexokinase in citrus guard cells controls stomatal aperture and reduces transpiration. *Frontiers in Plant Science* 6:1114.

Lundin, B., M. Hansson, B. Schoefs, A. V. Vener and C. Spetea. 2007. The Arabidopsis PsbO2 protein regulates dephosphorylation and turnover of the photosystem II reaction centre D1 protein. *The Plant Journal* 49:528–39.

Ma, Q., L.-J. Yue, J.-L. Zhang, G.-Q. Wu, A.-K. Bao and S.-M. Wang. 2012. Sodium chloride improves photosynthesis and water status in the succulent xerophyte *Zygophyllum xanthoxylum. Tree Physiology* 32:4–13.

Ma, Q.-H. 2008. Genetic engineering of cytokinins and their application to agriculture. *Critical Reviews in Biotechnology* 28:213–32.

Magalhães, A. P., N. Verde, F. Reis et al. 2015. RNA-seq and gene network analysis uncover activation of an ABA-dependent signalosome during the cork oak root response to drought. *Frontiers in Plant Science* 6:1195.

Mahajan, S., and N. Tuteja. 2005. Cold, salinity and drought stresses: An overview. *Archives of Biochemistry and Biophysics* 444:139–58.

Maréchaux, I., M. K. Bartlett, L. Sack et al. 2015. Drought tolerance as predicted by leaf water potential at turgor loss point varies strongly across species within an Amazonian forest. *Functional Ecology* 29:1268–77.

Marjanovic, Z., N. Uehlein, R. Kaldenhoff et al. 2005. Aquaporins in poplar: What a difference a symbiont makes! *Planta* 222:258–68.

Markesteijn, L., and L. Poorter. 2009. Seedling root morphology and biomass allocation of 62 tropical tree species in relation to drought- and shade-tolerance. *Journal of Ecology* 97:311–25.

Martínez-Vilalta, J. 2014. Carbon storage in trees: Pathogens have their say. *Tree Physiology* 34:215–7.

Maurel, C., Y. Boursiac, D.-T. Luu, V. Santoni, Z. Shahzad and L. Verdoucq. 2015. Aquaporins in plants. *Physiological Reviews* 95:1321–58.

Maurel, C., L. Verdoucq, D. T. Luu and V. Santoni. 2008. Plant aquaporins: Membrane channels with multiple integrated functions. *Annual Review of Plant Biology* 59:595–624.

McAdam, S. A. M., and T. J. Brodribb. 2016. Linking turgor with ABA biosynthesis: Implications for stomatal responses to vapour pressure deficit across land plants. *Plant Physiology* 171:2008–16.

McAdam, S. A. M., F. C. Sussmilch and T. J. Brodribb. 2016. Stomatal responses to vapour pressure deficit are regulated by high speed gene expression in angiosperms. *Plant, Cell & Environment* 39:485–91.

McCormick, A. J., M. D. Cramer and D. A. Watt. 2008. Changes in photosynthetic rates and gene expression of leaves during a source-sink perturbation in sugarcane. *Annals of Botany* (London) 101:89–102.

McDowell, N., W. T. Pockman, C. D. Allen et al. 2008. Mechanisms of plant survival and mortality during drought: Why do some plants survive while others succumb to drought? *New Phytologist* 178:719–39.

McDowell, N. G. 2011. Mechanisms linking drought, hydraulics, carbon metabolism, and vegetation mortality. *Plant Physiology* 155:1051–9.

McDowell, N. G., H. D. Adams, J. D. Bailey, M. Hess and T. E. Kolb. 2006. Homeostatic maintenance of ponderosa pine gas exchange in response to stand density changes. *Ecological Applications* 16:1164–82.

McDowell, N. G., N. Phillips, C. Lunch, B. J. Bond and M. G. Ryan. 2002. An investigation of hydraulic limitation and compensation in large, old Douglas-fir trees. *Tree Physiology* 22:763–74.

McKiernan, A. B., B. M. Potts, T. J. Brodribb et al. 2016. Responses to mild water deficit and rewatering differ among secondary metabolites but are similar among provenances within *Eucalyptus* species. *Tree Physiology* 36:133–47.

Meinzer, F. C., D. M. Johnson, B. Lachenbruch, K. A. McCulloh and D. R. Woodruff. 2009. Xylem hydraulic safety margins in woody plants: Coordination of stomatal control of xylem tension with hydraulic capacitance. *Functional Ecology* 23:922–30.

Meinzer, F. C., D. R. Woodruff, D. E. Marias, K. A. McCulloh and S. Sevanto. 2014. Dynamics of leaf water relations components in co-occurring iso- and anisohydric conifer species. *Plant, Cell & Environment* 37:2577–86.

Mencuccini, M. 2003. The ecological significance of long-distance water transport: Short-term regulation, long-term acclimation and the hydraulic costs of stature across plant life forms. *Plant, Cell & Environment* 26:163–82.

Méndez-Alonzo, R., H. Paz, R. C. Zuluaga, J. A. Rosell and M. E. Olson. 2012. Coordinated evolution of leaf and stem economics in tropical dry forest trees. *Ecology* 93:2397–406.

Miller, G. A. D., N. Suzuki, S. Ciftci-Yilmaz and R. O. N. Mittler. 2010. Reactive oxygen species homeostasis and signalling during drought and salinity stresses. *Plant, Cell & Environment* 33:453–67.

Mitchell, P. J., S. A. M. McAdam, E. A. Pinkard and T. J. Brodribb. 2016. Significant contribution from foliage-derived ABA in regulating gas exchange in *Pinus radiata*. *Tree Physiology*. doi: 10.1093/treephys/tpw092.

Mokany, K., R. J. Raison and A. S. Prokushkin. 2006. Critical analysis of root:shoot ratios in terrestrial biomes. *Global Change Biology* 12:84–96.

Molinari, H. B. C., C. J. Marur, J. C. B. Filho et al. 2004. Osmotic adjustment in transgenic citrus rootstock Carrizo citrange (*Citrus sinensis* Osb. × *Poncirus trifoliata* L. Raf.) overproducing proline. *Plant Science* 167:1375–81.

Morabito, D., and G. Guerrier. 2000. The free oxygen radical scavenging enzymes and redox status in roots and leaves of *Populus* × *Euramericana* in response to osmotic stress, desiccation and rehydration. *Journal of Plant Physiology* 157:74–80.

Moshelion, M., O. Halperin, R. Wallach, R. A. M. Oren and D. A. Way. 2015. Role of aquaporins in determining transpiration and photosynthesis in water-stressed plants: Crop water-use efficiency, growth and yield. *Plant, Cell & Environment* 38:1785–93.

Mott, K. A., and D. F. Parkhurst. 1991. Stomatal responses to humidity in air and helox. *Plant, Cell & Environment* 14:509–15.

Movahedi, A., J. Zhang, P. Gao et al. 2015. Expression of the chickpea *CarNAC3* gene enhances salinity and drought tolerance in transgenic poplars. *Plant Cell, Tissue and Organ Culture* 120:141–54.

Niinemets, Ü. 1999. Research review. Components of leaf dry mass per area – thickness and density – alter leaf photosynthetic capacity in reverse directions in woody plants. *New Phytologist* 144:35–47.

Niinemets, Ü. 2001. Global-scale climatic controls of leaf dry mass per area, density, and thickness in trees and shrubs. *Ecology* 82:453–69.

Niinemets, Ü. 2010. Responses of forest trees to single and multiple environmental stresses from seedlings to mature plants: Past stress history, stress interactions, tolerance and acclimation. *Forest Ecology and Management* 260:1623–39.

Noctor, G., A. Mhamdi and C. H. Foyer. 2014. The roles of reactive oxygen metabolism in drought: Not so cut and dried. *Plant Physiology* 164:1636–48.

Noctor, G., S. Veljovic-Jovanovic, S. Driscoll, L. Novitskaya and C. H. Foyer. 2002. Drought and oxidative load in the leaves of C3 plants: A predominant role for photorespiration? *Annals of Botany* 89:841–50.

Novaes, E., M. Kirst, V. Chiang, H. Winter-Sederoff and R. Sederoff. 2010. Lignin and biomass: A negative correlation for wood formation and lignin content in trees. *Plant Physiology* 154:555–61.

O'Brien, M. J., S. Leuzinger, C. D. Philipson, J. Tay and A. Hector. 2014. Drought survival of tropical tree seedlings enhanced by non-structural carbohydrate levels. *Nature Climate Change* 4:710–14.

O'Leary, B., J. Park and W. C. Plaxton. 2011. The remarkable diversity of plant PEPC (phosphoenolpyruvate carboxylase): Recent insights into the physiological functions and post-translational controls of non-photosynthetic PEPCs. *Biochemical Journal* 436:15–34.

Oren, R., J. S. Sperry, G. G. Katul et al. 1999. Survey and synthesis of intra- and interspecific variation in stomatal sensitivity to vapour pressure deficit. *Plant, Cell & Environment* 22:1515–26.

Ortas, I., D. Ortakçi and Z. Kaya. 2002. Various mycorrhizal fungi propagated on different hosts have different effect on citrus growth and nutrient uptake. *Communications in Soil Science and Plant Analysis* 33:259–72.

Osonubi, O., K. Mulongoy, O. O. Awotoye, M. O. Atayese and D. U. U. Okali. 1991. Effects of ectomycorrhizal and vesicular-arbuscular mycorrhizal fungi on drought tolerance of four leguminous woody seedlings. *Plant and Soil* 136:131–43.

Pang, T., L. Guo, D. Shim et al. 2015. Characterization of the transcriptome of the xerophyte *Ammopiptanthus mongolicus* leaves under drought stress by 454 pyrosequencing. *PLOS ONE* 10:e0136495.

Parniske, M. 2008. Arbuscular mycorrhiza: The mother of plant root endosymbioses. *Nature Reviews Microbiology* 6:763–75.

Paszkowski, U., and T. Boller. 2002. The growth defect of *lrt1*, a maize mutant lacking lateral roots, can be complemented by symbiotic fungi or high phosphate nutrition. *Planta* 214:584–90.

Pedranzani, H., R. Sierra-de-Grado, A. Vigliocco, O. Miersch and G. Abdala. 2007. Cold and water stresses produce changes in endogenous jasmonates in two populations of *Pinus pinaster* Ait. *Plant Growth Regulation* 52:111–6.

Perdiguero, P., M. D. C. Barbero, M. T. Cervera, C. Collada and Á Soto. 2013. Molecular response to water stress in two contrasting Mediterranean pines (*Pinus pinaster* and *Pinus pinea*). *Plant Physiology and Biochemistry* 67:199–208.

Perez-Martin, A., C. Michelazzo, J. M. Torres-Ruiz et al. 2014. Regulation of photosynthesis and stomatal and mesophyll conductance under water stress and recovery in olive trees: Correlation with gene expression of carbonic anhydrase and aquaporins. *Journal of Experimental Botany* 65:3143–56.

Perrone, I., G. Gambino, W. Chitarra et al. 2012. The grapevine root-specific aquaporin *VvPIP2;4N* controls root hydraulic conductance and leaf gas exchange under well-watered conditions but not under water stress. *Plant Physiology* 160:965–77.

Peterson, R. L., H. B. Massicotte and L. H. Melville. 2004. Mycorrhizas: Anatomy and cell biology. Ottawa: NRC Research Press.

Poorter, H., J. Buhler, D. Dusschoten, J. Climent and J. A. Postma. 2012. Pot size matters: A meta-analysis of the effects of rooting volume on plant growth. *Functional Plant Biology* 39:839–50.

Pospíšilová, J. 2003. Participation of phytohormones in the stomatal regulation of gas exchange during water stress. *Biologia Plantarum* 46:491–506.

Poyatos, R., D. Aguadé, L. Galiano, M. Mencuccini and J. Martínez-Vilalta. 2013. Drought-induced defoliation and long periods of near-zero gas exchange play a key role in accentuating metabolic decline of Scots pine. *New Phytologist* 200:388–401.

Pucholt, P., P. Sjödin, M. Weih, A. C. Rönnberg-Wästljung and S. Berlin. 2015. Genome-wide transcriptional and physiological responses to drought stress in leaves and roots of two willow genotypes. *BMC Plant Biology* 15:1–16.

Rajasekaran, L. R., and T. J. Blake. 1999. New plant growth regulators protect photosynthesis and enhance growth under drought of jack pine seedlings. *Journal of Plant Growth Regulation* 18:175–81.

Raveh, E. 2008. Partial root-zone drying as a possible replacement for 'Verdelli' practice in lemon production. *Acta Horticulturae* 2:537–41.

Raveh, E. 2012. Assessing salinity tolerance in citrus: Latest developments. In: *Advances in Citrus Nutrition*, ed. A.K. Srivastava, 425–33. The Netherlands: Springer-Verlag.

Raveh, E., and P. S. Nobel. 1999. CO_2 uptake and water loss accompanying vernalization for detached cladodes of *Opuntia ficus*. *International Journal of Plant Sciences* 160:92–7.

Regier, N., S. Streb, C. Cocozza et al. 2009. Drought tolerance of two black poplar (*Populus nigra* L.) clones: Contribution of carbohydrates and oxidative stress defence. *Plant, Cell & Environment* 32:1724–36.

Rensink, W. A., and C. R. Buell. 2005. Microarray expression profiling resources for plant genomics. *Trends in Plant Science* 10:603–9.

Ridolfi, M., M. L. Fauveau, P. Label, J. P. Garrec and E. Dreyer. 1996. Responses to water stress in an ABA-unresponsive hybrid poplar (*Populus koreana×trichocarpa* cv. Peace) I. Stomatal function. *New Phytologist* 134:445–54.

Rivero, R. M., M. Kojima, A. Gepstein et al. 2007. Delayed leaf senescence induces extreme drought tolerance in a flowering plant. *Proceedings of the National Academy of Sciences of the United States of America* 104:19631–6.

Rolland, F., E. Baena-Gonzalez and J. Sheen. 2006. Sugar sensing and signaling in plants: Conserved and novel mechanisms. *Annual Review of Plant Biology* 57:675–709.

Rotenberg, E., and D. Yakir. 2010. Contribution of semi-arid forests to the climate system. *Science* 327:451–4.

Sade, N., A. Gebremedhin and M. Moshelion. 2012. Risk-taking plants: Anisohydric behavior as a stress-resistance trait. *Plant Signaling & Behavior* 7:767–70.

Sade, N., M. Gebretsadik, R. Seligmann, A. Schwartz, R. Wallach and M. Moshelion. 2010. The role of tobacco aquaporin1 in improving water use efficiency, hydraulic conductivity, and yield production under salt stress. *Plant Physiology* 152:245–54.

Sakr, S., G. Alves, R. Morillon et al. 2003. Plasma membrane aquaporins are involved in winter embolism recovery in walnut trees. *Plant Physiology* 133:630–41.

Sala, A., F. Piper and G. Hoch. 2010. Physiological mechanisms of drought-induced tree mortality are far from being resolved. *New Phytologist* 186:274–81.

Sala, A., D. R. Woodruff and F. C. Meinzer. 2012. Carbon dynamics in trees: Feast or famine? *Tree Physiology* 32:764–75.

Salleo, S., A. Nardini, F. Pitt and M. A. L. Gullo. 2000. Xylem cavitation and hydraulic control of stomatal conductance in laurel (*Laurus nobilis* L.). *Plant, Cell & Environment* 23:71–9.

Santner, A., and M. Estelle. 2009. Recent advances and emerging trends in plant hormone signalling. *Nature* 459:1071–8.

Sathyan, P., R. J. Newton and C. A. Loopstra. 2005. Genes induced by WDS are differentially expressed in two populations of aleppo pine (*Pinus halepensis*). *Tree Genetics & Genomes* 1:166–73.

Schachtman, D. P., and J. Q. Goodger. 2008. Chemical root to shoot signaling under drought. *Trends in Plant Science* 13:281–7.

Schenk H. J., and S. Espino. 2011. Nighttime sap flow removes air from plant hydraulic systems. In: *8th International Workshop on Sap Flow*. Volterra, Italy.

Scholander, P. F., E. D. Bradstreet, E. A. Hemmington, H. T. Hammel. 1965. Sap pressure in vascular plants: Negative hydrostatic pressure can be measured in plants. *Science* 148:339–46.

Schraml, C., and H. Rennenberg. 2000. Sensitivity of different ecotypes of beech trees (*Fagus sylvatica* L.) to drought stress. *Forstwiss Centralbl* 119:51–61.

Schultz, H. R. 2003. Differences in hydraulic architecture account for near-isohydric and anisohydric behaviour of two field-grown *Vitis vinifera* L. cultivars during drought. *Plant, Cell & Environment* 26:1393–405.

Scoffoni, C., A. D. McKown, M. Rawls and L. Sack. 2012. Dynamics of leaf hydraulic conductance with water status: Quantification and analysis of species differences under steady state. *Journal of Experimental Botany* 63:643–58.

Scoffoni, C., M. Rawls, A. McKown, H. Cochard and L. Sack. 2011. Decline of leaf hydraulic conductance with dehydration: Relationship to leaf size and venation architecture. *Plant Physiology* 156:832–43.

Scoffoni, C., C. Vuong, S. Diep, H. Cochard and L. Sack. 2013. Leaf shrinkage with dehydration: Coordination with hydraulic vulnerability and drought tolerance. *Plant Physiology* 164:1772–88.

Secchi, F., M. E. Gilbert and M. A. Zwieniecki. 2011. Transcriptome response to embolism formation in stems of *Populus trichocarpa* provides insight into signaling and biology of refilling. *Plant Physiology* 157:1419–29.

Secchi, F., C. Lovisolo, N. Uehlein, R. Kaldenhoff and A. Schubert. 2007. Isolation and functional characterization of three aquaporins from olive (*Olea europaea* L.). *Planta* 225:381–92.

Secchi, F., and M. A. Zwieniecki. 2011. Sensing embolism in xylem vessels: The role of sucrose as a trigger for refilling. *Plant, Cell & Environment* 34:514–24.

Secchi, F., and M. A. Zwieniecki. 2014. Down-regulation of plasma intrinsic protein1 aquaporin in poplar trees is detrimental to recovery from embolism. *Plant Physiology* 164:1789–99.

Seelig, H.-D., R. J. Stoner and J. C. Linden. 2012. Irrigation control of cowpea plants using the measurement of leaf thickness under greenhouse conditions. *Irrigation Science* 30:247–57.

Shinozaki, K., and K. Yamaguchi-Shinozaki. 2007. Gene networks involved in drought stress response and tolerance. *Journal of Experimental Botany* 58:221–7.

Sibout, R., and G. Guerrier. 1998. Solute incompatibility with glutamine synthetase in water-stressed *Populus nigra*. *Environmental and Experimental Botany* 40:173–8.

Sinclair, T. R., N. M. Holbrook and M. A. Zwieniecki. 2005. Daily transpiration rates of woody species on drying soil. *Tree Physiology* 25:1469–72.

Sofo, A., B. Dichio, C. Xiloyannis and A. Masia. 2005. Antioxidant defences in olive trees during drought stress: Changes in activity of some antioxidant enzymes. *Functional Plant Biology* 32:45–53.

Speirs, J., A. Binney, M. Collins, E. Edwards and B. Loveys. 2013. Expression of ABA synthesis and metabolism genes under different irrigation strategies and atmospheric VPDs is associated with stomatal conductance in grapevine (*Vitis vinifera* L. cv Cabernet Sauvignon). *Journal of Experimental Botany* 64:1907–16.

Sperry, J. S. 2000. Hydraulic constraints on plant gas exchange. *Agricultural and Forest Meteorology* 104:13–23.

Spreer, W., S. Ongprasert, M. Hegele, J. N. Wünsche and J. Müller. 2009. Yield and fruit development in mango (*Mangifera indica* L. cv. Chok Anan) under different irrigation regimes. *Agricultural Water Management* 96:574–84.

Sreenivasulu, N., V. T. Harshavardhan, G. Govind, C. Seiler and A. Kohli. 2012. Contrapuntal role of ABA: Does it mediate stress tolerance or plant growth retardation under long-term drought stress? *Gene* 506:265–73.

Swarup, R., G. Parry, N. Graham, T. Allen and M. Bennett. 2002. Auxin cross-talk: Integration of signalling pathways to control plant development. *Plant Molecular Biology* 49:411–26.

Syvertsen, J. P., and Y. Levy. 1982. Diurnal changes in citrus leaf thickness, leaf water potential and leaf to air temperature difference. *Journal of Experimental Botany* 33:783–9.

Taiz, L., and E. Zeiger. 2010. *Plant Physiology*, 5th ed. Sunderland, MA: Sinauer Associates.

Taneda, H., and J. S. Sperry. 2008. A case-study of water transport in co-occurring ring- versus diffuse-porous trees: Contrasts in water-status, conducting capacity, cavitation and vessel refilling. *Tree Physiology* 28:1641–51.

Tardieu, F., and T. Simonneau. 1998. Variability among species of stomatal control under fluctuating soil water status and evaporative demand: Modelling isohydric and anisohydric behaviours. *Journal of Experimental Botany* 49:419–32.

Tatarinov, F., E. Rotenberg, K. Maseyk, J. Ogée, T. Klein and D. Yakir. 2016. Resilience to seasonal heat wave episodes in a Mediterranean pine forest. *New Phytologist* 210:485–96.

Tcherkez, G., E. Boex-Fontvieille, A. Mahe and M. Hodges. 2012. Respiratory carbon fluxes in leaves. *Current Opinion in Plant Biology* 15:308–14.

Tognetti, V. B., P. E. R. Mühlenbock and F. Van Breusegem. 2012. Stress homeostasis – the redox and auxin perspective. *Plant, Cell & Environment* 35:321–33.

Torres-Ruiz, J. M., A. Diaz-Espejo, A. Perez-Martin and V. Hernandez-Santana. 2014. Role of hydraulic and chemical signals in leaves, stems and roots in the stomatal behaviour of olive trees under water stress and recovery conditions. *Tree Physiology* 35:415–24.

Tran, L.-S. P., T. Urao, F. Qin et al. 2007. Functional analysis of AHK1/ATHK1 and cytokinin receptor histidine kinases in response to abscisic acid, drought, and salt stress in Arabidopsis. *Proceedings of the National Academy of Sciences of the United States of America* 104:20623–8.

Tschaplinski, T. J., and G. A. Tuskan. 1994. Water-stress tolerance of black and eastern cottonwood clones and four hybrid progeny. II. Metabolites and inorganic ions that constitute osmotic adjustment. *Canadian Journal of Forest Research* 24:681–7.

Tsuchihira, A., Y. T. Hanba, N. Kato, T. Doi, T. Kawazu, M. Maeshima. 2010. Effect of overexpression of radish plasma membrane aquaporins on water-use efficiency, photosynthesis and growth of Eucalyptus trees. *Tree Physiology* 30: 417–30.

Tsuchisaka, A., and A. Theologis. 2004. Unique and overlapping expression patterns among the Arabidopsis 1-amino-cyclopropane-1-carboxylate synthase gene family members. *Plant Physiology* 136:2982–3000.

Turner, N. C., and M. M. Jones. 1980. Turgor maintenance by osmotic adjustment: A review and evaluation. In: *Adaptation of Plants to Water and High Temperature Stress*, ed. N. C. Turner and P. J. Kramer, 87–103. New York: Wiley Interscience.

Tyree, M. T. 1999. Water relations and hydraulic architecture. In *Handbook of Functional Plant Ecology*, ed. F. I. Pugnaire and F. Valladares, 221–68. New York: Marcel Dekker.

Tyree, M. T., and J. S. Sperry. 1989. Vulnerability of xylem to cavitation and embolism. *Annual Review of Plant Physiology and Plant Molecular Biology* 40:19–36.

Umezawa, T., M. Okamoto, T. Kushiro et al. 2006. CYP707A3, a major ABA 8′-hydroxylase involved in dehydration and rehydration response in *Arabidopsis thaliana*. *The Plant Journal* 46:171–82.

Urao, T., B. Yakubov, R. Satoh et al. 1999. A transmembrane hybrid-type histidine kinase in Arabidopsis functions as an osmosensor. *The Plant Cell* 11:1743–54.

van Dongen, J. T., K. J. Gupta, S. J. Ramirez-Aguilar, W. L. Araujo, A. Nunes-Nesi and A. R. Fernie. 2011. Regulation of respiration in plants: A role for alternative metabolic pathways. *Journal of Plant Physiology* 168:1434–43.

Vanstraelen, M., and E. Benková. 2012. Hormonal interactions in the regulation of plant development. *Annual Review of Cell and Developmental Biology* 28:463–87.

Vélez-Bermúdez, I. C., T.-N. Wen, P. Lan and W. Schmidt. 2016. Isobaric tag for relative and absolute quantitation (iTRAQ)-based protein profiling in plants. In: *Plant proteostasis: Methods and Protocols*, ed. L. M. Lois and R. Matthiesen, 213–21. New York: Springer.

Verbruggen, N., and C. Hermans. 2008. Proline accumulation in plants: A review. *Amino Acids* 35:753–9.

Vicca, S., A. K. Gilgen, M. Camino Serrano et al. 2012. Urgent need for a common metric to make precipitation manipulation experiments comparable. *New Phytologist* 195:518–22.

Vincent, D., A. Ergul, M. C. Bohlman et al. 2007. Proteomic analysis reveals differences between *Vitis vinifera* L. cv. Chardonnay and cv. Cabernet Sauvignon and their responses to water deficit and salinity. *Journal of Experimental Botany* 58:1873–92.

Wang, L., H. Y. Su, L. Y. Han, C. Q. Wang, Y. L. Sun and F. H. Liu. 2014. Differential expression profiles of poplar *MAP kinase kinases* in response to abiotic stresses and plant hormones, and overexpression of *PtMKK4* improves the drought tolerance of poplar. *Gene* 545:141–8.

Wang, Y. Z., H. Z. Wang, R. F. Li, Y. Ma and J. H. Wei. 2011. Expression of a SK2-type dehydrin gene from *Populus euphratica* in a *Populus tremula* x *Populus alba* hybrid increased drought tolerance. *African Journal of Biotechnology* 10:9225–32.

Ward, E. J., R. Oren, D. M. Bell et al. 2013. The effects of elevated CO_2 and nitrogen fertilization on stomatal conductance estimated from 11 years of scaled sap flux measurements at Duke FACE. *Tree Physiology* 33:135–51.

Watkinson, J. I., A. A. Sioson, C. Vasquez-Robinet et al. 2003. Photosynthetic acclimation is reflected in specific patterns of gene expression in drought-stressed loblolly pine. *Plant Physiology* 133:1702–16.

West, A. G., K. R. Hultine, T. L. Jackson and J. R. Ehleringer. 2007. Differential summer water use by *Pinus edulis* and *Juniperus osteosperma* reflects contrasting hydraulic characteristics. *Tree Physiology* 27:1711–20.

Wheeler, J. K., B. A. Huggett, A. N. Tofte, F. E. Rockwell and N. M. Holbrook. 2013. Cutting xylem under tension or supersaturated with gas can generate PLC and the appearance of rapid recovery from embolism. *Plant, Cell & Environment* 36:1938–49.

Wolf, S., E. Werner, A. Christof et al. 2013. Contrasting response of grassland versus forest carbon and water fluxes to spring drought in Switzerland. *Environmental Research Letters* 8:035007.

Woodruff, D. R., F. C. Meinzer and B. Lachenbruch. 2008. Height-related trends in leaf xylem anatomy and shoot hydraulic characteristics in a tall conifer: Safety versus efficiency in water transport. *New Phytologist* 180:90–9.

Wu, Q.-S., X.-H. He, Y.-N. Zou, C.-Y. Liu, J. Xiao and Y. Li. 2012. Arbuscular mycorrhizas alter root system architecture of *Citrus tangerine* through regulating metabolism of endogenous polyamines. *Plant Growth Regulation* 68:27–35.

Wu, Q.-S., Y. Levy and Y. N. Zou. 2009. Arbuscular mycorrhizae and water relations in citrus. *Tree and Forest Science and Biotechnology* 3:105–12.

Wu, Q.-S., and R.-X. Xia. 2006. Arbuscular mycorrhizal fungi influence growth, osmotic adjustment and photosynthesis of citrus under well-watered and water stress conditions. *Journal of Plant Physiology* 163:417–25.

Wu, Q.-S., R.-X. Xia and Y.-N. Zou. 2006. Reactive oxygen metabolism in mycorrhizal and non-mycorrhizal citrus (*Poncirus trifoliata*) seedlings subjected to water stress. *Journal of Plant Physiology* 163:1101–10.

Wu, Q.-S., R.-X. Xia, Y.-N. Zou and G.-Y. Wang. 2007. Osmotic solute responses of mycorrhizal citrus (*Poncirus trifoliata*) seedlings to drought stress. *Acta Physiologiae Plantarum* 29:543.

Wu, Q.-S., and Y. N. Zou. 2009. Mycorrhizal influence on nutrient uptake of citrus exposed to drought stress. *The Philippine Agricultural Scientist* 92:33–8.

Xiao, X., X. Xu and F. Yang. 2008. Adaptive responses to progressive drought stress in two *Populus cathayana* populations. *Silva Fennica* 42: 705–19.

Xu, Z., Y. Jiang, B. Jia and G. Zhou. 2016. Elevated-CO_2 response of stomata and its dependence on environmental factors. *Frontiers in Plant Science* 7:657.

Yang, S.-J., Y.-J. Zhang, M. Sun, G. Goldstein and K.-F. Cao. 2012. Recovery of diurnal depression of leaf hydraulic conductance in a subtropical woody bamboo species: Embolism refilling by nocturnal root pressure. *Tree Physiology* 32:414–22.

Yin, C., B. Duan, X. Wang and C. Li. 2004. Morphological and physiological responses of two contrasting *Poplar* species to drought stress and exogenous abscisic acid application. *Plant Science* 167:1091–7.

Zalloni, E., M. de Luis, F. Campelo et al. 2016. Climatic signals from intra-annual density fluctuation frequency in Mediterranean pines at a regional scale. *Frontiers in Plant Science* 7:579.

Zawaski, C., and V. B. Busov. 2014. Roles of gibberellin catabolism and signaling in growth and physiological response to drought and short-day photoperiods in *Populus* trees. *PLOS ONE* 9:e86217.

Zeppel, M. J. B., S. P. Harrison, H. D. Adams et al. 2015. Drought and resprouting plants. *New Phytologist* 206:583–9.

Zhang, L., M. Y. Liu, G. R. Qiao, J. Jiang, Y. C. Jiang and R. Y. Zhuo. 2013. Transgenic poplar "NL895" expressing *CpFATB* gene shows enhanced tolerance to drought stress. *Acta Physiologiae Plantarum* 35:603–13.

Zhou, S., M. Li, Q. Guan et al. 2015. Physiological and proteome analysis suggest critical roles for the photosynthetic system for high water-use efficiency under drought stress in *Malus*. *Plant Science* 236:44–60.

Zhou, Y., F. Gao, R. Liu, J. Feng and H. Li. 2012. De novo sequencing and analysis of root transcriptome using 454 pyrosequencing to discover putative genes associated with drought tolerance in *Ammopiptanthus mongolicus*. *BMC Genomics* 13:266.

Zhu, L.-H., A. van de Peppel, X.-Y. Li and M. Welander. 2004. Changes of leaf water potential and endogenous cytokinins in young apple trees treated with or without paclobutrazol under drought conditions. *Scientia Horticulturae* 99:133–41.

Zimmermann, U., H. Schneider, L. H. Wegner and A. Haase. 2004. Water ascent in tall trees: Does evolution of land plants rely on a highly metastable state? *New Phytologist* 162:575–615.

Zufferey, V., H. Cochard, T. Ameglio, J. L. Spring and O. Viret. 2011. Diurnal cycles of embolism formation and repair in petioles of grapevine (*Vitis vinifera* cv. Chasselas). *Journal of Experimental Botany* 62:3885–94.

6 Plant Response to Mineral Nutrient Deficiency

Danqiong Huang and Wenhao Dai

CONTENTS

6.1 INTRODUCTION

It has been well documented that 17 elements are essential for plant growth and development, of which carbon (C), hydrogen (H), and oxygen (O) obtained from the air and water are not considered as minerals and the other 14 elements plants acquired from the soil are considered plant nutrients or mineral elements. According to the amount of each element needed for a healthy plant, mineral elements can be divided into two groups: macro- and micro-elements. Macro-elements, also called macronutrients or major nutrients, are required for plant growth and development in relatively large amounts (>1000 ppm). Six macronutrients include nitrogen (N), phosphorus (P), potassium

(K), calcium (Ca), magnesium (Mg), and sulfur (S). Micro-elements, also called micronutrients, are generally required by a plant in a small quantity (<100 ppm). These micronutrients include iron (Fe), manganese (Mn), zinc (Zn), copper (Cu), boron (B), molybdenum (Mo), chlorine (Cl), and nickel (Ni). In general, different species have their preference for nutrition requirement and very few soils have the capacity to supply all essential elements; therefore, mineral elements need to be supplemented through fertilization and other methods of soil improvement. The regular amount of each of the individual elements for healthy growth and development of plant species has been well documented (Barker and Pilbeam, 2015; Fernández-Escobar et al., 2016). Deficiency of one or several essential elements will result in defective growth, leading to a significant economic loss such as poor yield and quality. Plants respond to nutrient deficiency by showing specific symptoms. Such responses to nutrient deficiency are usually genotype dependent. For example, woody species such as apple, silver maple, pin oak, and some poplar and citrus species are very sensitive to iron deficiency caused by soil conditions such as high pH, poor aeration, or alkalinity. While other species, such as hackberry, some ash, and stone fruit species are tolerant to iron deficiency. Due to the specific characteristics woody species possess, such as a large body (crown and root system) and perennial nature, there are significant differences in nutrition management between woody and herbaceous species, although they share basic nutrition biology including similar deficiency symptoms and molecular regulations of nutrient uptake and transport. In this chapter, the roles of mineral nutrients in plant growth and development, physiological and molecular responses to nutrient deficiency, and the general management of nutrient deficiency in woody plants are reviewed. A case study that summarizes how a woody plant (*Populus tremula*) employs mechanisms in response to iron deficiency is included.

6.2 ROLES OF MINERAL NUTRIENTS IN PLANT GROWTH AND DEVELOPMENT

Plant nutrients refer to those elements that plants need to complete the life cycle. These elements play their individual or combined roles in the different stages of plant growth and development (Maathuis, 2009; Hänsch and Mendel, 2009). Some plant species need more elements than others. Some plant species are more sensitive to nutrient deficiency, while others are tolerant of nutrient deficiency. To keep healthy growth and development, plants have to acquire an appropriate amount of each of the individual micronutrients. A sufficient supply of these nutrients, particularly certain micronutrients, helps plants uptake, transport, and distribute the elements. A few elements are required for regulating a metal homeostasis network involving mobilization, intracellular trafficking, and storage of nutrients in plants. Deficiency of these nutrients can be caused by various factors including limited supply and low solubility/availability in the soil. Nutrient deficiency will cause abnormal growth and development (Maathuis, 2009; Hänsch and Mendel, 2009).

6.2.1 Macronutrients

Of the six macronutrients, N, P, and K are three primary elements that are required by plants. Often time, the soil cannot supply plants enough N, P, and K for a healthy growth and development. Deficiency of the three primary elements is frequently observed if they are not timely supplemented to the soil. The other three macronutrients, Ca, Mg, and S, are called second macronutrients and are required in moderate quantities. In general, the soil contains sufficient amounts of the second macronutrients for plant growth and development; however, the availability of macronutrients that plants can use is largely affected by soil condition and environment.

6.2.1.1 Nitrogen (N)

The element N is a primary building block of amino acids, a constituent of protein. N is also a component of vitamins and chlorophyll. N is involved in all plant physiological processes directly involved in photosynthesis. Plants absorb N as nitrate (NO_3^-) and ammonium (NH_4^+). Plants developed two mechanisms for acquiring N corresponding to the two ionic forms of N. Acquisition of N takes place on the plasma membrane and is performed by the plasma membrane-localized transporter in the root. N uptake is mediated by the high affinity H^+ coupled symporter, while ammonium uptake is mediated by the transporter functioning as the ion channel or possibly NH3/H^+ symporter (Krapp, 2015). Dry plant tissue contains N at the range of 1.5 to 4.0%. N is very mobile in plants; therefore, the N deficiency is evidenced on the old and lower leaves by uniform yellowing. Uncorrected N deficiency can eventually cause leaf scorching and drop. Shoots and branches are shorter and fewer in number. Different to herbaceous plants that are regularly suffer N deficiency, woody species are very efficient at recycling N and N deficiency occurs occasionally (Braun, 2012). Excess N causes excessive shoot growth, resulting in succulent tissues and delayed maturation, causing reduced resistance to biotic and abiotic stresses and reduced winter hardiness.

6.2.1.2 Phosphorus (P)

The element P is a vital component of nucleic acids that form the backbone of DNA. P is a major constituent of cellular membranes (phospholipids), and it plays an important role in energy transfer during metabolic processes in plant cells. P promotes root growth and development and enhances water use efficiency that helps improve plant tolerance to drought and cold stresses. Plants absorb P as PO_4^{3-}, HPO_4^{2-}, and H_2 PO_4^-. Plants use specialized transporters at the root/soil interface to extract P from the rhizosphere solution and then transport it across the membranes for further compartmentation. The P concentration in dry tissue is about 0.14 to 0.32%. Most soils contain high amount of P, but >80% of the P is tightly held in the soil. The soil pH affects the P solubility. In alkaline soil (pH > 7.5), the HPO_4^{2-} quickly react with calcium (Ca) and magnesium (Mg) to form less soluble compounds, while in acidic soil, the $H_2PO_4^-$ combines with Al and Fe to form less soluble compounds. P is mobile in plants. P deficiency symptoms include stunted plant, bronze or purple petioles of older leaves, followed by leaf yellowing, especially at the time of flowering and fruiting. Leaves will be smaller than normal and may be distorted and dropped earlier. Shoots appear to be normal length but are smaller in diameter. P deficiency is seldom observed in woody plants growing in the field because the large and deep root system helps plants uptake P from the subsoil where P is rich. Excess P can reduce plant growth because P can combine with Ca, Fe, and some micronutrients forming insoluble complexes.

6.2.1.3 Potassium (K)

The element K is involved in photosynthesis and respiration. It is required for stomata opening and closing and involved in modification of nutrient and water uptake and activation of enzymes. K is essential for protein synthesis and playing a role in increasing plant resistance to diseases and other stresses. Plants absorb K as K^+. K is one of the most abundant cations in plants. The plant root can effectively uptake K from the soil solution. K is then distributed into the aerial parts in the plant. Uptake of K is achieved by the K channel or secondary transporter by which K^+ is translocated from the root to the shoot or can be reversed from the shoot to the root via phloem. The cycling of K^+ in the plant provides a constant supply of cations to accompany anions (NO^{3-}) on their way to the shoot. The concentration of K in dry tissues is about 1 to 4%. K is mobile in plants; therefore, K deficiency symptoms are evident on the older leaves first with a marginal yellowing, followed by scorching and leaf drop. K deficiency is occasionally seen on woody plants with symptoms of necrotic spot, interveinal chlorosis and necrosis, or marginal necrosis on the old leaves. In some cases, crinkled and rolled-upward leaves and died-back shoots may be seen. Excess K will restrict plant growth. K can be easily leached by rain and irrigation water from the plant and recycled many times by a large tree.

6.2.1.4 Calcium (Ca)

The element Ca is a constituent of the cell wall and cell membrane and acts as a secondary messenger. It modifies the permeability of membranes and regulates nutrient uptake by roots and movement across membranes. Ca is involved in regulating certain enzyme systems (activating enzyme systems and neutralizing organic acids) in carbohydrate production and transportation. Ca plays an important role in root development and functioning. The ionic form absorbed by plants is Ca^{2+}. Ca enters the root through Ca^{2+} permeable channels, is stored in the vacuole, and translocated into other organs through xylem loading. The concentration of Ca in dry tissues is about 1%. Ca is not mobile in plants. Its deficiency symptoms occur first in the growing points, followed by the young leaf. The symptoms include yellow, necrotic, and distorted leaves and death of the growing points (bud or root). In woody species, low Ca has been linked to weak and/or broken stems. Ca deficiency in citrus is indicated by color fading at the leaf margin and the area between the main veins during the winter months. Ca deficiency restricts tree growth and decreased fruit production and produces poor quality fruits. Due to low mobility, fast-growing tissues suffer Ca^{2+} deficiency more frequently, resulting in bitter-pit symptoms in apple (Witney et al., 1991). Ca is rarely deficient in field soil because the subsoil contains sufficient calcium for plant growth. Excess Ca may raise the soil pH causing reduced plant growth.

6.2.1.5 Magnesium (Mg)

The element Mg is a key element of chlorophyll. Mg is involved in carbohydrate production and transportation. It acts as an enzyme activator regulating a few enzyme systems. Mg also acts as a regulator of the balance of cation and anion in cells. Together with K, Mg plays a role in regulating cell turgor. Plants absorb Mg as Mg^{2+}. The concentration of Mg in dry tissues is about 0.35 to 0.5%. Mg is very mobile in plants. Its deficiency symptoms can be seen on the older leaves with slight yellowing before dropping. In woody plants, Mg deficiency symptoms are most severe on the oldest leaves with a broad interveinal and/or marginal chlorosis. Excess Mg may cause the imbalance of Ca and K, resulting in reduced plant growth.

6.2.1.6 Sulfur (S)

The element S is a constituent of amino acids including cysteine and methionine. It is also a component of protoplasm. S promotes nodule formation in legume species and is known to help fixation of N in legumes; therefore, it plays a role in root growth and development. The ionic form that plants absorb is SO_4^{2-} (sulfate). Under certain conditions, such as flooding, the inorganic form SO_4^{2-} can be reduced to FeS, FeS_2 and H_2S, an organic form that cannot be transported in plants. Plants acquire sulfate through the specific sulfate transporters that are energized by the H^+ gradient. Sulfate can be translocated to the shoot via xylem and stored in vacuoles. The concentration of S in dry tissues is about 0.5%. Sulfate is immobile in plants. Its deficiency symptoms start with light green or yellow young leaves, then showing up on the older leaves with dark veins. In woody species, S deficiency is likely associated with severe stunting, reduced leaf size, reduced bud breaking, and less branching. Shoots are also shorter than normal ones. Sulfate deficiency is rare in field production because plants can obtain S from most complete fertilizers and air pollutants. Excess S could create a very low pH in some soils, influencing the plant's use of other nutrients.

6.2.2 Micronutrients

6.2.2.1 Iron (Fe)

The element Fe is involved in biosynthesis and functional maintenance of chlorophyll in plants. Fe is an important component in the ferredoxin/thioredoxin system and acts as an activator of many biochemical processes, such as the oxidation-reduction process. Accumulated evidence shows that iron functions as a co-factor involved in the formation of δ-aminolevulinic acid (ALA) that intermediates chlorophyll synthesis in higher plants. The ionic form absorbed by plants is Fe^{2+}. In the soil, Fe is predominant in the ferric oxidation state (Fe^{3+}) that is not available for plants to uptake. Dry plant tissues contain about 40 to 400 ppm. Fe is immobile in plants. The typical Fe deficiency symptom is interveinal chlorosis with the veins remaining green on young leaves. Many tree species including apple, citrus, banana, grape, pines, and poplars are susceptible to Fe deficiency. Symptoms on trees first occur on young leaves in the early growing stage and continue to develop on all leaves during the growing season. Shoot growth may be normal but shoots are smaller in diameter. In severe cases, leaves turn brown and gradually die, and eventually, branches or even an entire tree may die. Excess Fe may reduce Mn absorption.

6.2.2.2 Manganese (Mn)

The element Mn functions as an activator part in certain enzyme systems. Mn is involved in chlorophyll synthesis and required for CO_2 assimilation in photosynthesis and N metabolism. In plants, Mn is essential for phosphorus and Mg uptake. The ionic form absorbed by plants is Mn^{2+}. The concentration of Mn in dry tissue is about 25 to 100 ppm. Mn is immobile in plants. Deficiency symptoms appear on newly expanding leaves first by showing interveinal chlorosis. Shoot growth will be less than normal. As the deficiency progresses, new leaves become tiny. In some species such as *Ficus microcarpa,* new leaves are always uniformly chlorotic and defoliation may occur in severe cases. The availability of Mn is largely affected by the soil pH. When the pH increases >6.3, Mn becomes insoluble. Excess Mn may cause chlorosis, premature leaf abscission, and internal bark necrosis, especially in acid soils.

6.2.2.3 Zinc (Zn)

The element Zn an important component or regulator co-factor of enzymes and acts as an activator of a series of enzyme systems including oxidoreductases, transferases, hydrolases, lyases, isomerases, and ligases. Zn is essential for chlorophyll production, carbohydrate formation, and starch formation. It also influences plant hormone proteins and functions in cell elongation and seed development. Zn can be soluble or insoluble in the soil depending on the soil conditions including soil composition and soil pH (Kiekens, 1995). The ionic form of Zn absorbed by plants is Zn^{2+}. Zn is immobile in plants. Uptake of Zn by plants is a complex membrane-crossing process that is mediated by ZIP family transport proteins (Guerinot, 2000; Broadley et al., 2007; Suzuki et al., 2006). Zn deficiency is very common in high pH soil where $ZnOH^+$ and $Zn(OH)_2$ are dominant. Translocation of Zn from the root to the xylem is achieved either through the apoplast or symplast pathway. The concentration of Zn in dry tissues is about 15 to 60 ppm. Zn deficiency can disrupt photosynthesis and sugar transformation. Zn deficiency is uncommon in woody plants. Young leaves are affected first, showing smaller and narrower leaves than normal (little leaf) and rosetting leaves due to reduced internodes. Excess Zn may cause Fe deficiency and reduced Mn absorption.

6.2.2.4 Copper (Cu)

The element Cu is an important component of enzymes catalyzing several crucial pathways in photosynthesis, respiration, and protein and carbohydrates metabolism. Copper is involved in plant reproduction and plays a role in increasing sugar, intensifying color, and improving flavor of fruits and vegetables, and enhancing seed production.

Plants absorb Cu as Cu^{2+}. The concentration of Cu in dry tissue is about 6 to 40 ppm. Cu is immobile in plants. Deficiency symptoms can be evidenced on young leaves include yellowing of leaves and dieback of shoots. Although Cu deficiency in woody species is uncommon, young leaves are affected first with "wither tip" caused by dieback of the growing point. The stunted growth of shoots may result in a compact appearance of the tree. In severe cases, young shoots may die in the late spring or early summer. Excess Cu may cause Fe deficiency and reduced Mn and Mo absorption.

6.2.2.5 Boron (B)

The element B is essential for germination of pollen grains and formation of pollen tubes. B aids fruit setting and seed development. B regulates metabolism of carbohydrates and is also essential for cell division and maintaining the cell wall structure. It promotes maturity and is involved in translocation of Ca, sugars, and plant hormones. B can be available for plant uptake as BO_3^{2-} and $B_4O_7^{2-}$.

The concentration of B in dry tissue is about 25 to 50 ppm. B is immobile in plants. Symptoms of B deficiency appear on young leaves first. Deficiency is often seen in sandy and high pH soils. B deficiency is rare in woody species. Symptoms include small and dark green leaves and rosetted leaves caused by shortened internodes. In severe cases, chlorotic and necrotic leaves develop and the growing point may die. B deficiency may cause reduced fruit quality. Excess B may cause the same problems as B deficiency.

6.2.2.6 Chloride (Cl)

Chloride (Cl) is involved in photosynthesis, and protein and carbohydrate metabolism through the regulation of a series of enzymes. It acts as a counter anion balancing essential cations, such as K^+ and H^+ and plays a role in stabilizing membrane potential and regulating pH gradients across the membranes. Cl acts as an osmoticum in the vacuole balancing the osmotic pressure of the cell. Cl promotes the use of ammonium nitrogen and may be involved in fixation of N. Plants absorb Cl as Cl^- from soil solution, rainfall, and irrigation water. The concentration of Cl^- in the dry tissue is about 2000 to 20000 ppm and is much higher than other micronutrients. Cl is immobile in plants. Under Cl deficiency, interveinal chlorosis is first seen on young leaves. As the symptom progresses, curl or wilt leaf tips and scorched leaf margins occur. Overall plant growth is reduced. Cl can accumulate more in older tissues than in newly matured leaves. Excess Cl is toxic to plants. Woody plant Cl toxicity causes scorched leaves.

6.2.2.7 Molybdenum (Mo)

The element Mo plays a major role in the utilization of N and protein synthesis. Mo promotes the activity of nitrate reductase that reduces nitrate (NO^{3-}) to ammonium (NH^{4+}) in plants. Mo enhances the formation of legume nodules. It is also involved in the conversion of inorganic phosphates to organic forms. The ionic form plants absorb is MoO_4^{2-}. The concentration of Mo in the dry tissue is about 2 ppm. Mo is mobile in plants. Symptoms of Mo deficiency are similar to N deficiency. In severe cases, Mo deficiency may cause stunted growth, and marginal and interveinal chlorosis, although very few areas of the world are deficient in Mo. Excess Mo may encourage Fe deficiency.

6.2.2.8 Nickel (Ni)

The element Ni was classified as an essential nutrient in 1989. Ni is a component of the ureases in plants. The enzyme urease catalyzes urea N into ammonia that is usable by plants. Plant urases also promote N in legumes and improve disease tolerance in plants. Plants absorb Ni as Ni^+. The concentration of Ni in dry tissue is about 10 to 25 ppm. Ni is a mobile element in plants. Symptoms of Ni deficiency include distortion of foliage and small leaves that typically shows up on old leaves. Ni deficiency can cause reduced growth and yield. In woody plants, deficiency symptoms of Ni can be seen on the new growth as a rosetting appearance caused by shortened internodes. Deficiency of Ni may also result in reduced shoot growth or death of shoots and branches. A typical symptom of Ni in pecan is a mouse-ear that can be corrected by one-time spraying of Ni in the fall (Wood et al., 2004).

6.3 PHYSIOLOGICAL AND MOLECULAR RESPONSES OF WOODY PLANTS TO NUTRIENT DEFICIENCY

6.3.1 Physiological Responses to Nutrient Deficiency

Deficiency of one or multiple nutrients can cause a series of physical changes in plants. These changes may impact plant growth and development at any stage, from seed germination, vegetative growth, flower initiation, pollination, and fruit setting and ripening, and could eventually affect yield and quality. Plants have developed their abilities to alleviate and correct nutrient deficiency through various mechanisms (Maathuis, 2009). One of the most common strategies is to increase the total absorptive surface of the root system because the majority of nutrients are absorbed by the root system. Plants are able to develop a root system through modifying root morphology and physiology. Such a root system helps plants adapt to specific soil conditions, such as compact texture, high or low pH, drought, and waterlogging, to combat the nutrient deficiency. In *Arabidopsis*, N deficiency increased primary and lateral root length, P deficiency produced a shallower root system with more branches, and K deficiency generated shorter roots (Gruber et al., 2013). In woody plants, significantly increased branches and root/shoot ratio were observed under N and/or P deficiency (Tranbarger et al., 2003; Trubat et al., 2006; Afrousheh et al., 2010). In poplars, deficiency of P increased the activity level of acid phosphatases, phosphoenolpyruvate carboxylase, and malate dehydrogenase, which are all involved in P mobilization (Gan et al., 2016). Low N in plants caused reduced levels of amino acids and decreased the activities of nitrate reductase and glutamate synthase, which are related to N assimilation (Gan et al., 2016). K deficiency resulted in early chlorophyll degradation and decreased photosynthesis in mature leaves of cotton plants (Zhao et al., 2001). It is interesting that a differential response to nutrient deficiency was observed between male and female Manchurian poplar (*Populus cathayana*) (Zhang et al., 2014). The male trees suffering N and K deficiencies had higher photosynthesis, higher activities of glutamate dehydrogenase and peroxidase, and lower accumulation of plastoglobules (thylakoid-associated lipid droplets) in chloroplasts than the females, indicating that gender might play a role in response to N and P deficiency in *P. cathayana*. Similarly, Yang et al. (2015) reported that under K deficiency conditions, male poplars performed better than females by maintaining a higher K^+ content and K^+/Na^+ ratio. Mg deficiency disrupts the loading of sucrose into phloem, causing impaired carbon metabolism and decline of chlorophyll and carbon fixation rate. In a pine species (*Pinus radiate*), Mn deficiency reduced photosynthesis associated with closure of the stomata and reductions in the residual conductance; therefore suppressing plant growth by 15 to 25% in height (Laing et al., 2000). The physiological response of woody plants to S deficiency is rarely documented. Gilbert et al. (1997) reported that sulfate deprivation in young wheat plants has an early effect on CO_2 assimilation, Rubisco enzyme activity, and protein abundance. S deficiency also depressed root hydraulic conductivity.

Research on physiological responses of micronutrients in woody plants mainly focused on Fe and Zn. Under Fe deficiency, the activity of Fe^{3+}-chelate reductase in leaf mesophyll protoplasts is decreased. A decrease of usable Fe for chlorophyll formation could change the structure and composition of photosynthetic membranes in chloroplasts, consequently leading to chlorotic leaves. In poplar trees, the content of chlorophyll and carotenoid was significantly reduced under iron deficiency (Huang and Dai, 2015a,b). In several plant species, the number of thylakoid membranes and other photosynthetic units, such as grana and stroma lamellae deceased, resulting in a reduced efficiency of photosynthesis system II (PS II) under Fe deficiency (Spiller and Terry, 1980; Andaluz et al., 2006). Such structural and compositional changes impair the electron transport chain and the light harvesting system; therefore, the entire photosynthetic system is dramatically inhibited. Plants could employ various mechanisms to alleviate the stress caused by nutrient deficiency. For example, synthesis of ethylene and auxin was induced by Fe deficiency in plants (Landsberg, 1996; Romera et al., 1999). Plants responded to Fe deficiency by changing their morphology, particularly in the root (Romera and Alcántara, 1994; Landsberg, 1996; Schmidt and Bartels, 1996; Schmidt et al., 2000). Differences in response to nutrient deficiency were also found in different plant species. In non-graminaceous plants (grass species), Fe deficiency induces the release of reducing and chelating substances (reductants and chelators) from the root system and increase proton excretion in the rhizosphere, which enhance the reduction of Fe^{3+} to Fe^{2+}, facilitating the transport of Fe^{2+} across the plasma membranes. In woody plants (most of them are non-graminaceous), such as quince, pear, and olive trees, the stress of Fe deficiency resulted in an increased Fe^{3+} reducing capacity in the root system although the chlorophyll content in the leaf decreased (de la Guardia and Alcantara, 2002; Huang and Dai, 2015a,b). Similar events were observed in European aspen in which the contents of chlorophyll and carotenoids and the ratio of chlorophyll a to b were significantly decreased under Fe starvation (Huang and Dai, 2015a,b).

Under Zn deficiency, decreases in carbonic anhydrase activity, photochemical activity of chloroplasts, chlorophyll content, and chloroplast structure were reported in crops (Sharma et al., 1982; Ohki, 2006). By controlling generation and detoxification of reactive oxygen species (ROS), Zn deficiency interrupts the maintenance of integrity and function of cellular membranes and increases the leakage of organic compounds (Cakmak and Marschner, 1988).

Much research has focused on how plants uptake nutrients from the soil and transport between cells within the plant and how the deficiency of individual mineral elements affect primary photosynthesis, respiration, carbohydrate partitioning, and other biological pathways in the plant. Some research also elucidated how nutrient deficiency alters allocation of carbohydrates in the plant. Hermans et al. (2006) provided a review on how plants respond to the deficiency of the major nutrients (N, P, K, and Mg). Figure 6.1 illustrates a model of plant response to nutrient shortage by allocating biomass to specific organs. Deficiencies of N and P improve the ability of plants to acquire these elements, stimulate carbohydrate accumulation in the leaf, and allocate more carbon to the root; therefore, increasing the ratio of root to shoot (R:S) and consequently altering root morphology. Similar responses of plants to P deficiency were also reported by Sánchez-Calderón et al. (2006) and Niu et al. (2013). However, deficiencies of Mg and K induce sugar accumulation in the leaf, but not in the root because phloem transport is impaired in K- and Mg-deficient plants (Hermans et al., 2006, and references therein). Mechanisms of differential responses of plants to the deficiency of different nutrients have not been well explained. Recent advances in plant physiology, molecular biology, biotechnology, and relevant omics-technologies will certainly facilitate a better picture of how plants respond to nutrient deficiency.

6.3.2 Molecular Responses to Nutrient Deficiency

Molecular responses of plants to nutrient deficiency have been well researched in herbaceous plants, mainly in *Arabidopsis* and a few crop species. Such research covers a wide range of biological events responsible for acquisition, transport, and storage of various mineral elements and their impacts on photosynthesis, respiration, and partitioning of carbohydrates in plants. Recently, efforts have been made to understand how plants sense and signal the shortage of nutrients to improve or enhance their abilities to uptake and transport nutrients and to keep a healthy homeostasis in plant cells and tissues; however, considerable interest still focuses on the genes that regulate transporters of mineral elements. For example, in *Arabidopsis*, under the N deficiency condition, transport of ammonium (NH^+) is mediated by the transporters in the AMT/MEP/Rh (AMT) superfamily (Ludewig et al., 2007), while the transport of nitrate (NO_3^-) is mediated by the transporters in two other families, NPF (Nitrate Transporter 1/Peptide Transporter family, previously the NRT1/PTR family) and NRT2. Six AMT genes, 53 members in the NPF family and seven members in the NRT2 family have been identified. Research confirmed that under N limitation, five AMT genes were up-regulated, of which three *AtAMT* genes expressed mainly in root tips and epidermal cells, playing a role in direct absorption of NH^+ from the soil, and one *AtAMT* gene located in endodermis and cortex transporting NH^+ into cells through apoplastic pathway. Similarly, two *AtNRT2* genes that responded for nitrate uptake from the soil and one *AtNRT2* gene mediated the apoplastic absorption of NO_3^- are important for the sustaining growth of the N-deprived plants. Loss of function of these genes significantly reduced the growth of seedlings or adult plants. In *Populus trichocarpa*, 14 putative AMT genes have been annotated (Couturier et al., 2007). Five AMT genes encoding the functional ammonium transporters were confirmed by restoring ammonium absorption in the yeast strain that is defective in the endogenous ammonium transporters, and two AMT genes responsed positively to N deficiency. The orthologous *NRT2* gene expressed at a higher level in nitrate-fed *P.* × *canescens* than in ammonium-fed plants (Ehlting et al., 2007). In another two poplar species, *P. simonii* and *P.* × *euramericana,* the deficiency of P and N increased the expression levels

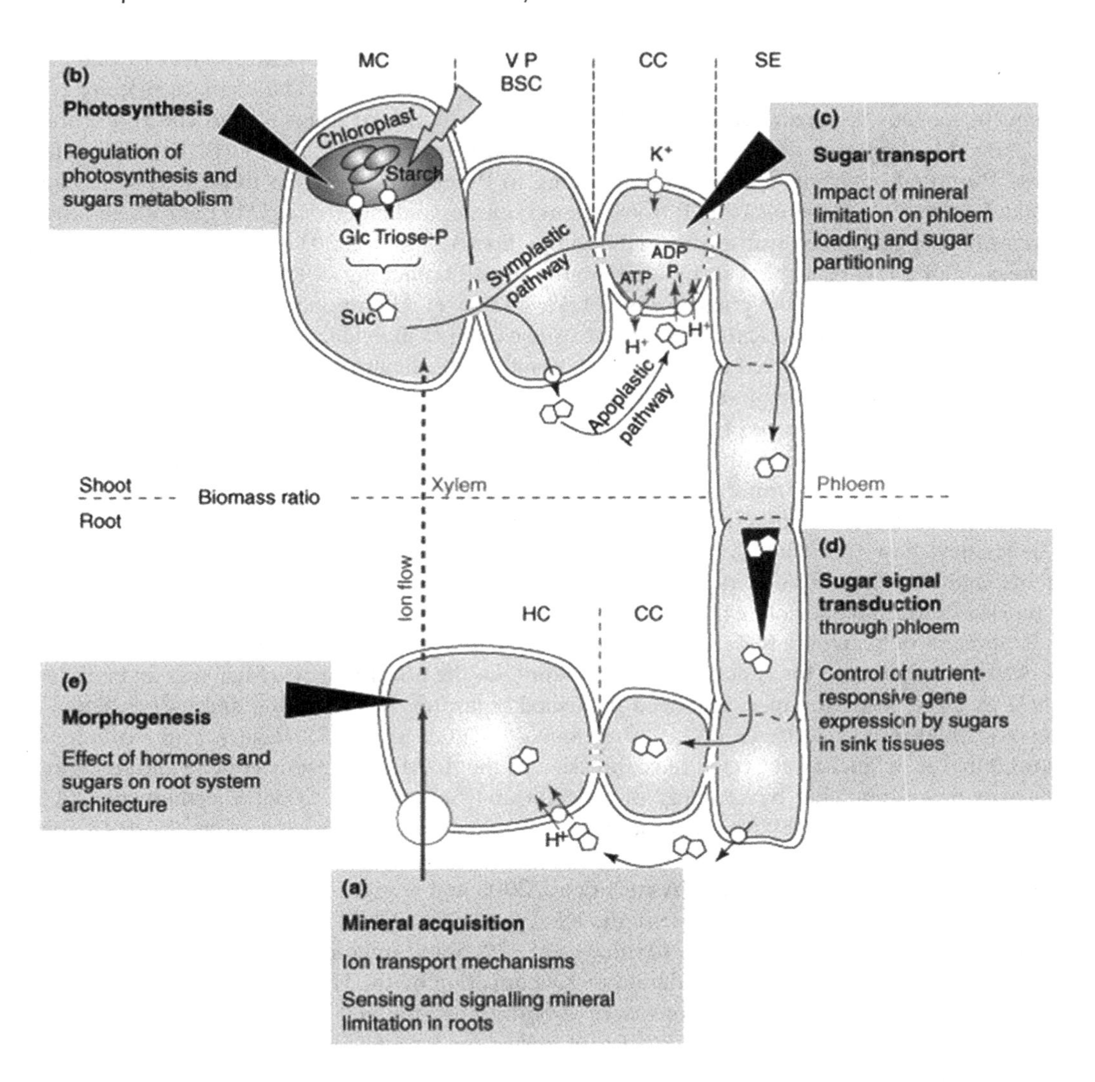

FIGURE 6.1 Plant responses to nutrient shortage by allocating biomass to specific organs. MC: mesophyll cell; VP: vascular parenchyma; BSC: bundle sheath cells; CC: companion cell; SE: sieve element; *and* HC: heterotrophic cell (Reprinted with permission from Hermans et al., 2006. *Trends in Plant Sci.*11: 610–617. https://doi.org/10.1016/j.tplants.2006.10.007).

of *PHT1;5*, *PHT1;9*, *PHT2;1*, *AMT2;1* and *NR* in the root, while decreasing the expression levels of *PHT1;9*, *PHO1;H1*, *PHO2*, *AMT1;1* and *NRT2;1* in the leaf (Gan et al., 2016). In peach (*Prunus persica*), two putative *NRT* transporters (*PpNrt2.1 and PpNrt2.2*) were identified (Nakamura et al., 2007). Application of nitrate to the root of N-deprived seedlings quickly induced the expression of *PpNrt2.1* within 0.5 h, but gradually induced the expression of *PpNrt2.2*; however, application of ammonium to the root of N-deprived seedlings repressed the expression of the *PpNrt2.1* and *PpNrt2.2* genes (Nakamura et al., 2007).

Phosphorus deficiency is not common in woody plants, particularly in tree species, because perennial woody plants usually have a large and deep root system can uptake P from the P-enriched subsoil. However, occasional P deficiency occurs in woody plants grown in soils rich in various cations, such as Ca^{2+} and Al^{3+}, because P can combine with these cations forming insoluble complexes. Plants have developed various mechanisms to uptake P under the P deficiency condition (Vance et al., 2003; Lambers et al., 2010; Aziz et al., 2014). Responses of plants to phosphorus deficiency are involved in a series of biological reactions and changes that are controlled by various genes and gene regulators to regulate P uptake from the soil and transport in the plant. Manipulation of these genes could improve the P-use-efficiency (PUE) of the plant in the P-limited environment. Genes responding to P deficiency include those that help plants uptake P from the soil by releasing P from organic sources and mobilizing P in the soil, and those responsible for phosphate transporters (PTs) in plants (Aziz et al., 2014 and references therein). Expression of a few genes that stimulate the growth and branching of the root system is induced by P deficiency, such as the phosphorus uptake 1 (*Pup 1*). Hammond et al. (2004) provided a review on how plants genetically responded to phosphorus deficiency. The review summarized that various genes are involved in responses of plants to P deficiency, including genes for P acquisition, tissue P economy, and various transcription factors including several cis-regulatory elements in the promoter regions of the genes. A recent review by Ha and Tran (2014) discussed plant responses to P deficiency and the molecular mechanisms regulating these responses. The review provides details on how transcription factors (TF), such as the phosphate starvation response 1 (PHR1), basic helix-loop-helix (*bHLH*), and C2H2 zinc finger TF, regulate plant response to P deficiency; how the PHO1 family and related proteins regulate the P transfer from roots to shoots; and how the purple acid phosphatases (PAPs) catalyze hydrolysis of phosphomonoesters to release Pi available for plants.

Responses of plants to K deficiency can be regulated by K^+ transporters and channels, two mechanisms plants use for K acquisition depending on the affinity to K^+ (Maathuis and Sanders, 1997). A large body of papers has been published regarding K acquisition and its regulation in plants (Ashley et al., 2005; Li et al., 2018; Santa-Maria et al., 2018). Research in *Arabidopsis* showed that K deficiency activated the expression of the *AtHAK5* transporter gene and its expression was down-regulated by resupply of K (Shin and Schachtman, 2004; Gierth et al., 2005); however, Rubio et al. (2000) reported that K^+-starvation decreased the expression of *AtHAK5* that might be regulated by NH_4^+. The response of the *HAK* genes in other species including barley, tomato, and rice was also reported (Ashley et al., 2005 and references therein; Yang et al., 2014). Other transporter genes, such as those in the KEA family and CHX family also demonstrated the responses to K^+ deficiency in plants (Ashley et al., 2005 and references therein). Li et al. (2018) reviewed the recent progress in regulation of K acquisition by the HAK/KUP/KT transporters in plants, indicating that such regulations were at both transcriptional and post-translational levels. They summarized that the HAK/KUP/KT transporter genes regulated plant K acquisition through a signal cascade and K^+ starvation can trigger a series of reactions including increases in ethylene, ABA, and ROS and changes in the membrane potential of root cells. In woody species, Han et al. (2018) reported on 111 and 181 genes differentially expressed in the male and female poplar plant (*Populus cathayana*), responding to K^+ deficiency. These genes are involved in a broad spectrum of biological pathways, such as photosynthesis, secondary metabolism, signal transduction, call wall formation, and plant responses to stresses, as well as regulation of transmembrane transport in plant cells.

Plant responses to deficiency of micronutrients are well documented (Hansch and Mendel, 2009; Barker and Pilbeam, 2006). In woody species, research has mainly focused on Fe deficiency; therefore, plant response to Fe deficiency will be thoroughly discussed in this chapter.

6.3.3 Response to Fe Deficiency in Woody Plants

Fe chlorosis is one of the major problems in calcareous soils, especially in arid and semi-arid regions. Worldwide, about 30% of the cultivated soils are calcareous in which Fe chlorosis limits agricultural production (Mori, 1999). Many tree species including apple, citrus, banana, grape, pines, and poplars are susceptible to Fe deficiency (Tagliavini and Rombola, 2001). The primary symptom of Fe chlorosis is interveinal chlorosis, resulting in a bright yellow leaf with a network of a green veins. Symptoms first occur on the young leaves in the early growing season and continue to develop on all leaves during the growing season. In severe cases, leaves turn brown and gradually die, and eventually, branches, or even the entire tree may die.

Considerable research has been done to control Fe chlorosis in plants. Current practices for correcting Fe chlorosis focus on 1) using species or cultivars tolerant to Fe deficiency; 2) creating ideal soil/root environments to improve Fe availability; 3) applying chelated Fe to the soil and/or to the plant directly. Lack of relevant germplasm and long breeding time make it difficult to develop cultivars tolerant to Fe deficiency *via* traditional breeding, especially for woody species. Creating an ideal soil/root environment, such as acidifying the soil and applying organic compounds to the soil to make Fe more available is not always practical and desirable for large crop fields and landscapes. Soil application of Fe elements has variable results, relatively high cost and sometimes slow response by the plants. Foliar spray and trunk injection may cause temporary leaf and trunk injury and a short-lasting effect. Compared to studies on annual crop species, limited research has been done for correcting tree chlorosis. So far, there is no satisfactory method to control tree chlorosis throughout the entire life span of trees either in natural stands or in urban forestry.

6.3.3.1 Iron Uptake, Transport, and Metabolism in Plants

Fe is an essential nutrient for plants, involved in biosynthesis, and the structure and functional maintenance of chlorophyll (Abadia, 1992). Fe is found as a component in many plant proteins in which the Fe-S proteins and the heme proteins are two major Fe-containing families. The ferredoxin/thioredoxin system plays a key role in regulating many pathways in the chloroplast (Malkin and Niyogi, 2000). The heme proteins known as the cytochromes are mainly involved in biological electron transfer in both respiration and photosynthesis processes. Cytochromes are also found to be involved in other plant processes, such as nitrate reduction and nitrogen fixation (Malkin and Niyogi, 2000).

It has been widely accepted that Fe deficiency causes a decrease in chlorophyll content and alteration in chlorophyll structure, resulting in plant chlorosis. Accumulated evidence shows that Fe plays an important role as a co-factor in chlorophyll formation *via* involvement in the pathway of formation of ALA, a biosynthetic intermediate in chlorophyll synthesis in higher plants (Pushnik et al., 1984). Studies showed that ALA formation was strongly inhibited by Fe-chelators, reflecting the requirement for Fe in ALA synthesis.

The structure and composition of photosynthetic membranes in chloroplasts changed under conditions of Fe deficiency. In several plant species, the number of thylakoid membranes and other photosynthetic units such as granal and stromal lamellae decreased (Spiller and Terry, 1980), and the efficiency of photosynthesis system II (PS II) lowered when they suffered Fe deficiency (Belkhodja et al., 1998; Bertamini et al., 2002). These structural and compositional changes dramatically affected the entire photosynthetic system of higher plants by impairing the electron transport chain and the light harvesting system (Abadia et al., 2000; Andaluz et al., 2006).

6.3.3.2 Iron Availability in the Soil

In many soils, the quantity of Fe is sufficient; however, most Fe is immobile and insoluble and cannot be utilized by plants. In calcareous or alkaline soils, Fe predominantly exists in the ferric form (Fe^{3+}), mainly in the inorganic state where little is available for plant uptake (Guerinot and Yi, 1994). The low Fe availability is primarily caused by the low solubility of Fe-oxides that is largely controlled by the pH of soil solution (Uren, 1984). Reduction of Fe^{3+} to Fe^{2+} increases the solubility of Fe, but further studies found that Fe^{2+} is not stable in the aerobic environment suitable for healthy growth of most plants. Therefore, the major concern is not only how much Fe^{3+} is reduced but also how long reduced Fe^{2+} is available to plants. These findings resulted in the conclusion that the major role of Fe nutrition is played by Fe forms available to plant roots rather than the amount of inorganic Fe (Guerinot and Yi, 1994; Briat et al., 1995).

6.3.3.3 Plant Responses to Iron Deficiency

The mechanism of Fe acquisition by higher plants can be distinctly classified into two strategies (Romheld and Marschner, 1986; Romheld, 1987; Mori, 1999). Non-graminaceous higher plants, known as Strategy I plants, acquire Fe from the soil through the co-ordinated action of H^+-ATPases, membrane-bound Fe^{3+}-chelate reductases, and Fe^{2+}-specific cation transporters (Guerinot and Yi, 1994). Under Fe deficiency, strategy I plants induce a number of mechanisms to alleviate the stress. These mechanisms include release of H^+, release of reducing and chelating substances (reductants and chelators), and enhancement of proton excretion in the rhizosphere to increase their ferric reduction capacity at the root surface to reduce Fe^{3+} to Fe^{2+} and transport Fe (II) through plasma membrane (Santi et al., 2005; Santi and Schmidt, 2008). In *Arabidopsis*, H^+-ATPase (AHA) plays a role in Fe-deficiency response with its activity being found in Fe-deficient roots (Colangelo and Guerinot, 2004). Evidence showed that increased reductase activities under Fe deficiency stresses enhanced reduction of Fe^{3+} to soluble Fe^{2+} and consequently increased iron uptake by plants (Romheld, 1987). In strategy II species (graminaceous species), plants produce iron-chelating mugineic acids (MAs) that belong to phytosiderophores (PSs) to the root surface to chelate Fe^{3+} under iron deficiency. The complex of Fe^{3+}-PSs is then transported into roots via YS1 PS (Yellow Stripe family) transporter (Curie et al., 2009). It is interesting that Fe^{2+} could be taken up in addition to Fe^{3+}-PSs complexes in rice without H^+-ATPase and Fe^{3+}-chelate reductase activity, which was often found in Strategy I plants (Ishimaru et al., 2006; Cheng et al., 2007).

The iron reduction has been well characterized (Schmidt, 1999). An externally oriented reductase has been found to be responsible for the reduction of ferric chelates. Localized on the root-cell plasma membrane, the reductase is responsible for transferring electrons for cytoplasmic reducing equivalents to extracellular ferric chelates (Bienfait et al., 1983; Buckhout et al., 1989; Holden et al., 1991; Schmidt, 1999). Evidence showed that the activity of ferric reductase in root tips increased under Fe deficiency, leading to an intensified reduction of Fe^{3+} to Fe^{2+} (Romheld and Marschner, 1986; Tagliavini and Rombola, 2001). A positive correlation between Fe^{3+} reduction capacity (RC) and the resistance to Fe deficiency was found in some annual plants, such as soybean (Jolley et al., 1992), dry bean (Ellsworth et al., 1997), and some woody species (de la Guardia and Alcantara, 2002).

The process of transportation of Fe from roots to leaves is involved in another reduction/oxidation process. Predominant Fe^{3+} state in plants indicates that the reduced Fe^{2+} was oxidized to the Fe^{3+} state before it was transported to aerial plant parts *via* the xylem. In this process, citrate plays a role as a long distance transporter (Fe^{3+}-citrate complex) (Olson et al., 1981). However, the Fe^{3+}-citrate compound has to be reduced to Fe^{2+} state before Fe can cross the leaf plasma membrane. Again, the Fe^{3+}-chelate reductase is involved in the process. Researchers observed that an increase of apoplastic pH induced by nitrate inhibited the activity of Fe^{3+}-chelate reductase, causing the decrease of Fe uptake rate into the cell (Kosegarten et al., 1999, 2001; Nikolic and Romheld, 1999).

6.3.3.4 Iron Homeostasis Regulation Genes in Plants

Molecular mechanisms of plants in response to Fe deficiency involve the regulation of ferric reductase (*FRO* genes) for uptake, iron transporters (*IRT* genes) for transport, and their regulators (*bHLH* genes). In general, genes in the *FRO* family encode Fe^{3+}-chelate reductase, reducing Fe^{3+} to soluble Fe^{2+}. Genes in the *IRT* family belonging to the ZIP (ZRT, IRT-like protein) family encode the Fe transporter proteins that are essential for iron uptake and transport in plants. The basic helix-loop-helix (*bHLH*) genes encode proteins, playing an important role in regulating these genes by binding to specific DNA sequences affecting the rate of transcription.

6.3.3.4.1 The FRO Gene Family

The *FRO* gene encodes the Fe^{3+}-chelate reductase, reducing Fe^{3+} to the soluble form Fe^{2+}. As a member of the flavocytochromes family, this enzyme has a function of transporting electrons across membranes (Robinson et al., 1999). The activity of ferric-chelate reductase at the plasma membrane of root epidermal cells increases when Fe is deficient (Bienfait et al., 1983). The *FRO2* is expressed in the epidermal cells of Fe-deficient roots and is thought to be the main Fe^{3+} chelate reductase in roots. A loss-of-function mutant (*frd1*) of *FRO2* showed a defect in Fe uptake when Fe was supplied as a Fe^{3+}-chelate and the plant was extremely chlorotic when grown on Fe-deficient medium (Connolly et al., 2003). The *AtFRO2* gene isolated from *Arabidopsis* is expressed in Fe-deficient roots. The activity of Fe^{3+}-chelate reductase decreased in loss-of-function mutations of *AtFRO2*, causing plant chlorosis and poor growth. Further evidence confirmed that low Fe-inducible ferric chelate reduction was caused by the decreased *AtFRO2* gene products (Robinson et al., 1999). Over-expression of *AtFRO2* in soybean resulted in enhanced root ferric reductase activity and tolerance to Fe-deficiency induced chlorosis (Vasconcelos et al., 2006).

The *PsFRO1* gene isolated from pea also encodes a Fe^{3+}-chelate reductase (Waters et al., 2002). Expression of *PsFRO1* in plants correlated with Fe^{3+}-chelate reductase activity. The *LeFRO1* encoding a Fe^{3+}-chelate reductase protein was isolated from tomato (Li et al., 2004). Expression of *LeFRO1* in yeast increased Fe^{3+}-chelate reductase activity. *LeFRO1* protein was targeted on the plasma membrane according to transient expression analysis. The results of RT-PCR indicated that the *LeFRO1* gene could express in roots, leaves, cotyledons, flowers, and young fruits. The expression level of *LeFRO1* in roots was highly related to Fe status. *LeFRO1* localizes on the plasma membrane in onion epidermal cells and confers Fe^{3+} reductase activity when expressed in yeast. In addition to *FRO2*, seven other *FRO* genes (*FRO1*, *FRO3*, *FRO4*, *FRO5*, *FRO6*, *FRO7*, and *FRO8*) were identified via genome annotation in *Arabidopsis* (Mukherjee et al., 2005). The expression of *FRO* genes in various locations in a plant suggests that different sets of *FRO* proteins are participating in Fe uptake in different plant tissues. The *FRO2*, *FRO3*, and *FRO5* genes are expressed in roots while *FRO3* is predominantly expressed in the vascular cylinder of roots, suggesting a role in Fe re-absorption from the root apoplast, while the *FRO6*, *FRO7*, and *FRO8* genes are shoot-specific. Similar tissue-specific expression of *FRO* genes was also reported by Mukherjee et al. (2005). Mukherjee et al. (2005) also found that the promoter of *FRO6* has multiple light-responsive elements. When exposed to light, a *FRO6* promoter-driven reporter gene was activated (Feng et al., 2006). Jain et al. (2014) reviewed the roles of *FRO* family metalloreductases in iron and copper homeostasis. They reported that in addition to the reduction function, *FRO* genes, such as *FRO7*, were involved in subcellular compartmentalization/distribution of Fe in the chloroplast (*FRO7*) and mitochondrion (*FRO3* and *FRO8*), as well as playing a role in photosynthesis. Muhammad et al. (2018) recently reported that rice *FRO* genes (*OsFRO1* and *OsFRO7*) showed a tissue-specific expression and their expression was regulated by several abiotic stresses, such as drought, heat, salt, and metals, as well as a change in phytohormones. Limited research has been reported regarding the response of the *FRO* genes to Fe deficiency in woody species. Huang and Dai (2013) cloned three putative *FRO* genes. The expression of the three genes was decreased in response to Fe deficiency in an Fe deficient susceptible line of *Populus tremula*.

6.3.3.4.2 The IRT Gene Family

The *IRT* gene encoding the Fe transporters protein is essential for Fe transport in plants (Vert et al., 2001, 2002). The *IRT1* and *IRT2* genes isolated from *Arabidopsis* are members of the ZIP (Zrt-, Irt-like proteins) metal transporter family. The *IRT1* gene was identified by functional complementation of an Fe uptake mutant of yeast. The *IRT2* gene was most similar in amino acid sequence to *IRT1*; however, *irt2* mutants showed no symptoms of Fe-deficiency and over-expression of *IRT2* did not appear to substitute for the loss of *IRT1*; thus, *IRT2* must play a different role than *IRT1*as an Fe transporter. A research study confirmed that *IRT2* was not involved in Fe absorption from the soil, but it was involved in intracellular Fe trafficking, playing a role in co-regulation with *FRO2* and *IRT1* (Vert et al., 2009). Further studies showed that as a divalent cation transporter, expression of the *IRT1* and *IRT2* genes were specifically detected in the root of the Fe deficient plants. The expression of the *IRT* genes was in response to Fe deficiency. Under the Fe deficiency condition, the level of mRNA increased, which enhanced the transportation of reduced Fes from the root surface into root cells through the plasma membrane. It was proved that *IRT1* expressed in the external cell layers of the root, specifically in response to Fe starvation. An *IRT1* knock-out *Arabidopsis* mutation (chlorotic) had a severe growth defect in the soil and resupply of soluble iron corrected the symptom of Fe deficiency. Expression of the *IRT2* gene was enhanced in roots as an altered response to Fe iron deficiency (Henriques et al., 2002). The *IRT* genes (*LeIRT1* and *LeIRT2*) cloned from tomato specifically expressed in roots; however, unlike in *Arabidopsis*, *LeIRT1* and *LeIRT2* expressed in both Fe sufficient and Fe deficient root, and Fe deficiency seemed to induce the expression of *LeIRT1* (Eckhardt et al., 2001). In woody plants, Qu et al. (2005) tested the tolerance of Fe deficiency in the *LeIRT2*-transgenic plants of *Malus robusta*. The result showed that the transgenic plants had a higher chlorophyll content in the leaf and a greater net photosynthetic rate than the control that even showed a severe leaf chlorosis. Another research study showed that the expression of *MxIRT1* was strongly enhanced in the root of *Malus xiaojinensis* under the Fe limited condition (Li et al., 2006). A similar result was found in a *Populus* species in which the expression of *PtIRT1* was dramatically increased under Fe deficiency (Huang and Dai, 2015b). Besides, *MxIRT1* and *PtIRT1*could complement Fe uptake activity in a yeast (*Saccharoomyces cerevisiae*) mutant strain DEY1453 (*fet3fet4*). A recent research found that 10 of 21 ZIP genes were induced by the deficiency of Cu, Zn, Mn, or Fe in *Populus trichocarpa* plants (Zhang et al., 2017), indicating that transporters from the ZIP family members are capable of transporting other metals including Zn, Mn, Zn, and Cd in plants.

6.3.3.4.3 Transcription Factors in Fe Homeostasis

It has been recognized that a small group of transcription factors regulate/activate expression of a series of genes that are closely related to environmental stresses, such as Fe deficiency, salinity, and drought, etc. Transcription factors that bind to promoter regulatory elements are often inducible and respond to various stress stimuli. Genes in the FRO and IRT families are regulated by a number of Fe-deficiency inducible transcription factors, such as *FER* (Ling et al., 2002), *FIT* (Colangelo and Guerinot, 2004; Jakoby et al., 2004; Bauer et al., 2007), *bHLH* (Yuan et al., 2005; Walker and Connolly, 2008; Wang et al., 2007; Yuan et al., 2008; Zheng et al., 2010), *IDEF* (Kobayashi et al., 2007; Ogo et al., 2008), *PYE* (Long et al., 2010), and *BTS* (Long et al., 2010). Additional genes that are involved in Fe homeostasis in plants have been recently summarized (Connorton et al., 2017 and references therein).

Several *bHLH* genes related to Fe deficiency were characterized in woody plants. In *Malus xiaojinesis*, three *bHLH* genes, *MxbHLH01*, *MxIRO2* and *MxFI*, were characterized. The *MxbHLH01* was specifically expressed in the root and its expression was up-regulated by Fe deficiency (Xu et al., 2011). The *MxIRO2* gene was induced by Fe deficiency in both the root and leaf and it might be interacting with other transcription factors to control the expression of genes related to Fe absorption (Yin et al., 2013). The *MxFIT* gene was up-regulated in the root under Fe deficiency, while almost no expression was detected in the leaf irrespective of Fe supply (Yin et al., 2014). Research also found that transgenic *Arabidopsis* plants with *MxFIT* showed a stronger resistance to Fe deficiency with an increase of both *AtIRT1* and *AtFRO2* transcripts in the root under the Fe deficient condition. In *Populus*, the *PtFIT* was up-regulated in the root under the Fe deficient condition (Huang and Dai, 2015a). One *PtFIT* transgenic *Populus* line accumulated more Fe under the Fe sufficient condition. Under Fe deficiency, the same line showed an enhanced Fe deficiency tolerance with a higher chlorophyll level and a greater ratio of Chla to Chlb than the control plant (Huang and Dai, 2015a).

6.4 MANAGEMENT OF MINERAL NUTRIENTS DEFICIENCY IN WOODY PLANTS

Considerable research has been done to manage nutrition deficiency in plant species. Methods for correcting nutrient deficiency include (1) field and foliage application of fertilizers; (2) creating ideal soil/root environments to improve nutrient availability; (3) using species or cultivars tolerant to nutrient deficiency. As discussed in this chapter, response of plants to nutrition deficiency is a complex physio-biological process that is regulated by the plant itself and the environmental conditions around the plant. These environmental conditions include the soil conditions such as soil structure, soil moisture, the level and type of minerals, organic matter, soil biota, and soil pH. Any activities that can improve soil conditions such as crop rotation, use of organic materials, and adjusting soil pH, will enhance the availability of nutrients in the soil for the plant. For example, research showed that the availability of most micronutrients increased as the soil pH decreased with the exception of Mo (Narwal et al., 1983; Wang et al., 2006); therefore, acidifying the soil and applying organic compounds to the soil that create an ideal soil/root environment can make more multiple nutrients available to the plant. Application of fertilizer to the soil and/or the plant foliage is still a dominant method to manage nutrient deficiency, particularly in crop production. Fertilization can be used to quickly and temporarily correct the deficiency of certain nutrients, particularly in so-called starving soil or for a specific nutrient.

Foliar application of nutrients seems more efficient than soil application, particularly for seasonal nutrient deficiency in woody species, due to the fact that woody plants have a large body and deep root system and plants may take time to uptake and transport nutrient(s) to the growing point. For example, foliar spraying of $FeSO_4$ on apple trees in the early growing season can quickly correct Fe chlorosis that is most severe on new leaves (Tagliavini et al., 2000). Foliar application can address the issue that some micronutrients are easily fixed on the surface of clay particles and might be chelated with other elements, decreasing their availability in the soil. The balance of individual elements in the soil and the plant should be considered due to the ion interaction. For instance, high N with the absence of other minerals may cause plant stress through an increased requirement for other nutrients for its rapid growth. A high level of P in the soil can sometimes enhance the symptoms of Zn deficiency because of the co-precipitation of insoluble $ZnO_3\ (PO_4)_2$ (Foth and Ellis, 1997). Loneragan and Webb (1993) found that cations, such as Ca, K, and Mg, decrease the plant absorption of Zn.

There are significant challenges in managing nutrient deficiency in woody plants due to their perennial nature, large body, and deep root system. For example, soil application of fertilizers has variable results, such as slow response and no response, due to the unknown conditions in the deep soil where only the deep roots of woody species can reach. Soil fertilization of trees is also difficult in urban areas for various reasons, mainly a limited area of soil exposure. It seems that foliar spray and trunk injection may be applicable, but their effect of correcting nutrient deficiency is only limited to a short time and several elements, such as Fe. Such applications should be repeated every year or every other year. Environmental conditions, particularly temperature and air humidity during the spraying day, might be a key factor affecting the result. Frequent trunk applications may also cause injury to the plant. Management of nutrient deficiency has been reported as early as 1948 when Hoagland (1948) reported using foliar spraying of $FeSO_4$ with Zn impurities to treat little leaf in peach, mottle leaf in citrus orchards, rosetting in pecan, and stunted pineapple growth. Although much research has been done for correcting plant nutrient deficiency, there is still no satisfactory method to control all nutrient deficiencies throughout the entire life span of trees either in natural stands or in urban forestry. Understanding mechanisms of plants responding to nutrient deficiencies will help manage these stresses and improve the productivity of woody plants under different nutrient deficiency conditions.

Development of plant materials that are resistant/tolerant to nutrient deficiency appears to be the best way to manage nutrient deficiencies in woody species; however, the long juvenile period and highly heterozygous genetic background of woody species, particularly for tree species, make it difficult to achieve such a goal. Efforts have been continuously made to select and develop woody plant materials that are tolerant to nutrient deficiency; however, it is a time- and labor-consuming process that needs a large amount of effort from breeders and other personnel. With advances in the fields of molecular genetics and biotechnology of woody species, more and more plant materials that are resistant to nutrient deficiency will be identified and developed, and that will help improve productivity and the aesthetic value of all woody species.

REFERENCES

Abadia, J. 1992. Leaf responses to Fe deficiency: A review. *J. Plant Nutri.* 15:1699–1713.

Abadia, J., F. Morales, and A. Abadia. 2000. Photosystem II efficiency in low chlorophyll, iron deficient leaves. *Plant Soil* 215:183–192.

Afrousheh, M., M. Ardalan, H. Hokmabadi, and M. Afrousheh. 2010. Nutrient deficiency disorders in *Pistacia vera* seedling rootstock in relation to eco-physiological, biochemical characteristics and uptake pattern of nutrients. *Sci. Hort.* 124:141–148.

Andaluz, S., A. F. López-Millán, J. De las Rivas, E. M. Aro, J. Abadia, and A. Abadia. 2006. Proteomic profiles of thylakoid membranes and changes in response to iron deficiency. *Photosynth. Res.* 89:141–155.

Ashley, M. K., M. Grant, and A. Grabov. 2005. Plant response to potassium deficiencies: A role for potassium transport proteins. *J. Exp. Bot.* 57:425–436. https://doi.org/10.1093/jxb/erj034.

Aziz, T., M. Sabir, M. Farooq, M. A. Maqsood, H. R. Ahmad, and E. A. Warraich. 2014. Phosphorus deficiency in plants: Responses, adaptive mechanisms, and signaling. In: *Plant Signaling: Understanding the Molecular Cross-Talks*, ed. K. H. Hakim et al. Springer, India. DOI: 10.1007/978-81-322-1542-4-7.

Barker, A. V., and D. J. Pilbeam. 2015. *Handbook of Plant Nutrition*. CRC Press/Taylor & Francis, Boca Raton, FL.

Bauer, P., H. Q. Ling, and M. L. Guerinot. 2007. FIT, the FER-like iron deficiency induced transcription factor in *Arabidopsis*. *Plant Physiol. Biochem.* 45:260–261.

Belkhodja, R., F. Morales, R. Quilez, A. F. Lopez-Millan, A. Abadia, and J. Abadia. 1998. Iron deficiency causes changes in chlorophyll fluorescence due to the reduction in the dark of the photosystem II acceptor side. *Photosynth. Res.* 56: 265–276.

Bertamini, M., K. Muthuchelia, and N. Nedunchezhan. 2002. Iron deficiency induced changes on the donor side of PS II in field grown grapevine (*Vitis vinifera* L. cv. Pinot noir) leaves. *Plant Sci.* 162:599–605.

Bienfait, H. F., R. J. Bino, A. M. van der Bliek, J. F. Duivenvoorden, and J. M. Fontaine. 1983. Characterization of ferric reducing activity in root of Fe deficient *Phaseolus vulgaris*. *Physiol. Plant.* 59:196–202.

Braun, L. 2012. *Nitrogen Uptake and Assimilation in Woody Crops*. University of Minnesota, Minneapolis and Saint Paul.

Briat, J. F., I. Fobis-Loisy, N. Grignon, S. Lobreaux, N. Pascal, G. Savino, S. Thoiron, N. von Wiren, and O. Van wuytswinkel. 1995. Cellular and molecular aspects of iron metabolism in plants. *Biol. Cell* 84:69–81.

Broadley, M. R., P. J. White, J. P. Hammond, I. Zelko, and A. Lux. 2007. Zinc in plants. *New Phytol.* 173:677–702.

Buckhout, T. J., P. F. Bell, D. G. Luster, and R. L. Chaney. 1989. Iron-stress induced redox activity in tomato (*Lycopersicum esculentum* Mill.) is localized on the plasma membrane. *Plant Physiol.* 90:208–213.

Cakmak, I., and H. Marschner. 1988. Increase in membrane permeability and exudation in roots of zinc deficient plants. *J. Plant Physiol.* 132:356–361.

Cheng, L., F. Wang, H. Shou, F. Huang, L. Zheng, F. He, J. Li, F.-J. Zhao, D. Ueno, J. F. Ma, and P. Wu. 2007. Mutation in nicotianamine aminotransferase stimulated the Fe(II) acquisition system and led to iron accumulation in rice. *Plant Physiol.* 145:1647–1657.

Colangelo, E. P., and M. L. Guerinot. 2004. The essential basic helix-loop-helix protein *FIT1* is required for the iron deficiency response. *Plant Cell* 16:3400–3412.

Connolly, E. L., N. H. Campbell, N. Grotz, C. L. Prichard, and M. L. Guerinot. 2003. Overexpression of the *FRO2* ferric chelate reductase confers tolerance to growth on low iron and uncovers posttranscriptional control. *Plant Physiol.* 133:1102–1110.

Connorton, J. M., J. Balk, and J. Rodriguez-Celma. 2017. Iron homeostasis in plants-a brief overview. *Metallomics* 9:813–823. DOI: 10.1039/c7mt00136c.

Couturier, J., B. Montanini, F. Martin, A. Brun, D. Blaudez, and M. Chalot. 2007. The expanded family of ammonium transporters in the perennial poplar plant. *New Phytol.* 174:137–150.

Curie, C., G. Cassin, D. Couch, F. Divol, K. Higuchi, M. L. Jean, J. Misson, A. Schikora, P. Czernic, and S. Mari. 2009. Metal movement within the plant: Contribution of nicotianamine and *yellow stripe 1-like* transporters. *Ann. Bot. Lond.* 103:1–11.

de la Guardia, M. D., and E. Alcántara. 2002. A comparison of ferric-chelate reductase and chlorophyll and growth ratios as indices of selection of quince, pear and olive genotypes under iron deficiency stress. *Plant Soil* 241:49–56.

Eckhardt, U., A. M. Marques, and T. J. Buckhout. 2001. Two iron-regulated cation transporters from tomato complement metal uptake-deficient yeast mutants. *Plant Mol. Biol.* 45:437–448.

Ehlting, B., P. Dluzniewska, H. Dietrich, A. Selle, M. Teuber, R. Haensch, U. Nehls, A. Polle, J. P. Schnitzler, H. Rennenberg, and A. Gessler. 2007. Interaction of nitrogen nutrition and salinity in Grey poplar (*Populus tremula × alba*). *Plant Cell Environ.* 30:796–811.

Ellsworth, J. W., V. D. Jolley, D. S. Nuland, and A. D. Blaylock. 1997. Screening for resistance to iron deficiency chlorosis in dry bean using iron reduction capacity. *J. Plant Nutr.* 20:1489–1502.

Feng, H., F. An, S. Zhang, Z. Ji, H. Q. Ling, and J. Zuo. 2006. Light-regulated, tissue-specific, and cell differentiation-specific expression of the *Arabidopsis* Fe(III)-Chelate reductase gene AtFRO6. *Plant Physiol.* 140:1345–1354.

Fernández-Escobar, R., M. Guerreiro, M. Benlloch, and M. Benlloch-González. 2016. Symptoms of nutrient deficiencies in young olive trees and leaf nutrient concentration at which such symptoms appear. *Sci. Hort.* 209:279–285.

Foth, H. D., and B. G. Ellis. 1997. *Soil Fertility.* CRC Press, Boca Raton, FL.

Gan, H., Y. Jiao, J. Jia, X. Wang, H. Li, W. Shi, C. Peng, A. Polle, and Z. Luo. 2016. Phosphorus and nitrogen physiology of two contrasting poplar genotypes when exposed to phosphorus and/or nitrogen starvation. *Tree Physiol.* 36(1): 22–38.

Gierth, M., P. Maser, and J. I. Schroeder. 2005. The potassium transporter AtHAK5 functions in K^+ deprivation-induced high-affinity K^+ uptake and AKT1 K^+ channel contribution to K^+ uptake kinetics in Arabidopsis roots. *Plant Physiol.* 137:1105–1114.

Gilbert, S., D. T. Clarkson, M. Cambridge, H. Lambers, and M. J. Hawkesford. 1997. Sulphate-deprivation has an early effect on the content of ribulose 1,5-bisphosphate carboxylase/oxygenase and photosynthesis in young leaves of wheat. *Plant Physiol.* 115:1231–1239.

Gruber, B. D., R. F. H. Giehl, S. Friedel, and N. von Wiren. 2013. Plasticity of the *Arabidopsis* root system under nutrient deficiencies. *Plant Physiol.* 163:161–179.

Guerinot, M. L. 2000. The ZIP family of metal transporters. *Biochim. Biophys. Acta* 1465: 190–198.

Guerinot, M. L., and Y. Yi. 1994. Iron: Nutritious, noxious, and not readily available. *Plant Physiol.* 104: 815–820.

Ha, S., and L. S. Tran. 2014. Understanding plant responses to phosphorus starvation for improvement of plant tolerance to phosphorus deficiency by biotechnological approaches. *Crit. Rev. Biotech.* 34: 16–30. https://doi.org/10.3109/07388551.2013.783549.

Hammond, J. P., M. R. Broadley, and P. J. White. 2004. Genetic responses to phosphorus deficiency. *Ann. Bot.* 94:323–332.

Han, Q., H. Song, Y. Yang, H. Jiang, and S. Zhang. 2018. Transcriptional profiling reveals mechanisms of sexually dimorphic responses of *Populus cathayana* to potassium deficiency. *Physiol. Plant.* 162: 301–315. DOI: 10.1111/ppl.12636.

Hänsch, R., and R. R. Mendel. 2009. Physiological functions of mineral micronutrients (Cu, Zn, Mn, Fe, Ni, Mo, B, Cl). *Curr. Opin. Plant Bio.* 12:259–266. DOI: 10.1016/j.pbi.2009.05.006.

Henriques, R., J. Jasik, M. Klein, E. Martinoia, U. Feller, J. Schell, M. S. Pais, and C. Koncz. 2002. Knock-out of *Arabidopsis* metal transporter gene *IRT1* results in iron deficiency accompanied by cell differentiation defects. *Plant Mol. Biol.* 50:587–597.

Hermans, C., J. P. Hammond, P. J. White, and N. Verbruggen. 2006. How do plants respond to nutrient shortage by biomass allocation? *Trends Plant Sci.* 11:610–617. https://doi.org/10.1016/j.tplants.2006.10.007.

Hoagland, D. R. 1948. *Lectures on the Inorganic Nutrition of Plants*, 2nd edition. Chronica Botanica Company, Waltham, MA.

Holden, M. J., D. G. Luster, R. L. Chaney, T. J. Buckout, and C. Robinson. 1991. Fe (III)-chelate reductase activity of plasma membranes isolated from tomato (*Lycopersicum esculentum* Mill.) roots. *Plant Physiol.* 97:537–544.

Huang, D., and W. Dai. 2013. *Characterization of Ferric Reductase Oxidase (FRO) Genes in Populus tremula L.* ASHS Annual Conference, Palm Desert, CA.

Huang, D., and W. Dai. 2015a. Molecular characterization of the basic helix-loop-helix (*bHLH*) genes that are differentially expressed and induced by iron deficiency in *Populus*. *Plant Cell Rep.* 34:1211–1124.

Huang, D., and W. Dai. 2015b. Two iron-regulated transporter (*IRT*) genes showed differential expression in poplar trees under iron or zinc deficiency. *J. Plant Physiol.* 186–187:59–67.

Ishimaru Y., M. Suzuki, T. Tsukamoto, K. Suzuki, M. Nakazono, T. Kobayashi, Y. Wada, S. Watanabe, S. Matsuhashi, M. Takahash, et al. 2006. Rice plants take up iron as an Fe3+-phytosiderophore and as Fe2+. *Plant J* 45:335–346.

Jain, A., G. T. Wilson, and E. Connolly. 2014. The diverse roles of FRO family metalloreductases in iron and copper homeostasis. *Front. Plant Sci.* 5:100. DOI: 10.3389/fpls.2014.00100.

Jakoby, M., H. Y. Wang, W. Reidt, B. Weisshaar, and P. Bauer. 2004. FRU(HLH029) is required for induction of iron mobilization genes in *Arabidopsis thaliana*. *FEBS Lett.* 577:528–534.

Jolley, V. D., D. J. Fairbanks, W. B. Stevens, R. E. Terry, and J. H. Orf. 1992. Root iron-reduction capacity for genotypic evaluation of iron efficiency in soybean. *J. Plant Nutr.* 15:1679–1690.

Kiekens, L. 1995. Zinc. In: *Heavy Metals in Soils*, 2nd edition, ed. B. J. Alloway. Blackie Academic and Professional, London.

Kobayashi, T., Y. Ogo, R. N. Itai, H. Nakanishi, M. Takahashi, S. Mori, and N. K. Nishizawa. 2007. The transcription factor IDEF1 regulates the response to and tolerance of iron deficiency in plants. *Proc. Natl. Acad. Sci. U.S.A.* 104:19150–19155.

Kosegarten, H. U., B. Hoffmann, and K. Mengel. 1999. Apoplastic pH and Fe (III) reduction in intact sunflower leaves. *Plant Physiol.* 121:1069–1079.

Krapp, A. 2015. Plant nitrogen assimilation and its regulation: A complex puzzle with missing pieces. *Curr. Opin. Plant Bio.* 25:115–122.

Laing, W., D. Greer, O. Sun, P. Beets, A. Lowe, and T. Payn. 2000. Physiological impacts of Mg deficiency in *Pinus radiata*: Growth and photosynthesis. *New Phytol.* 146:47–57.

Lambers, H., M. Brundrett, J. Raven, and S. Hopper. 2010. Plant mineral nutrition in ancient landscapes: High plant species diversity on infertile soils in linked to functional diversity for nutritional strategies. *Plant Soil* 334:11–31.

Landsberg, E. C. 1996. Hormonal regulation of iron-stress response in sunflower roots: A morphological and cytological investigation. *Protoplasma* 194:69–80.

Li, L., X. Cheng, and H. Q. Ling. 2004. Isolation and characterization of Fe(III)-chelate reductase gene *LeFRO1* in tomato. *Plant Mol. Biol.* 54:125–136.

Li, P., J. L. Qi, L. Wang, Q. N. Huang, Z. H. Han, and L. P. Yin. 2006. Functional expression of *MxIRT1*, from *Malus xiaojinensis*, complements an iron uptake deficient yeast mutant for plasma membrane targeting via membrane vesicles trafficking process. *Plant Sci.* 171:52–59.

Li, W., G. Xu, A. Alli, and L. Yu. 2018. Plant HAK/KUP/KT K^+ transporters: Function and regulation. *Semi. Cell Develop. Biol.* 74:133–141. https://doi.org/10.1016/j.semcdb.2017.07.009.

Ling, H. Q., P. Bauer, Z. Bereczky, B. Keller, and M. W. Ganal. 2002. The tomato fer gene encoding a bHLH protein controls iron-uptake responses in roots. *Proc. Natl. Acad. Sci. U.S.A.* 99:13938–13943.

Loneragan, J. F., and M. J. Webb. 1993. Interactions between zinc and other nutrients affecting the growth of plants. In: *Zinc in Soils and Plants*, ed. A. D. Robson. Kluwer Academic Publishers, Dordrecht, The Netherlands, 119–134.

Long, T. A., H. Tsukagoshi, W. Busch, B. Lahner, D. E. Salt, and P. N. Benfey. 2010. The bHLH transcription factor POPEYE regulates response to iron deficiency in *Arabidopsis* roots. *Plant Cell* 22:2219–2236.

Ludewig, U., B. Neuhduser, and M. Dynowski. 2007. Molecular mechanisms of ammonium transport and accumulation in plants. *FEBS Lett.* 581:2301–2308.

Maathuis, F. J. M. 2009. Physiological functions of mineral macronutrients. *Curr. Opin. Plant Bio.* 12:250–258. https://doi.org/10.1016/j.pbi.2009.04.003.

Maathuis, F. J. M., and D. Sanders. 1997. Regulation of K^+ absorption in plant root cells by external K^+: Interplay of different plasma membrane K^+ transporters. *J. Exp. Bot.* 48:451–458.

Malkin, R., and K. Niyogi. 2000. Photosynthesis. In: *Biochemistry and Molecular Biology of Plants*, ed. B. B. Buchanan et al. American Society of Plant Physiologists, Rockville, MD, 568–628.

Mori, S. 1999. Iron acquisition by plants. *Curr. Opin. Plant Biol.* 2:250–253.

Muhammad, I., X. Q. Jing, A. Shalmani, M. Ali, S. Yi, P. F. Gan, W. Q. Li, W. T. Liu, and K. M. Chen. 2018. Comparative in silico analysis of ferric reduction oxidase (*FRO*) genes expression patterns in response to abiotic stresses, metal and hormone applications. *Molecules* 23, 1163. https://doi.org/10.3390/molecules23051163.

Mukherjee, I., N. H. Campbell, J. S. Ash, and E. L. Connolly. 2005. Expression profiling of the Arabidopsis ferric chelate reductase (*FRO*) gene family reveals differential regulation by iron and copper. *Planta* 223:1178–1190. DOI: 10.1007/s00425-005-0165-0.

Nakamura, Y., Y. Umemiya, K. Masuda, H. Inoue, and M. Fukumoto. 2007. Molecular cloning and expression analysis of cDNA encoding a putative *Nrt2* nitrate transporter from peach. *Tree Physiol.* 27:503–510.

Narwal, R. P., B. R. Singh, and A. R. Panhwar. 1983. Plant availability of heavy metals in a sludge treated soil: I. Effect of sewage sludge and soil pH on the yield and chemical composition of rape. *J. Environ. Qual.* 12:358–365.

Nikolic, M., and V. Romheld. 1999. Mechanism of Fe uptake by the leaf symplast: Is Fe inactivation in leaf a cause of Fe deficiency chlorosis? *Plant Soil* 215:229–237.

Niu, Y. F., R. S. Chai, G. L. Jin, H. Wang, C. X. Tang, and Y. S. Zhang. 2013. Responses of root architecture development to low phosphorus availability: A review. *Annu. Bot.* 112:391–408. DOI: 10.1093/aob/mcs285.

Ogo, Y., T. Kobayashi, R. N. Itai, H. Nakanishi, Y. Kakei, M. Takahashi, S. Toki, S. Mori, and N. K. Nishizawa. 2008. A novel NAC transcription factor IDEF2 that recognizes the *iron deficiency-responsive element 2* regulates the genes involved in iron homeostasis in plants. *J. Biol. Chem.* 283:13407–13417.

Ohki, K. 2006. Effect of zinc nutrition on photosynthesis and carbonic anhydrase activity in cotton. *Physiol. Plant.* 38:300–304.

Olson, R. A., R. B. Clark, and J. H. Bennett. 1981. The enhancement of soil fertility by plant roots. *Amer. Sci.* 69:368–384.

Pushnik, J. C., G. W. Miller, and J. H. Manwaring. 1984. The role of iron in higher plant chlorophyll biosynthesis maintenance and chloroplast biogenesis. *J. Plant Nutr.* 7:733–758.

Qu, S. C., J. Y. Zhang, J. M. Tao, Y. S. Qiao, and Z. Zhang. 2005. Physiological reaction on resistance to iron deficiency in transgenic lines of *Malus robusta* by introducing *LeIRT2* gene. *Sci. Agric. Sin.* 38:1024–1028.

Robinson, N. J., C. M. Procter, E. L. Connolly, and M. L. Guerinot. 1999. A ferric-chelate reductase for iron uptake from soils. *Nature* 397:694–697.

Romera, F. J., and E. Alcántara. 1994. Iron-deficiency stress responses in cucumber (*Cucumis sativus* L.) roots: A possible role for ethylene? *Plant Physiol.* 105:1133–1138.

Romera, F. J., E. Alcántara, and M. D. De la Guardia. 1999. Ethylene production by Fe-deficient roots and its involvement in the regulation of Fe-deficiency stress responses by strategy I plants. *Ann. Bot.* 83:51–55.

Romheld, V. 1987. Different strategies for iron acquisition in higher plants. *Physiol. Plant.* 70:231–234.

Romheld, V., and H. Marschner. 1986. Mobilization of iron in the rhizophere of different plant species. *Adv. Plant Nutr.* 2:155–204.

Rubio, F., G. E. Santa-Maria, and A. Rodriguez-Navarro. 2000. Cloning of *Arabidopsis* and barley cDNAs encoding HAK potassium transporters in root and shoot cells. *Physiol. Plant.* 109:34–43.

Sánchez-Calderón, L., J. López-Bucio, A. Chacón-López, A. Gutiérrez-Ortega, and E. Hernández-Abreu. 2006. Characterization of low phosphorus insensitive mutants reveals a crosstalk between low phosphorus-induced determinate root development and the activation of genes involved in the adaptation of Arabidopsis to phosphorus deficiency. *Plant Physiol.* 140:879–889.

Santa-Maria, G. E., S. Oliferuk, and J. I. Moriconi. 2018. KT-HAK-KUP transporters in major terrestrial photosynthetic organisms: A twenty years tale. *J. Plant Physiol.* 226:77–90.

Santi, S., S. Cesco, Z. Varanini, and R. Pinton. 2005. Two plasma membrane H^+-ATPase genes are differentially expressed in iron-deficient cucumber plants. *Plant Physiol. Biochem.* 43:287–292.

Santi, S., and W. Schmidt. 2008. Laser microdissection-assisted analysis of the functional fate of iron deficiency-induced root hairs in cucumber. *J. Exp. Bot.* 59:697–704.

Schmidt, W. 1999. Mechanisms and regulation of reduction-based iron uptake in plants. *New Phytol.* 141:1–26.

Schmidt, W., and M. Bartels. 1996. Formation of root epidermal transfer cells in *Plantago. Plant Physiol.* 110:217–225.

Schmidt, W., J. Tittel, and A. Schikora. 2000. Role of hormones in the induction of iron deficiency responses in *Arabidopsis* roots. *Plant Physiol.* 122:1109–1118. DOI: https://doi.org/10.1104/pp.122.4.1109.

Sharma, C. P., P. N. Sharma, S. S. Bisht, and B. D. Nautiyal. 1982. Zinc deficiency induces changes in cabbage (*Brassica oleracea* var. Capitat). In: *Proceedings of the Ninth international Plant Nutrition Colloquium*, ed. A. Scaife. Commonwealth Agricultural Bureau, Warwick, England, 601–606.

Shin, R., and D. P. Schachtman. 2004. Hydrogen peroxide mediates plant root cell response to nutrient deprivation. *Proceedings of the National Academy of Sciences* 101:8827–8832.

Spiller, S., and N. Terry. 1980. Limiting factors in photosynthesis II. Iron stress diminishes photochemical capacity by reducing the number of photosynthetic units. *Plant Physiol.* 65:121–125.

Suzuki, M., M. Takahashi, T. Tsukamoto, S. Watanabe, S. Matsuhashi, J. Yazaki, N. Kishimoto, S. Kikuchi, H. Nakanishi, S. Mori, and N. K. Nishizawa. 2006. Biosynthesis and secretion of mugineic acid family phytosiderophores in zinc-deficient barley. *Plant J.* 48:85–97.

Tagliavini, M., J. Abadía, A. D. Rombolà, A. Abadía, C. Tsipouridis, and B. Mayangoni. 2000. Agronomic means for the control of iron deficiency chlorosis in deciduous fruit trees. *J. Plant Nutr.* 23:2007–2022. https://doi.org/10.1080/01904160009382161.

Tagliavini, M., and A. D. Rombola. 2001. Iron deficiency and chlorosis in orchard and vineyard ecosystems. *Euro. J. Agron.* 15:71–92.

Tranbarger, T. J., Y. Al-Ghazi, B. Muller, B. Teyssendier Del La Serve, P. Doumas, and B. Touraine. 2003.Transcription factor genes with expression correlated to nitrate-related root plasticity of *Arabidopsis thaliana. Plant Cell Environ.* 26:459–469. https://doi.org/10.1046/j.1365-3040.2003.00977.x.

Trubat, R., J. Cortina, and A. Vilagrosa. 2006. Plant morphology and root hydraulics are altered by nutrient deficiency in *Pistacia lentiscus* (L.). *Trees* 20:334–339. https://doi.org/10.1007/s00468-005-0045-z.

Uren, N. C. 1984. Forms, reactions and availability of iron in soils. *J. Plant Nutr.* 7:165–176.

Vance, C. P., C. Uhde-Stone, and D. L. Allan. 2003. Phosphorus acquisition and use: Critical adaptations by plants for securing a nonrenewable resource. *New Phytol.* 157:423–447.

Vasconcelos, M. W., H. Eckert, V. Arahana, G. Graef, M. A. Grusak, and T. Clemente. 2006. Molecular and phenotypic characterization of transgenic soybean expressing the *Arabidopsis* ferric chelate reductase gene, FRO2. *Planta* 224:1116–1128. DOI: 10.1007/s00425-006-0293-1.

Vert, G., M. Barberon, E. Zelazny, M. Seguela, J. F. Briat, and C. Curie. 2009. *Arabidopsis* IRT2 cooperates with the high-affinity iron uptake system to maintain iron homeostasis in root epidermal calls. *Planta* 229:1171–1179.

Vert, G., J. F. Briat, and C. Curie. 2001. *Arabidopsis* IRT2 gene encodes a root-periphery iron transporter. *Plant J.* 26:181–189.

Vert, G., N. Grotz, F. Dedaldechamp, F. Gaymard, M. L. Guerinot, J. F. Briat, and C. Curie. 2002. *IRT1*, an *Arabidopsis* transporter essential for iron uptake from the soil and for plant growth. *Plant Cell* 14:1223–1233.

Walker, E. L., and E. L. Connolly. 2008. Time to pump iron: Iron-deficiency-signaling mechanisms of higher plants. *Curr. Opin. Plant Biol.* 11:530–535. https://doi.org/10.1016/j.pbi.2008.06.013.

Wang, A. S., J. S. Angle, R. L. Chaney, T. A. Delorme, and R. D. Reeves. 2006. Soil pH effects on uptake of Cd and Zn by *Thlaspi caerulescens. Plant Soil* 281:325–337.

Wang, H. Y., M. Klatte, M. Jakoby, H. Baumlein, B. Weisshaar, and P. Bauer. 2007. Iron deficiency-mediated stress regulation of four subgroup Ib *BHLH* genes in *Arabidopsis thaliana. Planta* 226:897–908.

Waters, B. M., D. G. Blevins, and D. J. Eide. 2002. Characterization of *FRO1*, A pea ferric-chelate reductase involved in root iron acquisition. *Plant Physiol.* 129:85–94.

Witney, G. W., M. M. Kushad, and J. A. Barden. 1991. Induction of bitter pit in apple. *Sci. Hort.* 47:173–176. https://doi.org/10.1016/0304-4238(91)90039-2.

Wood, B. W., C. C. Reilly, and A. P. Nyczepir. 2004. Mouse-ear of pecan: A nickel deficiency. *HortScience* 39:1238–1242.

Xu, H. M., Y. Wang, F. Chen, X. Z. Zhang, and Z. H. Han. 2011. Isolation and characterization of the iron-regulated MxbHLH01 gene in *Malus xiaojinensis. Plant Mol. Biol. Rept* 29:936–942.

Yang, T., S. Zhang, Y. Hu, F. Wu, Q. Hu, G. Chen, J. Cai, T. Wu, N. Moran, L. Yu., and G. Xu. 2014. The role of a potassium transporter OsHAK5 in potassium acquisition and transport from roots to shoots in rice at low potassium supply levels. *Plant Physiol.* 166:945–959.

Yang, Y., H. Jiang, M. Wang, H. Korpelainen, and C. Li. 2015. Male poplars have a stronger ability to balance growth and carbohydrate accumulation than do females in response to a short-term potassium deficiency. *Physiol. Plant.* 155:400–413.

Yin, L., Y. Wang, M. Yuan, X. Zhang, H. Pan, X. Xu, and Z. Han. 2013. Molecular cloning, polyclonal antibody preparation, and characterization of a functional iron-related transcription factor *IRO2* from *Malus xiaojinensis. Plant Physiol. Biochem.* 67:63–70.

Yin, L., Y. Wang, M. Yuan, X. Zhang, X. Xu, and Z. Han. 2014. Characterization of *MxFIT*, an iron deficiency induced transcriptional factor in *Malus xiaojinensis. Plant Physiol. Biochem.* 75:89–95.

Yuan, Y., H. Wu, N. Wang, J. Li, W. Zhao, J. Du, D. Wang, and H. Q. Ling. 2008. FIT interacts with AtbHLH38 and AtbHLH39 in regulating iron uptake gene expression for iron homeostasis in *Arabidopsis. Cell Res.* 18:385–397.

Yuan, Y., J. Zhang, D. W. Wang, and H. Q. Ling. 2005. AtbHLH29 of *Arabidopsis thaliana* is a functional ortholog of tomato FER involved in controlling iron acquisition in strategy I plants. *Cell Res.* 15:613–621.

Zhang, H., S. Zhao, D. Li, X. Xu, and C. Li. 2017. Genome-wide analysis of the ZRT, IRT-like protein family and their responses to metal stress in *Populus trichocarpa. Plant Mol. Bio. Rept* 35:534–549.

Zhang, S., H. Jiang, H. Zhao, H. Korpelainen, and C. Li. 2014. Sexually different physiological responses of *Populus cathayana* to nitrogen and phosphorus deficiencies. *Tree Physiol.* 34:343–354.

Zhao, D. L., D. M. Oosterhuis, and C. W. Bednarz. 2001. Influence of potassium deficiency on photosynthesis, chlorophyll content, and chloroplast ultrastructure of cotton plants. *Photosynthetica* 39:103–109.

Zheng, L., Y. Ying, L. Wang, F. Wang, J. Whelan, and H. Shou. 2010. Identification of a novel iron regulated basic helix-loop-helix protein involved in Fe homeostasis in *Oryza sativa*. *BMC Plant Biol.* 10:166.

7 Plant Response to Salinity Stress

Qi Zhang and Wenhao Dai

CONTENTS

7.1 INTRODUCTION

Salinity is a major abiotic stress, severely and adversely affecting crop productivity and functionality. The U.S. Salinity Laboratory classified soils into four categories: (1) normal soil, (2) saline soil, (3) sodic soil, and (4) saline-sodic soil (Table 7.1) (US Salinity Laboratory Staff, 1954). Globally, approximately 955 million ha of salt-affected soils (saline, sodic, and saline-sodic) is the result of natural soil-forming processes (i.e. primary salinization) and 77 million ha is due to human activities (i.e. secondary salinization), such as improper irrigation methods and poor irrigation quality, fertilization, and applications of deicer (Carrow and Duncan, 1998; Metternicht and Zinck, 2003; Pessarakli, 1999). With changing climate, poor irrigation and cultural practices, and increasing food demand, the area of salt-affected soil is expanding at an annual rate of 10% (Joshi et al., 2016; Munns and Gilliham, 2015). For example, salt-affected soils in North Dakota, USA have increased from 1.05 million ha in 1987 to 2.35 million ha in 2010 (Brennan and Ulmer, 2010; Seelig, 2000). National Land and Water Resources Audit (2000) reported that 5.7 million ha are at risk or affected by dry land salinity in Australia, and the area will increase to 17 million ha in 50 years. Approximately 50% of these arable lands will become salt-affected by 2050, if no reclamation actions are taken (Jamil et al., 2011).

TABLE 7.1
Classification of Salt-Affected Soils by U.S. Salinity Lab

Class	Electrical Conductivity of Soil-water Saturated Paste (EC_e[a], dS m^{-1})	Exchangeable Sodium Percentage (ESP[b], %)	Sodium Absorption Ratio (SAR[c])	Soil pH
Normal	<4	<15	<12	<8.5
Saline	≥4	<15	<12	<8.5
Sodic	<4	≥15	≥12	>8.5
Saline-sodic	≥4	≥15	≥12	<8.5

[a] EC_e = total dissolved salts (TDS) or total soluble salts (TSS) [in parts per million (ppm) or mg g^{-1}]/640. 1 dS m^{-1} = 1 mmhos cm^{-1} = 1000 u mhos cm^{-1} = 0.1 S m^{-1}.

[b] ESP = Exchangeable Na on cation exchange capacity (CEC) sites/CEC. Unite: cmol kg^{-1} or meq 100 g^{-1}.

[c] $SAR = \frac{Na}{\sqrt{(Ca + Mg)/2}}$. Unit: meq L^{-1} or mmol L^{-1}.

7.2 EFFECTS OF SALINITY AND SODICITY ON SOIL PROPERTIES

7.2.1 Chemical Properties

The most common types of cations and anions detected in salt-affected soils are Ca^{2+}, Mg^{2+}, Na^+, K^+, Cl^-, SO_4^{2-}, HCO_3^-, NO_3^-, and CO_3^{2-} (Carrow and Duncan, 1998). Neutral soluble salts, such as NaCl, $MgCl_2$, $CaCl_2$, Na_2SO_4, $MgSO_4$, and $CaSO_4$, are dominant in saline soils (UC-NAR, 2016). Sodium is generally the dominant cation under saline conditions. Large amount of Ca^{2+} and Mg^{2+} is also commonly detected in saline soils due to a relative low soil pH (<8.5). In sodic soils, neutral soluble salts are absent. Instead, $NaHCO_3$ and Na_2CO_3 are two primary salts. Limited amount of Ca^{2+} and Mg^{2+} are seen in sodic soils due to their low solubility (i.e. precipitation) at a high soil pH (>8.5). Abrol et al. (1980) reported a positive relationship between pH and ESP in the saturated soil water paste in sodic soils, but not in saline soils. Soil pH may provide a sense of soil sodicity; however, it cannot be used as an accurate index of soil sodicity as other factors, such as soil type and climate, also affect the soil pH of a specific location (Abrol et al., 1980).

7.2.2 Physical and Hydraulic Properties

High exchangeable Na^+ in salt-affected soils results in adverse changes of soil's physical and hydraulic properties, including (1) increased smaller/larger pore ratio and discontinuous soil pores; (2) decreased water infiltration, percolation, and drainage; therefore, increasing the possibility of waterlogging; (3) lowered soil O_2 content and diffusion rate; and (4) increased soil hardness (Carrow and Duncan, 1998).

Negatively charged clay particles are stacked together into domains or tactoids by cations. Divalent cations, Mg^{2+} and Ca^{2+}, form thinner diffuse double layers of electrical charges of the clay than monovalent cation, Na^+; thus, a low ESP allows domains to flocculate together (Carrow and Duncan, 1998). Several domains, together with sand and silt, aggregate to form larger soil structure units by organic gels and polysaccharides, Fe/Al/Mn oxides, and divalent or trivalent cations (e.g. Ca^{2+} and Al^{3+}). When the ESP exceeds 15% (i.e. sodic), Na^+ starts to replace Ca^{2+} and Mg^{2+} on the CEC site, resulting in slaking of the large structural unit followed by soil dispersion. Clay and organic colloids are easily migrated by water and plug pore channels which hold air and water. The infiltration rate is highly reduced when EC is low but SAR (or ESP) is high (i.e. sodic).

The negative effects of Na in water infiltration may be alleviated (diluted) by the large amount of ions in water and soil under the high EC condition (i.e. high EC and high SAR or ESP, the saline-sodic condition).

7.3 PROBLEMS WITH HIGH SALINITY IN WOODY PLANTS

7.3.1 Responses of Woody Plants to Salinity Stress

Salinity adversely affects plant growth and development. For example, Everitt (1983a,b) reported reduced seed germination and radical growth of blackbrush (*Acacia rigidula*), guajillo (*A. berlandieri*), guayacan (*Porlieria angustifolia*), retama (*Parkinsonia aculeata*), and twisted acacia (*A. schaffneri*) under saline conditions (NaCl at 0–10 g L^{-1}). Bessa et al. (2017) observed reduced shoot biomass of seedlings of a few woody species including *Myracrodruon urundeuva*, *Handroanthus impetiginosus*, *Bauhinia unguculata*, *Erythrina velutina*, *Mimosa caesalpiniifolia*, and *Luetzelburgia auriculata* when they were exposed to salinity (1.2–8.4 dS m^{-1}). Shoot length of one-year-old grafted pear (*Pyrus communis*) was reduced by approximately 30% after a four-month NaCl treatment (5.0 dS m^{-1}) (Musacchi et al., 2006). Reduced shoot and root growth of citrus (*Citrus* spp.) seedlings has been well documented (Storey and Walker, 1999; Ferguson and Grattan, 2005). For the halophytes (plants that can complete their lifecycles at 200 to 300 mM NaCl), such as the mangrove species of *Acanthus ilicifolius*, *Aegiceras corniculatum* and *Avicennia marina*, seedling growth was affected at much higher rates of more than 25 ppt (approximately 45.5 dS m^{-1}) (Ye et al., 2005).

Under drought- or salinity-induced water deficiency, plants undergo anatomical changes in xylem by decreasing the ratio of vessel lumen to cell wall thickness to reduce cavitation and cell collapse in xylem (Hacke et al., 2001). Diameter of stem increment and vessel lumen was reduced in a salt-sensitive species (*P.* × *canescens*) after they were treated with 50 mM NaCl in a hydroponic system for six weeks (Chen and Polle, 2010). Janz et al. (2012) reported an increased number of vessels per area and decreased vessel diameter in *P.* × *canescens* when subjected to two-week salt stress with 25 mM NaCl or after additional two-week salt stress with 100 mM NaCl. Similar results were reported by Arend and Fromm (2007), Beniwal et al. (2010), and Schreiber et al. (2011). Such anatomical changes led to false year ring marks in woody plants (Chen and Polle, 2010). Although a salt-tolerant species, *P. euphratica*, showed similar anatomical changes only after a long-term (two months) saline exposure at a much higher concentration (150 mM NaCl) compared to *P.* × *canescens* (Chen and Polle, 2010), both species showed chemical changes in wood composition after two-weeks of 25 mM NaCl followed by another two-weeks at 100 mM NaCl (Janz et al., 2012). Cell wall pH decreased in both species, but to a higher extend in the sensitive one (Janz et al., 2012). The lignin:cellulose ratio increased by 6.8% in *P. euphratica* and 13.7% in *P.* × *canescens*, and the lignin:protein ratio increased by 9.4% in *P.* × *canescens* only. Janz et al. (2012) concluded that the cell wall carbohydrate composition contributed, at least partially, to these anatomical changes.

Yield loss in citrus species under the saline condition is mostly due to lower fruit number rather than size (Storey and Walker, 1999). Salinity induced fruit loss is also reported in other woody fruits including grape (*Vitis vinifera*) (Shani and Ben-Gal, 2005; Mohammadkhani et al., 2014; Walker et al., 2014), blueberry (*Vaccinium ashei*) (Wright et al., 1992; Machado et al., 2014), peach (*Prunus persica*) (Massi et al., 1998; Nasr et al., 1977), and plum (*Prunus domestica*) (Hoffman et al., 1989; Nasr et al., 1977). Furthermore, salinity negatively influences fruit quality, such as color density, anthocyanin concentration, pH, and total soluble solids in grapes (Stevens et al., 2011; Walker et al., 2014).

Unlike agricultural crops in which salinity tolerance is measured with a quantifiable growth and yield, salinity tolerance of landscape plants (ornamental shrubs, trees, and flowers) is determined by their aesthetic values. Niu et al. (2007) evaluated salinity tolerance (1.6 to 9.4 dS m^{-1}) of big bend bluebonnet (*Lupinus havardii*) and Texas bluebonnet (*L. texensis*). Although shoot growth of both species started to decline at 3.7 dS m^{-1}, leaf damage was observed at 5.7 dS m^{-1} in big bend bluebonnet and at 7.6 dS m^{-1} in Texas bluebonnet (Niu et al., 2007). Eight out of the 13 rose (*Rosa* spp.) cultivars had reduced shoot growth and total number of buds and flowers at 4.4 dS m^{-1}; however, the visual injury (leaf burning, necrosis, and discoloration) was mostly seen at 6.4 dS m^{-1} (Niu et al., 2013). Results from Niu et al. (2007) and Niu et al. (2013) showed that visual quality of landscape plants may not be correlated with plant growth. Niu et al. (2007) suggested that reduced growth of landscape plants under salinity may even be desirable because of low management requirement.

7.3.2 Factors Influencing the Severity of Salinity Stress

Large interspecific and intraspecific differences of salinity tolerance exist in plants, including woody species. For instance, swamp tupelo (*Nyssa sylvatica* var. *biflora*) seedlings showed reduced height, basal diameter, and stem and root biomass at 2 ppt of sea salt (approximately 3.6 dS m^{-1}); while the growth of buttonbush (*Cephalanthus occidentalis*) was not significantly affected at the same salt concentration (McCarron et al., 1998). As the salinity increased to 10 ppt (~18.2 dS m^{-1}), the growth of buttonbush was inhibited and the total mortality of swamp tupelo was observed. During a three-week saline exposure ($NaCl:CaCl_2{\cdot}2H_2O$, 2:1 molar ratio, 0 to 12 dS m^{-1}), no leaf damage was detected in eucalyptus (*Eucalyptus cmaldulensis*) seedlings (Leksungnoen et al., 2014). Conversely, big leaf maple (*Acer macrophyllum*) and canyon maple (*A. grandidentatum*) seedlings showed increased leaf damage and decreased leaf area under salinity (Leksungnoen et al., 2014). Under saline irrigation ($NaCl:CaCl_2$, 1:1 W:W; 1.2, 5.0, and 10.0 dS m^{-1}), Texas betony (*Stachys coccinea*) was the most salt tolerant among the six *Lamiaceae* species tested, with a visual scale of 3.7 (0 = dead, 1 = poor quality, >90% foliar damage, 2 = moderate quality, 50% to 90% foliar damage, 3 = high quality, <50% foliar damage, 4 = good quality, minimal foliar damage, and 5 = excellent with no foliar damage) and 56% reduction of dry weight at 10 dS m^{-1} compared with non-treated plants (Wu et al., 2016). Russian sage (*Perovskia atriplicifolia*) and spotted dead nettle (*Lamium maculatum*) grew well at 5.0 dS m^{-1} and showed an average visual scale of 2.4 (moderate to high quality) and low mortality rates (10% to 30%) at 10 dS m^{-1} (Wu et al., 2016), whereas, bugleweed (*Ajuga reptans*), Mexican oregano (*Poliomintha longiflora*), and cherry skullcap (*Scutellaria suffrutescens*) were the most sensitive to salinity stress with 0% survival rate at 10 dS m^{-1} (Wu et al., 2016). Mohammadkhani et al. (2014) determined salinity tolerance (NaCl, 0 to 100 mM) of nine grape (*Vitis*) genotypes in which H6 (*V. vinifera* 'GharaUzum' × *V. riparia* 'Kober 5BB') showed the least growth reduction (shoot and root length and biomass) compared to 'Aghshani', 'Anghootkeh', 'Shirazi', 'Silvestris' (*V. Labrusca* sub. Silvestris), H4 (*V. vinifera* 'Jighjigha' × *V. riparia* 'Gloire'), 'Zardkeh', and 'Kajhave'. Similarly, Niu et al. (2013) reported intraspecific differences in salinity tolerance ($NaCl:CaCl_2$, 2:1 molar ratio; 1.4–6.4 dS m^{-1}) in 13 rose genotypes during a saline exposure for 7 or 10 weeks. 'Belinda's Dream', 'Caldwell Pink', 'Little Buckaroo', and 'Sea Foam' were more salt-tolerant with no visual damage under salinity, followed by 'Folksinger', 'Quietness', 'Basye's Blueberry', 'Rise N Shine', and 'Carefree Beauty', while 'Winter Sunset', 'Iceberg', 'Marie Pavie', and 'The Fairy' were salt-sensitive based on severe visual damage and mortality (Niu et al., 2013). Shoot and root biomass, shoot height, and root length of 'Badami-Rize-Zarand' and 'Badami-e-Sefid' pistachio (*Pistacia vera*) rootstocks were both reduced when subjected to NaCl solutions (50, 100, and 150 mM) for 3 weeks (Rahneshan et al., 2018). However, the reduction rate in the growth indices was much lower in 'Badami-e-Sefid' than in 'Badami-Rize-Zarand' (Rahneshan et al., 2018). Relative salinity tolerance of most commonly used woody plants are reviewed by Francois and Maas (1999), Niknam and McComb (2000), Tanji and Kielen (2002), Cassaniti et al. (2012), and Grieve et al. (2012). Because of the interspecific and intraspecific differences in woody plants, fruit trees and vine crops can use plants with high salinity tolerance as rootstocks to improve stress tolerance (Francois and Maas, 1999). Stevens et al. (1996) reported a similar shoot and root growth rate of 'Chardonnay', 'Sultana', 'Ruby Cabernet', and 'Shiraz' vines on their own roots or on 'Ramsey' rootstock under the saline irrigation (25 mM; NaCl:$CaCl_2$, 17:1 molar ratio) applied from budburst to harvest; however, the 'Ramsey' rootstock reduced Cl^- tissue contents. Field grapevines of 'Sultana', grown either on their own roots or grated to 'Ramsey', '1103 Paulsen', 'J17-49', or four hybrids between *V.tis champini*, *V. berlandieri*, and *V. vinifera* were drip-irrigated with saline water (0.40, 1.75, and 3.50 dS m^{-1}) for five years (Walker et al., 2002). Yield (bunches per shoot, bunches per vine, weight per bunch, and weight per berry and total yield) were higher for 'Sultana' on 'Ramsey', '1103 Paulsen', and 'R2' than that on own roots and the other root stocks (Walker et al., 2002). Zrig et al. (2016) reported lower shoot growth of one-year-old sweet almond (*Prunus amygdalus*) 'Mazzetto' grafted onto 'GF677' (*P. amygdalus* × *P. persica*) than that onto 'Garnem' (*P. amygdalus* 'Garfi' × *P. persica* 'Nemared') under the saline condition (NaCl, 25–75 mM) for four weeks. The result from Zrig et al. (2016) indicates that the enhancement of salinity tolerance of grafted plants depends on rootstock selection.

In addition to the genetic differences, many other factors influence plant responses to salinity. Plants exposed to lower salinity levels show less damage than those exposed to high salinity levels (Bessa et al., 2017; Mohammadkhani et al., 2014; Niu et al., 2007, 2013; Rahneshan et al., 2018). Roses irrigated with a salt solution for seven weeks produced more flower buds, higher biomass, and lower visual damage than those irrigated with the salt solution for 10 weeks, indicating increasing damage along with the increasing exposure time of the salinity stress (Niu et al., 2013). Wright et al. (1992) irrigated rabbiteye blueberries (*Vaccinium ashei*) with NaCl and Na_2SO_4 at 0, 25, or 100 mM. They found that NaCl-treated rabbiteye blueberries had higher biomass than Na_2SO_4-treated ones. Research found that $CaCl_2$ was more detrimental to *Rosa rubiginosa* than NaCl when applied at the same molar concentration (0, 25, 50, 100, 150, and 200 mM) due to a higher reduction in photosynthetic activity and leaf water content (Hura et al., 2017). The results of Wright et al. (1992) and Hura et al. (2017) suggest that plant damage is related to salt type. Survival rate of buttonbush seedlings started to decrease two months after being irrigated with saline water at 10 ppt, whereas, a sharp decline of survival rate was observed in the plants under the combined saline-flooding (10 ppt) condition three weeks after the treatment (McCarron et al., 1998). Stevens et al. (2011) applied River Murray water (EC = 0.5 dS m^{-1}, control) and saline water (River Murray water mixed with NaCl, EC = 3.5 dS m^{-1}) to grapevines during one of the four growth stages, pre-flowering, berry formation, berry ripening, and postharvest, for six seasons. In one-year-old wood, Na^+ and Cl^- content with salt exposure during pre-flowering or postharvest were 1.4 and twice than that during berry formation and ripening, respectively, indicating that salinity tolerance that is partially associated with the tissue Na^+ and Cl^- content varies in growth stages. Woody plants, in general, have no salt tolerance at germination, low tolerance at the seedling stage, and high tolerance at the juvenile stage (4 to 6 months) (Niknam and McComb, 2000). There is no solid evidence that salt tolerance of woody plants would further increase as it ages after the juvenile stage (Niknam and McComb, 2000). For instance, the threshold salinity level causing growth reduction of two-year-old olive (*Olea europaea*) 'Arbequina' decreased from 6.7 dS m^{-1} in 1999 to 4.7 dS m^{-1} in 2000, and 3.0 dS m^{-1} in 2001 (Aragüés et al., 2005). Catlin et al. (1993) reported no differences in the threshold level of 'Santa Rosa' plum (*P. salicina*) on 'Marianna 2624' rootstock (*P. Cerasifera* × *P. munsoniana*) over a six-year field evaluation. Most research on salinity tolerance of woody plants are conducted in over a short term (weeks or months) and based on vegetative growth of young plants. A few studies have encompassed long-term evaluation of responses to salinity in woody plants. Boland et al. (1997) determined Na^+ content in the trunks of eight-year-old apricot trees (*Prunus armeniace*) that had been grown in an orchard with a high saline (~30 dS m^{-1}) and water table, two 30-year-old pear trees with or without salt damage, and orange (*Citrus sinensis*) trees (planted in 1960) irrigated with non-saline water (0.4 dS m^{-1}) or saline (2.8 dS m^{-1}) for nine seasons (1982–1991) and then non-saline water thereafter until being sampled in 1995. The results showed that the trunk wood acted as a store to minimize Na^+ translocation to leaf. Restriction of Na^+ transportation to the leaf was inhibited once the storage capacity of the trunk was exceeded; thus, injury occurred, and the severity of leaf injury was positively correlated with the Na^+ content in the trunk tissue. Bañuls et al. (1991)

observed salinity-calcium interactions in citrus plants, in which supplemental Ca^{2+} alleviated salinity damage in plant growth and defoliation of the two-year-old 'Navel' orange (*C. sinensis*) grafted on 'Cleopatra' mandarin (*C. reticulate*) and 'Troyer' citrange (*C. sinensis* × *Poncirus trifoliata*). Supplemental Ca^{2+} reduced negative effects of Na_2SO_4 (100 mM) in rabbiteye blueberries; however, it showed no effects on the NaCl-treated plants which were more damaged than those treated with Na_2SO_4 (Wright et al., 1992). Mozafari et al. (2018) reported that nanoparticles of iron (0.8 ppm) and potassium silicate (1 and 2 mM) elevated salinity damage in the grape 'Khoshnaw'. Research of Bañuls et al. (1991), Wright et al. (1992), and Mozafari et al. (2018) showed that application of micronutrients to stressed plants may help minimize salinity damage. As we have discussed before, the functionality of woody plants (i.e. evaluation criteria such as fruit production, vegetative growth, and aesthetics) need to be taken into consideration when determining salinity tolerance. Because of these complex results of salinity, plant tolerance to salinity can only be evaluated and compared as a 'relative' term.

7.4 MECHANISMS OF SALINITY DAMAGE

7.4.1 Osmotic Stress

In salt-affected soils (saline, sodic, and saline-sodic), total soil water content is not affected; however, plant-available water is decreased proportionally due to reduced water potential in the soil (Carrow and Duncan, 1998; Munns and Tester, 2008). Plant water stress or water deficiency induced by salinity is called osmotic stress or physiological drought. Osmotic stress begins once salt concentration in a growing medium is above a threshold level, roughly 4 dS m^{-1} for most plants (Munns and Tester, 2008). Symptoms of osmotic stress are similar to those of drought stress, primarily decreased growth and expansion of new leaves and shoots and delayed development. Salinity causes increased osmolality in growing media, resulting in decreased cell turgor in plant cells. Plant growth (i.e. cell division and cell elongation) only occurs when a positive turgor pressure (one of the major components determining plant water potential) occurs (Taiz, 1984). Using alga *Chara corallina* as a model plant, cell growth did not grow until turger pressure was above 0.3 MPa (Proseus et al., 2000). Growth decreased immediately once turgor pressure of growing cells was reduced. Furthermore, growth rate did not return to its original rate even after turgor pressure was recovered to its original value (Proseus et al., 2000). The relationship between relative yield or biomass and salinity is often described using the following equation: $Y = 100 - B(EC - A)$, in which Y is relative yield or biomass; A is the threshold value at which salinity damage is seen; B is the slope of growth reduction; and EC is the average root zone salinity (Ferguson and Grattan, 2005). Ferguson and Grattan (2005) suggested that the equation only reflects the damage from osmotic stress on woody plants. Growth reduction caused by salinity-induced ionic stress is above what is predicted based on osmotic effects alone.

Osmotic stress mainly regulates water movement and photosynthesis (Munns and Termaat, 1986). McCarron et al. (1998) reported reduced stomatal conductance in buttonbush and swamp tupelo exposed to 21 ppt saline water during the second day of the surge. In the salt-treated *Taxodium distichum*, reduced hydraulic efficiency was associated with increased woody density and resistance to xylem cavitation, suggesting that stressed plants may trade-off their growth to strengthen xylem and reduce xylem cavitation (Stiller, 2009). Salinity reduced photosynthesis, stomatal conductance, and water potential in eucalyptus, canyon maple, and big leaf maple (Leksungnoen et al., 2014). The inhibition of the aforementioned physiological indices was lower in the salt-tolerant eucalyptus than in the salt-sensitive canyon and bigleaf maple. In the study of Bessa et al. (2017), salinity caused reduced stomatal conductance, net assimilation rate and transpiration, and increased intrinsic water use efficiency (net assimilation rate/stomatal conductance), but had limited effects on the efficiency of photosystem II and tissue carbohydrate content. No relationship was observed between net assimilation rate or transpiration and the level of salinity tolerance; however, *Myracrodruon urundeuva*,

Handroanthus impetiginosus, and *Erythrina velutina* which had the lowest percentage of growth reduction among the six species evaluated in the experiment had a higher ability for stomatal control (Bessa et al., 2017). Bessa et al. (2017) suggested stomatal closure, which helps avoid water loss through transpiration and ion transportation to the photosynthetic tissue under the saline condition, was primarily responsible for the reductions in carbon assimilation.

In addition to reduced stomatal conductance, reduced photosynthesis observed under salinity may also be related to dehydrated and destroyed mesophyll cells and chlorophyll (Chl) and inhibited enzyme activities. Chlorophyll content was reduced in NaCl-treated *Picea mariana*, *P. glauca*, and *Pinus banksiana* seedlings (Croser et al., 2001). Total carotenoids, Chl a and Chl b decreased with increasing NaCl concentrations (0 to 300 mM) in the seven populations of *Picea abies* evaluated (Schiop et al., 2015). Chlorophyll a, Chl b, total Chl (i.e. Chl a+b), and carotenoid levels were reduced in two pistachio cultivars when treated with NaCl, with a higher reduction in the salt sensitive cultivar (Rahneshan et al., 2018). Percival (2005) suggested that Chl had the potential to be a quick, reliable, and inexpensive procedure to estimate whole plant salt tolerance, as a positive relationship was observed between leaf Chl content and salt tolerance of 30 woody perennial species when subjected to salinity. Chl was not affected by salinity in sweet almond 'Mazzetto' grafted to 'GF677' and 'Garnem' (Zrig et al., 2016). Total Chl and the Chl a/b ratio decreased in the leaves of 'Mazzetto'/'GF677' along with increasing salinity level, but not in 'Mazzetto'/'Garnem'. Changes in total Chl and Chl composition may be related to reduced activities in enzymes such as Rubisco and phosphoenolpyruvate carboxylase, Chl biosynthesis, and Chl degradation (Kumar et al., 2017). Carotenoids increased proportionally compared to total Chl at 75 mM in 'Mazzetto'/'Garnem' but not in 'Mazzetto'/'GF677'. Similarly, 'Mazzetto'/'Garnem' had a higher anthocyanin to total Chl ratio than 'Mazzetto'/'GF677'. Zrig et al. (2016) suggested that the rootstock affected the rate of Chl turnover or biosynthesis. Decreased Chl contents and enzyme activities (e.g. Rubisco) resulted in a reduced carboxylation rate which led to oxidative stress (Koyro et al., 2013).

7.4.2 Ion Toxicity and Imbalance

Another major stress induced by salinity is ion toxicity and imbalance (i.e. ionic stress), caused by ion accumulation and an inability for ion tolerance (Munns and Tester, 2008). It occurs at a later stage and has less severe effects (especially at low to moderate salinity levels) on plant growth and development compared to osmotic stress. Symptoms of ionic stress are mostly seen as leaf chlorosis and accelerated senescence or death of old leaves.

The dominant type of salt in most salt-affected soils is NaCl (Carrow and Duncan, 1998), although in some regions around the world including parts of India, Egypt and the US (e.g. California and North Dakota). SO_4^{2-} salts such as Na_2SO_4 and $MgSO_4$ are at a higher level than Cl^- salts (Bañuelos et al., 1993; Franzen, 2013; Manchanda and Sharma, 1989).

Sodium is not an essential but a beneficial (i.e. functional) nutrient for plants (Hopkins and Hüner, 2004). Sodium is absorbed via the roots, translocated to shoots through transpiration, and accumulated in leaf blades. It is involved in the transport of pyruvate, a critical intermediate compound between the bundle-sheath and mesophyll cells in the C4 photosynthetic pathway. Sodium may stimulate plant growth, particularly when K^+ is deficient, by maintaining cell turgor (Gorham, 2007). It is known that K^+ can play an important role in plant growth and development as well as in the maintenance of osmotic adjustment and cell turgor (Marshner, 1995). In addition, it is the major cation in plants that counterbalances the negative charge of anions, and plays a crucial role in the activation of the enzymes involved in the metabolism and synthesis of proteins and carbohydrates, as well as in the regulation of stomata opening and closing (Marshner, 1995). Under saline conditions, Na^+ competes with K^+ for the same binding sites and functionality of enzyme activation and interferes with K^+ uptake by using its physiological transport systems (Kumar et al., 2017). Na also activates outward-rectifying K^+ channels by the depolarization of the plasma membrane, further increasing K^+ loss. Increased Na^+ and decreased K^+ (i.e. increased Na^+/K^+ ratio), along with increasing NaCl concentrations were observed in pistachio when exposed to salinity stress (Rahneshan et al., 2018). Furthermore, the Na^+/K^+ ratio was highly associated with salt tolerance of pistachio cultivars. Similar results have been also observed in many other salt-treated woody plants including *Myracrodruon urundeuva*, *Handroanthus impetiginosus*, *Bauhinia unguculata*, *Erythrina velutina*, *Mimosa caesalpiniifolia*, *Luetzelburgia auriculata*, *Nerium oleander*, *Picea abies*, *Vitis* spp., and other ornamental species (Bessa et al., 2017; Cassaniti et al., 2012; Kumar et al., 2017; Mohammadkhani et al., 2014; Mozafari et al., 2018; Schiop et al., 2015; Walker et al., 2004); thus, the Na^+/K^+ ratio is often used as an indicator of the plant response and tolerance to high salinity. Typical symptoms of Na^+ toxicity are leaf necrosis, marginal scorching, or abscission (dropping) of the leaves. The concentration at which Na^+ causes toxicity is not well defined. Greenway and Osmond (1972) reported reduced enzyme activities at 100 mM NaCl in an *in vitro* assay.

High Na^+ absorption may also influence the uptake of other nutrients. Rahneshan et al. (2018) reported reduced tissue Ca^{2+}, Mg^{2+}, Fe^{2+}, and P in pistachio cultivars under salinity. The concentration of the aforementioned ions decreased significantly along with increasing NaCl levels in the sensitive cultivar 'Badami-e-Sefid'; however, the reduction was only observed in the higher salt concentrations (100 and 150 mM NaCl) in the tolerant cultivar 'Badami-Rize-Zarand' (Rahneshan et al., 2018). Sodium chloride (0 to 60 mM) decreased leaf and root Ca^{2+} and Mg^{2+} contents in Navel orange and Clementine scions on Cleopatra mandarin and Troyer citrange rootstocks, whereas, P content showed limited responses to salinity treatments (Bañuls et al., 1990). Walker et al. (2004) reported that a four-year means of Ca^{2+} was only decreased in the petiole and laminae of 'Sultana' grape rooted on 'J17-69' at 3.5 dS m^{-1} among eight evaluated rootstocks including its own roots, and it was not affected by salinity when tested in the juice. Mg increased in the petiole of 'Sultana' rooted on 'Ramsey', '1103 Paulsen', 'J17-69', and R1 and R3 hybrids, and in the laminae of the rootstock of '1103 Paulsen' and 'J17-69'. Grape juice Mg^{2+} concentrations showed an increase in 'Sultata' in the rootstock of its own root, 'Ramsey', 'J17-69', and R1 and R3 hybrids. They also observed negative correlations between grape juice Ca^{2+} content and yield and the weight of one-year-old pruning wood under salinity; however, no relationships between Mg^{2+} content and yield and the growth of one-year-old wood were detected. Although limited variations were observed in the Ca^{2+} and Mg^{2+} contents between non-treated and salt-treated (6.4 dS m^{-1}) roses, salt-sensitive varieties did have lower Ca^{2+} and Mg^{2+} contents at 6.4 dS m^{-1} than the tolerant ones (Niu et al., 2013). In contrast, Wu et al. (2016) reported that leaf Ca^{2+} content, along with Na^+, increased significantly with increasing salinity levels in six woody species. The results from Rahneshan et al. (2018), Walker et al. (2004), Niu et al. (2013), and Wu et al. (2016) suggest that it is more likely to be the Na^+/Ca^{2+} ratio, rather than the Ca^{2+} concentration itself, that is associated with plant salt tolerance (Cramer, 2002). Calcium has multiple functions in plants, including maintaining cell membrane integrity, controlling the activity of aquaporins that facilitate water movement across cell membranes, and regulating cell signaling as a secondary messenger (Rahneshan et al., 2018; Cramer, 2002). Ca is also required to maintain or enhance the selective absorption of K^+ by plants at high Na^+ concentrations (Epstein, 1998). Decreased Mg^{2+} under the saline condition may result in reduced chlorophyll synthesis and enzyme catalysis (Rahneshan et al., 2018). Although P is not typically considered the most limiting nutrient for plant growth (Rahneshan et al., 2018), P is required for ATP generation and activation of stress defense mechanisms and is energy intensive.

Cl, taken up as Cl^-, is required to counter balance cations maintaining electrical neutrality across membranes (Heckman, 2007; Hopkins and Hüner, 2004; Wege et al., 2017). It is one of the principal osmotically active solutes in the vacuole and an essential cofactor for the activation of the oxygen-evolving enzyme associated with photosystem II and participates in stomatal opening (Heckman, 2007; Hopkins and Hüner, 2004; Wege et al., 2017). Walker et al. (2004) reported Na^+ and Cl^- accumulation in saline-irrigated 'Sultata' grapes in the field. Although a lot of research has associated tissue Na^+ accumulation with reduced yield (quality and quantity) and aesthetics, 'Sultata' grapes grafted on their own roots (salt-sensitive based on the measurement of total leaf area and weight of one-year-pruning-wood) had higher Cl^- than the ones on 'Ramsey' rootstock (salt-tolerant), whereas the leaf Na^+ concentrations were similar between the two types of grafted plants (Walker et al., 1997). Over a nine-year field evaluation of 'Valencia' orange trees (*Citrus sinensis*) on sweet orange (*C. sinensis*) rootstock in saline soil, leaf Na^+ content increased at a later stage (~4 years after the saline exposure) than the Cl^- content (~2 years after) (Prior et al., 2007). The results from studies such as Walker et al. (1997) and Prior et al. (2007) do not suggest Cl^- is more toxic than Na^+, rather that plant species have a higher capability of Na^+ regulation via sequestration from roots and stems (Munns and Tester, 2008). Citrus shows Cl^- toxicity, slight leaf bronzing and leaf tip yellowing followed by tip burn and necrosis when tissue Cl^- concentration is above 0.7% of dry weight or 250 mM (Ferguson and Grattan, 2005; Munns and Tester, 2008), while most plant species can tolerant Cl^- as high as 400 mM (Munns and Tester, 2008). High Cl^- accumulation may not inhibit plant growth directly, but it affects production and translocation of essential factors regulating plant growth (Munns et al., 1988). Bañuls et al. (1990) observed decreased NO_3^- content in the leaves and roots of NaCl (0 to 60 mM) treated Troyer citrange and Cleopatra mandarin with a higher reduction in Troyer citrange than in Cleopatra mandarin. Furthermore, tissue NO_3^- content was negatively correlated with Cl^- accumulation (Bañuls et al., 1990). The negative feedback regulation between NO_3^- and Cl^- has also been reported in several other plant species, such as avocado (*Persea americana*) (Bar et al., 1997; Wiesman, 1995), peanut (*Arachis hypogea*) (Leidi et al., 1992), citrus (Bar et al., 1997), cotton (*Gossypium hirsutum*) (Leidi et al., 1992), and kiwifruit (*Actinidia deliciosa*) (Smith et al., 1987). The role of Cl^--induced NO_3^- reduction under saline conditions is unclear. It is suggested that NO_3^- replaced by Cl^- could have been used for plant growth; however, the change of NO_3^- content was fairly small and limited changes had been detected in leaf organic acids (Smith et al., 1987).

7.4.3 Oxidative Stress

Under saline conditions, stomatal closure and Na^+ and Cl^- toxicity impair photosynthetic machinery (directly and indirectly), reducing a plants' ability to fully utilize light that was absorbed by photosynthetic pigments; thus, formation and accumulation of reactive oxygen species (ROS) increase, leading to oxidative stress (Demidchik, 2017). H_2O_2 is the most important, relatively stable, non-radical ROS. It is often used as an indicator of intracellular ROS status (Demidchik, 2017). H_2O_2 has dual functions: 1) positively regulates plant growth and defense to biotic stresses at low concentration and 2) negatively impairs lipid, protein, DNA, and RNA at a high concentration; thus, the balance between H_2O_2 production and its scavenging is critical (Sofo et al., 2015). Mozafari et al. (2018) reported that H_2O_2 increased with an increase of NaCl concentrations (0, 50, and 100 mM) in grape 'Khoshnaw'. High H_2O_2 content can cause lipid peroxidation, which is commonly quantified by malondialdehyde (MDA) content (Mozafari et al., 2018). Elevated MDA indicates cell membrane damage and disruption of cell membrane selective permeability, resulting in increased electrolyte leakage (EL) (Kumar et al., 2017; Mozafari et al., 2018). Reduced total protein content in NaCl-treated 'Khoshnaw' grape may result from protein degradation and denaturation, and inhabitation of protein biosynthesis caused by oxidative stress (Mozafari et al., 2018). Reactive oxygen species, including H_2O_2, contribute to photoinhibition and reduced photosynthetic activities due to the damage of proteins such as Rubisco and other enzymes in photosystem II (Nishiyama et al., 2011).

7.5 MECHANISMS OF SALINITY TOLERANCE

7.5.1 Osmotic Adjustment

To counteract salinity-induced osmotic stress, plants undergo osmotic adjustment, one of the critical defense mechanisms to salinity stress (Marcum, 2008). Osmotic adjustment may be achieved through accumulation of inorganic ions and/or organic osmolytes, resulting in reduced osmotic potential in plant cells, and then water moves inside plant cells. Organic osmolytes are also called compatible solutes, which are naturally occurring organic compounds, non-toxic, and compatible with metabolic activity even at a high concentration in cytosol (Khan et al., 2010; Marcum, 2008). The most commonly detected compatible solutes in plants are sugars, proline, glycine betaine, and amino acids. Total proline content increased from 0.31 mg g^{-1} fresh weight at 0 mM NaCl to 1.14 mg g^{-1} fresh weight at 100 mM in 'Khoshnaw' grape (Mozafari et al., 2018). Mohammadkhani et al. (2014) also observed increased soluble sugars, proline, and glycine betaine along with increasing salinity levels in grapes; however, the salt tolerance was not closely associated with compatible solutes. For instance, 'Kajhave' had a higher sugar content than the other genotypes at low concentrations, but lower sugar content than the other genotypes at high concentrations (Mohammadkhani et al., 2014). Other than osmotic adjustment, organic osmolytes also help stabilize the tertiary structure of proteins at low concentrations (Munns and Tester, 2008).

Under a saline condition, large amounts of inorganic ions such as Na^+ and Cl^- are available in the external medium and can therefore be used for osmotic adjustment, if they can be successfully isolated in vacuole (i.e. compartmentation) as seen in salt-tolerant plants (Munns and Tester, 2008). The ion contents in salt-sensitive plants, in contrast, well exceed required levels for osmotic adjustment; thus, resulting in ion toxicity. In correspondence with the increased Na^+ concentrations in vacuoles, other subcellular compartments must increase osmotic pressure, possibly by accumulating K^+ to sub-toxic levels (<100 mM) and organic osmolytes to maintain their volume (Munns and Tester, 2008). X-ray microanalysis of mesophyll cells in plum leaves revealed that salt-treated (NaCl:$CaCl_2$, 2:1 molar ratio) plants accumulated Cl^- in the vacuoles for osmotic adjustment. Li et al. (2008) also observed salt compartmentation in two salt-treated (NaCl, 100–400 mM) mangrove species, *Kandelia candel* and *Bruguiera gymnorhiza*. Interestingly, more Na^+ compartmented in *K. candel* showing a higher reduction in photosynthetic rate (80%) but more Cl^- in *B. gymnorhiza* showed a lower reduction in photosynthetic rate (60%). Li et al. (2008) suggested that the differences in photosynthetic rate reduction between the two species was associated with the ability of Na^+ and Cl^- to be excluded from the chloroplast, rather than the bulk leaf salt concentration.

As limited organic osmolytes are available in the soil, plants mostly increase organic osmolytes content via *de novo* production, which is metabolically expensive and physiologically slow. Production of each molar of compatible solutes costs approximately 30 to 80 molars of ATP (Raven, 1985; Oren, 1999). Thus, plants that accumulate high organic solutes under the saline condition may have low growth rate. For instance, Mohammadkhani et al. (2014) reported significant negative correlations ($R^2 = -0.954$) between growth factors and compatible solutes (soluble sugars, proline, and glycinebetaine). It is more efficient to use inorganic ions mostly Na^+, K^+, and Cl^- than organic osmolytes to counterbalance the rapid drop of osmotic potential induced by salinity (Munns and Tester, 2008). Casas et al. (1991) observed a similar level of cell turgor between NaCl-adapted cells and non-adapted ones in *Atriplex nummularia*, suggesting that there are mechanisms other than osmotic adjustment that contribute to salinity tolerance.

7.5.2 Salt Overly Sensitive (SOS) Pathway

Salt overly sensitive pathway is a complex that includes three components, SOS3, SOS2, and SOS1 (Figure 7.1) (Ji et al., 2013). The Na ion around the root triggers a Ca^{2+} increase in cytoplasm that is sensed by a calcineurin B-like protein, SOS3 (Munns and Tester, 2008). Subsequently, SOS3 activates a srine/threonine protein kinase SOS2, forming a SOS3-SOS2 complex. The complex phosphorylates the Na^+/H^+ antiporter SOS1 to stimulate its Na^+/H^+ exchange activity at the plasma membrane, pumping Na^+ out and back to the external media (Munns and Tester, 2008; Quintero et al., 2002; Qiu et al., 2002). SOS2 also interacts with and activates the vacuolar H^+/Ca^{2+} antiporters for Na^+ compartmentation, independently of SOS3 (Batelli et al., 2007). The ability of SOS2 to increase salt tolerance is associated with its regulation on vacuolar H^+-ATPase (Batelli et al., 2007).

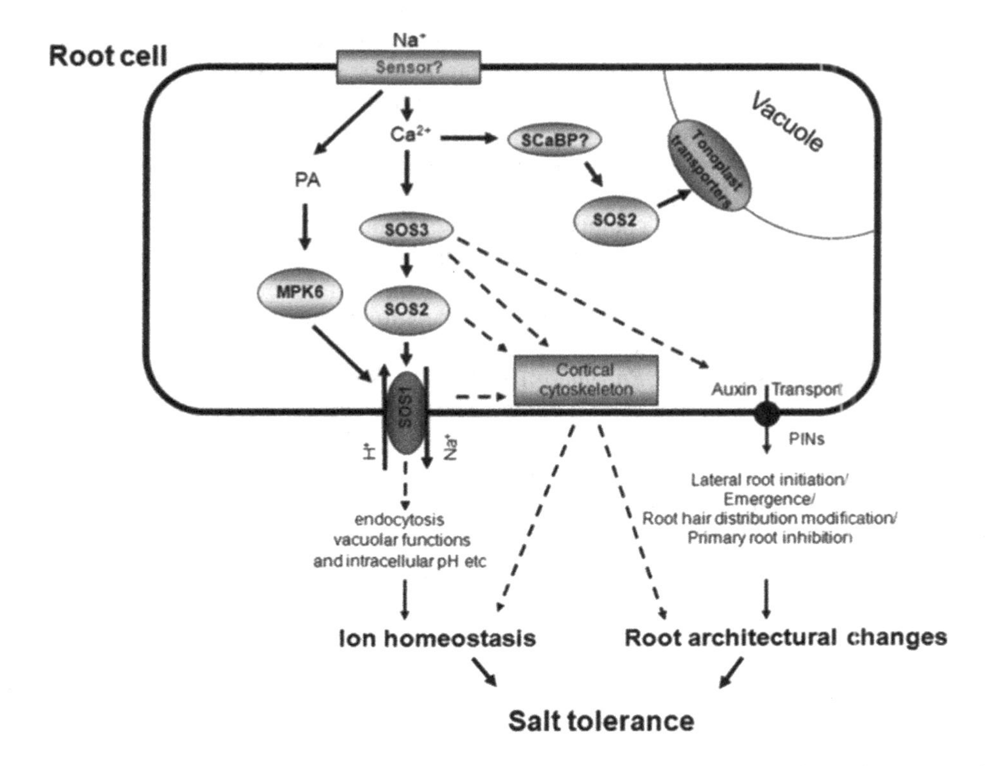

FIGURE 7.1 The current model of SOS signaling in root cortex/epidermal cells for maintenance of ion homeostasis, regulation of various cellular processes, and lateral root development based on current knowledge. Solid arrows indicate established, direct regulations, while dashed lines indicate suggested links between the components represented (Reprinted with permission from Ji et al., 2013).

Salt overly sensitive pathway regulates salt stress signaling in woody plants as well. Three SOS components, *PtSOS1*, *PtSOS2*, and *PtSOS3*, have been identified in poplar (*Populus trichocarpa*) (Tang et al., 2010). *PtSOS1*, *PtSOS2*, and *PtSOS3* interact cohesively with the same order as observed in *Arabidopsis*. Unlike the SOS in *Arabidopsis*, *PtSOS1* was specifically localized in the plasma membrane in poplar mesophyll cells, while *PtSOS2* was distributed throughout the cell and *PtSOS3* was predominantly targeted to the plasma membrane. Poplar plants with *PtSOS2* (*PtSOS2TD*) overexpression showed higher chlorophyll content and higher shoot fresh weight (47% to 94%) than the wild type plants under NaCl treatments (0 to 300 mM), particularly at higher concentrations (200 to 300 mM) (Yang et al., 2015). The improved salt tolerance in the transgenic plants was associated with a decreased Na^+ accumulation in the leaf resulting from enhanced Na^+/H^+ exchange activity and Na^+ efflux on the plasma membrane. Gao et al. (2016) isolated a Na^+/H^+ antiporter (SOS1) from *Chrysanthemum morifolium* (salt-sensitive) and *Crossostephium chinense*, *Artemisia japonica*, and *C. crassum* (salt-tolerant). The expression of SOS1 increased under salinity in all four species, but to a greater extent in the salinity tolerant plants.

7.5.3 Ion Secretion

Recretohalophytes are a small number of halophytes that have specialized cells including salt bladders (modified epidermal cells) or salt glands (modified trichrome) to secrete ions (Yuan et al., 2016). Eleven families (65 species) have been discovered to have salt secretion structures, including *Poaceae*, *Tamaricaceae*, *Plumbagenaceae*, *Conovolvulaceae*, *Verbenaceae*, *Sonnertiaceae*,

Acanthaceae, *Mysinaceae*, *Primulaceae*, *Frankeniaceae*, and *Scrophulariaceae* (Yuan et al., 2016). Salts are translocated from roots to leaves via the transpiration stream (Figure 7.2). Most ions are transported from mesophylls to salt glands through the basal penetration area that is not covered by cuticles. Sodium reflux is prevented by the transpiration stream, a valve structure that only opens inwardly, and the cuticle. Some Na^+ ions pass through plasmodesmata instead. Once inside the salt gland, Na^+ is either directly transported into the intercellular space (i.e. apoplastic) or through plasmodesmata between the salt gland component cells (i.e. symplastic) and then the Na^+ is stored in the collecting chamber. The ions are later excluded from the secreting pores due to high hydrostatic pressure.

Mangroves can exclude most of the salts from the root to minimize salinity stress. Scholander et al. (1962) evaluated salt balance in seven mangrove species and separated them into non-salt-secreting species including *Rhizophora mucronata*, *Bruguiera* prob. *exaristata*, *Sonneratia alba*, and *Lumnitzera littorea* and salt-secreting species including *Avicennia marina*, *Aegiceras corniulatum*, and *Aegialitis annulata* based on their ability to keep salts within physiologically acceptable levels. Total secretion rate increased from 2.0 mmol m^{-2} d^{-1} at 0 mM NaCl to 46.9 mmol m^{-2} d^{-1} at 865 mM NaCl in black mangrove (*Avicennia germinans*) to partially counteract the accumulation of salts transported into the leaves by transpiration (Sobrado and Greaves, 2000). Other ions, including K, S, Ca, Br, and Zn, also secreted from mangrove leaves along with Na^+

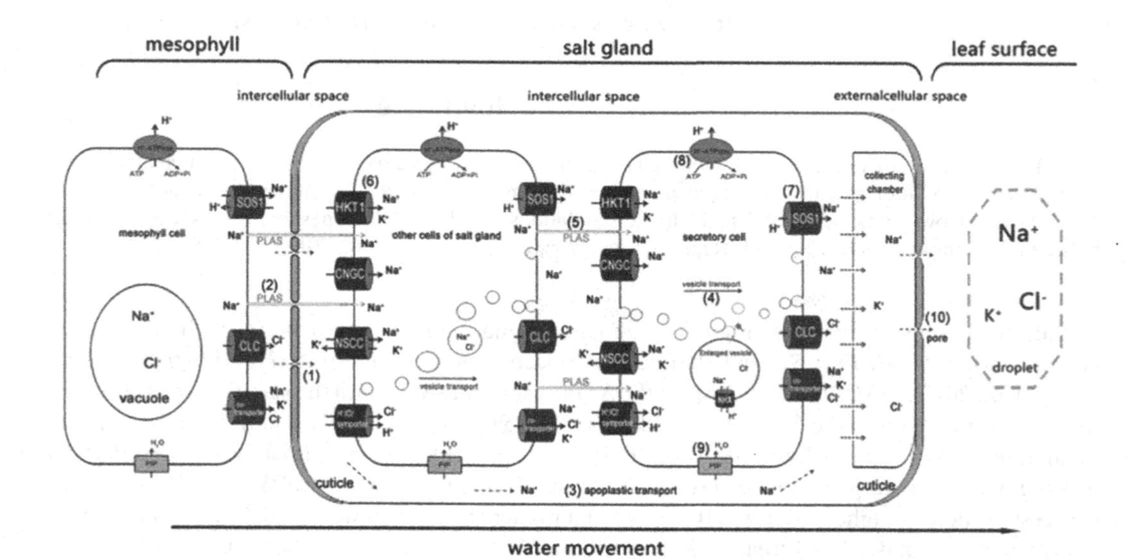

FIGURE 7.2 The possible pathway of Na^+ transport from the mesophyll to the salt gland, and the way in which Na^+ is secreted out. Na^+ is transported into the salt gland through the bottom penetration area, which is not covered with cuticles (1), and the plasmodesmata (2). In the salt gland, the ions can be directly transported into the intercellular space (3). The ions are wrapped in vesicles for transport from the cytosol to the plasma membrane and secreted out of the salt gland cells (4). Plasmodesmata play an important role in ion delivery among the salt gland component cells (5). Various ion transporters are widely involved in moving salt in (6) and out (7). H^+-ATPase participated in salt secretion by establishing electric potential difference and proton motive force across plasma membrane (8). Water channels (e.g., PIP) also take part in all the processes as a medium for ion transport (9). The ions are eventually exuded from the secreting pores at the top of the salt gland due to high hydrostatic pressure (10). Ion transporters responsible for influx (dark blue) and efflux (red) are asymmetrically distributed in the plasma membrane of salt gland cells. PLAS, plasmodesmata; HKT1, high-affinity K^+ transporter 1; CNGC, cyclic nucleotide-gated cation channel; NSCC, non-selective cationic channel; PIP, plasma membrane intrinsic protein; NHX, Na^+/H^+ antiporter; SOS1, Na^+/H^+ antiporter; CLC, chloride channel. (Reprinted from Yuan et al., 2016).

and Cl^-. However, the secretion amount from the aforementioned elements was much higher in the non-stressed plants (approximately 15% of the total secretion) compared to salt-stressed ones (<5% of the total secretion) (Sobrado and Greaves, 2000). Suárez and Medina (2008) reported that salt secretion rates of Na^+ and Cl^- increased linearly with NaCl concentrations (0 to 940 mM NaCl) in *A. germinans*. Despite the differences in the relative importance between Na^+ and Cl^- in ion secretion, Na^+ and Cl^- were the dominant ions in the total secretion, which help maintain favorable Na^+/K^+ and Cl^-/anion ratios in the leaf cells (Sobrado and Greaves, 2000; Suárez and Medina, 2008). Many factors, such as the salt concentration of the growth medium, light, temperature, oxygen pressure, and inhibitors of metabolism, influence ion secretion through the salt gland (Kozlowski, 1997).

Salt bladders are found in several genera of the Chenopodiaceae, including all *Atriplex* species (Schirmer and Breckle, 1982). Large amounts of ^{36}Cl and $^{35}SO_4$ accumulate in the stalk cell and peripheral cytoplasm of the large vacuolated bladder cell in *Atriplex* species after exposure to $K_2{}^{35}SO_4$ and $K^{36}Cl$ (Osmond et al., 1969). Light stimulated ion uptake to the bladder, particularly in young leaves (1 to 2 cm length), suggest that photosynthesis is a source of energy for ion secretion (Osmond et al., 1969). *A. halimus*, *A. prostrata*, and *Halimione portulacoides* all produce salt bladders. Both young and old bladders of *A. halimus* and *H. portulacoides* were effective in Na^+ and Cl^- accumulation contributing to their high salt tolerance, whereas *A. prostrata* had fewer bladders in old leaves than young ones, resulting in low salt tolerance (Freitas and Breckle, 1992). Removal of salt bladders affected *Atriplex* growth under the non-stress condition, yet inhibited the ability of Na^+ and Cl^- sequestration and K^+ retention which led to a salt-sensitive phonotype, reduced fresh weight, and increasing Na^+ accumulation under the stress (Kiani-Pouya et al., 2017). Removal of salt bladders also induced metabolic changes, particularly in gamma-aminobutyric acid (a non-protein amino acid known to modulate anion fluxes across the plasma membrane), proline, sucrose, inositol, and ion transport across cellular membranes (Kiani-Pouya et al., 2017).

7.5.4 Tolerance to Oxidative Stress

The activity of antioxidant enzymes, including superoxide dismutase (SOD), glutathione reductase (GR), catalase (CAT), and ascorbate peroxidase (APX), increased with increased salinity level, indicating a defense response to oxidative stress induced by salinity (Kumar et al., 2017; Mozafari et al., 2018). Superoxide dismutase is the front line of defending ROS, removing superoxide radicals – specifically, dissimulating $O_2^{\bullet-}$ radicals into H_2O_2 and O_2, which are precursors to other ROS and are generated in different subcellular compartments (Alscher et al., 2002). Catalase further detoxifies H_2O_2 to water and oxygen in peroxisomes and mitochondria. Ascorbate peroxidase and GR also removes H_2O_2 but through another pathway, the ascorbate-glutathione cycle. Carotenoids (which also contribute to photosynthesis), anthocyanins, and other non-enzymatic compounds and enzymes minimize oxidative stress by binding and scavenging. Mozafari et al. (2018) observed increased proline content in grape under salinity. Kumar et al. (2017) reported a 20% to 40% increase of total phenolic compounds and total flavonoids in the leaves of salt stressed oleander compare to the control plants. Kumar et al. (2017) suggests that secondary metabolites such as phenolic compounds and flavonoids (one type of phenolic) are activated only after the primary ROS scavenging system weakened in plants. Recent research suggests that increased oxidative enzyme activity has been attributed to the transcriptional activation of the SOS genes (Kumar et al., 2017). For instance, overexpressing *PtSOS2* in *populus* increased CAT activity, thus reducing salt-induced oxidative stress (Yang et al., 2015).

7.5.5 Dilution

Rhizophora mucronata was able to maintain a constant salt concentration at 510 to 560 meq L^{-1} when exposed to salinity at 17 meq Cl d^{-1}, due to its sufficient water uptake to 'dilute' salt concentration in cell sap (Levitt, 1980). This mechanism is observed in some moderately salt-resistant glacophytes with a rapid growth rate. In contrast, *Atriplex* species (Greenway et al., 1966) and *P. euphraica* (Chen and Polle, 2010) avoid the increase in salt concentration by increasing leaf succulence. The leaf cells, particularly the parenchyma, enlarge due to an increase of both water and dry matter to prevent salt content reaching a toxic concentration (Weissenböck, 1969).

7.6 MOLECULAR INFORMATION OF SALINITY TOLERANCE

Understanding the molecular and genetic basis of stress response is important for plant improvement. Comparing genetic variations in stress tolerance between and within closely related species might be a potentially powerful strategy for molecular breeding. Nevertheless, study gene responses to salinity just in salt-tolerant plants themselves can also increase knowledge of adaptation and defence mechanisms (Gu et al., 2004). Identifying genes involved in particular defense pathways and their genomic profiling provides more information on stress tolerance, especially for such stresses as salinity that are complex quantitative traits controlled by multiple genes (Chinnusamy et al., 2005).

Beritognolo et al. (2011) compared transcriptional changes in the leaves of two *Populus alba* genotypes, 6K3 (salt-sensitive) and 14P11 (salt-tolerant), when subjected to salt stress (200 mM NaCl) using cNDA microarrays containing 6,340 different genes. Among the commonly shared transcripts, 71 genes related to carbohydrate metabolism, energy metabolism, and photosynthesis were decreased and 13 genes related to metabolism and biosynthesis of proteins and macromolecules were increased. 21 genes increased specifically by salinity only in 14P11 and were associated with stress response, cell development, cell death, and catabolism; whereas, 11 genes increased specifically by salinity only in 6K3 and were involved in protein biosynthesis, metabolism of macromolecules, cell organization, and biogenesis. Furthermore, transcriptional responses to salinity varied in sexuality in some plant species. Photosynthesis-related genes were mainly up-regulated in male plants of *Populus yunnanensis* but down-regulated in female ones under salinity (50 mM NaCl for 7 d) (Jiang et al., 2012). The differences in salt tolerance between male and female poplars are related to the genes located in autosomes rather than in the female-specific region of the W chromosome (Jiang et al., 2012). Gu et al. (2004) profiled transcript changes in *Populus euthratica* (salt-tolerant) during salt stress (300 mM NaCl; 0.5–72 h) and recovery (1–72 h after removal of 300 mM NaCl). Among the transcripts up-regulated significantly by salinity were ionic and osmotic homeostasis elements, metabolism regulators, Rubisco activase, and photorespiration-related glycolate oxidase. During recovery, sucrose synthase, ABC transporter, calmodulin, Pop3 peptide, and aquaporin played important roles. Transcript analysis during salt exposure and recovery stage reveal a more 'complete' picture of plant responses to salinity, helping identify novel genes that deserve special attention in future studies.

REFERENCES

Abrol, I.P., R. Chhabra, and P.K. Gupta. 1980. A fresh look at the diagnostic criteria for sodic soils. pp. 142–147. In: *International Symposium on Salt Affected Soils*. Central Soil Salinity Reserch Institute, Karnal. Feb. 18–21, 1980.

Alscher, R.G., N. Erturk, and L.S. Heath. 2002. Role of superoxide dismutases (SODs) in controlling oxidative stress in plants. *J. Exp. Biol.* 531:1331–1341.

Aragüés, R., J. Puy, A. Royo, and J.L. Espada. 2005. Three-year field response of young olive trees (*Olea europaea* L., cv. Arbequina) to soil salinity: Trunk growth and leaf ion accumulation. *Plant Soil* 271:265–271.

Arend, M. and J. Fromm. 2007. Seasonal change in the drought response of wood cell development in poplar. *Tree Physiol.* 27:985–992.

Bañuelos, G.S., R. Mead, and G.J. Hoffman. 1993. Accumulation of selenium in wild mustard irrigated with agricultural effluent. *Agric. Ecosyst. Environ.* 43:119–126.

Bañuls, J., F. Legaz, and E. Primo-Millo. 1990. Effect of salinity on uptake and distribution of chloride and sodium in some citrus scion-rootstock combinations. *J. Hort. Sci.* 65:715–724.

Bañuls, J., F. Legaz, and E. Primo-Millo. 1991. Salinity-calcium interactions on growth and ionic concentration of citrus pants. *Plant Soil* 133:39–46.

Bar, Y., A. Apelbaum, U. Kafkafi, and R. Goren. 1997. Relationship between chloride and nitrate and its effect on growth and mineral composition of avocado and citrus plants. *J. Plant Nutri.* 20:715–731.

Batelli, G., P.E. Verslues, F. Agius, Q. Qiu, H. Fujii, S. Pan, K.S. Schumaker, S. Grillo, and J.K. Zhu. 2007. SOS2 promotes salt tolerance in part by interacting with the vacuolar H^+-ATPase and upregulating its transport activity. *Mol. Cell. Biol.* 27:7781–7790.

Beniwal, R.S., R. Langenfeld-Heyser, and A. Polle. 2010. Ectomycorrhiza and hydrogel protect hybrid poplar from water deficit and unravel plastic responses of xylem anatomy. *Environ. Exp. Bot.* 69:189–197.

Beritognolo, I., A. Harfouche, F. Brilli, F. Prosperini, M. Gaudet, M. Broschè, F. Salani, E. Kuzminsky, P. Auvinen, L. Paulin, J. Kangasjärvi, F. Loreto, R. Valentini, G.S. Mugnozza, and M. Sabatti. 2011. Comparative study of transcriptional and physiological responses to salinity stress in two contrasting *Populus alba* L. genotypes. *Tree Physiol.* 31:1335–1355.

Bessa, M.C., C.F. Lacerda, A.V. Amorim, A.M.E. Bezerra, and A.D. Lima. 2017. Mechanisms of salt tolerance in seedlings of six woody native species of the Brazillian semi-arid. *Rev. Ciênc. Agron.* 48:157–165.

Boland, A.M., P. Jerie, and E. Maas. 1997. Long-term effects of salinity on fruit trees. *Acta Hort.* 2:599–606.

Brennan, J. and M. Ulmer. 2010. *Salinity in the Northern Great Plains.* Natural Resources Conservation Service, Bismarck, ND.

Carrow, R.N. and R.R. Duncan. 1998. *Salt-Affected Turfgrass Sites: Assessment and Management.* Ann Arbor Press, Chelsea, MI.

Casas, A.M., R.A. Bressan, and P.M. Hasegawa. 1991. Cell growth and water relations of the halophyte, *Atriplex nummularia* L., in response to NaCl. *Plant Cell Rep.* 10:81–84.

Cassaniti, C., D. Romano, and T.J. Flowers. 2012. The response of ornamental plants to saline irrigation water. In: *Irrigation – Water Management, Pollution and Alternative Strategies.* I. Garcia-Garizabal (ed.). InTech. ISBN: 978-953-51-0421-6. Available from: http://www.intechopen.com/books/irrigation-water-management-pollution-and-alternative-strategies/response-of-ornamental-plants-to-saline-irrigation-water.

Catlin, P.B., G.J. Hoffman, R.M. Mead, and R.S. Johnson. 1993. Long-term response of mature plum trees to salinity. *Irrig. Sci.* 13:171–176.

Chen, S. and A. Polle. 2010. Salinity tolerance of *Populus. Plant Biol. (Stuttg.)* 12:317–333.

Chinnusamy, V., A. Jagendorf, and J.K. Zhu. 2005. Understanding and improving salt tolerance in plants. *Crop Sci.* 45:437–448.

Cramer, G.R. 2002. Sodium-calcium interactions under salinity stress. pp. 205–227. In: *Salinity. Environment-Plants-Molecules.* A. Läuchli and U. Lüttge (eds.). Kluwer, Dordrecht.

Croser, C., S. Renault, J. Franklin, and J. Zwiazek. 2001. The effect of salinity on the emergence and seedling growth of *Picea mariana, Picea glauca* and *Pinus banksiana. Environ. Pollut.* 115:9–16.

Demidchik, V. 2017. Reactive oxygen species and their role in plant oxidative stress. pp. 64–96. In: *Plant Stress Physiology.* S. Shabala (ed.). CABI, Boston, MA.

Division of Agriculture and Natural Resources, University of California (UC-ANR). 2016. Salinity Management. Available from https://ucanr.edu/sites/Salinity/Salinity_Management/ (Accessed on June 11, 2018).

Epstein, E. 1998. How calcium enhances plant salt tolerance. *Science* 280:1906–1907.

Everitt, J.H. 1983a. Seed germination characteristics of three woody plant species from south Texas. *J. Range Manag.* 36:246–249.

Everitt, J.H. 1983b. Seed germination characteristics of two woody legumes (retama and twisted acacia) from south Texas. *J. Range Manag.* 36:411–414.

Ferguson, L. and S.R. Grattan. 2005. How salinity damages citrus: Osmotic effects and specific ion toxicities. *HortTechnology* 15:95–99.

Francois, L.E. and E.V. Maas. 1999. Crop response and management of salt-affected soils. pp. 169–201. In: *Handbook of Plant and Crop Stress.* 2nd Ed. M. Pessarakli (ed.). Marcel Dekker, Inc, New York.

Franzen, D. 2013. *Managing saline soils in North Dakota.* North Dakota State University Extension Service, Fargo.

Freitas, H. and S.W. Breckle. 1992. Importance of bladder hairs for salt tolerance of field-grown *Atriplex* species from a Portuguese salt marsh. *Flora* 187:283–297.

Gao, J., J. Sun, P. Cao, L. Ren, C. Liu, S. Chen, F. Chen, and J. Jiang. 2016. Variation in tissue Na+ content and the activity of *SOS1* genes among two species and two related genera of Chrysanthemum. *BMC Plant Biol.* 16:98. doi:10.1186/s12870-016-0781-9.

Gorham, J. 2007. Sodium. pp. 569–583. In: *Handbook of Plant Nutrition*. A.V. Barker and D.J. Pilbeam (eds.). CRC Press, Taylor & Francis Group, LLC, Boca Raton, FL.

Greenway, H., A. Gunn, and D.A. Thomas. 1966. Plant responses to saline-substrates. VIII. Regulation of ion concentrations in salt-sensitive and halophytic species. *Aust. J. Biol. Sci.* 19:741–756.

Greenway, H. and C.B. Osmond. 1972. Salt responses of enzymes from species differing in salt tolerance. *Plant Physiol.* 49:256–259.

Grieve, C.M., S.R. Grattan, and E.V. Maas. 2012. Plant salt tolerance. pp. 405–459. In: *ASCE Manual and Reports on Engineering Practice*. 2nd Ed. No. 71. Agricultural Salinity Assessment and Management. W.W. Wallender and K.K. Tanji (eds.). ASCE, Reston, VA.

Gu, R., S. Fonseca, L.G. Puskás, L. Hackler, Jr., A. Zvara, D. Dudits, and M.S. Pais. 2004. Transcript identification and profiling during salt stress and recovery of *Populus euphratica*. *Tree Physiol.* 24:265–276.

Hacke, U.G., J.S. Sperry, W.T. Pockman, S.D. Davis, and K.A. McCulloch. 2001. Trends in wood density and structure are linked to prevention of xylem implosion by negative pressure. *Oecologia* 126:457–461.

Heckman, J.R. 2007. Chloride. pp. 279–292. In: *Handbook of Plant Nutrition*. A.V. Barker and D.J. Pilbeam (ed.). CRC Press, Taylor & Francis Group, LLC, Boca Raton, FL.

Hoffman, G.J., P.B. Catlin, R.M. Mead, R.S. Johnson, L.E. Francois, and D. Goldhamer. 1989. Yield and foliar injury response of mature plum trees to salinity. *Irrig. Sci.* 10:215–229.

Hopkins, W.G. and N.P.A. Hüner. 2004. *Introduction to Plant Physiology*. 3rd Ed. John Wiley & Sons, Inc., Hoboken, NJ.

Hura, T., B. Szewczyk-Taranek, K. Hura, K. Nowak, and B. Pawlowska. 2017. Physiological responses of *Rosa rubiginosa* to saline environment. *Water Air Soil Pollut.* 228:81. doi: 10.1007/s11270-017-3263-2.

Jamil, A., S. Riaz, M. Ashraf, and M.R. Foolad. 2011. Gene expression profiling of plants under salt stress. *Crit. Rev. Plant Sci.* 30:435–458.

Janz, D., S. Lautner, H. Wildhagen, K. Behnke, J. Schnitzler, H. Rennenberg, J. Fromm, and A. Polle. 2012. Salt stress induces the formation of a novel type of 'pressure wood' in two *Populus* species. *New Phytol.* 194:129–141.

Ji, H., J.M. Pardo, G. Batelli, M.J. Van Oosten, R.A. Bressan, and X. Li. 2013. The salt overly sensitive (SOS) pathway: Established and emerging roles. *Mol. Plant* 6:275–286.

Jiang, H., S. Peng, S. Zhang, X. Li, H. Korpelainen, and C. Li. 2012. Transcriptional profiling analysis in Populus synnanensis provides insights into molecular mechanisms of sexual differences in salinity tolerance. *J. Exp. Bot.* 63:3709–3726.

Joshi, R., B. Singh, A. Bohra, and V. Chinnusamy. 2016. Salt stress signaling pathways: Specificity and cross talk. pp. 51–78. In: *Managing Salt Tolerance in Plants: Molecular and Genomic Perspectives*. S.H. Wani and M.A. Hossain (eds.). CRC Press, Boca Raton, FL.

Khan, S.H., N. Ahmad, F. Ahmad, and R. Kumar. 2010. Naturally occurring organic osmolytes: From cell physiology to disease prevention. *IUBMB Life* 62:891–895.

Kiani-Pouya, A., U. Roessner, N.S. Jayasinghe, A. Lutz, T. Rupasinghe, N. Bazihizina, J. Bohm, S. Alharbi, R. Hedrich, and S. Shabala. 2017. Epidermal bladder cells confer salinity stress tolerance in the halophyte quinoa and Atriplex species. *Plant Cell Environ.* 40:1900–1915.

Koyro, H.W., T. Hussain, B. Huchzermeyer, and M.A. Khan. 2013. Photosynthetic and growth responses of a perennial halophytic grass *Panicum turgidum* to increasing NaCl concentrations. *Environ. Exp. Bot.* 91:22–29.

Kozlowski, T.T. 1997. Responses of woody plants to flooding and salinity. *Tree Physiol.* Monograph 1:1–29.

Kumar, D., M.Al. Hassan, M.A. Naranjo, V. Agrawal, M. Boscaiu, and O. Vicente. 2017. Effects of salinity and drought on growth, ionic relations, compatible solutes and activation of antioxidant systems in oleander (*Nerium oleander* L.). *PLOS ONE* 12:e0185017. https://doi.org/10.1371/journal.pone.0185017.

Leidi, E.O., M. Silberbush, M.I.M. Soares, and S.H. Lips. 1992. Salinity and nitrogen nutrition studies on peanut and cotton plants. *J. Plant Nutri.* 15:591–604.

Leksungnoen, N., R.K. Kjelgren, R.C. Beeson, Jr., P.G. Johnson, G.E. Cardon, and A. Hawks. 2014. Salt tolerance of three tree species differing in native habitats and leaf traits. *HortScience* 49:1194–1200.

Levitt, J. 1980. *Responses of Plants to Environmental Stresses. II. Water, Radiation, Salt, and Other Stresses*. Academic Press, New York.

Li, N., S. Chen, X. Zhou, C. Li, J. Shao, R. Wang, E. Fritz, A. Hüttermann, and A. Polle. 2008. Effect of NaCl on photosynthesis, salt accumulation and ion compartmentation in two mangrove species, *Kandelia candel* and *Bruguiera gymnorhiza*. *Aquat. Bot.* 88:303–310.

Machado, R.M.A., D.R. Bryla, and O. Vargas. 2014. Effects of salinity induced by ammonium sulfate fertilizer on root and shoot growth of highbush blueberry. *Acta Hort.* 407–414.

Manchanda, H.R. and S.K. Sharma. 1989. Tolerance of chloride and sulphate salinity in chickpea (*Cicer arietinum*). *J. Agric. Sci. Camb.* 113:407–410.

Marcum, K.B. 2008. Physiological adaptations of turfgrasses to salinity stress. pp. 407–428. In: *Handbook of Turfgrass Management and Physiology*. M. Pessarakli ed. CRC Press, Boca Raton, FL.

Marshner, H. 1995. *Mineral Nutrition of Higher Plants*. 2nd Ed. Academic Press, San Diego, CA.

Massi, R., R. Gucci, and M. Tattini. 1998. Salinity tolerance in four different rootstocks for peach. *Acta Hort.* 465:363–369.

McCarron, J.K., K.W. McLeod, and W.H. Conner. 1998. Flood and salinity stress of wetland woody species, buttonbush (*Cephalanthus occidentalis*) and swamp tupelo (*Nyssa sylvatica* var. *biflora*). *Wetlands* 18:165–175.

Metternicht, G.I. and J.A. Zinck. 2003. Remote sensing of soil salinity: Potentials and constraints. *Remote Sens. Environ.* 85:1–20.

Mohammadkhani, N., R. Heidari, N.Abbaspour, and F. Rahmani. 2014. Evaluation of salinity effects on ionic balance and compatible solute contents in nine grape (*Vitis* L.) genotypes. *J. Plant Nutri.* 37:1817–1836.

Mozafari, A., A.G. Asl, and N. Ghaderi. 2018. Grape response to salinity stress and role of iron nanoparticle and potassium silicate to mitigate salt induced damage under in vitro conditions. *Physiol. Mol. Biol. Plants* 24:25–35.

Munns, R., P.A. Gardner, M.L. Tonnet, and H.M. Rawson. 1988. Growth and development in NaCl-treated plants II. Do Na^+ or Cl^- concentrations in dividing or expanding tissues determine growth in barley? *Aust. J. Plant Physiol.* 15:529–540.

Munns, R. and M. Gilliham. 2015. Salinity tolerance of crops – What is the cost? *New Phytol.* 208:668–673.

Munns, R. and A. Termaat. 1986. Whole-plant responses to salinity. *Aust. J. Plant Physiol.* 13:143–160.

Munns, R. and M. Tester. 2008. Mechanisms of salinity tolerance. *Annu. Rev. Plant Biol.* 59:651–681.

Musacchi, S., M. Quartieri, and M. Tagliavini. 2006. Pear (*Pyrus communis*) and quince (*Cydonia oblonga*) roots exhibit different ability to prevent sodium and chloride uptake when irrigated with saline water. *Eur. J. Agron.* 24:268–275.

Nasr, T.A., E.M. El-Azab, and M.Y. El-Shurafa. 1977. Effect of salinity and water table on growth and tolerance of plum and peach. *Sci. Hort.* 7(3):225–235.

National Land and Water Resources Audit. 2000. *Australian Dryland Salinity Assessment 2000: Extent, Impacts, Processes, Monitoring and Management Options*. National Land and Water Resources Audit, Canberra, Australia.

Niknam, S.R. and J. McComb. 2000. Salt tolerance screening of selected Australian woody species – A review. *Forest Ecol. Manag.* 139:1–19.

Nishiyama, Y., S.I. Allakhverdiev, and N. Murata. 2011. Protein synthesis is the primary target of reactive oxygen species in the photoinhibition of photosystem II. *Physiol. Plant.* 142:35–46.

Niu, G., D.S. Rodriguez, L. Aguiniga, and W. Mackay. 2007. Salinity tolerance of *Lupinus havardii* and *Lupinus texensis*. *HortScience* 42:526–528.

Niu, G., T. Starman, and D. Byrne. 2013. Responses of growth and mineral nutrition of garden roses to saline water irrigation. *HortScience* 48:756–761.

Osmond, C.B., U. Lüttge, K.R. West, C.K. Pallaghy, and B. Shacher-Hill. 1969. Ion absorption in *Atriplex* leaf tissue. II. Secretion of ions to epidermal bladders. *Aust. J. Biol. Sci.* 22:797–814.

Oren, A. 1999. Bioenergetic aspects of halophilism. *Microbiol. Mol. Biol. Rev.* 63:334–348.

Percival, G.C. 2005. Identification of foliar salt tolerance of woody perennials using chlorophyll fluorescence. *HortScience* 40:1892–1897.

Pessarakli, M. 1999. Soil salinity and sodicity as particular plant/crop stress factors. pp. 1–15. In: *Handbook of Plant and Crop Stress*. M. Pessarakli (ed.). 2nd Ed. Marcel Dekker, Inc., New York.

Prior, L.D., A.M. Grieve, K.B. Bevington, and P.G. Slavich. 2007. Long-term effects of saline irrigation water on 'Valencia' orange trees: Relationships between growth and yield, and salt levels in soil and leaves. *Aust. J. Agric. Res.* 58:349–358.

Proseus, T.E., G.L. Zhu, and J.S. Boyer. 2000. Turgor, temperature and the growth of plant cells: Using *Chara corallina* as a model system. *J. Exp. Bot.* 51:1481–1494.

Qiu, Q.S., Y. Guo, M.A. Dietrich, K.S. Schumaker, and J.K. Zhu. 2002. Regulation of SOS1, a plasmamembrane Na^+/H^+ exchanger in *Arabidopsis thaliana*, by SOS2 and SOS3. *Proc. Natl. Acad. Sci. U.S.A.* 99, 8436–8441.

Quintero, F.J., M. Ohta, H. Shi, J.K. Zhu, and J.M. Pardo. 2002. Reconstitution in yeast of the *Arabidopsis* SOS signaling pathway for Na^+ homeostasis. *Proc. Natl. Acad. Sci. U.S.A.* 99, 9061–9066.

Rahneshan, Z., F. Nasibi, and A.A. Moghadam. 2018. Effects of salinity stress on some growth, physiological, biochemical parameters and nutrients in two pistachio (*Pistacia vera* L.) rootstocks. *J. Plant Interact.* 13:73–82.

Raven, J.A. 1985. Regulation of pH and generation of osmolarity in vascular plants: A cost-benefit analysis in root growth response to a salinity gradient. *J. Exp. Bot.* 62:69–77.

Schiop, S.T., M. Al Hassan, A.F. Sestras, M. Boscaiu, R.E. Sestras, and O. Vicente. 2015. Identification of salt stress biomarkers in Romanian Carpathian populations of *Picea abies* (L.) Karst. *PLOS ONE* 10:e0135419. http://doi.org/10.1371/journal.pone.0135419. PMID:26287687.

Schirmer, U. and S. Breckle. 1982. The role of bladders for salt removal in some Chenopodiaceae (mainly *Atriplex* species). pp. 215–232. In: *Tasks for Vegetation Science*. D. Sen and K. Rajpurohit (eds.). Dr. W. Junk, The Hague.

Scholander, P.F., H.T. Hammel, E. Hemmingsen, and W. Garey. 1962. Salt balance in mangroves. *Plant Physiol.* 37:722–729.

Schreiber, S.G., U.G. Hacke, A. Hamann, and B.R. Thomas. 2011. Genetic variation of hydraulic and wood anatomical traits inhybrid poplar and trembling aspen. *New Phytol.* 190:150–160.

Seelig, B.D. 2000. Salinity and sodicity in North Dakota soils. North Dakota State Univ. Extension Service Bulletin 57.

Shani, U. and A. Ben-Gal. 2005. Long-term response of grapevines to salinity: Osmotic effects and ion toxicity. *Am. J. Enol. Vitic.* 56:148–154.

Smith, G.S., C.J. Clark, and P.T. Holland. 1987. Chlorine requirement of kiwifruit (*Actinidia deliciosa*). *New Phytol.* 106:71–80.

Sobrado, M.A. and E.D. Greaves. 2000. Leaf secretion composition of the mangrove species *Avicennia germinans* (L.) in relation to salinity: A case study by using total-reflection X-ray fluorescence analysis. *Plant Sci.* 159:1–5.

Sofo, A., A. Scopa, M. Nuzzaci, and A. Vitti. 2015. Ascorbate peroxidase and catalase activities and their genetic regulation in plants subjected to drought and salinity stresses. *Int. J. Mol. Sci.* 16:13561–13578.

Stevens, R.M., G. Harvey, and G. Davies. 1996. Separating the effects of foliar and root salt uptake on growth and mineral composition of four grapevine cultivars on their own roots and on 'Ramsey' rootstock. *J. Amer. Soc. Hort. Sci.* 121:569–575.

Stevens, R.M., G. Harvey, and D.L. Partington. 2011. Irrigation of grapevines with saline water at different growth stages: Effects on leaf, wood and juice composition. *Aust. J. Grape Wine Res.* 17:239–248.

Stiller, V. 2009. Soil salinity and drought alter wood density and vulnerability to xylem cavitation of baldcypress (*Taxodium distichum* (L.) Rioch.) seedlings. *Environ. Exp. Bot.* 67:164–171.

Storey, R. and R.R. Walker. 1999. Citrus and salinity. *Sci. Hortic.* 78:39–81.

Suárez, N. and E. Medina. 2008. Salinity effects on leaf ion composition and salt secretion rate in *Avicennia germinans* (L.) L. *Braz. J. Plant Physiol.* 20:131–140.

Taiz, L. 1984. Plant cell expansion: Regulation of cell wall mechanical properties. *Annu. Rev. Plant Physiol.* 35:585–657.

Tang, R., H. Liu, Y. Bao, Q. Lv, L. Yang, and H. Zhang. 2010. The woody plant poplar has a functionally conserved salt overly sensitive pathway in response to salinity stress. *Plant Mol. Biol.* 74:367–380.

Tanji, K.K. and N.C. Kielen. 2002. *Agricultural Drainage Water Management in Arid and Semi-Arid Areas*. Food and Agriculture Organization of the United Nations, Rome.

United States Salinity Laboratory Staff. 1954. Diagnosis and improvement of saline and alkali soils. *Agriculture Handbook No. 60*. United States Department of Agriculture.

Walker, R.R., D.H. Blackmore, P.R. Clingeleffer, and R.L. Correll. 2002. Rootstock effects on salt tolerance of irrigated field-grown grapevines (*Vitis vinifera* L. cv. Sultana) 1. Yield and vigour inter-relationships. *Aust. J. Grape Wine Res.* 8:3–14.

Walker, R.R., D.H. Blackmore, P.R. Clingeleffer, and R.L. Correll. 2004. Rootstock effects on salt tolerance of irrigated field-grown grapevines (*Vitis vinifera* L. cv. Sultana) 2. Ion concentrations in leaves and juice. *Aust. J. Grape Wine Res.* 10:90–99.

Walker, R.R., D.H. Blackmore, P.R. Clingeleffer, and D. Emanuelli. 2014. Rootstock type determines tolerance of Chardonnay and Shiraz to long-term saline irrigation. *Aust. J. Grape Wine Res.* 20:496–506.

Walker, R.R., D.H. Blackmore, P.R. Clingeleffer, and F. Iacono. 1997. Effect of salinity and Ramsey rootstock on ion concentrations and carbon dioxide assimilation in leaves of drip-irrigated, field-grown grapevines (*Vitis vinifera* L. cv. Sultana). *Aust. J. Grape Wine Res.* 3:66–74.

Wege, S., M. Gilliham, and S.W. Henderson. 2017. Chloride: Not simply a 'cheap osmoticum', but a beneficial plant macronutrient. *J. Exp. Bot.* 68:3057–3069.

Weissenböck, G. 1969. The effect of the salt content of the soil upon the morphology and ion accumulation of halophytes. *Flora (Jena)* B158:369–389.

Wiesman, Z. 1995. Rootstock and nitrate involvement in 'Ettinger' avocado response to chloride stress. *Sci. Hort.* 62:33–43.

Wright, G.C., K.D. Patten, and M.C. Drew. 1992. Salinity and supplemental calcium influence growth of rabbiteye and southern highbush blueberry. *J. Amer. Soc. Hort. Sci.* 117:749–756.

Wu, S., Y. Sun, G. Niu, G.L.G. Pantoja, and A.C. Rocha. 2016. Responses of six *Lamiaceae* landscape species to saline water irrigation. *J. Environ. Hort.* 34:30–35.

Yang, Y., R. Tang, C. Jiang, B. Li, T. Kang, H. Liu, N. Zhao, X. Ma, L. Yang, S. Chen, and H. Zhang. 2015. Overexpression of the *PtSOS2* gene improves tolerance to salt stress in transgenic poplar plants. *Plant Biotechnol. J.* 13:962–973.

Ye, Y., N.F. Tam, C. Lu, and Y. Wong. 2005. Effects of salinity on germination, seedling growth and physiology of three salt-secreting mangrove species. *Aquat. Bot.* 83:193–205.

Yuan, F., B. Leng, and B. Wang. 2016. Progress in studying salt secretion from the salt glands in recretohalophytes: How do plants secrete salt? *Front. Plant Sci.* 7:977. doi: 10.3389/fpls.2016.00977.

Zrig, A., H.B. Mohamed, T. Tounekti, H. Khemira, M. Serrano, D. Valero, and A.M. Vadel. 2016. Effect of rootstock on salinity tolerance of sweet almond (cv. Mazzetto). *S. Afr. J. Bot.* 102:50–59.

8 Plant Response to Low Temperature

Guo-Qing Song

CONTENTS

8.1 INTRODUCTION

8.1.1 Low Temperatures in Plant Life Cycles

There is great diversity of tolerance to extreme low and high temperatures among plant species (Strimbeck et al. 2015). The lowest temperature at which a plant can survive is genetically controlled and is often affected by many factors, for example, natural environment, acclimation, and plant developmental stage (Larcher 2003; Kreyling, Schmid, and Aas 2015). Deciduous fruits such as apples, grapes, pears, stone fruits (e.g. peaches, nectarines, apricots, and cherries) and berries (blueberries, cranberries, pomegranate, and kiwi) are of importance for human health due to their unique nutritional values packed with fiber, sugars, minerals, vitamins, anti-oxidants, etc. and medicinal values. Most of the cultivated deciduous fruit plants originate and are grown in temperate climates where their ability to survive depends on a tolerance to low temperatures. Thus, their survival is based on plant species and genotypes. The hardiness zones defined by the U.S. Department of Agriculture is often a reference for choosing appropriate fruit crops and cultivars to grow at specific production sites. In annual growth cycles of deciduous fruit crops, warm temperatures in spring break bud dormancy and initiate a new season of plant growth and development (i.e. flowering, fruiting, and growing) under conditions of increasing photoperiod and warm temperatures through summer to early fall. As fall continues, deciduous plants become dormant in response to the shortening photoperiod and decreasing temperatures, and they remain dormant in winter and early spring, during which exposure to freezing temperatures maintains dormancy and chilling hours build up until release of dormancy for the next round of plant growth cycle (Rohde and Bhalerao 2007; Byrne and Bacon 1992). During this annual growth cycle, temperatures and light are the key environmental factors that guide seasonal plant growth and development (Saure 1985).

8.1.2 Photoperiod, Low Temperature, and Plant Dormancy

Photoperiod regulates flowering time of many herbaceous plants (Fornara, de Montaigu, and Coupland 2010; Greenup et al. 2009). For deciduous woody plants, flowering time often relies on plant dormancy but does not necessarily rely on photoperiod, although short photoperiod and low temperatures are considered the main environmental signals to induce flower bud formation and plant dormancy in fall and early winter (Olsen 2010; Wilkie, Sedgley, and Olesen 2008). Low temperature plays a major role during winter dormancy, when freezing temperatures (≤0°C) prevent the buds from bursting and protect the plants from freezing damage, and chilling temperatures (>0°C) promote the build-up of chemical signals needed for releasing dormancy and forcing the buds to burst in spring (Kramer, Leinonen, and Loustau 2000; Chuine et al. 2016). Chilling does not usually result in damage to dormant tissues but can cause damage when it occurs to un-acclimated leaf or flower tissues. In fact, insufficient winter chilling can reduce bud breaking and is a main concern for production of fruit in temperate zones. Freezing that occurs either before plant acclimation or during dormancy releasing can cause plant injury or death and is therefore the major concern for sustainable fruit production (Melke 2015; Olsen 2010, 2003; Atkinson, Brennan, and Jones 2013; Luedeling et al. 2011). Studies on low temperature tolerance in higher plants have been conducted for more than 100 years (Strimbeck et al. 2015). A recent review with a focus on proteomics and metabolomics on plants that can naturally survive some of the lowest temperatures on Earth shows that changes in carbohydrate and compatible solute concentrations, membrane lipid composition, and proteins have important roles in plant survival in extreme freezing temperatures (Strimbeck et al. 2015).

Flowering plants are thought to be descendants of woody plants originally restricted to warm habitats (Sinnott and Bailey 1915; Wing and Boucher 1998; Feild et al. 2004; Zanne et al. 2014). During evolution, some ancestral plants gained new structural and functional traits and inhabited novel cold environments. For example, plants possessing transport networks of small safe conduits or having the ability to shut down hydraulic function by dropping leaves during freezing have successfully adapted to a vast range of freezing prone environments (Zanne et al. 2014). In contrast to the evolution pathways of herbaceous plants with small conduits, the evolutionary pathway to deciduous plants in freezing environments was largely *via* a shift to seasonal leaf shedding only after exposure to freezing, evolving into the freezing tolerance related trait (Zanne et al. 2014).

Climate change during the past 40 years has caused a shift to an earlier onset of the growing season of trees (e.g. 2.3 days/decade in temperate Europe) (Chuine et al. 2016; Root et al. 2003; Parmesan and Yohe 2003; Atkinson, Brennan, and Jones 2013). The reduced winter chill is often associated with insufficient chilling hours. Warmer weather sometimes leads to fruit and nut trees flowering out of season. The increased temperature fluctuation during plant blooming time turns seasonal frost into a danger, which often causes freezing injuries to flowers and young fruits. Apparently, climate change is occurring and its potential impact on production of deciduous fruit crops needs to be addressed (Luedeling et al. 2011; Atkinson, Brennan, and Jones 2013; Melke 2015).

Plant breeding to develop low chilling requirements and freezing tolerant cultivars is considered to be a long-term solution to mitigate both reduced winter chill and increase freezing damage, and to secure deciduous fruit production (Atkinson, Brennan, and Jones 2013). Many studies have yielded considerable insight into the molecular and biophysical mechanisms and functional genomics of plant dormancy and freezing tolerance (Anderson 2015; Strimbeck et al. 2015). This review will focus on recent studies on transcriptomic analysis of genes and networks involving in low temperature responses (i.e. dormancy, vernalization, and freezing tolerance) of deciduous fruit crops.

8.2 CHILLING REQUIREMENT AND PLANT FLOWERING

The widely used terminologies to describe different plant dormancy physiological statuses include eco-dormancy, para-dormancy, and endo-dormancy (Lang et al. 1987). Endo-dormancy (winter dormancy) affects freezing tolerance and flowering of deciduous fruit crops and is essential for their survival over winter and for reproduction in the following year (Anderson 2015; Olsen 2006; Liu, Zhu, and Abbott 2015; Zinn, Tunc-Ozdemir, and Harper 2010). Under inductive low temperatures in the fall, plants are acclimated to develop freezing tolerance and accumulation of effective chilling hours is stimulated (Ouellet and Charron 2013). Sufficient chilling accumulation gives plants full vernalization, which is a prerequisite to breaking bud dormancy and enabling a normal flowering time and vegetative growth.

8.2.1 Variations of Chilling Requirements in Deciduous Fruit Crops

Winter dormancy and chilling requirements are genetically controlled and environmentally affected (Olsen 2006, 2010; Liu, Zhu, and Abbott 2015). Chilling requirements vary among fruit crops, for example, cranberries (1,200 to 1,400 chill hours (CH chilling hours calculated based on the hours at temperatures between 1.1–12.2°C using Utah model)), hazelnuts (400 to 1,400 CH), sweet cherries (300 to 1,000 CH), sour cherries (700 to 1,000 CH), apples (200 to 1,000 CH), pears (500 to 15,00 CH), peaches (100 to 1,000 CH), blueberries (150 to 1,200 CH), raspberries (700 to 800 CH), blackberries (200 to 500 CH), grapes (>100 CH), almonds (200 to 400 CH), walnut (600 to 700 CH), strawberries (200 to 300 CH), kiwifruits (600 to 800 CH), figs (100 to 200 CH), pomegranates (100 to 200 CH), pecans (400 to 500 CH), jujubes (200 to 400 CH), quinces (300 to 500 CH), pistachios (450 to 1,200 CH), plums (200 to 900 CH), persimmons (200 to 400 CH). The ranges of CHs for each fruit and nut crop show that the chilling requirements often vary widely from one variety to another within one crop. Choosing varieties with the right chilling requirement is essential for successful fruit and nut production.

8.2.2 Chilling-Regulated Flowering in Deciduous Fruit Crops

Flowering is a prerequisite for fruiting and plays a significant role in the life cycle of flowering plants (Angiosperms). A wealth of information on plant flowering (e.g. behaviors, regulation factor, and molecular mechanism) in literature has suggested that a network of flowering pathway genes controls seasonal flowering. To date, flowering mechanisms have been well known in *Arabidopsis thaliana*, rice, and cereals (Fornara, de Montaigu, and Coupland 2010; Amasino 2010; Greenup et al. 2009; Trevaskis et al. 2007; Lee and Lee 2010; Wellmer and Riechmann 2010; Pin et al. 2010; Song et al. 2015; Michaels 2009; Huijser and Schmid 2011; Higgins, Bailey, and Laurie 2010; Cockram et al. 2007); in contrast, those mechanisms in woody plants remain to be revealed (Jameson and Clemens 2015; Wilkie, Sedgley, and Olesen 2008; Jia et al. 2014; Zhang et al. 2010).

Photoperiod is the primary environmental factor that regulates the flowering time of many annual plants. *A. thaliana* transcription factor *CONSTANS* (*CO*) and *Flowering Locus T* (*FT*) are key regulators in the photoperiodic flowering pathway (Golembeski et al. 2014). The *CO* activates the *FT* to promote flowering under inductive long-day conditions in *A. thaliana*. *CO-FT* pathway, equivalent to rice (*Oryza sativa*) *OsCO* (*Hd1*)-*OsFT* (*Hd3a*) and wheat (*Triticum aestivum*) *CO-TaFT* [*VERNALIZATION3* (*VRN3*)] (Griffiths et al. 2003; Yan et al. 2006), is often considered a conserved pathway in all flowering plants.

Temperature/vernalization is necessary for normal flowering of winter annual and deciduous perennial plants. The *A. thaliana* MADs-box gene *Flowering Locus C* (*FLC*) in winter annual *A. thaliana* ecotypes plays a central role in their vernalization pathway. The *A. thaliana FRIGIDA* (*FRI*)-*FLC-FT* regulatory loop seems to be conserved in the family *Brassicaceae*. In vernalization-requiring wheat and barley, a *VRN1-VRN2- TaFT* (*VRN3*) loop regulates vernalization-mediated flowering (Greenup et al. 2009); however, this loop was recently found to be unique to core Pooideae (Woods et al. 2016). It is interesting to point out that *VRN1* is a *FRUITFUL-LIKE* MADS BOX gene and *VRN2* is a *CONSTANS-LIKE* gene (*COL*) (Preston and Kellogg 2008; Yan et al. 2003, 2004; Hemming et al. 2008). In woody plants, neither a functional *FLC-LIKE* nor a functional *VRN2-LIKE* has been documented. However, based on the studies in a peach (*Prunus persica*) evergrowing mutant, a cluster of six MADS-box transcription factors were termed as *DORMANCY ASSOCIATED MADS-BOX* (*DAM*) genes because their roles in dormancy are analogous to *FLC* in *A. thaliana* or *VRN2* in cereals, and the loss of all or part of four of the six *DAMs* resulted in the non-dormant evergrowing mutant (Bielenberg et al. 2008; Wang et al. 2002). The *DAM* genes show high similarities to *A. thaliana AGAMOUS-LIKE 24* (*AGL24*) and *SHORT VEGETATIVE PHASE* (*SVP*) genes (Jimenez et al. 2009; Jimenez, Reighard, and Bielenberg 2010; Sasaki et al. 2011). Functional analysis of the *DAMs* through reverse genetics has not been reported in peach. Taken together, the *CO-FT* regulatory module in the photoperiodic flowering pathway is conserved in flowering plants and the *FLC/VRN2/DAM-FT* regulatory module in vernalization/chilling-mediated flowering pathway varies among divergent plant species (Böhlenius et al. 2006; Woods et al. 2016).

Studies on *DAM* or *DAM*-like genes (herein *DAMs*) is well summarized in a recent review (Horvath 2015). Differential expression analysis *DAMs* in several deciduous fruit crops (Table 8.1) has suggested that *DAMs* likely play a significant role in regulating endo-dormancy responses in some perennials (Horvath 2015). To date, the *DAM5* and *DAM6* are the top candidates that show analogous roles to *FLC* or *VRN2* in peach (Jimenez, Reighard, and Bielenberg 2010). Six tandemly arrayed *PmDAMs* of Japanese apricot (*Prunus mume*) are similar to the *DAMs* in peach. The result of ectopic expression of the apricot *PmDAM6* in poplar (*Populus tremula* × *Populus tremuloides*) suggests that the *PmDAM6* has analogous roles to *FLC* or *VRN2*; meanwhile, *PmDAM4*, *PmDAM5*, and *PmDAM6* show genetic associations with chilling requirement in apricots (Sasaki et al. 2011). On the other hand, several *A. thaliana DAMs* (e.g. *AGL24* and *SVP*) in addition to *FLC* are known to be functional in regulating plant flowering (Gregis et al. 2006; Jimenez et al. 2009; Jimenez, Reighard, and Bielenberg 2010; Fornara, de Montaigu, and Coupland 2010; Mateos et al. 2015), and while *DAMs* are found in most dicot plant species, only a few seem to have a role in endo-dormancy responses (Horvath 2015). Overall, more direct evidence is still needed to support a valid *DAM*(s)-*FT* module in regulating vernalization mediated flowering in deciduous fruit and nut crops.

Regardless of the upstream *CO* or *COL* gene in photoperiodic flowering pathway and *FLC*, *VRN2*, or *DAMS* in vernalization pathway, the downstream *FT* or *FT*-like is generally considered the primary florigen and its function in proposed floral regulatory pathways (*e.g.,* Greenup et al., 2009 and Fornara et al. 2010) is conserved in flowering plants. In *A. thaliana* and cereal crops, *FLC* and *VRN2* negatively regulate *FT* and *FT*-like genes. Overexpression of *FT* or *FT*-like genes promotes flowering. Functional analyses of *FT* and *FT*-like genes have been reported in numerous studies for 20 plant species of 15 families (Pin et al. 2010; Wickland and Hanzawa 2015). These include several woody plant species (Table 8.1). For example, overexpression of a blueberry (*Vaccinium corymbosum*) *FT*-like gene (*VcFT*) was able to reverse the photoperiodic and chilling requirements and to promote plant flowering (Song et al. 2013). Phenotypically, overexpression of *VcFT* (about 2,896-fold increase in leaf tissues) caused early and continuous flowering in *in vitro* shoots and in one-year-old 'Aurora' plants. However, our observation on two- and three-year-old plants showed that the *VcFT*-overexpressing plants did not flower normally, and the majority of the flower buds

TABLE 8.1
Literature Survey on Genes that Regulate Flowering, Dormancy, and Freezing Tolerance in Deciduous Fruit and Nut Crops

Deciduous Fruit and Nut Crops	Genes	References
Apple	Ectopic and over-expression apple *FT* genes	Trankner et al. (2010), Mimida et al. (2011), Li et al. (2010), Kotoda et al. (2010)
	DAM-like genes	Mimida et al. (2015), Kumar et al. (2016), Allard et al. (2016), Falavigna et al. (2014)
	Ectopic expression of a peach *PpCBF1* gene in apple	Wisniewski et al. (2011), Wisniewski, Norelli, and Artlip (2015)
Apricot (*Prunus mume*)	*DAM* genes	Sasaki et al. (2011)
	CBF genes in apricot	Zhang et al. (2013)
Bilberry (*Vaccinium myrtillus*)	Ectopic expression of a bilberry *CBF* gene in *A. thaliana*	Oakenfull, Baxter, and Knight (2013)
Blueberry (*Vaccinium corymbosum*)	Overexpression of a blueberry *FT* gene (*VcFT*)	Song et al. (2013), Walworth et al. (2016), Gao et al. (2016)
	Overexpression of a blueberry *DDF1* gene (*VcDDF1*)	Walworth et al. (2012), Song and Gao (2017)
Fig (*Ficus carica* L.)	Ectopic expression of a fig *FT* gene (*FcFT1*)	Ikegami et al. (2013)
Grapes	Characterization of four *CBF/DREB1*-like genes from wild grape (*Vitis riparia*) cultivated grape (*Vitis vinifera*) in *A. thaliana*	Xiao et al. (2006, 2008)
	DEREB/CBF genes in grapes *Vitis amurensis* and *Vitis vinifera*	Zandkarimi et al. (2015), Xin et al. (2013)
	Overexpression a *Vitis vinifera* C-repeat binding protein 4 gene (*VvCBF4*)	Tillett et al. (2012)
	Vitis vinifera CONSTANS-like genes (*VvCO* and *VvCOL1*),	Almada et al. (2009)
	Vitis vinifera FT (*VvFT*) and endodormancy	Vergara et al. (2016)
Kiwifruit	Kiwifruit flowering- and dormancy-related genes	Varkonyi-Gasic et al. (2013), Voogd, Wang, and Varkonyi-Gasic (2015)
Peach	Genes related to dormancy and freezing tolerance	Takemura et al. (2015), Bai et al. (2013)
	Ectopic expression of a citrus *FT* in pear gene (*CiFT*)	Matsuda et al. (2009)
Physic nut (*Jatropha curcas* L.)	*AP2/ERF* genes	Tang et al. (2016), Wang et al. (2013)
Plum (*Prunus domestica* L)	Ectopic expression of a poplar (*Populus trichocarpa*) *FT* gene in plum.	Srinivasan et al. (2012)
Sweet cherry (*Prunus avium* L.)	Ectopic expression of a sweet cherry *CBF* in *Arabidopsis*	Kitashiba et al. (2004)
	Quantitative trait loci (QTLs) associated with temperature requirements for bud dormancy release and flowering	Castede et al. (2015)

did not open under greenhouse conditions without chilling. Vernalized *VcFT*-overexpressing plants can flower normally. These results suggest that overexpression of *VcFT* is not capable of fulfilling all the roles of vernalization in blueberry (Walworth, Chai, and Song 2016; Gao et al. 2016). More interesting, compared to unvernalized flower buds, vernalized flower buds did not show differential expression of the *VcFT* (Song and Chen 2018).

8.2.3 Primary Dormancy Genes in Deciduous Fruit Crops

For some deciduous *Prunus* fruit crops (e.g. peach and apricot), *DAMs* are primary candidate genes that play a significant role in plant dormancy responses and chilling requirements (Horvath 2015). However, there is a high potential that the primary dormancy gene could be out of the scope of the known genes, such as *FLC, VRN,* or *DAMs* in deciduous fruit crops due to their genetic diversity. For example, a recent study in kiwifruit (*Actinidia* spp.) showed that three *SOC1*-like genes of kiwifruit had no impact on establishment of winter dormancy and induction of precocious flowering, but one of them reduced the duration of dormancy in the absence of winter chilling (Voogd, Wang, and Varkonyi-Gasic 2015). In poplar (*Populus tremula* x *Populus alba* Institut National de la Recherche Agronomique 717-IB4), *EARLY BUD-BREAK 1* (*EBB1*) has a major role in timing vegetative budbreak after winter dormancy (Yordanov et al. 2014). Comparative transcriptome analysis of a low-chilling requirement pear and a high-chilling requirement pear (*Pyrus pyrifolia*) suggest that the APETALA2/ethylene response (AP2/ERF) transcription factors are involved in endo-dormancy maintenance and in the transition from endo-dormancy to eco-dormancy (Takemura et al. 2015). In grapevine (*Vitis vinifera*), a MADS BOX gene *VvFLC2*, in contrast to the *VvSVP* genes and *PmDAM6*-like genes, could participate in bud dormancy (Diaz-Riquelme et al. 2012). In blueberries, compared to unvernalized flower buds, full vernalization caused differential expression of orthologues of 88.5% of known *A. thaliana* flowering pathway genes and 85.5% of known *A. thaliana* MADS-box genes (Song and Chen 2018). Orthologues of *FLC* and other MADS-box genes that show a similar expression pattern to *FLC* in response to vernalization in *A. thaliana* were not found in blueberry, suggesting that neither *FLC*- nor *DAM*s-mediated vernalization likely works in blueberry. Taken together, the key dormancy gene does not seem to be conserved in deciduous fruit crops.

8.3 ACCLIMATION, DORMANCY, AND FREEZING TOLERANCE

Low temperatures induce cold acclimation or dormancy. This is an adaptive mechanism that has evolved to enhance plant tolerance to cold or freezing. Since plant freezing tolerance depends on numerous factors, such as natural environment, plant species/genotypes, plant developmental stages, acclimation state, organs, and tissues (Strimbeck et al. 2015), studies on genetics and functional genomics of cold acclimation are anticipated to lead to improved cold/freezing tolerance through breeding or genetic engineering.

8.3.1 The APETALA2/Ethylene Response (AP2/ERF) Transcription Factors

The AP2/ERFs play a significant role in plant responses to several abiotic stresses (e.g. cold, dehydration, and high salinity) (Mizoi, Shinozaki, and Yamaguchi-Shinozaki 2012) that have inspired many recent studies on genome-wide analysis of the AP2/ERF in several plant species (Table 8.1). The AP2/ERFs include DREB1 (dehydration responsive element-binding factor 1) DREB1and DREB2. The CBF/DREB1 (C-repeat-binding factor/DREB1) genes belong to a large family of the AP2/ERF transcription factors and have a conserved DNA binding domain recognizing the dehydration-responsive element/C-repeat (DRE/CRT) *cis*-acting element in the promoters of their target genes (Liu et al. 1998; Mizoi, Shinozaki, and Yamaguchi-Shinozaki 2012; Qin, Shinozaki, and Yamaguchi-Shinozaki 2011). It appears that all plant species tested undergo cold acclimation through a universal process that belongs at least partially to the CBF/DREB1-mediated cold-response pathway (Thomashow et al. 2001; Miller, Galiba, and Dubcovsky 2006; Navarro et al. 2011; Mizoi, Shinozaki, and Yamaguchi-Shinozaki 2012; Carvallo et al. 2011). This CBF/DREB1 pathway has been well documented in *A. thaliana* (Liu et al. 1998; Gilmour et al. 1998; Stockinger, Gilmour, and Thomashow 1997; Jaglo-Ottosen et al. 1998). The DREB2 group of the AP/ERF transcription factors function in both drought- and heat-stress responses (Mizoi, Shinozaki, and Yamaguchi-Shinozaki 2012; Sun et al. 2014).

8.3.2 CBF/DREB1 Genes and Plant Freezing Tolerance

The potential of using *CBF* genes (*CBF1, CBF2,* and *CBF3*) to increase plant freeze tolerance has been demonstrated in both herbaceous and woody plants. This has laid the foundation for the utility of the CBF/DREB1 pathway genes to enhance freezing tolerance of commercially important crops. However, modulating expression of the CBF/DREB1 pathway genes for enhanced tolerance to abiotic stress is often associated with undesirable impacts on plant growth and development. For example, the constitutive expression of a peach (*Prunus persica*) *CBF1* gene (*PpCBF1*) in apple rootstock resulted in improved freezing tolerance and altered plant growth and dormancy (Wisniewski et al. 2011). A slight increase in freezing tolerance in non-cold-acclimated vines and dwarf phenotypes, which are similar to that reported for *CBFs* overexpression in *A. thaliana* and other species, were observed in transgenic grape vines overexpressing a grapevine *CBF* (*VvCBF4*) (Tillett et al. 2012). Ectopic expression of a *CBF1* orthologue from European bilberry (*Vaccinium myrtillus*) enhanced freeze tolerance of *A. thaliana* plants (Oakenfull, Baxter, and Knight 2013). Collectively, beyond freezing tolerance, the over-expression of *CBFs* is often associated with reduced plant growth (*CBF1, CBF3,* and *CBF4*) or altered developmental processes such as flowering time, leaf senescence, and plant longevity (*CBF2* and *CBF3*) (Gilmour et al. 2000; Kasuga et al. 1999; Benedict et al. 2006; Tillett et al. 2012; Achard et al. 2008; Sharabi-Schwager et al. 2010; Jaglo-Ottosen et al. 1998; Haake et al. 2002). The occurrence of these additional phenotypic changes (e.g. dwarfism), which are sometimes desirable traits for crop/fruit production, is often interpreted by the complexity of the *CBF*s-mediated low-temperature regulatory networks (Tillett et al. 2012; Park et al. 2015; Sharabi-Schwager et al. 2010; Novillo et al. 2011).

Of the major CBF/DREB1 transcription factors, relatively few studies have been conducted on DRE1E_ARATH [designated as *DWARF AND DELAYED FLOWERING 1* (*DDF1*)] and DRE1F_ARATH (designated as *DDF2*) (Sakuma et al. 2002; Magome et al. 2004, 2008; Lehti-Shiu et al. 2015; Kang et al. 2011). *A. thaliana* plants overexpressing the *DDF1* showed dark-green leaves, dwarfism, and late flowering; meanwhile the plants displayed enhanced tolerance to cold, drought, heat and high salinity (Magome et al. 2004, 2008; Kang et al. 2011).

In woody fruit crops, the constitutive expression of the *CBF1* and *CBF4* or their orthologues enabled similar phenotype changes to those observed in *A. thaliana*, suggesting that the *CBF/DREB1* mediated-freezing tolerance is conserved in woody fruit crops (Wisniewski, Norelli, and Artlip 2015; Tillett et al. 2012; Oakenfull, Baxter, and Knight 2013; Jin et al. 2009). Ectopic expression of a *Muscadinia rotundifolia CBF2* resulted in dwarfism and late flowering of transgenic *A. thaliana* plants (Wu et al. 2017). To date, few studies have been done on overexpression of *DDF1, DDF2,* or their orthologues in crops. In our previous study, overexpression of a blueberry-derived *CBF* (*BB-CBF*) enhanced cold tolerance in leaves and dormant buds but not in flower tissues of a southern highbush blueberry cultivar; meanwhile, no dwarfism was observed (Walworth et al. 2012). The *BB-CBF* was initially designated as an orthologue of *CBF2* (Polashock et al. 2010), so there was not much surprise when no dwarfing was shown because overexpression of *CBF2* is mainly dedicated to delayed leaf senescence and extended plant longevity in *A. thaliana* (Sharabi-Schwager et al. 2010). Recently, we found that the *BB-CBF* (herein renamed as *VcDDF1*) shows a higher similarity to *A. thaliana DDF1* than to the *CBF2* (Song and Gao 2017).

8.3.3 Low-Temperature Regulatory Networks

As revealed in *A. thaliana*, the low-temperature regulatory network is more complicated than the CBF-CRT/DRE regulatory module (Park et al. 2015; Thomashow 2010). For example, acclimation (cold) induced differential expression of 2,637 cold-regulated (COR) genes, but in contrast, overexpression of *CBF1*, *CBF2*, and *CBF3* can only alter 171 (6.5%) COR genes (Park et al. 2015). Accordingly, it was pointed out that freezing tolerance occurs with cold acclimation and is only partially dependent on the CBF regulatory module (Park et al. 2015). Accordingly, changes of freezing tolerance induced by manipulation of CBF pathway genes could vary among plant species (El Kayal et al. 2006; Navarro et al. 2011; Carvallo et al. 2011).

For deciduous fruit crops, many efforts have been focusing mainly on developing transgenic plants for freezing tolerance using *CBF* or *CBF* orthologues (Table 8.1). The CBF-CRT/DRE regulatory module has not been well studied through forward genetics or reverse genetics. In our recent studies, transcriptomic responses to the overexpression of the blueberry *DDF1* (herein, *VcDDF1*-OX) showed differentially expressed COR genes (compared to non-transgenic plants), which contribute to the increase in freezing tolerance (Song and Gao 2017).

8.4 POTENTIAL ROLES OF PHYTOHORMONES IN PLANT FLOWERING, DORMANCY, AND FREEZING TOLERANCE

Phytohormones are involved in plant flowering, dormancy, and freezing tolerance. In *A. thaliana*, it has been well documented that expression of phytohormone genes play roles in cold acclimation (Shi, Ding, and Yang 2015; Shi et al. 2012), dormancy (Kendall et al. 2011) and plant flowering (Fornara, de Montaigu, and Coupland 2010; Bernier 2013). Based on our studies in blueberry, a schematic diagram was drawn to describe the impact of low temperatures on plant flowering, dormancy, and freezing tolerance (Figure 8.1).

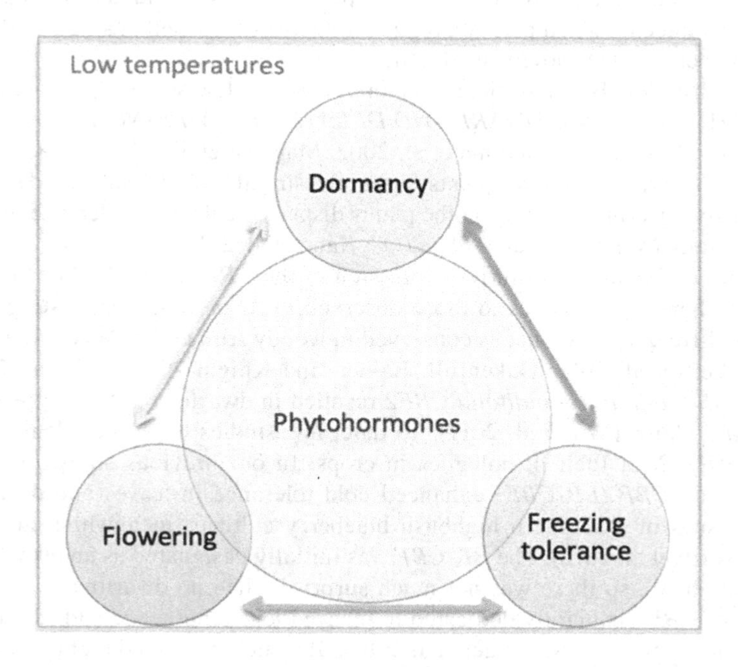

FIGURE 8.1 The potential interactions of phytohormone genes in plant flowering, dormancy and freezing tolerance.

8.4.1 Phytohormones in Plant Cold Acclimation

The delayed flowering of *A. thaliana* plants caused by the overexpression of *CBF1,2,3* was attributed to the increased expression of *FLC* and repressed expression of *SOC1* (Seo et al. 2009), suggesting an interaction of *CBFs* with vernalization-mediated flowering pathway genes. The gibberellin (GA) pathway interacts with the flowering pathway through the *SOC1* (Fornara, de Montaigu, and Coupland 2010; El-Showk, Ruonala, and Helariutta 2013). Thus, the delayed flowering and growth retardation caused by the overexpression of *CBF1,2,3* are due to the changes of DELLA proteins, the negative regulators in GA signaling (Achard et al. 2008; Sharabi-Schwager et al. 2010). Similarly, the dwarfism of *A. thaliana* plants was caused by the overexpression of *DDF1* through reduced bioactive GA (Magome et al. 2008; Kang et al. 2011). In woody plants, overexpression of *VcDDF1* altered expression of 21 floral genes, 88 phytohormone genes (i.e. 11 for IAA, 23 for GA, 5 for cytokinin, and 49 for ethylene), and 725 blueberry COR genes in unvernalized flower buds (Figure 8.2) (Song and Gao 2017). Cold acclimation affects expression of *CBFs* and phytohormone genes (Shi, Ding, and Yang 2015; Shi et al. 2012; Yang et al. 2005). The *CBFs* were also found to act in parallel with a low-temperature signaling pathway in the regulation of dormancy in *A. thaliana* seeds through *GA2OXIDASE6* (*GA2OX6*) and *DELAY OF GERMINATION1* (*DOG1*) expression (Kendall et al. 2011).

8.4.2 Phytohormones in Plant Dormancy

In woody perennials plants, induction and release of bud dormancy is a complex process (Arora, Rowland, and Tanino 2003). During bud dormancy induction, low temperatures and short days induce growth cessation, enable cold acclimation for enhancing freezing tolerance, and fulfill chilling requirements. Several phytohormones (e.g. ABA, ethylene, GA, and auxin) are involved in bud formation, cold acclimation, bud dormancy, and flowering (Wen et al. 2016; Wisniewski, Norelli, and Artlip 2015; Wang et al. 2015; Gao et al. 2016; Niu et al. 2014; Yordanov et al. 2014; Baba et al. 2011;

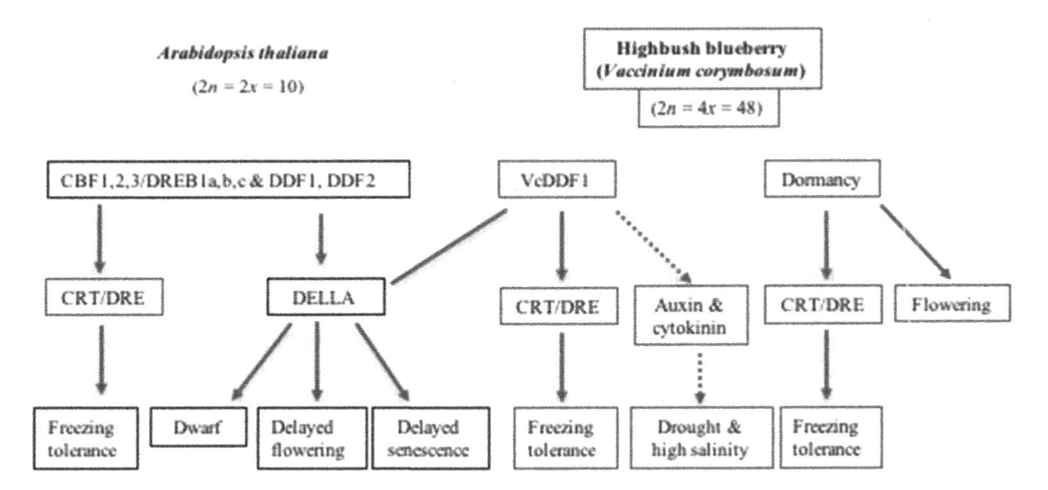

FIGURE 8.2 Schematic diagram illustrating the effect of overexpression of a *DDF1* orthologue of blueberry (*VcDDF1*) on plant freezing tolerance but not on plant growth (dwarf) and development (delayed flowering and senescence) and the effect of full vernalization on blueberry flowering and freezing tolerance. As a comparison, the diagram illustrating the effect of the overexpression of *CBF/DREB1* on freezing tolerance, plant growth, and development in *A. thaliana* was derived from the previous reports. The overexpression of *VcDDF1* showed little effect on DELLA proteins but could affect plant tolerance to drought and high salinity through the altered gene expression in auxin and cytokinin pathways (Song and Gao 2017). Black arrows show positive effect. Black line shows no significant effect. Black dash arrows show projected positive effect.

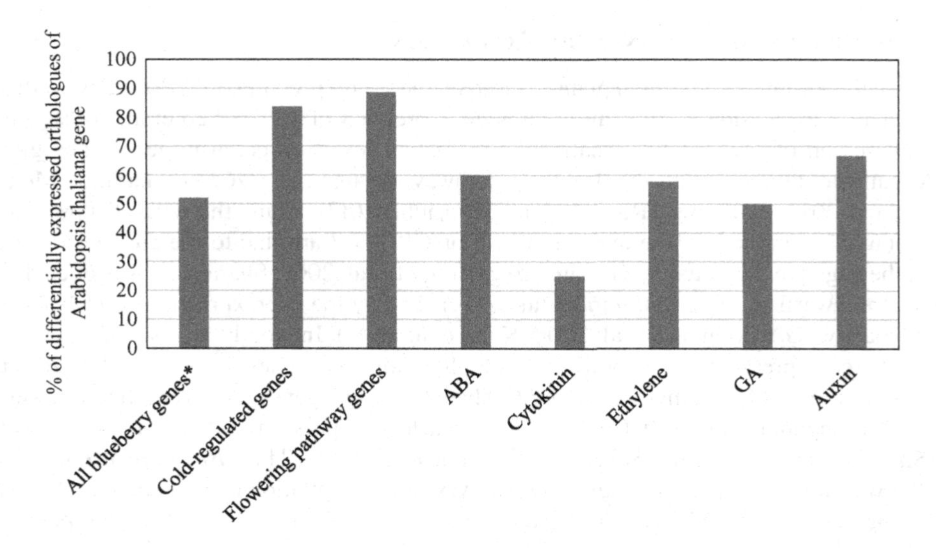

FIGURE 8.3 Effect of vernalization on percentages of differentially expressed (DE) orthologues of *A. thaliana* genes in vernalized blueberry flower buds (in comparison to non-vernalized flower buds). GA: gibberellic acid. ABA: abscisic acid. *DE genes including *A. thaliana* genes and genes annotated to other species (Song and Chen 2018).

Leida et al. 2012; Rios et al. 2014). More recently, our transcriptome analysis of vernalized and unvernalized flower buds of blueberry has revealed that vernalized flower buds showed differential expression (compared to unvernalized floral buds) of flowering pathway genes, phytohormone genes, and blueberry COR genes. (Figure 8.3) (Song and Chen 2018).

8.4.3 Phytohormones in Plant Flowering

Phytohormones play important roles in regulating plant development and stature formation. Of these hormones, GA plays a significant role in regulating plant flowering time and in determining plant stature. Mutations resulting in reduced GA biosynthesis or increased GA degradation often produce dwarf plants with delayed plant flowering (Fornara, de Montaigu, and Coupland 2010; Busov et al. 2006; Hartweck 2008; Daviere and Achard 2013; Ji et al. 2014). Other phytohormone genes, e.g. auxin (Yamaguchi et al. 2013; Krizek 2011), cytokinin (Osugi and Sakakibara 2015; Bernier 2013; Ren et al. 2009), ethylene (Achard et al. 2007), brassinosteroid (Domagalska et al. 2007; Kutschera and Wang 2012), jasmonic acid (Ishiguro et al. 2001), nitric oxide (Takahash and Morikawa 2014), peptide hormone (Wang and Irving 2011), and salicylic acid (Martinez et al. 2004; Rivas-San Vicente and Plasencia 2011)], could also affect plant flowering and plant size (Bernier et al. 1993; Bernier 2013). Overexpression of the *VcFT* in the flowering pathway is associated with differential expression of 33 known flowering pathway genes, 110 phytohormone pathway genes (i.e. three for ABA, 26 for IAA, six for cytokinin, 39 for ethylene, and 36 for GA), and 822 blueberry COR genes in leaf tissues (Gao et al. 2016; Walworth, Chai, and Song 2016).

8.5 CONCLUSION REMARKS

Deciduous woody crops (fruits and nuts) have evolved very successful winter dormancy (i.e. endodormancy and eco-dormancy) to acquire freezing tolerance to survive in winter, but also to gain sufficient chilling units for the fitness of the next season's plant growth and development. Winter dormancy is genetically controlled and environmentally regulated. To date, the molecular mechanism of freezing tolerance and endo-dormancy is not fully understood in woody plants. With the global warming and temperature changes in winter, insufficient chilling units and freezing damage are two major low temperature stresses threatening the production of deciduous fruits. Comparative transcriptome analysis has provided new insights into the genes and gene networks that respond to low temperatures during plant cold acclimation and dormancy. Functional analyses of individual cold/freeze tolerance gene(s) through genetic engineering to improve freezing tolerance have been demonstrated in both herbaceous and woody plants. In this review, recent studies have indicated that three low-temperature-related gene networks, including dormancy/vernalization, plant flowering, and freezing tolerance, interact with each other in deciduous woody plants (Figure 8.1). Manipulation of individual gene(s) through genetic engineering in any of these three networks usually ends up with a broad impact on gene interactions.

REFERENCES

Achard, P., M. Baghour, A. Chapple, et al. 2007. The plant stress hormone ethylene controls floral transition via DELLA-dependent regulation of floral meristem-identity genes. *Proc Natl Acad Sci U S A* 104 (15):6484–9.

Achard, P., F. Gong, S. Cheminant, M. Alioua, P. Hedden, and P. Genschik. 2008. The cold-inducible CBF1 factor-dependent signaling pathway modulates the accumulation of the growth-repressing DELLA proteins via its effect on gibberellin metabolism. *Plant Cell* 20 (8):2117–29.

Allard, A., M. C. Bink, S. Martinez, et al. 2016. Detecting QTLs and putative candidate genes involved in budbreak and flowering time in an apple multiparental population. *J Exp Bot* 67 (9):2875–88.

Almada, R., N. Cabrera, J. A. Casaretto, S. Ruiz-Lara, and E. Gonzalez Villanueva. 2009. VvCO and VvCOL1, two CONSTANS homologous genes, are regulated during flower induction and dormancy in grapevine buds. *Plant Cell Rep* 28 (8):1193–203.

Amasino, R. 2010. Seasonal and developmental timing of flowering. *Plant J* 61 (6):1001–13.

Anderson, J. V. 2015. *Advances in Plant Dormancy*. Switzerland: Springer International Publishing.

Arora, R., L. J. Rowland, and K. Tanino. 2003. Induction and release of bud dormancy in woody perennials: A science comes of age. *HortScience* 38:11.

Atkinson, C. J., R. M. Brennan, and H. G. Jones. 2013. Declining chilling and its impact on temperate perennial crops. *Environ Exp Bot* 91:48–62.

Baba, K., A. Karlberg, J. Schmidt, et al. 2011. Activity-dormancy transition in the cambial meristem involves stage-specific modulation of auxin response in hybrid aspen. *Proc Natl Acad Sci U S A* 108 (8):3418–23.

Bai, S. L., T. Saito, D. Sakamoto, A. Ito, H. Fujii, and T. Moriguchi. 2013. Transcriptome analysis of Japanese pear (Pyrus pyrifolia Nakai) flower buds transitioning through endodormancy. *Plant Cell Physiol* 54 (7):1132–51.

Benedict, C., J. S. Skinner, R. Meng, et al. 2006. The CBF1-dependent low temperature signalling pathway, regulon and increase in freeze tolerance are conserved in Populus spp. *Plant Cell Environ* 29 (7):1259–72.

Bernier, G. 2013. My favourite flowering image: The role of cytokinin as a flowering signal. *J Exp Bot* 64 (18):5795–9.

Bernier, G., A. Havelange, C. Houssa, A. Petitjean, and P. Lejeune. 1993. Physiological signals that induce flowering. *Plant Cell* 5 (10):1147–55.

Bielenberg, D. G., Y. Wang, Z. G. Li, et al. 2008. Sequencing and annotation of the evergrowing locus in peach [*Prunus persica* (L.) Batsch] reveals a cluster of six MADS-box transcription factors as candidate genes for regulation of terminal bud formation. *Tree Genet Genom* 4 (3):495–507.

Böhlenius, H., T. Huang, L. Charbonnel-Campaa, et al. 2006. CO/FT Regulatory Module controls timing of flowering and seasonal growth cessation in trees. *Science* 312 (5776):1040–3.

Busov, V., R. Meilan, D. W. Pearce, et al. 2006. Transgenic modification of gai or rgl1 causes dwarfing and alters gibberellins, root growth, and metabolite profiles in Populus. *Planta* 224 (2):288–99.

Byrne, D. H., and T. A. Bacon. 1992. Chilling estimation: Its importance and estimation. *Texas Horticult* 18 (8):2.

Carvallo, M. A., M. T. Pino, Z. Jeknic, et al. 2011. A comparison of the low temperature transcriptomes and CBF regulons of three plant species that differ in freezing tolerance: Solanum commersonii, Solanum tuberosum, and Arabidopsis thaliana. *J Exp Bot* 62 (11):3807–19.

Castede, S., J. A. Campoy, L. Le Dantec, et al. 2015. Mapping of candidate genes involved in bud dormancy and flowering time in sweet cherry (Prunus avium). *PLOS ONE* 10 (11):e0143250.

Chuine, I. C., M. Bonhomme, J.-M. Legave, et al. 2016. Can phenological models predict tree phenology accurately in the future? The unrevealed hurdle of endodormancy break. *Glob Change Biol* 22 (10):17.

Cockram, J., H. Jones, F. J. Leigh, et al. 2007. Control of flowering time in temperate cereals: Genes, domestication, and sustainable productivity. *J Exp Bot* 58 (6):1231–44.

Daviere, J. M., and P. Achard. 2013. Gibberellin signaling in plants. *Development* 140 (6):1147–51.

Diaz-Riquelme, J., J. Grimplet, J. M. Martinez-Zapater, and M. J. Carmona. 2012. Transcriptome variation along bud development in grapevine (Vitis vinifera L.). *BMC Plant Biol* 12 (1):181.

Domagalska, M. A., F. M. Schomburg, R. M. Amasino, R. D. Vierstra, F. Nagy, and S. J. Davis. 2007. Attenuation of brassinosteroid signaling enhances FLC expression and delays flowering. *Development* 134 (15):2841–50.

El Kayal, W., M. Navarro, G. Marque, G. Keller, C. Marque, and C. Teulieres. 2006. Expression profile of CBF-like transcriptional factor genes from Eucalyptus in response to cold. *J Exp Bot* 57 (10):2455–69.

El-Showk, S., R. Ruonala, and Y. Helariutta. 2013. Crossing paths: Cytokinin signalling and crosstalk. *Development* 140 (7):1373–83.

Falavigna, V. D., D. D. Porto, V. Buffon, M. Margis-Pinheiro, G. Pasquali, and L. F. Revers. 2014. Differential transcriptional profiles of dormancy-related genes in apple buds. *Plant Mol Biol Rep* 32 (4):796–813.

Feild, T. S., N. C. Arens, J. A. Doyle, T. E. Dawson, and M. J. Donoghue. 2004. Dark and disturbed: A new image of early angiosperm ecology. *Paleobiology* 30 (1):82–107.

Fornara, F., A. de Montaigu, and G. Coupland. 2010. SnapShot: Control of flowering in Arabidopsis. *Cell* 141 (3): 550–550.e1–2.

Gao, X., A. E. Walworth, C. Mackie, and G. Q. Song. 2016. Overexpression of blueberry FLOWERING LOCUS T is associated with changes in the expression of phytohormone-related genes in blueberry plants. *Hortic Res* 3 (1):16053.

Gilmour, S. J., A. M. Sebolt, M. P. Salazar, J. D. Everard, and M. F. Thomashow. 2000. Overexpression of the Arabidopsis CBF3 transcriptional activator mimics multiple biochemical changes associated with cold acclimation. *Plant Physiol* 124 (4):1854–65.

Gilmour, S. J., D. G. Zarka, E. J. Stockinger, M. P. Salazar, J. M. Houghton, and M. F. Thomashow. 1998. Low temperature regulation of the Arabidopsis CBF family of AP2 transcriptional activators as an early step in cold-induced COR gene expression. *Plant J* 16 (4):433–42.

Golembeski, G. S., H. A. Kinmonth-Schultz, Y. H. Song, and T. Imaizumi. 2014. Photoperiodic flowering regulation in Arabidopsis thaliana. *Adv Bot Res* 72:1–28.

Greenup, A., W. J. Peacock, E. S. Dennis, and B. Trevaskis. 2009. The molecular biology of seasonal flowering-responses in Arabidopsis and the cereals. *Ann Bot* 103 (8):1165–72.

Gregis, V., A. Sessa, L. Colombo, and M. M. Kater. 2006. AGL24, SHORT VEGETATIVE PHASE, and APETALA1 redundantly control AGAMOUS during early stages of flower development in Arabidopsis. *Plant Cell* 18 (6):1373–82.

Griffiths, S., R. P. Dunford, G. Coupland, and D. A. Laurie. 2003. The evolution of CONSTANS-like gene families in barley, rice, and Arabidopsis. *Plant Physiol* 131 (4):1855–67.

Haake, V., D. Cook, J. L. Riechmann, O. Pineda, M. F. Thomashow, and J. Z. Zhang. 2002. Transcription factor CBF4 is a regulator of drought adaptation in Arabidopsis. *Plant Physiol* 130 (2):639–48.

Hartweck, L. M. 2008. Gibberellin signaling. *Planta* 229 (1):1–13.

Hemming, M. N., W. J. Peacock, E. S. Dennis, and B. Trevaskis. 2008. Low-temperature and daylength cues are integrated to regulate FLOWERING LOCUS T in barley. *Plant Physiol* 147 (1):355–66.

Higgins, J. A., P. C. Bailey, and D. A. Laurie. 2010. Comparative genomics of flowering time pathways using Brachypodium distachyon as a model for the temperate grasses. *PLOS ONE* 5 (4):e10065.

Horvath, D. 2015. Dormancy-associated MADS-BOX genes: A review. In *Advances in Plant Dormancy*, edited by J. V. Anderson. Switzerland: Springer International Publishing.

Huijser, P., and M. Schmid. 2011. The control of developmental phase transitions in plants. *Development* 138 (19):4117–29.

Ikegami, H., H. Nogata, Y. Inoue, et al. 2013. Expression of FcFT1, a FLOWERING LOCUS T-like gene, is regulated by light and associated with inflorescence differentiation in fig (Ficus carica L.). *BMC Plant Biol* 13 (1):216.

Ishiguro, S., A. Kawai-Oda, J. Ueda, I. Nishida, and K. Okada. 2001. The DEFECTIVE IN ANTHER DEHISCIENCE gene encodes a novel phospholipase A1 catalyzing the initial step of jasmonic acid biosynthesis, which synchronizes pollen maturation, anther dehiscence, and flower opening in Arabidopsis. *Plant Cell* 13 (10):2191–209.

Jaglo-Ottosen, K. R., S. J. Gilmour, D. G. Zarka, O. Schabenberger, and M. F. Thomashow. 1998. Arabidopsis CBF1 overexpression induces COR genes and enhances freezing tolerance. *Science* 280 (5360):104–6.

Jameson, P. E., and J. Clemens. 2015. Phase change and flowering in woody plants of the New Zealand flora. *J Exp Bot*. doi: 10.1093/jxb/erv472

Ji, S. H., M. A. Gururani, J. W. Lee, B. O. Ahn, and S. C. Chun. 2014. Isolation and characterisation of a dwarf rice mutant exhibiting defective gibberellins biosynthesis. *Plant Biol (Stuttg)* 16 (2):428–39.

Jia, T. Q., D. F. Wei, S. Meng, A. C. Allan, and L. H. Zeng. 2014. Identification of regulatory genes implicated in continuous flowering of longan (Dimocarpus longan L.). *PLOS ONE* 9 (12):e114568.

Jimenez, S., A. L. Lawton-Rauh, G. L. Reighard, A. G. Abbott, and D. G. Bielenberg. 2009. Phylogenetic analysis and molecular evolution of the dormancy associated MADS-box genes from peach. *BMC Plant Biol* 9 (1):81.

Jimenez, S., G. L. Reighard, and D. G. Bielenberg. 2010. Gene expression of DAM5 and DAM6 is suppressed by chilling temperatures and inversely correlated with bud break rate. *Plant Mol Biol* 73 (1–2):157–67.

Jin, W. M., J. Dong, Y. L. Hu, Z. P. Lin, X. F. Xu, and Z. H. Han. 2009. Improved Cold-resistant Performance in Transgenic Grape (Vitis vinifera L.) overexpressing cold-inducible transcription factors AtDREB1b. *HortScience* 44 (1):35–9.

Kang, H. G., J. Kim, B. Kim, et al. 2011. Overexpression of FTL1/DDF1, an AP2 transcription factor, enhances tolerance to cold, drought, and heat stresses in Arabidopsis thaliana. *Plant Sci* 180 (4):634–41.

Kasuga, M., Q. Liu, S. Miura, K. Yamaguchi-Shinozaki, and K. Shinozaki. 1999. Improving plant drought, salt, and freezing tolerance by gene transfer of a single stress-inducible transcription factor. *Nat Biotechnol* 17 (3):287–91.

Kendall, S. L., A. Hellwege, P. Marriot, C. Whalley, I. A. Graham, and S. Penfield. 2011. Induction of dormancy in Arabidopsis summer annuals requires parallel regulation of DOG1 and hormone metabolism by low temperature and CBF transcription factors. *Plant Cell* 23 (7):2568–80.

Kitashiba, H., T. Ishizaka, K. Isuzugawa, K. Nishimura, and T. Suzuki. 2004. Expression of a sweet cherry DREB1/CBF ortholog in Arabidopsis confers salt and freezing tolerance. *J Plant Physiol* 161 (10):1171–6.

Kotoda, N., H. Hayashi, M. Suzuki, et al. 2010. Molecular characterization of FLOWERING LOCUS T-like genes of apple (Malus × domestica Borkh.). *Plant Cell Physiol* 51 (4):561–75.

Kramer, K., I. Leinonen, and D. Loustau. 2000. The importance of phenology for the evaluation of impact of climate change on growth of boreal, temperate and Mediterranean forests ecosystems: An overview. *Int J Biometeorol* 44 (2):67–75.

Kreyling, J., S. Schmid, and G. Aas. 2015. Cold tolerance of tree species is related to the climate of their native ranges. *J Biogeogr* 42 (1):156–66.

Krizek, B. A. 2011. Auxin regulation of Arabidopsis flower development involves members of the AINTEGUMENTA-LIKE/PLETHORA (AIL/PLT) family. *J Exp Bot* 62 (10):3311–9.

Kumar, G., P. Arya, K. Gupta, V. Randhawa, V. Acharya, and A. K. Singh. 2016. Comparative phylogenetic analysis and transcriptional profiling of MADS-box gene family identified DAM and FLC-like genes in apple (Malusx domestica). *Sci Rep* 6 (1):20695.

Kutschera, U., and Z. Y. Wang. 2012. Brassinosteroid action in flowering plants: A Darwinian perspective. *J Exp Bot* 63 (10):3511–22.

Lang, G. A., J. D. Early, G. C. Martin, and R. L. Darnell. 1987. Endo-, para-, and ecodormancy: Physiological terminology and classification for dormancy research. *HortScience* 22:8.

Larcher, W. 2003. *Physiological Plant Ecology: Ecophysiology and Stress Physiology of Functional Groups*. 4th ed. Berlin; New York: Springer.

Lee, J., and I. Lee. 2010. Regulation and function of SOC1, a flowering pathway integrator. *J Exp Bot* 61 (9):2247–54.

Lehti-Shiu, M. D., S. Uygun, G. D. Moghe, et al. 2015. Molecular evidence for functional divergence and decay of a transcription factor derived from whole-genome duplication in Arabidopsis thaliana. *Plant Physiol* 168 (4):1717–34.

Leida, C., A. Conesa, G. Llacer, M. L. Badenes, and G. Rios. 2012. Histone modifications and expression of DAM6 gene in peach are modulated during bud dormancy release in a cultivar-dependent manner. *New Phytol* 193 (1):67–80.

Li, W. M., Y. Tao, Y. X. Yao, Y. J. Hao, and C. X. You. 2010. Ectopic over-expression of two apple Flowering Locus T homologues, MdFT1 and MdFT2, reduces juvenile phase in Arabidopsis. *Biol Plantarum* 54 (4):639–46.

Liu, Q., M. Kasuga, Y. Sakuma, et al. 1998. Two transcription factors, DREB1 and DREB2, with an EREBP/AP2 DNA binding domain separate two cellular signal transduction pathways in drought- and low-temperature-responsive gene expression, respectively, in Arabidopsis. *Plant Cell* 10 (8):1391–406.

Liu, Z., H. Zhu, and A. Abbott. 2015. Dormancy behaviors and underlying regulatory mechanisms: From perspective of pathways to epigenetic regulation. In *Advances in Plant Dormancy*, edited by J. V. Anderson. Switzerland, Zürich: Springer International Publishing.

Luedeling, E., E. H. Girvetz, M. A. Semenov, and P. H. Brown. 2011. Climate change affects winter chill for temperate fruit and nut trees. *PLOS ONE* 6 (5):e20155.

Magome, H., S. Yamaguchi, A. Hanada, Y. Kamiya, and K. Oda. 2004. dwarf and delayed-flowering 1, a novel Arabidopsis mutant deficient in gibberellin biosynthesis because of overexpression of a putative AP2 transcription factor. *Plant J* 37 (5):720–9.

Magome, H., S. Yamaguchi, A. Hanada, Y. Kamiya, and K. Oda. 2008. The DDF1 transcriptional activator upregulates expression of a gibberellin-deactivating gene, GA2ox7, under high-salinity stress in Arabidopsis. *Plant J* 56 (4):613–26.

Martinez, C., E. Pons, G. Prats, and J. Leon. 2004. Salicylic acid regulates flowering time and links defence responses and reproductive development. *Plant J* 37 (2):209–17.

Mateos, J. L., P. Madrigal, K. Tsuda, et al. 2015. Combinatorial activities of SHORT VEGETATIVE PHASE and FLOWERING LOCUS C define distinct modes of flowering regulation in Arabidopsis. *Genome Biol* 16:31.

Matsuda, N., K. Ikeda, M. Kurosaka, et al. 2009. Early flowering phenotype in transgenic pears (Pyrus communis L.) expressing the CiFT gene. *J Jpn Soc Hortic Sci* 78 (4):410–6.

Melke, A. 2015. The physiology of chilling temperature requirements for dormancy release and bud-break in temperate fruit trees grown at mild winter tropical climate. *J Plant Stud* 4 (2):47.

Michaels, S. D. 2009. Flowering time regulation produces much fruit. *Curr Opin Plant Biol* 12 (1):75–80.

Miller, A. K., G. Galiba, and J. Dubcovsky. 2006. A cluster of 11 CBF transcription factors is located at the frost tolerance locus Fr-A(m)2 in Triticum monococcum. *Mol Genet Genom* 275 (2):193–203.

Mimida, N., S.-I. Kidou, H. Iwanami, et al. 2011. Apple FLOWERING LOCUS T proteins interact with transcription factors implicated in cell growth and organ development. *Tree Physiol* 31 (5): 555–66.

Mimida, N., T. Saito, T. Moriguchi, A. Suzuki, S. Komori, and M. Wada. 2015. Expression of DORMANCY-ASSOCIATED MADS-BOX (DAM)-like genes in apple. *Biol Plantarum* 59 (2):237–44.

Mizoi, J., K. Shinozaki, and K. Yamaguchi-Shinozaki. 2012. AP2/ERF family transcription factors in plant abiotic stress responses. *Biochim Biophys Acta* 1819 (2):86–96.

Navarro, M., C. Ayax, Y. Martinez, et al. 2011. Two EguCBF1 genes overexpressed in Eucalyptus display a different impact on stress tolerance and plant development. *Plant Biotechnol J* 9 (1):50–63.

Niu, S., Q. Gao, Zhexin Li, X. Chen, and W. Li. 2014. The role of gibberellin in the CBF1-mediated stress-response pathway. *Plant Mol Biol Rep* 32 (4):852–63.

Novillo, F., J. Medina, M. Rodriguez-Franco, G. Neuhaus, and J. Salinas. 2011. Genetic analysis reveals a complex regulatory network modulating CBF gene expression and Arabidopsis response to abiotic stress. *J Exp Bot* 63 (1):293–304.

Oakenfull, R. J., R. Baxter, and M. R. Knight. 2013. A C-repeat binding factor transcriptional activator (CBF/DREB1) from European bilberry (Vaccinium myrtillus) induces freezing tolerance when expressed in Arabidopsis thaliana. *PLOS ONE* 8 (1):e54119.

Olsen, J. E. 2003. Molecular and physiological mechanisms of bud dormancy regulation. *Environ Stress Hortic Crops* 618:437–53.

Olsen, J. E. 2006. Mechanisms of dormancy regulation. *Proc Xth Int Symp Plant Bioregul Fruit Prod* 727:157–65.

Olsen, J. E. 2010. Light and temperature sensing and signaling in induction of bud dormancy in woody plants. *Plant Mol Biol* 73 (1–2):37–47.

Osugi, A., and H. Sakakibara. 2015. Q&A: How do plants respond to cytokinins and what is their importance? *BMC Biol* 13 (1):102.

Ouellet, F., and J.-B. Charron 2013. Cold acclimation and freezing tolerance in plants. In *eLS*. Chichester, UK: John Wiley & Sons, Ltd.

Park, S., C. M. Lee, C. J. Doherty, S. J. Gilmour, Y. Kim, and M. F. Thomashow. 2015. Regulation of the Arabidopsis CBF regulon by a complex low-temperature regulatory network. *The Plant Journal* 82 (2):193–207.

Parmesan, C., and G. Yohe. 2003. A globally coherent fingerprint of climate change impacts across natural systems. *Nature* 421 (6918):37–42.

Pin, P. A., R. Benlloch, D. Bonnet, et al. 2010. An antagonistic pair of FT homologs mediates the control of flowering time in sugar beet. *Science* 330 (6009):1397–400.

Polashock, J. J., R. Arora, Y. Peng, D. Naik, and L. J. Rowland. 2010. Functional identification of a C-repeat binding factor transcriptional activator from blueberry associated with cold acclimation and freezing tolerance. *J Am Soc Hortic Sci* 135 (1):40–8.

Preston, J. C., and E. A. Kellogg. 2008. Discrete developmental roles for temperate cereal grass VERNALIZATION1/FRUITFULL-like genes in flowering competency and the transition to flowering. *Plant Physiol* 146 (1):265–76.

Qin, F., K. Shinozaki, and K. Yamaguchi-Shinozaki. 2011. Achievements and challenges in understanding plant abiotic stress responses and tolerance. *Plant Cell Physiol* 52 (9):1569–82.

Ren, B., Y. Liang, Y. Deng, et al. 2009. Genome-wide comparative analysis of type-A Arabidopsis response regulator genes by overexpression studies reveals their diverse roles and regulatory mechanisms in cytokinin signaling. *Cell Res* 19 (10):1178–90.

Rios, G., C. Leida, A. Conejero, and M. L. Badenes. 2014. Epigenetic regulation of bud dormancy events in perennial plants. *Front Plant Sci* 5:247.

Rohde, A., and R. P. Bhalerao. 2007. Plant dormancy in the perennial context. *Trends Plant Sci* 12 (5):217–23.

Root, T. L., J. T. Price, K. R. Hall, S. H. Schneider, C. Rosenzweig, and J. A. Pounds. 2003. Fingerprints of global warming on wild animals and plants. *Nature* 421 (6918):57–60.

Sakuma, Y., Q. Liu, J. G. Dubouzet, H. Abe, K. Shinozaki, and K. Yamaguchi-Shinozaki. 2002. DNA-binding specificity of the ERF/AP2 domain of Arabidopsis DREBs, transcription factors involved in dehydration- and cold-inducible gene expression. *Biochem Biophys Res Commun* 290 (3):998–1009.

Sasaki, R., H. Yamane, T. Ooka, et al. 2011. Functional and expressional analyses of PmDAM genes associated with endodormancy in Japanese apricot. *Plant Physiol* 157 (1):485–97.

Saure, M. C. 1985. Dormancy release in deciduous fruit trees. *Hortic Rev* 7:239–300.

Seo, E., H. Lee, J. Jeon, et al. 2009. Crosstalk between cold response and flowering in Arabidopsis is mediated through the flowering-time gene SOC1 and its upstream negative regulator FLC. *Plant Cell* 21 (10):3185–97.

Sharabi-Schwager, M., A. Lers, A. Samach, C. L. Guy, and R. Porat. 2010. Overexpression of the CBF2 transcriptional activator in Arabidopsis delays leaf senescence and extends plant longevity. *J Exp Bot* 61 (1):261–73.

Shi, Y., Y. Ding, and S. Yang. 2015. Cold signal transduction and its interplay with phytohormones during cold acclimation. *Plant Cell Physiol* 56 (1):7–15.

Shi, Y., S. Tian, L. Hou, et al. 2012. Ethylene signaling negatively regulates freezing tolerance by repressing expression of CBF and type-A ARR genes in Arabidopsis. *Plant Cell* 24 (6):2578–95.

Sinnott, E. W., and I. W. Bailey. 1915. The evolution of herbaceous plants and its bearing on certain problems of geology and climatology. *J Geol* 23 (4):289–306.

Song, G. Q. and Chen, Q. 2018. Comparative transcriptome analysis of unvernalized, vernalized, and late pink buds reveals genes involved in vernalization-mediated flowering in blueberry.

Song, G. Q., and X. Gao. 2017. Transcriptomic changes reveal gene networks responding to the overexpression of a blueberry DWARF and DELAYED FLOWERING 1 gene in transgenic blueberry plants. *BMC Plant Biol* 17 (1):106.

Song, G. Q., A. Walworth, D. Y. Zhao, N. Jiang, and J. F. Hancock. 2013. The Vaccinium corymbosum FLOWERING LOCUS T-like gene (VcFT): A flowering activator reverses photoperiodic and chilling requirements in blueberry. *Plant Cell Rep* 32 (11):1759–69.

Song, Y. H., J. S. Shim, H. A. Kinmonth-Schultz, and T. Imaizumi. 2015. Photoperiodic flowering: Time measurement mechanisms in leaves. *Annu Rev Plant Biol* 66 (1):441–64.

Srinivasan, C., C. Dardick, A. Callahan, and R. Scorza. 2012. Plum (Prunus domestica) trees transformed with poplar FT1 result in altered architecture, dormancy requirement, and continuous flowering. *PLOS ONE* 7 (7):e40715.

Stockinger, E. J., S. J. Gilmour, and M. F. Thomashow. 1997. Arabidopsis thaliana CBF1 encodes an AP2 domain-containing transcriptional activator that binds to the C-repeat/DRE, a cis-acting DNA regulatory element that stimulates transcription in response to low temperature and water deficit. *Proc Natl Acad Sci U S A* 94 (3):1035–40.

Strimbeck, G. R., P. G. Schaberg, C. G. Fossdal, W. P. Schroder, and T. D. Kjellsen. 2015. Extreme low temperature tolerance in woody plants. *Front Plant Sci* 6:884.

Sun, J., X. Peng, W. Fan, M. Tang, J. Liu, and S. Shen. 2014. Functional analysis of BpDREB2 gene involved in salt and drought response from a woody plant Broussonetia papyrifera. *Gene* 535 (2):140–9.

Takahash, M., and H. Morikawa. 2014. Nitrogen dioxide accelerates flowering without changing the number of leaves at flowering in Arabidopsis thaliana. *Plant Signal Behav* 9 (10).

Takemura, Y., K. Kuroki, Y. Shida, et al. 2015. Comparative transcriptome analysis of the less-dormant Taiwanese pear and the dormant Japanese pear during winter season. *PLOS ONE* 10 (10):e0139595.

Tang, Y., S. Qin, Y. Guo, et al. 2016. Genome-wide analysis of the AP2/ERF gene family in physic nut and overexpression of the JcERF011 gene in rice increased its sensitivity to salinity stress. *PLOS ONE* 11 (3):e0150879.

Thomashow, M. F. 2010. Molecular basis of plant cold acclimation: Insights gained from studying the CBF cold response pathway. *Plant Physiol* 154 (2):571–7.

Thomashow, M. F., S. J. Gilmour, E. J. Stockinger, K. R. Jaglo-Ottosen, and D. G. Zarka. 2001. Role of the Arabidopsis CBF transcriptional activators in cold acclimation. *Physiol Plantarum* 112 (2):171–5.

Tillett, R. L., M. D. Wheatley, E. A. Tattersall, K. A. Schlauch, G. R. Cramer, and J. C. Cushman. 2012. The Vitis vinifera C-repeat binding protein 4 (VvCBF4) transcriptional factor enhances freezing tolerance in wine grape. *Plant Biotechnol J* 10 (1):105–24.

Trankner, C., S. Lehmann, H. Hoenicka, et al. 2010. Over-expression of an FT-homologous gene of apple induces early flowering in annual and perennial plants. *Planta* 232 (6):1309–24.

Trevaskis, B., M. N. Hemming, E. S. Dennis, and W. J. Peacock. 2007. The molecular basis of vernalization-induced flowering in cereals. *Trends Plant Sci* 12 (8):352–7.

Varkonyi-Gasic, E., S. M. Moss, C. Voogd, T. Wang, J. Putterill, and R. P. Hellens. 2013. Homologs of FT, CEN and FD respond to developmental and environmental signals affecting growth and flowering in the perennial vine kiwifruit. *New Phytol* 198 (3):732–46.

Vergara, R., X. Noriega, F. Parada, D. Dantas, and F. J. Perez. 2016. Relationship between endodormancy, FLOWERING LOCUS T and cell cycle genes in Vitis vinifera. *Planta* 243 (2):411–9.

Vicente, R.-S. M., and J. Plasencia. 2011. Salicylic acid beyond defence: Its role in plant growth and development. *J Exp Bot* 62 (10):3321–38.

Voogd, C., T. Wang, and E. Varkonyi-Gasic. 2015. Functional and expression analyses of kiwifruit SOC1-like genes suggest that they may not have a role in the transition to flowering but may affect the duration of dormancy. *J Exp Bot* 66 (15):4699–710.

Walworth, A. E., B. Chai, and G. Q. Song. 2016. Transcript profile of flowering regulatory genes in VcFT-overexpressing blueberry plants. *PLOS ONE* 11 (6):e0156993.

Walworth, A. E., L. J. Rowland, J. J. Polashock, J. F. Hancock, and G. Q. Song. 2012. Overexpression of a blueberry-derived CBF gene enhances cold tolerance in a southern highbush blueberry cultivar. *Mol Breed* 30 (3):1313–23.

Wang, D., Z. Gao, P. Du, et al. 2015. Expression of ABA metabolism-related genes suggests similarities and differences Between seed dormancy and bud dormancy of peach (Prunus persica). *Front Plant Sci* 6:1248.

Wang, H., Z. Zou, S. Wang, and M. Gong. 2013. Global analysis of transcriptome responses and gene expression profiles to cold stress of Jatropha curcas L. *PLOS ONE* 8 (12):e82817.

Wang, Y., L. L. Georgi, G. L. Reighard, R. Scorza, and A. G. Abbott. 2002. Genetic mapping of the evergrowing gene in peach [Prunus persica (L.) Batsch]. *J Hered* 93 (5):352–8.

Wang, Y. H., and H. R. Irving. 2011. Developing a model of plant hormone interactions. *Plant Signal Behav* 6 (4):494–500.

Wellmer, F., and J. L. Riechmann. 2010. Gene networks controlling the initiation of flower development. *Trends Genet* 26 (12):519–27.

Wen, L. H., W. J. Zhong, X. M. Huo, W. B. Zhuang, Z. J. Ni, and Z. H. Gao. 2016. Expression analysis of ABA- and GA-related genes during four stages of bud dormancy in Japanese apricot (Prunus mume Sieb. et Zucc). *J Hortic Sci Biotechnol* 91 (4):362–9.

Wickland, D. P., and Y. Hanzawa. 2015. The FLOWERING LOCUS T/TERMINAL FLOWER 1 gene family: Functional evolution and molecular mechanisms. *Mol Plant* 8 (7):983–97.

Wilkie, J. D., M. Sedgley, and T. Olesen. 2008. Regulation of floral initiation in horticultural trees. *J Exp Bot* 59 (12):3215–28.

Wing, S. L., and L. D. Boucher. 1998. Ecological aspects of the Cretaceous flowering plant radiation. *Annu Rev Earth Planet Sci* 26 (1):379–421.

Wisniewski, M., J. Norelli, and T. Artlip. 2015. Overexpression of a peach CBF gene in apple: A model for understanding the integration of growth, dormancy, and cold hardiness in woody plants. *Front Plant Sci* 6:85.

Wisniewski, M., J. Norelli, C. Bassett, T. Artlip, and D. Macarisin. 2011. Ectopic expression of a novel peach (Prunus persica) CBF transcription factor in apple (Malus x domestica) results in short-day induced dormancy and increased cold hardiness. *Planta* 233 (5):971–83.

Woods, D. P., M. A. McKeown, Y. Dong, J. C. Preston, and R. M. Amasino. 2016. Evolution of VRN2/Ghd7-like genes in vernalization-mediated repression of grass flowering. *Plant Physiol* 170 (4):2124–35.

Wu, J., K. M. Folta, Y. Xie, W. Jiang, J. Lu, and Y. Zhang. 2017. Overexpression of Muscadinia rotundifolia CBF2 gene enhances biotic and abiotic stress tolerance in Arabidopsis. *Protoplasma* 254 (1): 239–51.

Xiao, H., M. Siddiqua, S. Braybrook, and A. Nassuth. 2006. Three grape CBF/DREB1 genes respond to low temperature, drought and abscisic acid. *Plant Cell Environ* 29 (7):1410–21.

Xiao, H., E. A. Tattersall, M. K. Siddiqua, G. R. Cramer, and A. Nassuth. 2008. CBF4 is a unique member of the CBF transcription factor family of Vitis vinifera and Vitis riparia. *Plant Cell Environ* 31 (1):1–10.

Xin, H., W. Zhu, L. Wang, et al. 2013. Genome wide transcriptional profile analysis of Vitis amurensis and Vitis vinifera in response to cold stress. *PLOS ONE* 8 (3):e58740.

Yamaguchi, N., M. F. Wu, C. M. Winter, et al. 2013. A molecular framework for auxin-mediated initiation of flower primordia. *Dev Cell* 24 (3):271–82.

Yan, L., D. Fu, C. Li, et al. 2006. The wheat and barley vernalization gene VRN3 is an orthologue of FT. *Proc Natl Acad Sci U S A* 103 (51):19581–6.

Yan, L., A. Loukoianov, G. Tranquilli, M. Helguera, T. Fahima, and J. Dubcovsky. 2003. Positional cloning of the wheat vernalization gene VRN1. *Proc Natl Acad Sci U S A* 100 (10):6263–8.

Yan, L. L., A. Loukoianov, A. Blechl, et al. 2004. The wheat VRN2 gene is a flowering repressor down-regulated by vernalization. *Science* 303 (5664):1640–4.

Yang, T. W., L. J. Zhang, T. G. Zhang, H. Zhang, S. J. Xu, and L. Z. An. 2005. Transcriptional regulation network of cold-responsive genes in higher plants. *Plant Sci* 169 (6):987–95.

Yordanov, Y. S., C. Ma, S. H. Strauss, and V. B. Busov. 2014. EARLY BUD-BREAK 1 (EBB1) is a regulator of release from seasonal dormancy in poplar trees. *Proc Natl Acad Sci* 111 (27):10001–6.

Zandkarimi, H., A. Ebadi, S. A. Salami, H. Alizade, and N. Baisakh. 2015. Analyzing the expression profile of AREB/ABF and DREB/CBF genes under drought and salinity stresses in grape (Vitis vinifera L.). *PLOS ONE* 10 (7):e0134288.

Zanne, A. E., D. C. Tank, W. K. Cornwell, et al. 2014. Three keys to the radiation of angiosperms into freezing environments. *Nature* 506 (7486):89–92.

Zhang, H., D. E. Harry, C. Ma, et al. 2010. Precocious flowering in trees: The FLOWERING LOCUS T gene as a research and breeding tool in Populus. *J Exp Bot* 61 (10):2549–60.

Zhang, J., W. R. Yang, T. R. Cheng, H. T. Pan, and Q. X. Zhang. 2013. Functional and evolutionary analysis of two CBF genes in Prunus mume. *Can J Plant Sci* 93 (3):455–64.

Zinn, K. E., M. Tunc-Ozdemir, and J. F. Harper. 2010. Temperature stress and plant sexual reproduction: Uncovering the weakest links. *J Exp Bot* 61 (7):1959–68.

9 Plant Response to Oxidative Stress

Rolston St. Hilaire

CONTENTS

9.1 INTRODUCTION

Oxidative stress occurs when excess photon energy is redirected to the production of reactive of species (ROS) primarily in the electron transport chain systems in the chloroplast and mitochondria, and the capacity of the plant to remove ROS is overwhelmed (Asada, 1999). Woody plants have a complex antioxidant system to combat the effects of oxidative stress (Smirnoff, 1993). The production of ROS is one of the earliest plant biochemical responses to abiotic and biotic stresses known in plant cells (Anjum et al., 2011). Consequently, one of the immediate responses of the woody plant to oxidative stress is to remove the offending ROS species. If unchecked ROS cell concentrations rise and cause oxidative damage to cell membranes (lipid peroxidation), proteins, RNA and DNA molecules, and ultimately, the cell dies.

Under normal conditions, ephemeral changes in environmental conditions such as strong light, pathogens, moisture deficits, and extreme temperatures produce ROS and challenge steady state conditions of the plant. These changes only transiently disrupt plant function because woody plants are very adept at reducing the impact of ROS. However, Osakabe et al. (2012) observed that prolonged and repeated severe environmental stresses that affect growth and development engender long-lasting effects in woody plants because of their long growth periods. What has been observed is that plants growing in harsh environments show enhanced antioxidant activity that combats the effect of the constant generation of ROS (Duan et al., 2005). The balance between ROS production and antioxidant capability is important to plants because the ability of plants to mitigate the impact of various stresses and maintain their productivity may be related to the ROS scavenging efficiency of a plant's antioxidant systems (Xiao et al., 2008).

Many plant stress tolerance capabilities have been correlated to the ROS scavenging capabilities of antioxidants (Anjum et al., 2011; Bowen O'Connor et al., 2013). For example, Bowen-O'Connor et al. (2013) reported that in *Acer grandidentatum* (bigtooth maple), enhanced tolerance to prolonged light was correlated with elevated levels of the antioxidants xanthophyll, zeaxanthin and β-carotene. In contrasting populations of *Populus cathayana* (Manchurian poplar) from wet and dry climates, the enhanced drought tolerance of plants from the xeric environment correlated with higher peroxidase (POD) antioxidant activity (Xiao et al., 2008). However, Vickers et al. (2009) cautioned that it is often difficult to distinguish cause-and-effect relationships for ROS because ROS production is ephemeral and a plant's lipid and aqueous phase of its response network work together. This makes the direct measurement of ROS difficult.

While ROS can produce deleterious effects in plants, they can also serve as important signaling molecules (Foyer et al., 1995; Niyogi, 1999). For example, ROS activates genes that determine apoptosis which is the systematic death of a cell. Apoptosis is a mechanism that plants use to contain injury (Gravano et al., 2004). That cell death could produce a visible symptom induced by ROS signaling is evidence of the plant's defense mechanism at work and not an indicator of an injury to the plant (Gravano et al., 2004). The ROS hydrogen peroxide is known to be a diffusible signaling molecule in plant defense responses (Vranova et al., 2002). More importantly, ROS can activate genes that are directly responsible for the cell detoxification process (Gravano et al., 2004).

Mechanisms of ROS detoxification exist in all plants and can be categorized as enzymatic and non-enzymatic (Reddy et al., 2004; Smirnoff, 1993). Superoxide dismutase (SOD), catalase (CAT), ascorbate peroxidase (APX), POD and glutathione reductase (GR) are enzymes that are involved in pathways that scavenge ROS. Non-enzymatic antioxidant carotenoids such as β-carotene, α-tocopherol and xanthophylls, and flavanones, and plant osmolytes such as proline, ascorbic acid, and anthocyanins are known to mitigate the impact of ROS. Both β-carotene and α-tocopherol along with other carotenoids have been characterized as low molecular weight antioxidants (Bartosz, 1997). The role that enzymatic and non-enzymatic antioxidant systems play in mitigating the impact of ROS are co-occurring, and this could be very significant for woody plants.

Interestingly, the levels of SOD either slightly increased or decreased when anthocyanin production was photoinduced (Grace et al., 1995). Vitamin E (α- tocopherol) is a well-studied carotenoid antioxidant. Two decades ago, Ishii et al. (1996) showed anthocyanins had a greater activity than vitamin E. Thus, there is evidence that anthocyanins function as plant antioxidants (Chalker-Scott, 1999; Yamasaki et al., 1996).

9.2 MAJOR OXIDATIVE SPECIES

Efficient energy production in plants is facilitated through the reduction of O_2 (Vranova et al., 2002). And in plant cells, organelles and the cytosol are potential sources of ROS. Reactive oxygen species, namely hydrogen peroxide (H_2O_2), super oxide anions ($O_2^{\cdot-}$), and hydroxyl radicals ($OH^{\cdot}$) are produced as a consequence of the reduction of O_2. While the initiation of the reduction of O_2 requires energy, and the energy from excess light at PSII (photosystem II) is partly released through photochemical quenching, singlet oxygen ($^1O^2$), another ROS, is formed when O_2 accepts excess energy (Foyer, 2018).

Of the major oxidative species, $O_2^{\cdot-}$, H_2O_2, $^1O^2$, $OH^{\cdot}$, the $O_2^{\cdot-}$ anions are confined within biological membranes. The activity of SOD hastens the formation of H_2O_2 and O_2 from them ($2O_2^{\cdot-} + e^- + 2H^+ = H_2O_2 + O_2$; Vranova et al., 2002; Halliwell and Gutteridge, 1989). Additionally, the $O_2^{\cdot-}$ anion can autonomously reduce quinones and the ferric iron (Fe^{3+}) and cuprous iron (Cu^{2+}) complexes of enzymes (Vranova et al., 2002). In proteins that have an iron-sulfur center, the reduction of Fe^{3+} to Fe^{2+} leads to enzyme inactivation, and the accompanying loss of Fe^{3+} from the enzymes drives the Fenton reaction. In a process that is mediated by radicals, Fenton chemistry is at the heart of the potentially damaging non-enzymatic lipid peroxidation (Cakmak, 2000). While $O_2^{\cdot-}$ anions do not traverse biological membranes, they exist in equilibrium with the more reactive hydroperoxyl radical ($HO_2^{\cdot}$) that is formed through protonation. The hydroperoxyl radical can cross membranes and trigger lipid peroxidation by coopting hydrogen ions from polyunsaturated fatty acids and lipid hydroperoxidases (Halliwell and Gutteridge, 1989; Vranova et al., 2002).

Singlet oxygen is among the most damaging oxidative species and once formed, $^1O^2$ can react with other biological species to subsequently form endoperoxidases and hydroperoxidases (Choudhury et al., 2017; Foyer, 2018; Halliwell and Gutteridge, 1989). When formed in the chloroplasts $^1O^2$ is known to reprogram nuclear gene expression leading to chlorosis, programmed cell death, and other responses related to biotic and abiotic stress responses (Leister et al., 2017).

The H_2O_2 formed through dismutation can cross cell membranes (Vranova et al., 2002), and H_2O_2 formed in chloroplasts and diffusing to the nucleus can serve as an important stress signaling molecule. Singly, H_2O_2 may inactivate enzymes, but more importantly, the $OH^{\cdot}$ radical is formed during Fenton reactions (Halliwell and Gutteridge, 1989). Because plants do not have an enzymatic mechanism to detoxify $OH^{\cdot}$, the accumulation of this OH radical eventually causes cell death (Vranová et al., 2002).

While the chloroplast, mitochondria, peroxisome, and apoplast (Choudhury et al., 2017), are the major cell areas where ROS is produced during abiotic stress, ROS profiles are different at each site. Super oxide anions ($O_2^{\cdot-}$) that accumulate in the mitochondria during abiotic stress are typically converted to H_2O_2 by mitochondrial superoxide dismutase (MnSOD) (Huang et al., 2016). Consequently, each organelle has developed some specificity in detoxifying ROS.

9.3 OXIDATIVE STRESS EFFECT ON PLANT GROWTH AND DEVELOPMENT

Woody plant adaptive responses to the oxidative stresses present in the environment, such as radiation stress, extreme temperature, water and nutrient deficit, and pest and diseases have shaped vegetation types (Bussotti, 2008). Because existing oxidative stresses in the plant's evolutionary environment may have shaped the plant's responses to stresses, the effect of stresses on plant growth and development may be a result of the pre-existing competency to tolerate that stress. Different plant parts, for example roots vs. shoots, might also have differential responses to stress. Evidently, the constitutive presence of oxidative stress in the growth environment of woody plants implies there is a greater need for antioxidant protection (Bussotti, 2008; Fryer, 1992). For example, the antioxidants ascorbate and α-tocopherol were higher in sun leaves than in shade leaves (Fryer, 1992). In addition, the age of the woody plant tissue might modulate the impact of oxidative stress. In *Fagus sylvatica* (European beech) older leaves had a higher photoprotective capacity against foliar surface-based impacts of oxidative stress, such as those induced by light or pollutant uptake (Wieser et al., 2003).

On the other hand, the environment itself could alleviate the effects of stress and compensate for the deleterious effect of oxidative stress. This was evident in a study done on *Quercus robur* (English oak) and *Pinus pinaster* (cluster pine) (Schwanz et al., 1996), where growth in an environment with elevated CO_2 caused reductions in the activities of SODs in both species. Thus, the presence of high concentrations of CO_2 has the potential to mitigate the impact of oxidative stress in leaves, while providing the plant with the metabolic flexibility to increase antioxidative activity in response to further exposure to oxidative stress.

In addition to the environment, cultivar differences might dictate how an oxidative stress influences the growth and development of woody plants. Cultivar differences in response to drought-induced oxidative stress also were observed in *Olea europaea* (olive) cultivars grown under low-water availability conditions (Bacelar et al., 2007). In another study, the olive cultivars, 'Gaidourelia', 'Kalamon', 'Koroneiki' and 'Megaritiki' were subjected to drought stress, and photosynthetic performance, antioxidant activity, and oxidative stress damage were assessed (Petridis et al., 2012). While drought stress impaired photosynthetic function, and increased the accumulation of phenolic compounds, Petridis et al. (2012) concluded that the severity of the impact of oxidative stress created by water limitation, depended on the genotype and on the level and duration of the stress.

In general, oxidative stress reduces the growth and development of woody plants. In *Pistacia vera* (pistachios), drought-induced oxidative stress decreased vegetative growth, leaf relative water content, chlorophyll content, carotenoid content, and total soluble proteins (Khoyerdi et al., 2016). In the woody shrubs, *Pyracantha fortuneana* (Chinese firethorn) and *Rosa cymosa* (elder-flower rose) and the trees, *Broussonetia papyrifera* (paper mulberry), *Cinnamomum bodinieri*, *Platycarya longipes*, and *Platycarya tatarinowii*, malondialdehyde content, which is an indicator of membrane lipid peroxidation and ion leakage, gradually increased as oxidative stress increased (Liu et al., 2011).

9.4 PHYSIOLOGICAL RESPONSE TO OXIDATIVE STRESS

Physiological responses to oxidative stress may be expressed in different forms (Bussotti et al., 2007), and this may be due to the fact that different abiotic stresses, such as drought, heat, salinity and prolonged light, produce different ROS profiles that specify the acclimation response and help the plant tailor its response to the precise stress (Choudhury et al., 2017). Ozone is one of the common pollutants that induces oxidative stress in woody plants (Bussotti et al., 2007; Iglesias et al., 2006). The ozone that enters into the leaf through the stomata decays into ROS and triggers an oxidative burst (Iglesias et al., 2006). Consequently, leaf stippling, foliage reddening, leaf senescence, and subsequent new foliage production are all visible symptoms resulting from the plant's physiological responses to oxidative stress triggered by ozone stress (Bussotti et al., 2007). New foliage growth allows the plant to replace the old leaves with ones that are more efficient in light of the heightened oxidative stress (Bussotti et al., 2007). The maintenance of full photosynthetic physiological capacity in the face of oxidative stress is a worthy target for plants exposed to stress and not necessarily exhibiting visible symptoms.

The metabolic expense to the plant in maintaining photosynthetic capacity depends in part on the severity of the stress and the external environment of the woody plant. Photosynthesis was higher and stomatal conductance was lower in leaves of *Quercus robur* (English oak) grown under elevated (700 ± 50 μL L^{-1}) CO_2 conditions compared to normal (350 ± 30 μL L^{-1}) CO_2 levels (Schwanz et al., 1996). The availability of a better supply of photosynthates will limit the plants metabolic expense and engender a better physiological response to oxidative stress. In this same study, Schwanz et al. (1996) reported that in plants of English oak and *Pinus pinaster* (cluster pine) exposed to drought stress and elevated CO_2, the antioxidant activities of SOD in leaves was higher compared to that in leaves grown under normal CO_2 levels (Schwanz et al., 1996). Since SOD detoxifies $O_2^{\cdot-}$ ($2O_2^{\cdot-} + e^- + 2H^+ = H_2O_2 + O_2$), the authors concluded that the rate of $O_2^{\cdot-}$ production was increased in plants grown with elevated CO_2 and subjected to drought stress. If elevated CO_2 enhances the plant's fitness to cope with environmental stresses such as drought, then some ecological benefits are likely to be gained (Schwanz et al., 1996).

Woody plant physiological response to oxidative stress might involve non-enzymatic means. In *Populus × canescens* (grey poplar) exposed to Cd-induced oxidative stress, He et al. (2011), suggested that phenolics may be essential ROS scavengers because the accumulation of phenolics was greater in the tissues that had the strongest Cd accumulation.

For photosynthesis to occur, membranes that envelope the photosynthetic apparatus must be stable. Cell membrane stability and integrity in the face of abiotic stress is considered to be a good physiological indicator of oxidative stress resilience in plants (Liu et al., 2011). Peroxidation of the lipids in the cell membrane can be the result of ROS attack. Hence, the leakage of ions from the membrane, as measured by the production of malondialdehyde, can used be used as a proxy for oxidative stress (Smirnoff, 1993). Better values of the efficiency of PSII, as measured through the ratio of variable to maximal fluorescence (F_v/F_m), and lower malondialdehyde content would be good physiological indicators of resiliency to oxidative stress in plants.

Drought-induced oxidative stress is common among woody plants growing in natural stands or managed landscapes. Liu et al. (2011) suggested that the plants that accumulated proline and soluble sugars, especially under sustained drought had a better capacity to maintain their water status because of the higher capacity to osmotically adjust their cell contents. Interestingly, the accumulation of proline was positively correlated with SOD activity in five woody species (Liu et al., 2011). This suggests that non-enzymatic and enzymatic parameters might be working in tandem to curb oxidative stress.

A change in antioxidant enzyme activity is a common response to drought-induced oxidative stress. Progressive drought stress caused a gradual increase in the activities of SOD, CAT, POD, APX and GR in cuttings of *Populus kangdingensis* and *Populus cathayana*, originating from high and low altitudes in the eastern Himalaya respectively (Yang and Miao, 2010). However, the physiological response of those plants was tempered by their altitudinal origins.

In beech, sun exposed leaves of seedlings and adult canopy trees had more total ascorbate and α-tocopherol and lower amounts of photosynthetic pigments than leaves in the shaded part of the crown (Wieser et al., 2003). This physiological response to possible oxidative stress appears to be an investment in antioxidant compounds to protect the photosynthetic apparatus.

9.5 MOLECULAR RESPONSE TO OXIDATIVE STRESS

As was shown in a study of the environmental stresses of 59 African savanna tree species, unpacking the molecular mechanisms of woody species to environmental stress is important to the understanding of stress tolerances and how stress resistance might be improved (Prior et al., 1987). Molecular cascades are important because they signal the plant to produce enough of a substance to protect the plant from stress. Furthermore, plants, in response to those stresses, change their architecture on a molecular level to express phenotypes that are more tolerant of stress or counteract the stresses with antioxidants (Ahuja et al., 2010). The molecular response to environmental stress in woody plants, such as drought or cold is triggered by either a change in plant metabolism, ionic shifts, plant hormones, or ROS. Once the signal is perceived, the activated stress-responsive genes work to maintain the woody plant's steady state and protect protein and cellular membranes from damage (Osakabe et al., 2012). Molecular signaling and the subsequent physiological and biochemical pathways that are mobilized are important defenses against oxidative stresses that could potentially kill the plant (Osakabe et al., 2012). Unfortunately, very little is known about these relationships in woody plants.

In woody plants, oxidative stress causes the transcription of many genes (Teixeira et al., 2005; Zheng et al., 2015). The responses to oxidative stress, such as those caused by drought stress, are known to be regulated at the transcriptional and post-transcriptional levels through noncoding endogenous microRNAs (Esmaeili et al., 2017). Some of the gene products are multifunctional. For example, in *Eucalyptus gunnii* (cider gum), metallothionein genes up-regulated in response to cold stress code for products that bind metal ions, scavenge ROS, and repair plasma membranes (Keller et al., 2013). The Chinese white poplar's (*Populus tormentosa*) molecular defense mechanisms were studied in ten-week-old seedlings subjected to salinity stress. Gene expression patterns were profiled in response to the various salinity solutions using RNA-sequencing (Zheng et al., 2015). Compared with controls (no salinity) 4,903 genes were up regulated and 3,461 genes were down regulated after 24 hours of salinity treatment. The differentially expressed genes were highly enriched in hormone- and ROS-related biological processes leading Zheng et al. (2015) to conclude that H_2O_2 and hormones are responsible for a role in the salt stresses in poplar. In *Populus deltoides* (eastern cottonwood) subjected to copper (Cu) stress, Guerra et al. (2009) showed that the genes differentially expressed in Cu stress also included several enzymes involved in ROS scavenging such as peroxidases, Cu/Zn superoxide dismutases, and catalases that were down regulated. Similarly, down regulation of those genes was observed in other *Populus* taxa exposed to other heavy metal stress. And this lead Guerra et al. (2009) to conclude similar responses of these enzymes are expected in different poplar clones subjected to heavy metal stress.

While many genes encoding for different enzymes will be up regulated in response to stress, in specific woody plants, certain genes might contribute to an outsized role in oxidative stress reduction (Teixeira et al., 2005). In *Eucalyptus grandis* (flooded gum), mining of the FORESTs gene data bank revealed that a large gene family encoding for phospholipid hydroperoxide glutathione peroxidases, gene sequences encoding for superoxide dismutase isoforms, and the ubiquity of catalase transcripts suggest that these three enzymes and their isoforms have major importance in eucalyptus antioxidant metabolism (Teixeira et al., 2005).

Two eucalyptus genotypes designated as G1 and G2 were exposed to water limiting conditions and screened for differential gene expression in response to water treatment (Villar et al., 2011). After the whole sequencing set of the eucalyptus genome was screened, genotype G2 showed a larger number of specific contigs (10.5% of the *Eucalyptus spp* sequencing set) than G1 (5.5% of the *Eucalyptus spp* sequencing set). So, the larger set of genes activated in genotype G2 might have triggered more specific responses to water deficit. Because the genotype G2 maintained higher biomass growth during the dry season, Villar et al. (2011) concluded that the ability to express a broader set of genes in response to drought stress might be a key molecular adaptation to drought. In Poli and 58-861, two clones of black poplar (*Populus nigra*) originating from mesic and xeric environments respectively, the relative gene expression of SOD was 9.2- and 4.5-fold higher in roots of Poli than that of 58-861 under well-watered and water limiting conditions (Regier et al., 2009). This suggests that SOD activity might be enhanced in woody plant genotypes unaccustomed to water limiting conditions.

In an experiment where Poli and 58-861 were exposed to Cd heavy metal stress, the assessment gene expression of GR and APX using semi-quantitative RT-PCR did not show marked changes (Gaudet et al., 2011). While GR and APX are known to be involved in the oxidative stress response, Gaudet et al. (2011) concluded that the ROS detoxification in response to Cd stress was not activated. Because multigene families code for GR and APX, several enzyme isoforms exist, and this makes it important to test for the right isoform in oxidative stress studies (Gaudet et al., 2011).

9.6 RESISTANCE AND TOLERANCE TO OXIDATIVE STRESS

Coordinated growth and development activities, physiological and biochemical adjustment, and molecular responses contribute to woody plant resistance and tolerance to oxidative stress. In contrast to their herbaceous counterparts, woody plants develop responses to stresses eventually making them better able to mitigate the impact of oxidative stress. Schwanz et al. (1996) reasoned that the rapid development of drought stress in a tender herbaceous leaf might overwhelm the metabolic capacity of the plant. Seedlings of English oak (*Quercus robur*) that were drought-stressed for six weeks maintained predawn water potentials that were similar to well-watered plants (Schwanz et al., 1996). This slow adaptation to drought could involve osmotic adjustments of cell contents that help maintain cell turgor and consequently the production of metabolites that could combat the effect of oxidative stress. In many species, Liu et al. (2011) noted that under drought conditions, the accumulation of proline and other soluble sugars was associated with drought tolerance.

The pool of metabolites that can be mobilized or is present when the plant is exposed to oxidative stress is helpful in conferring resistance and tolerance to oxidative stress. The nonenzymatic function of those metabolites is well documented in the case of the carotenoid pool that can be mobilized if the plant is exposed to photo-oxidative stress. The carotenoid β-carotene scavenges singlet oxygen and limits the production of ROS from the light harvesting complexes of the photosynthetic apparatus (Telfer et al., 1994). Consequently, plants with a high β-carotene pool are expected to have enhanced tolerance to oxidative stress.

Another potential indicator of a woody plant's tolerance of oxidative stress is its age (Novak et al., 2005; Wieser et al., 2003). Older trees of *Picea abies* (Norway spruce) were less susceptible to ozone-induced oxidative stress than seedlings (Wieser et al., 2002). And Wieser et al. (2003) also showed this to be true for *Fagus sylvatica* (Beech) and cautioned researchers that age-related differences in leaf morphology mandates that leaf area-based antioxidant capacity data rather than dry-mass data is needed to truly decipher age-related differences in oxidative stress tolerance of woody plants.

Leaf functional traits can be drivers of how plants thrive in an oxidative stress environment (Bussotti, 2008). Leaves with high tissue density have a high photosynthesis capacity per surface unit and high water-use efficiency to acclimate well to oxidative stress because this functional leaf trait facilitates detoxification processes. These leaves are relatively unaffected by environmental changes, showed enhance resistance and tolerance to oxidative stress, and consequently are long-lived in the face of oxidative pressure (Bussotti, 2008).

The extent of the woody plant lipid peroxidation of its cell membranes when in a stressful environment is considered to be a good indicator of resistance and tolerance to oxidative stress

(Liu et al., 2011). While elucidation of their roles is still being conducted, the phytochemicals flavonoids and isoflavonoids, already known for antioxidant functions might be key players in preventing lipid peroxidation (Arora et al., 2000). Flavonoids and isoflavonoids could partition into cell membranes and decrease membrane fluidity and thereby allow the plant to resist the ill effects of oxidative stress-induced lipid peroxidation (Arora et al., 2000). Other phytochemicals such as carotenoids provide photoprotection and have a structural role in photosynthesis (Bowen-O'Connor et al., 2013). Equally important, carotenoids inhibit lipid peroxidation and stabilize membranes (Niyogi, 1999). Thus, plants rich in flavonoids, isoflavonoids, and carotenoids could potentially display resilience to oxidative stress.

While the abundance of antioxidant phytochemicals improves stress resistance, plants with better antioxidant activity at the chloroplast, a major source of ROS, have enhanced tolerance to oxidative stress. Plants with elevated expression of GR in the chloroplast had increased resistance to oxidative stress (Foyer et al., 1995).

In conclusion, woody plants develop responses to oxidative stresses more slowly than their herbaceous counterparts and this might allow them to better mitigate the impact of oxidative stress. What is clear is that woody plants immediately respond to the threat of oxidative stress that ROS generates. Inevitably, the capacity of the plant to mobilize enzymatic and nonenzymatic means to blunt the damaging effects of ROS correlates with the oxidative stress resistance and tolerance of the woody plant.

REFERENCES

Ahuja, I., R. C. de Vos, A. M. Bones, and R. D. Hall. 2010. Plant molecular stress responses face climate change. *Trends in Plant Science* 15:664–674.

Anjum, S. A., X. Y. Xie, L. C. Wang, M. F. Saleem, C. Man, and W. Lei. 2011. Morphological, physiological and biochemical responses of plants to drought stress. *African Journal of Agricultural Research* 6:2026–2032.

Arora, A., T. M. Byrem, M. G. Nair, and G. M. Strasburg. 2000. Modulation of liposomal membrane fluidity by flavonoids and isoflavonoids. *Archives of Biochemistry and Biophysics* 373:102–109.

Asada, K. 1999. The water–water cycle in chloroplasts scavenging of active oxygens and dissipation of excess photons. *Annual Review of Plant Physiology and Plant Molecular Biology* 50:601–639.

Bacelar, E. A., D. L. Santos, J. M. Moutinho-Pereira, J. I. Lopes, B. C. Gonçalves, T. C. Ferreira, and C. M. Correia. 2007. Physiological behaviour, oxidative damage and antioxidative protection of olive trees grown under different irrigation regimes. *Plant and Soil* 292:1–12.

Bartosz, G. 1997. Oxidative stress in plants. *Acta Physiologiae Plantarum* 19:47–64.

Bowen-O'Connor, C. A., D. M. VanLeeuwen, T. M. Sterling, G. Bettmann, and R. St. Hilaire. 2013. Variation in violaxanthin and lutein cycle components in two provenances of *Acer grandidentatum* L. exposed to short-term contrasting light. *Acta Physiologiae Plantarum* 35:541–548.

Bussotti, F. 2008. Functional leaf traits, plant communities and acclimation processes in relation to oxidative stress in trees: A critical review. *Global Change Biology* 14:2727–2739.

Bussotti, F., R. J. Strasser, and M. Schaub. 2007. Photosynthetic behavior of woody species under high ozone exposure probed with the JIP-test: A review. *Environmental Pollution* 147:430–437.

Cakmak, I. 2000. Possible roles of zinc in protecting plant cells from damage by reactive oxygen species. *New Phytologist* 146:185–205.

Chalker-Scott, L. 1999. Environmental significance of anthocyanins in plant stress responses. *Photochemistry and Photobiology* 70:1–9.

Choudhury, F. K., R. M. Rivero, E. Blumwald, and R. Mitter. 2017. Reactive oxygen species, abiotic stress and stress combination. *The Plant Journal* 90:856–886.

Duan, B. L., Y. W. Lu, C. Y. Yin, O. Junttila, and C. Y. Li. 2005. Physiological responses to drought and shade in two contrasting *Picea asperata* populations. *Physiologia Plantarum* 124:476–484.

Esmaeili, F., B. Shiran, H. Fallahi, N. Mirakhorli, H. Budak, and P. Martínez-Gómez. 2017. In silico search and biological validation of microRNAs related to drought response in peach and almond. *Functional and Integrative Genomics* 17:189–201.

Foyer, C. H. 2018. Reactive oxygen species, oxidative signaling and the regulation of photosynthesis. *Environmental and Experimental Botany* 154:134–142. https://doi.org/10.1016/j.envexpbot.2018.05.003.

Foyer, C. H., N. Souriau, S. Perret, M. Lelandais, K. J. Kunert, C. Pruvost, and L. Jouanin. 1995. Overexpression of glutathione reductase but not glutathione synthetase leads to increases in antioxidant capacity and resistance to photoinhibition in poplar trees. *Plant Physiology* 109:1047–1057.

Fryer, M. J. 1992. The antioxidant effects of thylakoid vitamin E (α-tocopherol). *Plant, Cell and Environment* 15:381–392.

Gaudet, M., P. Fabrizio, I. Beritognolo, V. Iori, M. Zacchini, A. Massacci, G. S. Mugnozza, and M. Sabatti. 2011. Intraspecific variation of physiological and molecular response to cadmium stress in *Populus nigra* L. *Tree Physiology* 31:1309–1318.

Grace, S., B. A. Logan, A. Keller, B. Deming-Adams, and W. Adams III. 1995. Acclimation of leaf antioxidants systems to light stress. *Plant Physiology* 108:36.

Gravano, E., F. Bussotti, R. J. Strasser, M. Schaub, K. Novak, J. Skelly, and C. Tani. 2004. Ozone symptoms in leaves of woody plants in open-top chambers: Ultrastructural and physiological characteristics. *Physiologia Plantarum* 121:620–633.

Guerra, F., S. Duplessi, A. Kohler, F. Martin, J. Tapia, P. Lebed, F. Zamudio, and E. González. 2009. Gene expression analysis of *Populus deltoides* roots subjected to copper stress. *Environmental and Experimental Botany* 67:335–344.

Halliwell, B. and J. M. C. Gutteridge. 1989. *Free radicals in biology and medicine*. Oxford, Clarendon Press.

He, J., J. Qin, L. Long, Y. Ma, H. Li, K. Li, X. Jiang, T. Liu, A. Polle, Z. Liang, and Z. Luo. 2011. Net cadmium flux and accumulation reveal tissue-specific oxidative stress and detoxification in *Populus×canescens*. *Physiologia Plantarum* 1:50–63.

Huang, S., O. Van Aken, M. Schwarzländer, K. Belt, and A. Millar. 2016. The roles of mitochondrial reactive oxygen species in cellular signaling and stress responses in plants. *Plant Physiology* 171:1551–1559.

Iglesias, D. J., Á. Calatayud, E. Barreno, E. Primo-Millo, and M. Talon. 2006. Responses of citrus plants to ozone: Leaf biochemistry, antioxidant mechanisms and lipid peroxidation. *Plant Physiology and Biochemistry* 44:125–131.

Ishii, G., M. Mori, and Y. Umemura. 1996. Antioxidants activity and food chemical properties of anthocyanins from the colored tuber flesh of potatoes. *Journal of the Japanese Society for Food Science and Technology* 43:962–966.

Khoyerdi, F. F., M. H. Shamshiri, and A. Estaji. 2016. Changes in some physiological and osmotic parameters of several pistachio genotypes under drought stress. *Scientia Horticulturae* 198:44–51.

Keller, G., P. B. Cao, H. San Clemente, W. El Kayal, C. Marque, and C. Teulières. 2013. Transcript profiling combined with functional annotation of 2,662 ESTs provides a molecular picture of *Eucalyptus gunnii* cold acclimation. *Trees* 27:1713–1735.

Leister, D., L. Wang, and T. Kleine. 2017. Organellar gene expression and acclimation of plants to environmental stress. *Frontiers in Plant Science* 8:387. doi: 10.3389/fpls.2017.00387.

Liu, C., Y. Liu, K. Guo, D. Fan, G. Li, Y. Zheng, R. Yang, and R. Yang. 2011. Effect of drought on pigments, osmotic adjustment and antioxidant enzymes in six woody plant species in karst habitats of southwestern China. *Environmental and Experimental Botany* 71:174–183.

Niyogi, K. K. 1999. Photoprotection revisited: Genetic and molecular approaches. *Annual Review of Plant Physiology and Plant Molecular Biology* 50:333–359.

Novak, K., M. Schaub, J. Fuhrer, J. M. Skelly, C. Hug, W. Landolt, P. Bleuler, and N. Kräuchi. 2005. Seasonal trends in reduced leaf gas exchange and ozone-induced foliar injury in three ozone sensitive woody plant species. *Environmental Pollution* 136:33–45.

Osakabe, Y., A. Kawaoka, N. Nishikubo, and K. Osakabe. 2012. Responses to environmental stresses in woody plants: Key to survive and longevity. *Journal of Plant Research* 125:1–10.

Petridis, A., I. Therios, G. Samouris, S. Koundouras, and A. Giannakoula. 2012. Effect of water deficit on leaf phenolic composition, gas exchange, oxidative damage and antioxidant activity of four Greek olive (*Olea europaea* L.) cultivars. *Plant Physiology and Biochemistry* 60:1–11.

Prior, J., J. Tuohy, and J. Whiting. 1987. Interrelationships between aspects of nitrogen metabolism and solute accumulation and the distribution of subtropical woody plants in southeast Africa. *New Phytologist* 107:427–439.

Reddy, A. R., K. V. Chaitanya, P. P. Jutar, and K. Sumithra. 2004. Differential antioxidative responses to water stress among five mulberry (*Morus alba* L.) cultivars. *Environmental and Experimental Botany* 52:33–42.

Regier, N., S. Streb, C. Cocozza, M. Schaub, P. Cherubini, S. C. Zeeman, and B. Frey. 2009. Drought tolerance of two black poplar (*Populus nigra* L.) clones: Contribution of carbohydrates and oxidative stress defence. *Plant, Cell and Environment* 32:1724–1736.

Schwanz, P., C. Picon, P. Vivin, E. Dreyer, J.-M. Cuehl, and A. Polle. 1996. Responses of antioxidative systems to drought stress in pendunculate oak and maritime pine as modulated by elevated CO_2. *Plant Physiology* 110:393–402.

Smirnoff, N. 1993. The role of active oxygen in the response of plants to water deficit and desiccation. *New Phytologist* 125:27–58.

Telfer, A., S. M. Bishop, D. Phillips, and J. Barber. 1994. Isolated photosynthetic reaction center of photosystem II as a sensitiser for the formation of singlet oxygen: Detection and quantum yield determination using a chemical trapping technique. *Journal of Biological Chemistry* 269:13244–13253.

Teixeira, F. K., L. Menezes-Benavente, V. C. Galvão, and M. Margis-Pinheiro. 2005. Multigene families encode the major enzymes of antioxidant metabolism in *Eucalyptus grandis* L. *Genetics and Molecular Biology* 28:529–538.

Vickers, C. E., J. Gershenzon, M. T. Lerdau, and F. Loreto. 2009. A unified mechanism of action for volatile isoprenoids in plant abiotic stress. *Nature Chemical Biology* 5:283–291.

Villar, E., C. Klopp, C. Noirot, E. Novaes, M. Kirst, C. Plomion, and J. M. Gion. 2011. RNA-Seq reveals genotype-specific molecular responses to water deficit in eucalyptus. *BMC Genomics* 12:538.

Vranova, E., D. Inze, and F. Van Breusegem. 2002. Signal transduction during oxidative stress. *Journal of Experimental Botany* 53:1227–1236.

Wieser, G., K. Hecke, M. Tausz, K.-H. Häberle, T. E. E. Grams, and R. Matyssek. 2003. The influence of microclimate and tree age on the defense capacity of European beech (*Fagus sylvatica* L.) against oxidative stress. *Annals of Forest Science* 60:131–135.

Wieser, G., K. Tegischer, M. Tausz, K. H. Häberle, T. E. E. Grams, and R. Matyssek. 2002. Age-effects on Norway spruce (Picea abies (L.) Karst.) susceptibility to ozone uptake. A novel approach relating stress avoidance to defense. *Tree Physiology* 22:583–590.

Xiao, X., X. Xu, and F. Yang. 2008. Adaptive responses to progressive drought stress in two Populus cathayana populations. *Silva Fennica* 42:705–719.

Yamasaki, H., H. Uefuji, and Y. Sakihama. 1996. Bleaching of red anthocyanin by superoxide radical. *Archives of Biochemistry and Biophysics* 332:183–186.

Yang, F., and L. F. Miao. 2010. Adaptive responses to progressive drought stress in two poplar species originating from different altitudes. *Silva Fennica* 44:23–37.

Zheng, L., Y. Meng, J. Ma, X. Zhao, T. Cheng, J. Ji, S. Shi, C. Meng, N. Deng, L. Chen, S. Shi, and Z. Jiang. 2015. Transcriptomic analysis reveals importance of ROS and phytohormones in response to short-term salinity stress in *Populus tomentosa*. *Frontiers in Plant Science* 6:678.

10 Plants Response to Heavy Metal Stress

Raj Narayan Roy and Bidyut Saha

CONTENTS

10.1 INTRODUCTION

The term heavy metals (HMs) refers to a group of metals and metalloids with atomic density heaver than 5 gm/cm^3, or five times or more, greater than water (Hawkes, 1997). HMs such as Fe, Mn, Mo, Ni, Zn and Cu are considered essential micronutrients because they are essential for normal growth and metabolic processes of plants (Vangronsveld and Clijsters, 1994). On the other hand, Cd, Pb, Cr and Hg are nonessential and can be highly toxic to a plant's system (Rai et al., 2004; Sebastiani et al., 2004).

Heavy metals are naturally present in the terrestrial and aquatic environments. Geologic and anthropogenic activities increase their concentrations and make these elements harmful to microbes, plants and animals. Natural phenomena such as weathering and volcanic eruptions and activities including mining and smelting of metals, burning of fossil fuels, use of fertilizers and pesticides in agriculture, production of batteries and other metal products in industries, sewage sludge and municipal waste disposal all significantly contribute to heavy metal pollution

(Raskin et al., 1994; Duffus, 2002; Shen et al., 2002; He et al., 2005). A number of HMs are bioaccumulative as they do not degrade in the environment nor easily metabolized. Such metals accumulate in the food chain through uptake at primary producer level and subsequent consumption at consumer level.

Heavy metal toxicity is one of the major abiotic stresses in plants. The level of heavy metal toxicity depends on several factors including the dose, route of exposure and chemical state. Age, genetic background, biochemical and physiological adaptation and nutritional status in individual plants also play important roles in HM toxicity (Verkleji, 1993). Plants are stationary and the roots of a plant are the primary site where heavy metal ions are contracted. In aquatic systems the whole plant body is exposed to these ions. HMs are also absorbed directly by the leaves when particles are deposited on the foliar structure. In recent years, there has been an increasing ecological and global public health concern associated with environmental contamination by these metals. Also, human exposure has risen dramatically as a result of an exponential increase of their use in several industrial, agricultural, domestic and technological applications (Bradl, 2002).

Herbaceous crop species such as cereals, vegetables and fodders are a large part of the work that has been done in relation to the effect of heavy metals on plants because of the importance of those crops from an economic point of view and their short life cycles make them preferable as experimental material; however, reports about the effect of HMs on woody plants is scarce. Woody plants should be an excellent tool for biomonitoring of the HMs due to their rapid growth, large biomass, profuse root apparatus and low impact on the food chain and human health (Estrabou et al., 2011; Tomašević et al., 2008). For example the accumulation of HMs in trees was studied in order to assess air quality and bioaccumulation fluxes (Bajpai et al., 2009; Tomaševic et al., 2004)

10.2 HEAVY METAL TOXICITY

The term "heavy metal" (HM) is in many people's minds means metals (or their derivatives) that are toxic to a living system. However, based on scientific facts, this is a feeling rather than a fact. Two important facts about HMs are: (1) the effect of any substance on a living system is related to the available concentration on its cells, thus no substance is always toxic, (2) a number of metals (Co, Cu, Fe, Mn, Mo, Ni and Zn) are essential for metabolism in the cell at low concentrations, and these are known as micronutrients, even though they are toxic at high concentrations. The presence of these metals show bell-shaped dose-response relationships (Marschner, 1995). Thus, a HM is not toxic per se, it is only toxic when its concentration in the plant exceeds a certain threshold ("it is the dose that makes the effect"). As far as we know, all of these plant micronutrients are transition elements. The plant can use metals in mineral form only, as there is no direct metal availability to plants (Appenroth, 2009). Micronutrients are indispensable for biogenesis, proper functioning of nucleic acid, chlorophyll and hydro carbonates as well as for stress resistance. Some heavy metals play an important role in plants cells as micronutrients (Rengel, 2004), while others have stimulating and provoking effects on plants even in trace concentrations (Nyitrai et al., 2007; Kovacs et al., 2009). There are two important facts about the heavy metals. First, heavy metals are generally non-toxic but when their quantity exceeds a certain limit they become toxic. Second, some of the HMs are essential for development and growth of plants such as Co, Cu, Mn, Fe, Ni, Zn and Mo (Rengel, 2004). On the basis of plant growth parameters, phytotoxicity of heavy metals are $As^{5+} < As^{3+} < Cr^{6+} < Zn^{2+} < Ni^{2+} < Cu^{2+} < Hg^{2+} < Cd^{2+}$ (Assche and Clijsters, 1990).

10.2.1 Physiological and Biochemical Mechanisms of HM Toxicity in Plants

Based on the chemical and physical properties of metals, different molecular mechanisms of HM toxicity in plants can be distinguished as follows.

10.2.1.1 Production of Reactive Oxygen Species (ROS) by Autoxidation and Fenton Reaction

The redox active HMs (Fe, Cu, Cr, Co) are directly involved with the redox reaction in cells. These HMs form $O_2^{\bullet-}$ and consequently H_2O_2 and $^{\bullet}OH$ via the Haber-Weiss and Fenton reactions (Schutzendubel and Polle, 2002). Exposure of plants to redox inactive heavy metals (Cd, Zn, Ni, Al, etc) also results in oxidative stress through a number of indirect mechanisms. The important mechanisms are interaction with the antioxidant defense system, disruption of the electron transport chain or induction of lipid peroxidation. The latter can be due to an HM-induced increase in lipoxygenase (LOX) activity.

ROS are more reactive than O_2 and cause severe toxic impacts on living systems. As spinoffs of different metabolic reactions, ROS are continually produced in different cellular parts of plants, particularly in mitochondria and chloroplasts (Rasmusson et al., 2007; del Rio et al., 2006; Navrot et al., 2007). Toxicity due to ROS can destroy the DNA structure, stimulating the oxidation of lipids and proteins as well as degradation of chlorophyll pigments (Schutzendubel and Polle, 2002). Heavy metals disturb the equilibrium between the cleaning and production of ROS such as salinity, droughts, ultraviolet (UV)-radiation, air pollution, extremes of temperature, pathogens and herbicides (Gill and Tuteja, 2010). Heavy metals have the capacity to create OH via the Haber/Fenton-Weiss reaction (Kehrer, 2000). Normally, the ROS is expeditiously cleaned by the antioxidant system (Dubey and Pandey, 2011). The hydroxyl radical ($OH^{\bullet-}$), superoxide radical ($O_2^{\bullet-}$), hydrogen peroxide (H_2O_2) and hydroxyl radical ($OH^{\bullet-}$) are the main mediators for peroxidative damage (Dubey and Pandey, 2011).

10.2.1.2 Blocking of Essential Functional Groups in Biomolecules

Heavy metal can strongly bind to oxygen, nitrogen and sulphur atoms. This binding affinity is associated by free enthalpy of the formation of the product of the HM and ligand with low solubility of these products. Due to these capabilities, HMs are capable of binding to cysteine residues of enzymes resulting in inactivation. For example, binding capability of Cd to sulfhydryl groups of structural proteins and enzymes leads to misfolding as well as inactivation and/or interference with redox-regulation (Hall, 2002; Dalcorso et al., 2008).

10.2.1.3 Composite Mechanism Leading to Membrane Disruption

Heavy metals may be responsible for membrane damage through diverse mechanisms. According to Merarg (1993) the chief mechanisms are (1) oxidation and cross-linking with protein thiols, (2) inhibition of key membrane protein such as H^+-ATPase and (3) changes in the composition and fluidity of membrane lipids. In response to HM stress accumulation of methyl glyoxal, a cytotoxic compound was found to increase in plants due to impairment of the glyoxalase system leading to oxidative stress by reducing the GSH content (Singla-Pareek et al., 2006; Hossain et al., 2009, 2010).

BOX

An interesting model related to step-wise action of HMs in plants was suggested by Fodor (2002).

1. There are interactions with other ionic components present at the locus of entry into the plant rhizosphere that subsequently have consequences for metabolism.
2. These result in animpact on the formation of reactive oxygen species (ROS) in the cell wall and an influence on the plasmalemma membrane system.
3. The metal ion reacts with all possible interaction partners within the cytoplasm, including proteins, other macromolecules and metabolites.

4. In this stage, factors influence homeostatic events including water uptake, transport and transpiration. Symptoms start to develop and at subsequent stages they become visible.
5. As a result, chlorophyll and, usually to a lesser degree, carotenoid contents decrease, having obvious consequences for photosynthesis and plant growth (Barcelo and Poschenrieder, 2004). Death of the plant cell occurs. This model has the advantage that visible effects are linked to metabolic events that are influenced by the metal ion of interest.

10.2.2 Effect of an Individual HM on a Plant System

Plants may uptake HMs in soluble form surrounding the root or when solubilized by root exudates. For normal growth and development, the plant requires certain HMs, and these metals become toxic to plants at excessive amounts (Blaylock and Huang, 2000). The effect may be both direct and indirect. The direct toxic effects of HMs include inhibition of cytoplasmic enzymes and damage to cell structures due to oxidative stress (Assche and Clijsters, 1990; Jadia and Fulekar, 1999). The indirect toxic effect is the replacement of essential nutrients at the cation exchange sites of plants (Taiz and Zeiger, 2002). Both direct and indirect toxic effects of HMs lead to impairment of plant growth, leading to death of the plant (Schaller and Diez, 1991). Metals such as Pb, Cd, Hg, and As have adverse effects even at very low concentrations, as they do not have any beneficial role in plant growth or growth medium.

10.2.2.1 Effects of Copper (Cu) on Plants

Though copper (Cu) is an essential metal for normal plant growth and development, it is also potentially toxic at supra-optimal concentration (Thomas et al., 1998). Exposure of plants to excess Cu generates oxidative stress and ROS leading to disturbance of metabolic pathways and damage to macromolecules (Mohnish and Kumar, 2015b; Hegedus et al., 2001; Stadtman and Oliver, 1991). Cu dust had adverse effects on various photosynthesis pigmentation secretions in many trees species leaves (Mohnish and Kumar, 2015b).

10.2.2.2 Effects of Mercury (Hg) on Plants

The available forms of Mercury (Hg) in the environment are Hg^{+2}, Hg° and methyl-Hg. However, in agricultural soil the predominant form is Hg^{+2} (Han et al., 2006). In soil, Hg mainly remains in solid phase through adsorption onto sulfides, clay particles and organic matters. Hg^{+2} can readily accumulate in higher and aquatic plants (Kamal et al., 2004; Israr et al., 2006). A high level of Hg^{+2} is highly phytotoxic to plant cells and can induce visible injuries and physiological disorders (Zhou et al., 2007). After binding with water channel proteins, Hg^{+2} can induce leaf stomata to close and thus create physical obstruction of water flow in the plant (Zhang and Tyerman, 1999). At elevated concentration, Hg^{+2} can also impair mitochondrial activity thus inducing oxidative stress by triggering the generation of ROS. This leads to the disruption of membrane lipids and cellular metabolism in plants (Messer et al., 2005)

10.2.2.3 Effects of Cadmium (Cd) on Plants

Cadmium (Cd) may interfere with the uptake, transport and use of several elements (Ca, Mg, P and K) and water by plants (Das et al., 1997). Inhibition of root Fe (III) reductase is induced by Cd and leads to Fe (II) deficiency that seriously affects photosynthesis. Absorption of nitrate and its transport from roots to shoots is reduced by Cd causing inhibition of nitrate reductase activity in the shoots (Hernandez et al., 1996). Nitrogen fixation and primary ammonia assimilation decreases in nodules of leguminous plants due to Cd treatments (Balestrasse et al., 2003). Permeability of plasma membrane can be affected by Cd toxicity causing a reduction in water content and thus influences the water balance of the plant (Costa and Morel, 1994). A toxic effect of Cd has been shown to be reduction of ATPase activity of the plasma membrane fraction of root (Fodor et al., 1995). Induction of lipid peroxidation causing alterations in the functionality of cell membranes by Cd was also reported (Fodor et al., 1995). Toxicity of Cd also includes inhibition of chlorophyll biosynthesis and decreasing of the activity of enzymes responsible for CO_2 fixation (De Filippis and Ziegler, 1993).

10.2.2.4 Effects of Manganese (Mn) on Plants

Mn is a vital mineral nutrient that performs a key role in several physiological processes, particularly photosynthesis (Kitao et al., 1997). Mn deficiency is a widespread problem, particularly in sandy soils, organic soils (with a pH above 6), heavily weathered and tropical soils. Common symptoms of Mn toxicity include necrotic brown spotting on leaves, petioles and stems (Wu, 1994). Manganese toxicity starts with chlorosis with brownish spotting of the older leaves and moves toward the younger leaves in the course of time (Bachman and Miller, 1995). Symptoms start at the leaf margins as spots, progressing to the interveinal areas to marginal and interveinal portion of leaves forming lesions and leading to death in acute cases (Elamin and Wilcox, 1986; Bachman and Miller, 1995). General leaf bronzing and shortening of internodes has also been documented (Crawford et al,. 1989). Another common symptom of youngest leaf, stem and petiole tissues associated with chlorosis and browning of these tissues is crinkle-leaf (Wu, 1994; Bachman and Miller, 1995). Cracking and brownish coloration of roots are common symptoms of Mn toxicity (Le Bot et al., 1990; Foy et al., 1995). Excess Mn is responsible for inhibition of chlorophyll synthesis through Mn-induced Fe deficiency (Clarimont et al., 1986; Horst, 1988).

10.2.2.5 Effect of Zinc (Zn) on Plants

For biosynthesis of chlorophyll in plants, Zinc (Zn) plays a vital role. Zn deficiency is responsible for leaf discoloration leading to chlorosis. Chlorosis due to zinc deficiency usually affects the base of the leaf. High levels of Zn in soil inhibit many plant metabolic functions resulting in retarded growth and causes senescence. Zinc toxicity in plants limits the growth of both root and shoot (Choi et al., 1996). The chlorosis may arise partly from an induced iron (Fe) deficiency as hydrated Zn^{+2} and Fe^{+2} ions have similar radii (Marschner, 1986). Excess Zn can also give rise to manganese (Mn) and Cu deficiencies in plant shoots. Such deficiencies have been ascribed to a hindered transfer of these micronutrients from root to shoot. This hindrance is because the Fe and Mn concentrations in plants grown in Zn rich media are greater in the the root in comparison to that of shoot (Ebbs and Kochin, 1997).

10.2.2.6 Effects of Cobalt (Co) on Plants

Cobalt complexes form by interacting with other elements. Phytotoxicity of cobalt depends on physicochemical properties,including their electronic structure and ion parameters, of these complexes and coordination environment. The uptake and distribution of Co in plants is species-specific and governed by different mechanisms (Kukier et al., 2004; Bakkaus et al., 2005). Information related to the phytotoxic effect of excess Co is scanty. Surplus Co limited the concentration of Fe, chlorophyll, protein and the activity of catalase in leaves. Exposure to excess Co significantly decreases water potential and transpiration rate (Chatterjee and Chatterjee, 2000).

10.3 HEAVY METAL EFFECT ON PLANT GROWTH AND DEVELOPMENT

Plants growing under natural conditions are exposed to various environmental factors that constitute their micro and macro environment. Any deviation in these factors from their optimum levels is deleterious and leads to stress resulting in impairment of plant growth. Elevated contents of both essential and non-essential HMs lead to toxic symptoms and growth inhibition in plants. Environmental quality (i.e. air, soil and water), microbial activities, plant growth etc. are affected by HM contamination (Järup, 2003; Wang et al., 2009). Toxicity of HMs in plants depends on the plant species, the concentration and type of metal, the chemical form in which the HM is present and soil composition and pH (Nagajyoti et al., 2010; Nagata et al., 2013). The molecular and physiological basis of plant interactions with the environment has attracted considerable interest in recent years. As the plants are sessile organisms, they may be constantly exposed to adverse environmental conditions during their life cycle that negatively affect their growth, development or productivity.

The negative influence of heavy metals on the growth and activities of soil microorganisms also indirectly affects the growth of plants. Reduction in the number of beneficial soil microorganisms due to high concentrations of HMs may lead to a decrease in organic matter decomposition resulting in lower soil fertility. Enzymatic activities are very much useful for plant metabolism but are hampered due to HMs' interference with activities of soil microorganisms. The plants under stress conditions are most likely to be adversely affected by high concentrations of HMs. Growth inhibition is associated with most HMs, while the tolerance limits for HM toxicity is specific for each species and even for a variety of plants (Broadley et al., 2007).

Plant growth and development are essential processes of life and propagation of the species. We cannot think about the existence of the animal kingdom on earth without the growth and development of plants. Growth is chiefly expressed as a function of genotype and environment, which consists of external and internal growth factors. The presence of HMs in the external environment leads to changes in growth and development patterns of the plant. The effect of HM toxicity on the growth of plants varies according to the particular HM involved in the process. Non-essential HMs such as Pb, Cd, Hg and As have adverse effects even at very low concentrations in the growth medium. On the other hand, essential HMs like Cu, Zn and Ni are essential for growth at very low concentration, although they become toxic at supra-optimal amounts (Sfaxi-Bousbih et al., 2009). HMs are available for plant uptake only in soluble form either present as soluble components in the soil solution or solubilized by root exudates (Blaylock and Huang, 2000). The ability of plants to accumulate essential metals means that they acquire other nonessential metals (Djingova and Kuleff, 2000). Some of the direct toxic effects include inhibition of cytoplasmic enzymes and damage to cell structures due to oxidative stress (Jadia and Fulekar, 2009). Replacement of essential nutrients at cation exchange sites plants is an example of indirect toxicity (Taiz and Zeiger, 2002). Negative influence of HMs on the growth and activities of soil microorganisms may also indirectly affect the growth of plants.

10.3.1 Effect of HMs on Growth of Root

The roots of higher plants are the first organ that comes into contact with toxic metal concentrations. This organ usually accumulates significantly higher metal levels than that of the shoots (Breckle, 1989). Under long-term exposure to HMs, the growth of roots is generally more reduced than shoots thus decreasing the root/top ratio. But the amount of metal uptake and the intensity of root growth inhibition are dependent on the plant species as well as the growth conditions (Clark, 1982; Breckle, 1989). Different metals may inhibit root growth by different mechanisms. Inhibition of root cell division by Zn (Fontes and Cox, 1998), inhibition of root cell elongation by Zn, Cu and Pb (Wainwright and Woolhouse, 1977; Lane et al., 1978) and the extension of the cell cycle by Zn (Powell et al., 1986) have been seen to be responsible for decreased root growth.

The search for the key mechanisms of HM-induced toxicity on root growth is very difficult because root tissue grows differently from other tissues to a large extent. As the distribution of HMs through root tissue is irregular thus toxicity effects and even modes of injury within roots differs considerably. Toxicity of HM affects the length of the primary root as well as the architecture of the entire root system (Breckle, 1989). At a supra-optimal concentration of Pb, Cu, Zn, Mn, Cd or Cr lateral root formation is enhanced giving rise to a denser and more compact root system, and root hair density is generally decreased (Breckle, 1989). These morphogenetic changes indicate the hormonal disbalance in roots caused by HM toxicity. But the changes in endogenous hormone levels underlying these morphogenetic alterations are virtually unexplored. The inhibition of root elongation may be established as a sensitive parameter for assessing the response of trees to HMs (Kahle, 1993).

Most of the Cu absorbed by the plant is retained in the roots (Wisniewski and Dickinson, 2003). Likewise, much of the Pb absorbed by plants accumulates in roots and only a small fraction is translocated to the aerial part of the plant (Patra et al., 2004).

A major portion of Pb is retained in the cell wall of roots in the form of carbonates. Two mechanisms – binding to ion exchange sites and extracellular precipitation are involved (Jarvis and Leung, 2002). However, the endoderm of the root acts as an additional barrier to Pb absorption and penetration to the interior of the stele and its transport to the aerial plant part (Weis and Weis, 2004). Fixation of Cr^{3+} by plant tissue has an effect on cell division leading to inhibition of root growth. It also disturbs the osmotic relations that narrow down Ca^{2+} transport through the plasma membrane to the cytoplasm (Liu et al., 1992; Barbosa et al., 2007). However, study of Liu et al. (1992) proves that the toxicity of Cr(VI) is greater than Cr(III) on cell division. Furthermore, Cr (VI) causes a delay in the cell cycle leading to an inhibition in cell division and thereby reduces root growth (Sundaramoorthy et al., 2010). The effect of different HMs on the growth of root woody plants is summarized in Table 10.1.

10.3.2 Effect of HMs on Growth of the Shoot

The growth of the plant stem significantly decreases due to HM toxicity. The effect of various HMs on the growth of shoots is influenced by environmental factors and the genotype of the subjected plant. The toxicity created due to the effect of supra-optimal Ni toxicity produces disorders in metabolic systems of plants resulting in quick inhibition of cell division (Yusuf et al., 2011). Cr strongly inhibits the incorporation of various metals such as Fe, P, Ca, Mg, K, Mn, Cu and Zn in different cellular constituents of the plant (Biddappa and Bopaiah, 1989). The result of specific interactions between Cr and P, Cr and Fe or Cr and Cu are responsible for the inhibitory effects of Cr on plant growth (Barbosa et al., 2007). Such interactions could be associated with the chemical properties of these metals, for example the charge (Cr^{3+} and Fe^{3+}) and the effective ionic radius (Cr and Cu).

The seedling length of *Thespesia populnea* is reduced with increased concentrations of lead. The reduction of height could be attributed to the adverse affect of metal treatments on cell elongation and expansion. There are toxic effects of Pb on seedling growth of *Albizia lebbeck* recorded at a concentration as low as 10 μmol/l. The growth of plant stems significantly decreased with an increased concentration of Pb due to its toxicity (Farooqi et al., 2009). Ni toxicity creates disorders in metabolic systems of plants resulting in a fast inhibition of cell division (Yusuf et al., 2011). There is reduction in plant height and biomass, and a decrease in chlorophyll may be due to an inhibitory effect on carotenoid, sugar, starch, and amino acid content (Manivasagaperumal et al., 2011). The effect of different HMs on the growth of shoot of woody plants is summarized in Table 10.2.

10.3.3 Effect of HMs on Leaves of Woody Plants

Many morphological and anatomical changes in leaves occur due to HM toxicity as it is an important site of photosynthesis and many biochemical reactions. Cd is responsible for essential nutrient deficiency and decreased concentrations of many macronutrients in plants (Siedlecka, 1995). It can impair the absorption of traces elements (Cu, Fe, Zn and Mn) by plants through competition (Sanitá di Toppi and Gabbrielli, 1999). Cd competes with Fe for the same absorption site in the plasma membrane causing Fe deficiency in plants, particularly in leaves, leading to chlorosis (Wong et al., 1984; Sanitá di Toppi and Gabbrielli, 1999). Excess Cu inhibits cellular elongation due to the increase in plasmalemma permeability and cell wall lignification thus affecting overall growth of the plant (Arduini et al., 1995). Leaf chlorosis promoted by Cr could be caused by the inhibition of Fe absorption or the reduction of N transport (Barbosa et al., 2007). High Cr concentration can hinder the photosynthetic process through disturbing the ultrastructure of chloroplast

TABLE 10.1
Effect of HMs on Growth of Root

HM	Plant Tested	Dose	Effect	References
Cu	*Pinus pinea* *Pinus pinaster*	1 μM 5 μM	Taproot elongation reduced, only short lateral primordia formed, No significant effect on mitotic index, Cell elongation more sensitive to Cu than cell division. Inhibited root growth of both species.	Arduini et al. (1995)
Pb	*Thespesia populnea*	10 μmol/l	63% inhibition of root length	Kabir et al. (2008)
		30 μmol/l	82% inhibition of root length	
		50 μmol/l	94% inhibition of root length	
		70 μmol/l	97% inhibition of root length	
Cd	*Thespesia populnea*	10 μmol/l	23% inhibition of root length	Kabir et al. (2008)
		30 μmol/l	75% inhibition of root length	
		50 μmol/l	84% inhibition of root length	
		70 μmol/l	99% inhibition of root length	
Pb	*Albizia lebbeck*	10 μmol/l	6% inhibition of root length	Faruuqi et al. (2011)
		30 μmol/l	18% inhibition of root length	
		50 μmol/l	25% inhibition of root length	
		70 μmol/l	27% inhibition of root length	
		90 μmol/l	73% inhibition of root length	
Cd	*Albizia lebbeck*	10 μmol/l	8% inhibition of root length	Faruuqi et al. (2011)
		30 μmol/l	20% inhibition of root length	
		50 μmol/l	35% inhibition of root length	
		70 μmol/l	59% inhibition of root length	
		90 μmol/l	88% inhibition of root length	
Zn	*Picea abies*	30–60 μM	Root elongation was greatly inhibited	Kahle (1993)
Cd	*Picea abies*	30–60 μM	Root elongation ceased after 5 days	Kahle (1993)
Hg	*Picea abies*	0.1 μM	Root elongation severely depressed	Kahle (1993)
		5 μM	Root elongation ceased completely, shrinkage observed 1–5 mm behind the root tip	
Cu	*Betula papyrifera*	60 μg/l	Radicle elongation was reduced by 25%	Patterson and Olsonj (1983)
Pb	*Fagus sylvatica*	44 ppm	Root hair formation strongly inhibited	Breckle (1991)
		283 ppm	Root hairs were completely lacking	
Ni & Cu	*Acer rubrum* *Pinus resinosa*	Ni (10 mg/1) & Cu (20 mg/l)	No. of lateral roots higher, primordia closer together, stunted or thickened; with bulbous root tips, root tissues discolored	Heale and Ormrod, 1982
Pb	*Fagus sylvatica*	44 ppm	Root elongation rates of seedlings were significantly reduced ~ 30%,	Breckle and Kahle (1992)
Pb & Cd	*Fagus sylvatica*	Pb (20 ppm) & Cd (1 ppm)	Considerable decrease in the contents of K, Ca, Mg, Fe, Mn, and Zn	Breckle and Kahle (1992)
Cd & Pb	*Picea sitchensis*	Cd (2.5 ppm) & Pb (48 ppm)	Seedlings grown was significantly reduced	Burton et al. (1984)
Pb	*Leucaena leucocephala*	25 ppm	35% decreased root length	Muhammad et al. (2008)
		50 ppm	40% decreased root length	
		75 ppm	55% decreased root length	
		100 ppm	70% decreased root length	
Cd	*Leucaena leucocephala*	25 ppm	35% decreased root length	Muhammad et al. (2008)
		50 ppm	60% decreased root length	
		75 ppm	67% decreased root length	
		100 ppm	78% decreased root length	
Zn	*Ailanthus altissima*	25 μM	No. of lateral root reduced ≥ 50%	Samuilov et al. (2014)
		100 μM	No. of lateral root reduced ≥ 86%	
		250 μM	No. of lateral root reduced ≥ 93%	

TABLE 10.2
Effect of HMs on Growth of Shoot of Woody Plants

HM	Plant Tested	Dose	Effect on Shoot Length	References
Pb	*Thespesia populnea*	25 µmol/l	Shoot length reduction	Kabir et al. (2008)
Cd	*Thespesia populnea*	10 µmol/l	20% inhibition	Kabir et al. (2008)
		30 µmol/l	28% inhibition	
		50 µmol/l	56% inhibition	
		70 µmol/l	68% inhibition	
Pb	*Albizia lebbeck*	10 µmol/l	2% inhibition	Farooqi et al. (2011)
		30 µmol/l	5% inhibition	
		50 µmol/l	8% inhibition	
		70 µmol/l	10% inhibition	
		90 µmol/l	27% inhibition	
Cd	*Albizia lebbeck*	10 µmol/l	14% inhibition	Faooqi et al. (2011)
		30 µmol/l	16% inhibition	
		50 µmol/l	26% inhibition	
		70 µmol/l	42% inhibition	
		90 µmol/l	47% inhibition	
Pb & Cu	*Lythrum salicaria*	500 mg/kg	Rapid and complete death of the stem	Nicholls and Mal (2003)
Pb	*Thespesia populnea*)	10 µ mol/l	Seedling growth	Kabir et al. (2008)
Ni	*Acer rubrum* *Cornus stolonifera* *Lonicera tatarica* *Pinus resinosa*mg	$\geq 10^{-1}$ mg·l	Growth and development were retarded	Heale and Ormrod (1982)
Cu	*Acer rubrum* *Cornus stolonifera* *Lonicera tatarica* *Pinus resinosa*mg	$\geq 20^{-1}$ mg·l	Growth and development of plants were retarded	Heale and Ormrod (1982)
Pb	*Leucaena leucocephala*	25 ppm	10% decreased	Muhammad et al. (2008)
		50 ppm	20% decreased	
		75 ppm	33% decreased	
		100 ppm	50% decreased	
Cd	*Leucaena leucocephala*	25 ppm	24% decreased	Muhammad et al. (2008)
		50 ppm	36% decreased	
		75 ppm	48% decreased	
		100 ppm	57% decreased	
Zn	*Ailanthus altissima*	100 µM	45% decreased	Samuilov et al. (2014)
		200 µM	53% decreased	
		500 µM	52% decreased	
Zn	*Plathymenia reticulata*	10^{-3} M $HgCl_2$ 10^{-2} M $HgCl_2$	Significant reduction 100%	Cardoso et al. (2015)

(Panda and Choudhury, 2005). Cr probably reduces the size of the peripheral parts of the antenna complex thus decreasing the chlorophyll *a:b* ratio (Shanker et al., 2003). The decrease in chlorophyll *a* can be correlated to the destabilization as well as degradation of proteins of the peripheral part. Decline in photosynthetic pigment content and changes in stomatal function adversely affect light and dark reactions (Mysliwa-Kurdziel et al., 2004). Ni toxicity may cause anarchy in metabolic systems of plants resulting in fast inhibition of cell division (Yusuf et al., 2011). The effect of HMs on leaves of some woody plants is presented in Table 10.3.

TABLE 10.3
Effect of HMs on Growth of Leaves of Woody Plants

HM	Plant Tested	Dose	Effect on Leaves	References
Zn	*Cajanus cajan*	1 mM	40% reduction in leaf area	Sheoran et al. (1990)
Pb	*Thespesia populnea*	10–25 μmol/l	Reduction in number and area of leaves	Kabir et al. (2008)
Pb & Cu	*Lythrum salicaria.*	500 mg/kg	Rapid and complete death of the leaves and stem	Nicholls and Mal (2003)
Pb	*Albizia lebbeck*	25 μmol/l	Number of leaves, leaf area and circumference were significantly reduced	Faruuqi et al. (2011)
Pb	*Fagus sylvatica*	6 ppm	Significant leaf area reduction	Breckle and Kahle (1992)
Cd	*Fagus sylvatica* L	0.3 ppm Cd	Significant leaf area reduction	Breckle and Kahle (1992)
Ni	*Acer rubrum* *Cornus stolonifera* *Lonicera tatarica* *Pinus resinosa*mg	10^{-1} mg·l	Marginal and interveinal chlorosis, alteration of leaf pigmentation, severe leaf drop, and abnormally small leaves.	Heale and Ormrod (1982)
Cu	*Acer rubrum* *Cornus stolonifera* *Lonicera tatarica* *Pinus resinosa*mg	20^{-1} mg·l	Marginal and interveinal chlorosis, alteration of leaf pigmentation, severe leaf drop, and abnormally small leaves.	Heale and Ormrod (1982)
Zn	*Ailanthus altissima*	100 μM 250 μM 500 μM	≥40% reduction of number of leaf ≥63% reduction of number of leaf ≥80% reduction of number of leaf	Samuilov et al. (2014)
Pb & Cu	*Lythrum salicaria*	500 mg/kg	Rapid and complete death of the leaves	Nicholls and Mal (2003)

10.3.4 Effect of HMs on Biomass Production

An increase in biomass production in terms of dry matter is the first prerequisite for normal growth and development in plants. Carbon compounds account for 80–90% of the total dry matter produced by plants. HMs are responsible for the impairment of biomass production at their supra-optimal concentration, but Cd seems to be the most toxic heavy metal to plants (Table 10.4). HMs may be easily absorbed and translocated to different plant parts resulting in reduction of the ability of the plant to absorb water and nutrients (Souza, 2007; Sharma and Dietz, 2009). Consequently, major changes in the function of roots and leaves will affect developmental processes such as formation of flowering, embryogenesis and seed formation.

10.4 PHYSIOLOGICAL RESPONSE TO HEAVY METAL STRESS

Heavy metals (HMs) are a very heterogeneous group of elements. Their effect on plant systems varies widely with their chemical properties (Figure 10.1). Due to non-degradable nature, HMs are persistent in nature and accumulate in soils as well as in plants. HMs at supra-optimal concentrations has been shown to interfere with various metabolic processes resulting in stunted growth and development of the plant even resulting in death of the plant in extreme cases (Tomar et al., 2000). The physiological activities of plants that HS interfere with include changes in certain organelles, disruption of membrane structure, photosynthesis, respiration, water relation, nutrient absorption, dry matter accumulation, germination of seeds and ultimately the yield of the plant (Wilkins, 1978; Lata, 1989; Hagemayor, 1993; Prasad, 1997; Linger et al., 2005; Panda, 2007). Levels of antioxidants in plants also interfere with HMs, causing reduction of the nutritive value of the product.

TABLE 10.4
Effect of HMs on Biomass Production of Woody Plants

HM	Plant Tested	Dose	Effect on Biomass (dry wt.)	References
Pb	*Thespesia populnea*	10 μmol/l	24% inhibition	Kabir et al. (2008)
		30 μmol/l	26% inhibition	
		50 μmol/l	31% inhibition	
		70 μmol/l	41% inhibition	
Cd	*Thespesia populnea*	10 μmol/l	23% inhibition	Kabir et al. (2008)
		30 μmol/l	45% inhibition	
		50 μmol/l	53% inhibition	
		70 μmol/l	89% inhibition	
Ni	*Glycine max*	>50 mM	Reduction in plant biomass	El-Shintinawy and El-Ansary (2000)
Pb	*Albizia lebbeck*	100 μmol/l	Significantly reduction	Faruuqi et al. (2011)
Ni	*Glycine max*	1 mM	Stunted growth	Prasad et al. (2005)
Pb	*Fagus sylvatica* L	18 ppm	Significant reduction	Breckle and Kahle (1992)
Cd	*Fagus sylvatica* L	3.6 ppm	Significant reduction	Breckle and Kahle (1992)
Pb	*Thespesia populnea*	10 μmol/l	Reduction in plant biomass	Kabir et al. (2008)
Cd	*Thespesia populnea*	50 μmol/l	Significant reduction	Kabir et al. (2008)
Pb	*Leucaena leucocephala*	25 ppm	20% inhibition	Muhammad et al. (2008)
		50 ppm	22% inhibition	
		75 ppm	27% inhibition	
		100 ppm	42% inhibition	
Cd	*Leucaena leucocephala*	25 ppm	20% inhibition	Muhammad et al. (2008)
		50 ppm	25% inhibition	
		75 ppm	30% inhibition	
		100 ppm	47% inhibition	
Cd	*Platanus occidentalis*	10–100 ppm	10.% inhibition	Carlson and Bazzaz (1977)
Cu	*Prunus cerasifera*	50 μM	No significance reduction	Lombardi and Sebastiani (2005)
		100 μM	Significance reduction	

Toxicities in plant physiology exerted by HMs can be grouped into four proposed mechanisms. These include (1) similarities with the nutrient cations, resulting in a competition for absorption at root surface, e.g. As and Cd compete with P and Zn, respectively, for their absorption, (2) direct interaction of heavy metals with the sulfhydryl group (-SH) of functional proteins that disrupts their structure and function and renders inactivation, (3) displacement of essential cations from specific binding sites that leads to a collapse of function and (4) generation of reactive oxygen species (ROS) that consequently damages the macromolecules (Sharma and Dietz, 2009; DalCorso et al., 2013).

10.4.1 Effect of HMs on Transpiration

Transpiration is a vital physiological process of plants that is essential for the uptake of mineral nutrients and maintenance of internal temperature. A number of vital transpiration processes are disturbed due to adverse effect of HMs. Adverse effects of Cd on stomatal function (Bazzaz et al., 1974; Kirkham, 1978), water transport (Lamoreaux and Chaney, 1977; Barcelo et al., 1988), cell wall elasticity (Baszynski et al., 1980), growth and photosynthesis (Siedlecka and Baszynski, 1993; Fodor et al., 1996) have been well documented. Poschenrieder et al. (1989) opined that Cd-treated plants showed lower leaf relative water content and higher stomatal resistance.

The build up of the concentration of abscisic acid (ABA) in response to Cd leads to stomatal closure (Poschenrieder et al., 1989). It was found that transpiration rates were reduced by 20% in

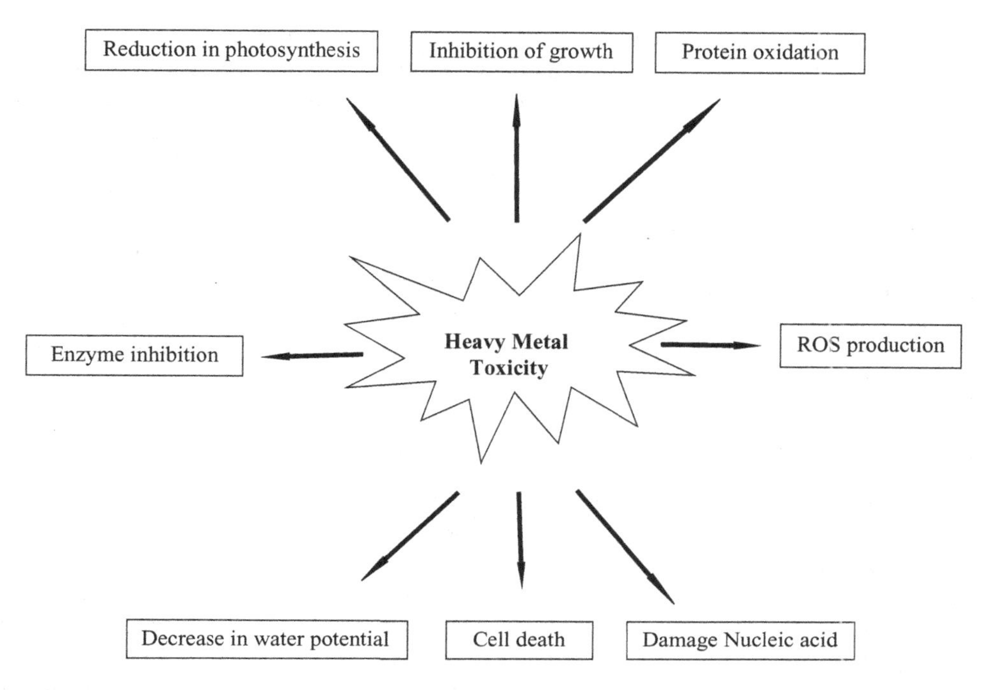

FIGURE 10.1 Adverse effect of Heavy Metal on plant.

ten-year-old beech trees (*Fagus sylvatica*) after three months of exposure to soil containing 2.5 ppm plant-available Cd (Lamoreaux and Chaney, 1978). Breckle and Kahle (1992) reported that the excised leaves of silver maple (*Acer saccharinum*) after 64 h exposure to 0.180 mM Cd exhibited 21% reduced transpiration. Reduction of the rate of transpiration was found to correlate with the concentration of the Cd solution and amount of time of exposure. Lamoreaux and Chaney (1978) also reported that higher plants grown on Cd contaminated environment showed disturbed water balance.

10.4.2 Effect of HM on Imbibitions

Imbibition of seed is a crucial process of seed germination. Three important phases of germination are (1) rapid imbibition due to the water potential gradient between the dry seed and external environment, (2) metabolic activity of the seed is restored, which characterizes the beginning of the germination process and (3) renewed hydration that concludes in the emergent of a radicle and finalization of germination (Nonogaki et al., 2010). Aquaporins play an important role in imbibition as well as the germination processes of seeds (Liu et al., 2007, 2013). Functionality of the aquaporin in seeds may be related to the water absorption and activation of the metabolic system leading to an elevated rate of germination (Liu et al., 2007). Mercury contact with aquaporin may oxidize the thiol groups of cysteine residues resulting in irreversible interruption of the activity of aquaporin (Aroca et al., 2012). Mercury chloride is mostly used as a tool to study the activity of aquaporin in water transport at various stages of plant development, due to its great efficiency in blocking the activity of these proteins (Przedpelska-Wasowicz and Wierzbicka, 2011).

10.4.3 Effect of HMs on Seed Germination

Germination of seed is the first physiological process of plant life. The process is regulated by hormonal interactions as well as environmental factors and occurs only when conditions become favorable, which can be severely affected by HMs (Iglesias and Babiano, 1997). Ni is reported to be toxic to most plant species. It affects amylase, protease and ribonuclease activity that impair the digestion and mobilization of food reserves in germinating seeds, thus retarding seed germination (Ahmad and Ashraf, 2011). Bae et al. (2016) studied the effect of different HMs on germination of seeds of various woody plants. They reported the degree of inhibition is dependent on the HM involved and its concentration, as well as which plant species were tested (Table 10.5). The germination rate of pigeon pea (*Cajanus cajan*) seed decreased by 20% in a 1.5 mM solution of Ni in proportion to the concentration of Ni (Rao and Sresty, 2000). Mercury has potentially toxic side effects on plant metabolism (Gaspar, 2011). At high concentrations due to generation of oxidative stress, important cellular components are damaged by this metal and lead to death of the embryo of the seed. On the other hand, at lower concentrations, mercury can alter the pattern of hydration and germination of seeds (Jain et al., 2008). The effects of various HMs on seed germination of many woody plants are presented in Table 10.5.

10.4.4 Effect of HMs on Mitochondrial Respiration

Plant respiration at supra-optimal concentrations of HMs is usually inhibited (Lösch, 2004). *In vitro* study on the effect of Pb on mitochondrial preparations from plant cells revealed a decrease in respiration rate with increasing Pb (Vangrosveld and Clijsters, 1994). Pb affects the flow of electrons of the electron transport system through uncoupling of oxidative phosphorylation (Miller et al., 1973). However, a stimulation of respiration is observed in whole plants (Lee et al., 1976), detached leaves (Lemoreaux and Chaney, 1978), isolated protoplasts (Parys et al., 1998) and mitochondria (Koeppe and Miller, 1970). However, in some plant species the presence of some HMs at lower concentrations increases respiration (Romanowska et al., 2002).

10.4.5 Effect of HMs on Membrane Functionality

Sanz et al. (2009) have studied the effects of HMs on root cell plasma membranes, the first selective barrier encountered when entering the plant, using rice as model plant. Distinctive effects of Cd and Ni on membrane function (i.e. Em and membrane permeability) were observed in the short term. They confirmed the pattern of Em changes caused by Cd and Ni using barley roots and have also followed the effects of both metals in the longer term in rice. The distinct effects caused by Cd and Ni are due to differences in cellular responses, triggered when entering the cytoplasm (i.e. an efficient detoxifying mechanism for Cd), more than to different direct effects on membranes (Sanz et al., 2009). Zn affects membrane permeability due to oxidation of proteins and lipids, and consequently it leads to solute wasting in the germination medium causing failure in nutrient availability to the growing axis, which impairs emergence and growth of the embryonic axis (Rahoui et al., 2010).

10.4.6 Effect of HMs in Water Relation

Heavy metals may restrict the movement of water from roots to upper parts of plants causing severe dehydration in shots (Haag-Kerwer et al., 1999; Chen et al., 2004). Toxic effects of heavy metals may impair functionality of stomata, movement of water through apoplast and symplast and water uptake etc. (Barcelo and Poschenrieder, 2004). Ni toxicity decreased the water content and transpiration rate in plants (Sheoran et al., 1990b; Bishnoi et al., 1993; Molas, 1997). The primary toxic impact of heavy metals is the reduced transpiration rate and shutting down of the stomatal aperture (Molas, 1997; Seregin and Ivanov, 2001). The elevation of abscisic acid levels in the leaf tissues results in closing stomata under Ni stress (Molas, 1997). Excess Ni reduced the area of the transpiring surface (leafs and blades) of plants (Molas, 1997; Chen et al., 2009). Rao and Sresty (2000) stated that Ni toxicity increased the production of ROS and caused peroxidative damage in membrane lipids. It was well documented that over-accumulation of lipid peroxidation resulted in

TABLE 10.5
Effect of HMs on the Germination of Seeds on Woody Plants

HM	Plant Tested	Dose	Effect % (Inhibition)	References
Pb	*Leucaena leucocephala*	25 ppm	8%	Muhammad et al. (2008)
		50 ppm	10%	
		75 ppm	14%	
		100 ppm	21%	
Pb	*Leucaena leucocephala*	25 ppm	15%	Muhammad et al. (2008)
		50 ppm	22%	
		75 ppm	24%	
		100 ppm	33%	
Zn	*Ailanthus altissima*	25 μM	29.%	Samuilov et al. (2014)
		100 μM	50%	
		250 μM	49%	
Hg	*Plathymenia reticulata*	10^{-3} M	Significant	Cardoso et al. (2015)
		10^{-2} M	100%	
Ni	*Cajanus cajan*	1.5 mM	20%	Rao and Sresty (2000)
Pb	*Thespesia Populnea*	10 μ mol/l	24%	Kabir et al. (2008)
		30 μ mol/l	46%	
		50 μ mol/l	70%	
		70 μ mol/l	74%	
Cd	*Thespesia populnea*	10 μ mol/l	44%	Kabir et al. (2008)
		30 μ mol/l	58%	
		50 μ mol/l	80%	
		70 μ mol/l	84%	
Pb	*Albizia lebbeck*	10 μmol/l	8%	Farooqi et al. (2011)
		30 μmol/l	21%	
		50 μmol/l	23%	
		70 μmol/l	31%	
		90 μmol/l	44%	
Cd	*Albizia lebbeck*	10 μmol/l	13%	Farooqi et al. (2011)
		30 μmol/l	31%	
		50 μmol/l	44%	
		70 μmol/l	50%	
		90 μmol/l	77%	
Cu	*Ambrosia artemisiifolia*	50 mg/kg	5%	Bae et al. (2016)
		100 mg/kg	27%	
		200 mg/kg	51%	
Cu	*Coronilla varia*	50 mg/kg	13%	Bae et al. (2016)
		100 mg/kg	41%	
		200 mg/kg	81%	
Cu	*Lotus corniculatus*	50 mg/kg	22%	Bae et al. (2016)
		100 mg/kg	44%	
		200 mg/kg	72%	
Cu	*Trifolium arvense*	50 mg/kg	37%	Bae et al. (2016)
		100 mg/kg	79%	
		200 mg/kg	87%	
Cd	*Ambrosia artemisiifolia*	5 mg/kg	1%	Bae et al. (2016)
		10 mg/kg	4%	
Cd	*Coronilla varia*	5 mg/kg	17%	Bae et al. (2016)
		10 mg/kg	23%	
Cd	*Lotus corniculatus*	5 mg/kg	26%	Bae et al. (2016)
		10 mg/kg	27%	
Cd	*Trifolium arvense*	5 mg/kg	34%	Bae et al. (2016)
		10 mg/kg	42%	

toxicity of heavy metals and oxidative damage (Pandolfini et al., 1992; Luna et al., 1994; Chaoui et al., 1997). The H_2O_2 quantity significantly increased in blades of wheat under Ni stress (Gajewska and Skłodowska, 2007). The primary toxic impact of heavy metals is reduced transpiration rate and shutting down of the stomatal aperture (Molas, 1997; Seregin and Ivanov, 2001). The elevation of ABA levels in leaf tissues that was responsible for stomata closing was observed under Ni stress (Molas, 1997).

10.4.7 Effect of HMs on Photosynthesis

Heavy metals affect both light and dark reactions of photosynthesis either directly or indirectly and influence the various physiological processes related to photosynthesis as well. HMs may directly impair the photosynthetic machinery due to their capability to bind various sensitive sites of photosynthetic apparatus. They can hamper some processes of light reaction through interfering with components of the PSII located on the thylakoid membranes. Excess Cu degrades granal stacks and stroma lamellae and causes an increase in the number and size of plastoglobuli and intrathylakoidal inclusions (Baszynski et al., 1988). Toxic effect of metal on phototropic organisms appears to be strongly related to an increase in the levels of lipid peroxidation and protein carbonylation, as well as the production of antioxidant defense systems (Gallego et al., 1996; Devi and Prosad, 2005). In higher plants, photosynthesis is indirectly reduced by heavy metal accumulation in leaves, which influences the functioning of stomata. Cu blocks the photosynthetic electron transport chain and degrades chlorophyll (Pätsikkä et al., 2002).

10.4.7.1 Effect of HMs on Photosynthetic Pigments

Different HMs are found to be inhibitors of chlorophyll biosynthesis particularly on d-aminolaevulinic acid (ALA)-dehydratase and protochlorophyllide reductase (De Filippis and Pallaghy, 1976; De Filippis et al., 1981). Excess Mg can be removed by Cu from chlorophyll, present in both antenna complexes and reaction centers, thus damaging the structure and function of chlorophyll and impairing the supply of Mg^{2+}, Fe^{2+}, Zn^{2+} and Mg^{2+} (Van Assche and Clijsters, 1990; Kupper et al., 2003).

10.4.7.2 Effect of HMs on Photosynthetic Enzymes

The water splitting enzymes at the oxidizing side of PSII and NADPH and oxidoreductase at the reducing side of PSI are potentially sensitive to HMs (Vangrosveld and Clijsters, 1994). Energy dependent enzymes of the carbon fixation pathway are affected by Zn, Cd and Hg (De-Filippis and Pallaghy, 1994). A majority of metals affect the key enzymes of the Calvin cycle, for example Rubisco and phosphofructokinase (Vangrosveld and Clijsters, 1994). Interaction of HMs with functional SH groups is generally seen as the mechanism of the inhibition of enzymes. A study on *Erythrina variegates* found decreased activity of Rubisco under Cd stress (Muthuchelian et al., 2001). Siborova (1988) the decreased activity might correlate with the formation of mercaptide by Cd with the thiol group of Rubisco. The photosynthesis of woody plants decreased under the Cd polluted atmosphere before the symptoms appeared (Huang and Zhang, 1986). Thus, the photosynthetic yield would be one of the indicators for air pollution.

10.4.7.3 Effect of HMs on Photosystem

PSII is more susceptible than PSI to HMs. The cytb6/f complex has also been suggested as an inhibitory site (Singh and Singh, 1987). The most apparent effect of the toxic action of HMs on PSII is inhibition of oxygen evolution accompanied by quenching of variable fluorescence (Hsu and Lee, 1988; Arellano et al., 1995). HM ions inhibit both the donor and the acceptor sides of PSII, but the most sensitive site of the metal inhibitor is located on the oxidation (donor) side of PSII (Cedeno-Maldonado and Swader, 1972; Vierke and Stuckmeier, 1977), where a reversible inhibition of Tyrz (the redox active tyrosine residue of the D1 protein) takes place. At higher metal concentration, primary quinine acceptor QA (Jegerschold et al., 1995), non-haem iron (Singh and

Singh, 1987; Jagerschold et al., 1995) and secondary quinine acceptor QB (Mohanty et al., 1989) are marked as the target sites of metal inhibitory action on the acceptor site of PSII. Beside the oxygen-evolving complex, cyt b559 is the most sensitive component. Yruela et al. (1996) found that Cu decreased the level of the photoreduced cytb559 and slowed down its rate of photoreduction, whereas Jegerschold et al. (1995) observed the conversion to low potential form at high Cu concentration. Burda et al. (2003) suggest that the primary target sites in photosystems II for Cu are tyrosine Z, high and low potential cytb559 and chlorophyll Z, and that these sites are the source of Cu-induced fluorescence quenching and oxygen evolution inhibition in PSII. Reports related to the effects of heavy metals on the photosynthesis of woody plants are represented in the Table 10.6.

10.5 RESISTANCE AND TOLERANCE TO HEAVY METAL STRESS

Resistance and tolerance to heavy metals (HMs) stress is the ability of a plant to survive in an environment that is toxic to other plants (Macnair et al., 2000). HM resistant/tolerant plants have developed a complex network of highly effective homeostatic mechanisms liable for controlling the uptake, amassing, trafficking and detoxification of metals. The recognized components of this network include metal transporters in charge of metal uptake and vacuolar transport; chelators involved in metal detoxification via buffering cytosolic metal concentrations and chaperones serving delivery and trafficking of metal ions (Clemens, 2001; Hall, 2002). However, HM tolerance and resistance of plants may be a result of two basic strategies: metal exclusion and metal accumulation (Baker, 1987).

Such adaptation of plants is the resultant interaction between its genotype and the concerned environment (Macnair et al., 2000). Genetic studies indicate that HM tolerance is governed by a small number of major genes along with some minor modified genes (Macnair et al., 2000; Schat et al., 2000). It is not clear whether tolerance confers a broad resistance to several different metals (co-tolerance) or involves a series of independent metal-specific mechanisms (multi tolerance) (Schat et al., 2000). However, evidence is favorable for multi tolerance, suggesting that specific mechanisms are involved for each metal present at a toxic concentration (Macnair et al., 2000). The important mechanisms are listed in Table 10.8.

TABLE 10.6
Effect of HMs on the Photosynthesis of Woody Plants

HM	Plant Tested	Dose	Effect	References
Ni	*Populus nigra*	200 μM	Stomatal conductance decreased significantly	Velikova et al. (2010)
Ni	*Phaseolus vulgaris*	200 mM	Increased the stomatal resistance	Rauser and Dumbroff (1981)
Cd	*Acer saccharinum*	0.090 mM	Net photosynthesis were reduced to 21%	Lamoreaux and Chaney (1978)
Cd	*Cajanus cajan*	10 mM	Photosynthesis reduced by 86%	Sheoran et al. (1990)
Ni	*Glycine max*	1 mM	Decreased photosynthetic electron transport activity	Prasad et al. (2005)
Cd	*Phaselous cocineus*	25 μM	Deficiency of photosynthetic apparatus	Skórzyńska-Polit et al. (1998)
Pb & Cd	*Fagus sylvatica*	Pb (20 ppm) Cd (1 ppm)	Ca, Mn, K, Zn & Fe decreased root and leaves.	Kahle and Breckle (1989)
Pb	*Picea abies*		Decreased Ca and Mn concentration	Godbold and Kettner (1991)

10.5.1 Restriction of Uptake and Transport of HMs

HM-tolerant plants restrict uptake and transport of HMs in various ways as described below.

10.5.1.1 Root Exudates

The first obstruction encountered against the entry of HMs is at root level. Root exudates play a variety of roles in metal tolerance (Marschner, 1995).

Root exudates in the form of extracellular carbohydrates such as mucilage and callose are very important for immobilization of HMs in the cell wall (Waraich et al., 2011; Revathi and Venugopal, 2013). This keeps free ions away from root tissue and therefore translocates ions to shoots, thus lessening phytotoxicity. Among a number of ions associated with roots, only a portion are absorbed by the cells.

10.5.1.2 Mycorrhizas

Ectomycorrhizas of higher plants are very much effective in ameliorating the effects of HM toxicity on the host plant. They restricts metal movement to the host roots from the external environment through diverse mechanisms that are noticeably species- and metal- specific (Jentschke and Godbold, 2000). Marked variations in responses to metals have also been observed both between fungal species and to different metals within a species (Hüttermann et al., 1999). The key mechanisms are absorption as well as adsorption of metals by hyphal sheath, reduced access to the apoplast due to the hydrophobicity of the fungal sheath and chelation by fungal exudates (Figure 10.2). These different exclusion mechanisms vary significantly between different plant/fungal interactions. Colpaert and van Assche (1992) reported on the retention of Zn by ectomycorrhizal fungi *Paxillus involutus* that results in reduction of Zn content of the host *Pinus sylvestris.*

10.5.1.3 The Cell Wall

The root cell wall is directly in contact with metals in the soil solution. Adsorption onto the cell wall must be of limited capacity and thus have a limited effect on metal activity at the surface of the plasma membrane. The heavy metal-tolerant *Silene vulgaris* ssp *humilis* accumulates a variety of metals in the epidermal cell walls as silicates or bonds to a protein (Bringezu et al., 1999). A significant portion is adsorbed by negatively charged groups in the cell wall of roots. Absorption ensures the accumulation of significant concentration of HM(s) in the roots and thus expression of a

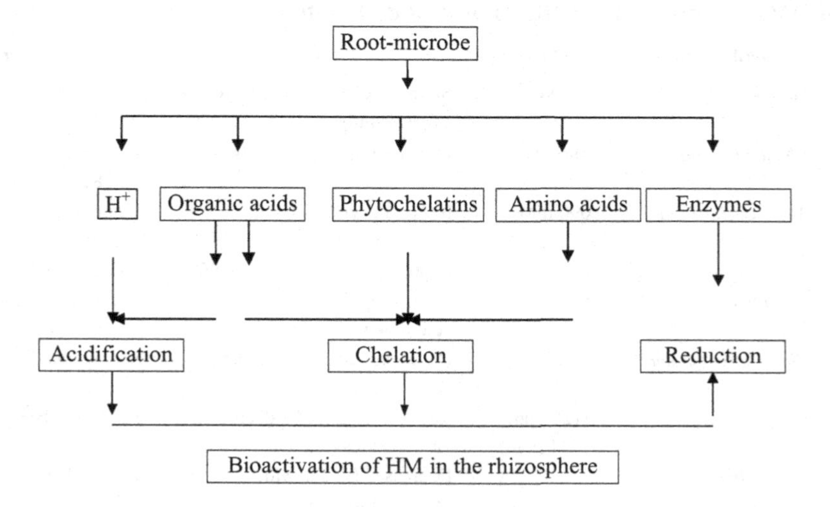

FIGURE 10.2 Probable process in Heavy Metal mobilization to the rhizosphere by root-microbes interaction.

limited concentration in shoots is guaranteed (Visioli and Marmiroli, 2013). However, pectins and histidines stand out for as agents for the immobilization of HMs in the cell wall (Watt and Evans, 1999; Baetz and Martinoia, 2014).

10.5.1.4 Plasma Membrane

Plasma membrane is the first 'living' structure of a plant that is prone to being rapidly affected by HMs leading to ion leakage. Such damage could result from various mechanisms including the oxidation and cross-linking of protein thiols, inhibition of key membrane proteins such as the H^+-ATPase. Changes to the composition and fluidity of membrane lipids also play a vital role in membrane integrity (Quartacci et al., 2001). Thus tolerance maybe correlated with the protection of plasma membrane integrity against HM damage to prevent increased leakage of solutes from cells (Meharg, 1993). However, the effects on membranes are metal-specific. In contrast to Cu, Zn protects membranes against oxidation generally without causing membrane leakage (Ernst et al., 1992; Cakmak, 2000). On the other hand, involvement of heat-shock proteins or metallothioneins, ensure membrane integrity by repairing that damage in the presence of HMs (Salt et al., 1998).

The role of the cell membrane in metal homeostasis is vital. It works by blocking or lessening entry into the cell or pumping out through efflux mechanisms. As many of these cations are essential, so complete exclusion is not possible, thus selective efflux may be more realistic (Silver, 1996). It appears that the metabolic penalty of possessing more specific uptake mechanisms for restricting the entry of toxin ions is the greater than that of having an inducible efflux system (Silver, 1996). However, exclusion or reduced uptake mechanisms in higher plants is quite limited. The best studied mechanism is the relation of toxicity to reduced uptake as an adapted tolerance (Meharg and Macnair, 1992).

10.5.2 Complexation and Compartmentation of HMs within the Plant Cell

10.5.2.1 Intracellular Sequestration

The main storage compartment of HMs in plant cells is the vacuole. Through vacuolar sequestration, HM ions are transported into the vacuole and thus reduce the levels of toxic metals in the cytosol (Soudek et al., 2014). A number of HMs particularly Zn, Cd, Mo and Ni are accumulated in vacuoles of the HM-tolerant plants (Brune et al., 1995; De, 2000). *Jatropha curcus* accumulates high concentrations of Cr in roots. The level of free Cr ions in the roots of this plant remains low as most of the Cr ions are either immobilized or compartmentalized in vacuoles or form Cr–phytochelatin complexes (Yadav, 2010). The Ni-hyperaccumulator plant *Alyssum serpytllifolium* can accumulate 72% of the cellular Ni in the vacuole (Brooks et al., 1980). The reactive interactions between metal ions and S, N and O make organic acids and amino acids potential ligands for metal chelation. Citrate has been proposed as the major ligand for Cd^{2+} at low Cd concentration within the cell (Wagner, 1993). It forms a Ni-citrate complex in the Ni-hyperaccumulating plant *Sebertia acuminate* (Sagner et al., 1998). Malate and oxalate are also implicated in metal tolerance, metal transport through xylem sap and vacuolar metal sequestration (Rauser, 1999). Thus, vacuolar compartmentalization is potentially an important mechanism for HM tolerance.

10.5.2.2 Formation of Metal Complex with Phytochelatins

Plant cells subjected to HMs rapidly synthesize phytochelatins (PCs). These consist of metal-binding polypeptides and act to sequester and detoxify excess metal ions. Biosynthesis of PC is primarily induced and regulated by a wide array of HMs (Cd, Pb, Zn, Sb, Ag, Ni, Hg, Cu, Sn, Au, Bi). However, the binding of metals to PCs has been evidenced in Cu, Zn, Pb and Cd only (Grill, 1989; Maitani et al., 1996). PCs have been detected in the roots of HM-sensitive *Acer pseudoplatanus* grown on soil having total Zn content 28 g/kg but at lower Zn concentration (0.3 g/kg) PC is not traceable (Grill et al., 1988). PCs production capability of *Acer* at its roots was distinctly higher than those of the high HM-tolerant plant *Silene cucubalus* from the same site. Although, the Zn content at the root of *Acer* was found to be four times higher than those at *S. cucubalus*. This supports the idea that metal-tolerant plants have additional tolerance mechanisms besides complexation of metal ions by PC (Grill et al., 1988). Maitani et al. (1996) showed that *Rubia tinctorum* root cultures initiate PC synthesis only after being exposed to a number of HMs such as Zn, Cu and Cd. PCs are enzymatically synthesized by PC synthase directly from glutathione (Grill et al., 1989; Cobbett, 2000). The incapability of the plant to synthesize these peptides results in growth inhibition or cell death in high HM-contaminant soil.

10.5.2.3 Complexing by Metallothioneins

A range of higher plants produce metallothioneins (MTs), one of the important metal-binding cysteine-rich peptides (Prasad, 1999). Four types of MTs – MT1, 2 MT2, MT3 and MT4 types have been recognized (Goldsbrough, 2000). However, MT mRNA was strongly induced in *Arabidopsis* seedlings by Cu, but only slightly by Cd and Zn (Zhou and Goldsbrough, 1994). When genes for MTI and MT2 from *Arabidopsis* were expressed in an MT-deficient yeast mutant, both genes complemented the mutation and ensured a high level of resistance to Cu. Van Vliet et al. (1995) showed that MT genes can be induced by Cu and that the expression of MT2 RNA is increased in a Cu-sensitive mutant of *Arabidopsis* that accumulates high concentrations of Cu. MT genes could clearly play a role in metal metabolism, although their precise function is not clear, they may have distinct functions for different metals (Hamer, 1986). The role of MTs in HM detoxification in plants is yet to be established (Schat et al., 2000). They could function as antioxidants although evidence is lacking (Dietz et al., 1999), while a role in plasma membrane repair is another possibility (Salt et al., 1998).

10.5.3 Heat Shock Proteins

All groups of living organisms expressed a higher quantity of heat shock proteins (HSPs) during their growth at supra-optimal temperature. They are also known to be expressed in response to a variety of stress conditions including HMs (Lewis et al., 1999). They act as molecular chaperones in normal protein folding, its assembly and in the protection as well as in repairing of proteins under stress conditions.

10.5.4 Proline

The accumulation of proline is an indicator of HM tolerance. This amino acid acts as an osmolyte by antioxidative, osmoprotection properties and metal chelator (Farago and Mullen, 1979). Proline takes part in the reconstruction of chlorophyll (Carpena et al., 2003), regulation of cytosolic acidity tolerance to stress by osmoregulation (Gajewska and Skłodowska, 2008), stabilization of protein synthesis (Kuznetsov and Shevyakova, 1997), stabilization of the macromolecules and organelles (John et al., 2008) to prevent denaturation of enzymes (Gajewska and Skłodowska, 2008) and also serves as source of nitrogen and energy in recovery growth (Chandrashekhar and Sandhyarani, 1996). The elevation of proline levels under abiotic stress may result due to the increase of de novo synthesis or decline of degradation (Kasai et al., 1998) as well as the effect of proline on membrane permeability (Pesci and Reggiani, 1992).

10.5.5 Plant Antioxidant System

Reactive oxygen species (ROS) e.g. $•O_2^-$, H_2O_2 and •OH are unavoidable byproducts of various stresses. The unwanted overproduction of ROS results in massive cellular damage (Miller et al., 2008). An antioxidant system is essentially required to overcome the cytotoxic effects of ROS. The plant antioxidant system consists of ROS-scavenging enzymes (superoxide dismutase, ascorbate peroxidase and catalase) as well as low-molecular-weight antioxidants such as metallothionein, carotenoids, glutathione, ascorbate, etc. (Table 10.7). Transgenic plants overexpress these

TABLE 10.7
Antioxidants Produced by Plant in Response to HM

Compounds	Target
Low molecular weight	*Antioxidants*
Ascorbate	O_2 ($^1\Delta_g$), •OH, O_2•, HO_2•
β-Carotene	O_2 ($^1\Delta_g$), RO_2•
α-Tocopherol	RO_2•
Glutathione	Nonspecific
Urate	O_2 ($^1\Delta_g$), metal
Flavonoid	•OH and HOCl
Phytochelatin	Metal
Enzyme antioxidants	*Reaction catalyzed*
Superoxide dismutase	$2O_2^{\bullet -} + 2H^+ \rightarrow H_2O_2 + O_2$
Catalase	$2H_2O_2 \rightarrow 2H_2O + O_2$
Glutathione peroxidase	H_2O_2 or ROOH + 2GSH → $2H_2O$ or ROH + GSSG
Ascorbate peroxidase	H_2O_2 + Ascorbate → H_2O + Monodehydroascorbate
Thioredoxin	Prot-S_2 + Prot'$(SH)_2$ → Prot$(SH)_2$ + Prot'-S_2
Peroxiredoxin	ROOH + R'$(SH)_2$ → ROH + R'S_2 + H_2O
Glutathione reductase	GSSG + NAD(P)H + H^+ → 2GSH + $NAD(P)^+$

antioxidant genes to maintain a high antioxidant capacity in cells. This capability of transgenic plants can be linked to increased tolerance against various adverse conditions particularly HM tolerance (Guo et al., 2009; Wang et al., 2010).

Phenolics include a wide range of compounds such as catechol, flavonols, cyanide, caffeic acid, lignins, ferulic acid, tannins, chlorogenic acid and capsaicin. The rise in phenolic content may be due to defensive functioning of these compounds against HM stress by metal chelation and ROS scavenging (Brown et al., 1998; Lavid et al., 2001). The level of phenol content increases under HM stress due to its antioxidative activity governed by the ability to chelate HM ions the inhibition of a superoxide-driven Fenton reaction (Arora et al., 1998) and membranes stabilization by decreasing membrane fluidity (Blokhin et al., 1998). Phenolic compounds can also protect cells from oxidative stress by phenol-coupled APX reactions (Polle et al., 1997).

At present, processes involved in dropping toxicity are of great interest. This knowledge is essential for the development of crops where there are highly contaminated soils through phytoremediation (Salt et al., 1998). There is no single mechanism that can account for tolerance to a wide range of metals (Table 10.8). Thus, breeding plants for broad phytoremediation purposes will involve a large number of genetic changes (Macnair et al., 2000). However, the increased application of molecular genetic techniques will have a huge impact on our understanding of HM tolerance. The increased availability of gene deletion mutants, or of plants over or under expressing certain key genes, will again provide valuable evidence in relation to tolerance mechanisms. Such information will allow us to construct detailed models of the various responses that occur when plants, both sensitive and tolerant, are subjected to HM stress.

10.6 MOLECULAR RESPONSE TO HEAVY METAL STRESS

Plants perceive stress signals during growth under adverse environmental conditions. The stress signal is essential for appropriate response to help them to deal with the stress condition. The stress signal initiates a coordinated complex physiological and biochemical process, including changes in global gene expression, protein modification as well as primary and secondary metabolite compositions to tide over the stress condition (Urano et al., 2010). In the last decade functional genomics

TABLE 10.8
Summary of Potential Mechanisms Involved in the Detoxification of and Tolerance to Specific Metals

Mechanism	Metal	Key Reference
Mycorrhizas	Zn, Cu, Cd	Jentschke and Godbold (2000)
Cell wall, exudates	Various, including Ni, Al	Salt et al. (2000), Ma et al. (1997)
Plasma membrane		
Reduced uptake	Arsenate Ni	Meharg and Macnair (1992), Arazi et al. (1999)
Phytochelatins	Cd	Cobbett (2000)
Metallothioneins	Cu	Murphy and Taiz (1997)
Organic acids, amino acids	Various	Rauser (1999)
Heat shock proteins	Various	Neumann et al. (1994)
Vacuolar compartmentation	Zn	Van der Zaal et al. (1999)

approaches have partially unraveled the complex mechanisms of molecular response to HMs. These approaches cover a range from stress perception and transduction through a cascade of signaling molecules to the expression modulation of genes related to plant stress response (Matsui et al., 2008). The function of newly identified stress-responsive non-coding RNA brings knowledge that will make easier understanding of complex responses to stress (Borsani et al., 2005). An understanding of the molecular mechanisms involved and the genetic basis are important aspects for developing plants as agents for the phytoremediation of contaminated sites (Salt et al., 1998; Cobbett, 2000). Using molecular techniques, some fast growing high-biomass nonaccumulators can be engineered to achieve some of the properties of the hyperaccumulators. Determination of the molecular mechanisms of metal accumulation will be key factors in reaching the destination.

10.6.1 Signal Transduction in Response to Heavy Metal Stress

Signaling plays a central role in HM stress responses. The plant activates a complex signal transduction system during HM stress to counteract the situation. After sensing the HM, the plant cell activates specific genes to neutralize the stress stimuli. Even gene expression patterns may change in response to toxic elements. A signal transduction cascade may be responsible for differential gene regulation. The cascade initiates with the synthesis of stress-related proteins and signaling molecules. These molecules ultimately act on the transcription of specific metal-responsive genes to counteract the stress (Maksymiec, 2007). Stress-related genes are activated by a number of signal transduction pathways. Different signaling pathways may be used to respond to different HMs (DalCorso et al., 2010). Ca-calmodulin system, hormones, ROS signaling and the mitogen-activated protein kinase (MAPK) phosphorylation cascade are well studied.

10.6.1.1 The Ca-Calmodulin System

Ca ions and calmodulin are well known second messengers of external stimuli that also participate in HM signaling (Suzuki et al., 2001). Actually, HM stress is responsible for significantly increasing the concentration of Ca in cells (DalCorso et al., 2008). It stimulates calmodulin-like proteins that interact with Ca ions. Their conformational change is in response to Ca binding. Calmodulin proteins regulate a variety of mechanisms involved in HM transport, metabolism and tolerance (Yang and Poovaiah, 2003). Excess HMs ensures the stability of Ca channels leading to increased calcium flux into the cell. Higher intracellular Ca levels are observed in plants exposed to Cd, inducing adaptive mechanisms that alleviate the toxic effects of the HM (Skórzyńska-Polit et al., 1998). In Ni and Pb toxicity, Ca-calmodulin system is also involved. Transgenic tobacco plants capable of expressing

NtCBP4 (*Nicotiana tabacum* calmodulin-binding protein), tolerate higher levels of Ni^{2+} but are hypersensitive to Pb^{2+}. This reflects the exclusion ability of Ni^{2+} but the accumulation of more Pb^{2+} than in wild-type plants (Arazi et al., 1999, 2000).

10.6.1.2 Mitogen-Activated Protein Kinase (MAPK) Cascade

Involvement of MAPK signaling in HM stress for different plant has been doccumented. MAPKinases are activated by ROS production induced upon HM stress. They convert the perception of HM into intracellular signals to the nucleus for initiation of appropriate responses. However, MAPK cascades are not specific to a single stress condition. Interaction between the different MAPKinase modules and the possible transcription factors activated by MAPKs can be identified by the use of functional protein microarrays or phosphoproteomics (Van Bentem et al., 2008; Popescu et al., 2009). Specific genes targeted by these transcription factors in their turn can be detected through different molecular strategies. Better knowledge of HM stress responses on plants and its regulation opens up future perspectives for study of the difficulty in signaling modules in plant responses due to the globally changing environment.

Heavy metal exposure for 2 to 60 min quickly increases mRNA as well as activity levels, and activation of MAPKs is transient. In *Arabidopsis* accumulation of ROS within 60 min of exposure to $CdCl_2$ activates very low concentrations (1 μM) of MPK3 and MPKS (Liu et al., 2010). Rao et al. (2011) expect MAPK cascades in rice as OsMKK4/OSMPK3. Increased transcript levels of *OsMPK3* in leaves as well as in roots of 14-day-old rice plants was observed at exposyre of 50 μM arsenite for 30 min. Elevated gene expression of *OsMKK4* was also observed in leaves and roots after 3 h exposure to arsenite. In *Medicago sativa* roots, transient activation of MAPKs (SIMK, SAMK, MKK2 and MMK3) was triggered within 10 min of exposure at a concentration of 100 μM $CuCl_2$. However, in *Arabidopsis* protoplasts, transient expression assays with HA-tagged SIMK, SAMK, MMK2 and MKK3, and a myc-tagged MAPKK (SIMKK), showed that SIMKK specifically activated SIMK and SAMK after exposure to100 μM $CuCl_2$ (Jonak et al., 2004). Opdenakker et al. (2012) showed that 24 h exposure of *Arabidopsis thaliana* seedlings to reasonable concentrations of Cu and Cd increased transcript levels of MAPKinases in a time-dependent manner.

These changes in gene expression seemed to be related to the production of H_2O_2 by these HMs, directly and fast by Cu (Fenton-Haber Weiss reactions) or indirectly and delayed by Cd (e.g., via NADPH oxidases). However, increased activity of ANP1, MPK3 and MPK6 within 10 min was found after application of 200 μM H_2O_2 to *Arabidopsis* protoplasts. Co-transfection of protoplasts with ANP1 and MPK3 or MPK6 revealed that ANP1 could further enhance the activity of MPK3 and MPK6 after H_2O_2 treatment (Kovtun et al., 2000). Gene expression of *OXI1* enhanced after 30 min in 7-day-old seedlings with a treatment of 10 mM H_2O_2 (Rentel et al., 2004). On the other hand *oxi1* knockout mutants failed to activate MPK3 and MPK6 after treatment with H_2O_2. Additionally, a primary root elongation based toxicity test showed that *oxi1* and *mpk6* knock-outs were more tolerant to excess Cu but not Cd. These findings suggest that OXI1 and MPK6 play important roles in the observed stress response following Cu exposure (Roelofs et al., 2008).

However, knowledge about the downstream signaling targets of MAPKs is scarce under HM stress. Roelofs et al. (2008) are of the opinion that all abiotic stresses are responsible for switching on more than one stress-responsive pathway. They have observed this in the overlap of transcription factors used by each stressor. Their speculation is that bZIP, MYB and MYC transcription factors could be downstream targets of MAPK signaling in plant metal stress.

10.6.1.3 Hormones in the Heavy Metal Response

Along with many physiological and developmental processes, plant hormones play a crucial role in adaptation to HM stress. Hormone synthesis is regulated in the presence of HMs (Peleg and Blumwald, 2011). Ethylene, JAs and SA play important roles in hormone synthesis.

HM effects on ethylene production in plants are both metal- and concentration-specific (Abeles et al., 1992; Thao et al., 2015). Ethylene production was enhanced in the presence of Cd in a number of plants including *Hordeum vulgare* (Vassilev et al., 2004), *Lycopersicon esculentum* (Iakimova et al., 2008), *Pisum sativum* (Rodríguez-Serrano et al., 2006, 2009), *Brassica juncea* (Masood et al., 2012; Asgher et al., 2014), *Glycine max* (Chmielowska-Bąk et al., 2013), *A. thaliana* (Schellingen et al., 2014) and *Triticum aestivum* plants (Khan et al., 2015). Long duration, 16 days, exposure to Cd decreased ethylene liberation in *A. thaliana* (Carrió-Seguí et al., 2015). Fascinatingly, a Cd-tolerant *H. vulgare* genotype showed a larger increase in ethylene emission after 15 days of Cd exposure as compared to a Cd-sensitive genotype (Cao et al., 2014). Up to 6 hr after exposure to excess Cu or Zn (25–500 μM), seven-days-old *A. thaliana* seedlings produced more ethylene than unexposed seedlings (Mertens et al., 1999). In contrast, no significant changes in ethylene emission were detected for *A. thaliana* seedlings *in vitro* grown in the presence of 25 or 50 μM Cu over nine days (Lequeux et al., 2010). These findings signify an effect of exposure time and/or plant age. Excess Cu (500 μM) can augment ethylene production in *Helianthus annuus* and *T. aestivum* leaf discs. On the other hand, exposure to 500 μM Cd only enhanced its emission in *T. aestivum* leaves (Groppa et al., 2003), pointing toward species-specificity of metal stress. Moreover, different parts of *A. thaliana* showed variation in stimulation of ethylene emission after exposure to excess Cu or Cd. However, the highest production rate was observed in inflorescences (Keunen et al., 2016). This response declined with increasing age of different plant parts.

On the other hand, Ni and Zn exposure led to higher ethylene release from *B. juncea* leaves (Khan and Khan, 2014). Also Fe (Yamauchi and Peng, 1995) and Li toxicity (Naranjo et al., 2003) may be linked to stress-induced ethylene production. However, the mechanistic basis is becoming increasingly clear. For example, Cu induced an increased expression of ACO1 and ACO3 genes in *Nicotiana glutinosa* (Kim et al., 1998). It has been suggested that up regulation of ACO genes serves as a good ethylene production indicator (Rudus et al., 2012). Nevertheless, ACC production by ACS covers the rate-limiting step in the ethylene biosynthesis pathway.

It was found that Cu can increase ACS transcript in *A. thaliana* at high levels (Weber et al., 2006). Activity of ACS increased in *B. juncea* plants exposed to Cd (Asgher et al., 2014), Ni or Zn (Khan and Khan, 2014), as well as in Cd-exposed *T. aestivum* plants (Khan et al., 2015). Transcript levels of ACS and ACO genes were rapidly enhanced in Cu-exposed *B. oleracea* (Jakubowicz et al., 2010), Al-exposed *A. thaliana* (Sun et al., 2010), Cr-exposed *Oryza sativa* (Trinh et al., 2014) and Hg-treated *O. sativa* (Chen et al., 2014b) as well as *M. sativa* plants (Montero-Palmero et al., 2014). In addition, Cd was able to enhance ACS gene expression in *G. max* (Chmielowska-Bąk et al., 2013) and ACS and/or ACO transcription in *H. vulgare* (Cao et al., 2014), *O. sativa* (Trinh et al., 2014) and *A. thaliana* plants (Weber et al., 2006; Schellingen et al., 2014).

Latter it was found that Cd-induced increase in ACC and ethylene biosynthesis was mainly attributed to up regulated ACS2 and ACS6 expression, as mutants lacking both isoforms failed to enhanced ethylene release when exposed to Cd (Schellingen et al., 2014). These enzymes are both phosphorylated by the MAPKs MPK3 and MPK6, increasing their half-life (Lin et al., 2009; Han et al., 2010; Skottke et al., 2011). Furthermore, MPK3 and MPK6 are able to induce ACS2 and ACS6 transcription via the transcription factor WRKY33 (Li et al., 2012). As MAPKs are clearly implicated in HM-induced signaling responses in plants, they might affect ethylene biosynthesis during metal stress (Opdenakker et al., 2012). However, most studies focused on ACS or ACO gene expression levels. Dorling et al. (2011) have pointed out the importance of the effects of HM stress on enzyme abundance, activity and post translational modifications. Plants exposed to Cd, Cu, Fe and Zn at a level of toxicity produce higher amounts of ethylene, but Co does not have the same effect (Maksymiec, 2007; Wise and Naylor, 1988). Cd and Cu stimulate ethylene synthesis by upregulating ACC synthase expression and activity (Pell et al., 1997).

JA – JA also plays a leading role as a signaling molecule in higher plants under different environmental stresses (Wasternack, 2014; Kamal and Komatsu, 2016). An externally small amount of JA is capable of enhancing plant tolerance against abiotic stresses (Alam et al., 2014; Chen et al., 2014), plant growth and gene expression (Creelman and Mullet, 1995; Cheong and Choi, 2003). Cu and Cd can induce the rapid accumulation of JA in *Phaseolus coccineus* (Maksymiec et al., 2005).

Cu has also been shown to have this effect in rice (Rakwal et al., 1996), as well as in *A. thaliana* (Maksymiec et al., 2005). It was found that supplementation of JA minimizes the accumulation of H_2O_2, MDA and NADPH oxidase leading to stabilization of biomolecules (Sirhindi et al., 2016). Ni stress caused negative impacts on soybean seedlings. However, coapplication of JA enhanced osmolytes, activity of antioxidant enzymes and gene expression of seedlings to struggle against the injurious effects of Ni (Sirhindi et al., 2016).

SA – SA is involved in HM stress responses. It was found that in the presence of Cd the level of SA in barley roots increases. Exogenous SA can also protect roots from lipid peroxidation caused by Cd toxicity (Metwally et al., 2003). The effect of SA on Cd-induced stress was investigated in plants. When SA and Cd were applied simultaneously in young maize (hybrid Norma) plants, less damage was manifest than without application of SA. However, SA treatment itself also caused oxidative stress and damage to the root system, and inhibited the phytochelatin synthase enzyme (Pal et al., 2002).

10.6.1.4 The Role of Reactive Oxygen Species

Production of ROS is one of the major consequences of HM stress causing widespread damage (Figure 10.3). Its activity as signaling molecules has also been reported. Heavy metals like Cd can produce ROS directly via the Fenton and Haber–Weiss reactions and indirectly by inhibiting antioxidant enzymes (Romero-Puertas et al., 2007). In particular, H_2O_2 acts as a signaling molecule in response to HMs and other stresses (Dat et al., 2000). H_2O_2 levels rise in response to Cu and Cd treatment in *A. thaliana* (Maksymiec and Krupa, 2006). H_2O_2 levels also go up upon Hg exposure in tomato (Cho and Park, 2000) and in response to Mn toxicity in barley (Cho and Park, 2000). This increase in H_2O_2 accumulation changes the redox status of the cell leading to production of antioxidants through activation of antioxidant mechanisms.

10.6.2 Modulation of Transcription Factors

The fate of plants under HM stress is dependent on the regulation of gene expression. Various stressors trigger large number of genes and several proteins in order to link the signaling pathways leading to stress tolerance (Umezawa et al., 2006; Valliyodan and Nguyen, 2006; Manavalan et al., 2009; Nakashima et al., 2009; Tran et al., 2010). These genes can be grouped into two categories: the regulatory genes and the functional genes (Tran et al., 2010). Regulatory genes can regulate various stress-responsive genes cooperatively and/or separately through encoding various transcription factors (TFs) ensuring constitution of a gene network. In addition, functional genes encode metabolic compounds, as those play a crucial role in HM stress. Such important compounds are amines, alcohols and sugars. Usually members of multi-gene families are responsible for master regulators, control and expression of gene clusters. A single TF can control the expression of many target genes. The specific binding capability of the TF to the cis-acting element in the promoters of its target genes makes it possible (Wray et al., 2003; Nakashima et al., 2009). The DNA-binding domain of TFs generally interacts with cis-regulatory elements in the promoters of its target genes. A protein-protein interaction domain helps in oligomerization of TFs with other regulators (Wray et al., 2003; Shiu et al., 2005). This regulatory system of transcription is known as "regulon" (Nakashima et al., 2009). Important TFs families responsible for stress response in plants are ABI3VP1, AREB/ABF, ARF, AtSR, AP2/EREBP, bHLH, bZIP, DREB1/CBF HB, ARID, CCAAT-HAP2, CCAAT- DR1, CCAAT-HAP3, CCAAT-HAP5, CPP, C2H2, C3H, C2C2-Dof, C2C2-YABBY, C2C2-CO-like, C2C2-Gata, E2F-DP, E2F-DP, EMF1, HSF, MADS, MYB, MYC, NAC, SBP, TUB, WRKY, etc. (Singh et al., 2002; Shiu et al., 2005; Shameer et al., 2009).

In plants, TFs may induce regulation of corresponding transcriptional processes due to HM treatments (LeDuc et al., 2006). The first FER regulatory gene involved in Fe uptake was reported in tomato (Liang et al., 2013). The functional analog of FER is FER-like Deficiency Induced Transcription factor (FIT) played a vital role under Fe deficiency in *Arabidopsis* (Yuan et al., 2005).

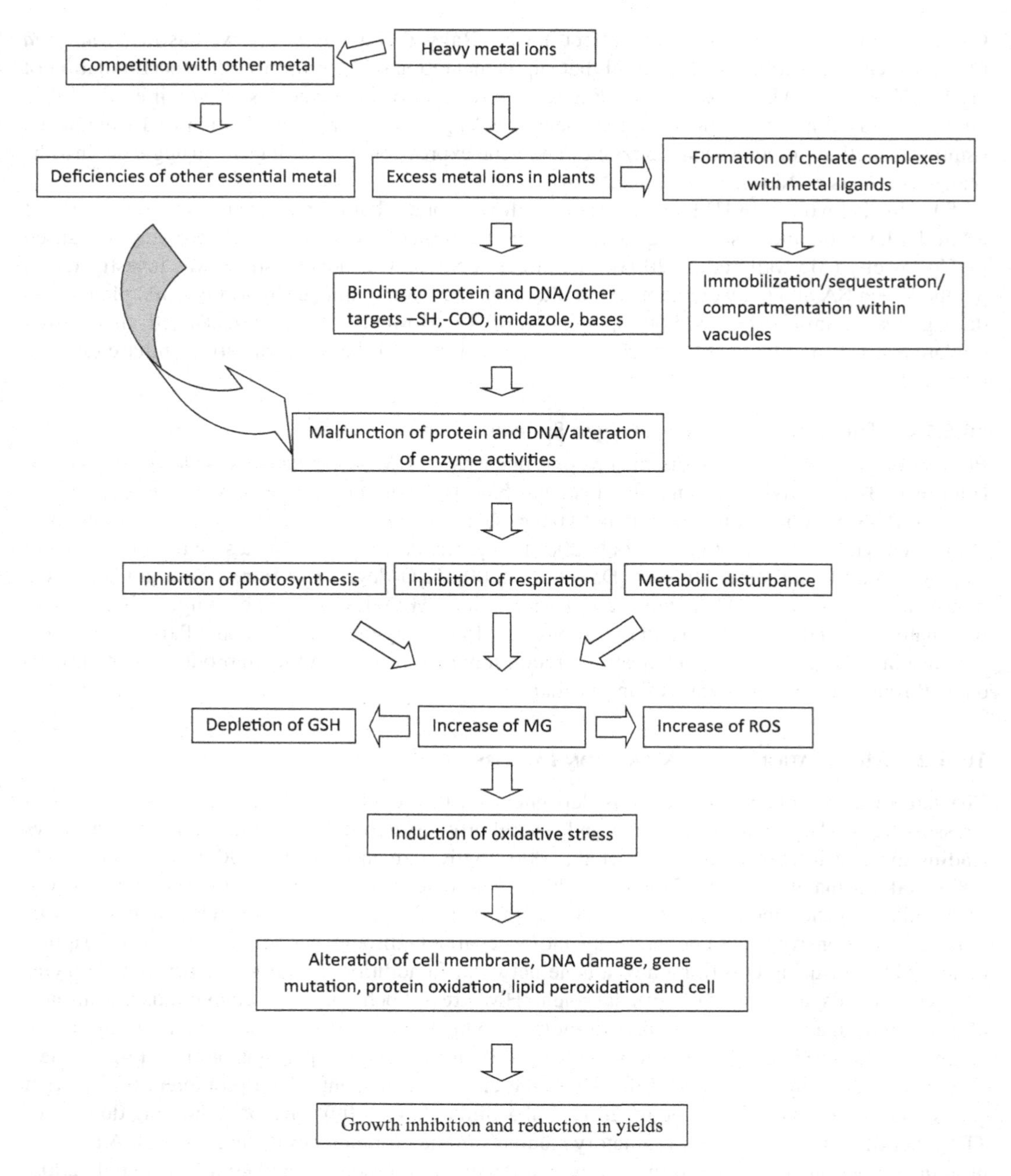

FIGURE 10.3 Possible molecular mechanism of response to HM including metal-mediated ROS induction responsible for damage of higher plants.

Beside these, other subgroups of bHLH family *viz.*, AtbHLH38, AtbHLH39, AtbHLH100 and AtbHLH101 are up regulated in roots and leaves of *Arabidopsis* under Fe deficiency (Wang et al., 2007; Yuan et al., 2008). AtbHLH38 or AtbHLH39 interacts with FIT to form hetero dimers causing direct activation of transcription factors for ferric chelate reductase and ferrous transporters. These two major genes regulate Fe uptake under deficient conditions (Varotto et al., 2002; Vert et al., 2002; Yuan et al., 2008). IRT1 has been reported as the most essential ferrous transporter in *Arabidopsis*. It is also able to transport Zn, Mn, Co, Ni and Cd. Thus, these metals are accumulated under Fe deficiency (Vert et al., 2002; Schaaf et al., 2006). In *Arabidopsis* expression of *FIT* with *AtbHLH38* or *AtbHLH39* promotes the expression of several other transporters *viz*., HMA3, (MTP3) and Iron Regulated Transporter2 (*IRT2*). This expression plays a regulatory role in maintaining Fe content under Cd exposure (Wu et al., 2012). At Cd stress, *A. thaliana* and *B. juncea* induce basic region leucine zipper (bZIP) and zinc finger transcription factors (Ramos et al., 2007). Two transcription factors, ERF1 and ERF5, belonging to the AP2/ERF superfamily have been reported to from in *A. thaliana* during exposure to Cd (Nakano et al., 2006; Herbette et al., 2006). Similar induction of TFs has been reported in *A. halleri* under Cd stress (Weber et al., 2006). Heavy metal stress may also be responsible for down regulation of the dehydration-responsive element-binding protein (DREB) transcription factor (concerned with cold and osmotic stress responsive genes) in the roots of *A. thaliana* (Nakashima and Yamaguchi-Shinozaki, 2006). DREB might have helped in normalizing osmotic potential to keep the toxic effects of HM away from plants. This may possible through reducing the flow of HM-contaminated water. Thus, it could be an acclimation response. A clear idea of the interrelated mechanisms regulating the expression of these genes is important to generate genetically improved crop plants for extreme environments under HM stress (Umezawa et al., 2006; Valliyodan and Nguyen, 2006; Nakashima et al., 2009).

10.6.3 Mobilization of Metal Ions to Extracellular Exudates and to the Cell Wall

Bioavailability of some metals to plants is restricted due to their low solubility in oxygenated water and strong binding ability to soil particles (da Silva et al., 2001). Plants have evolved diverse mechanisms to increase the bioavailability of micronutrient metals. The acidification of rhizosphere and exudation of carboxylates can enhance metal accumulation in plants. For HM mobilization, a HM has to be captured by root cells followed by bonding to the outer cell wall. Bioavailability of metal cations increases due to excretion of organic acids (e.g. malate, citrate) as metal chelators and rhizosphere pH decrease (Ross, 1994). Also, outside the root, organic acids may form an inconsumable complex with the HM and thus impair metal uptake. Citrate inhibits Cu uptake in *Arabidopsis* (Murphy et al., 1999). Plants can also influence their rhizospheric pH through proton pumps in the root cell membrane. Under iron deficit, the gene AHA2 encoded proton pump, an H+-ATPase is up regulated, causing enhanced proton efflux from the root (Fox and Guerinot, 1998). Enhanced proton influx is also found in an *Arabidopsis* mutant due to increased rhizosphere pH (Moffat, 1999). Grass exudates phytosiderophores, biosynthesizing them from nicotianamine. It can bind Fe and facilitate its uptake (Higuchi et al., 1994). A plant's microbe uptake of HMs can also be affected by rhizospheric bacteria. Hg uptake can be enhanced by bacteria (de Souza et al., 1999). Reduced metal uptake by mycorrhizae leads to enhanced tolerance (Frey et al., 2000; Rufyikiri et al., 2000). Certain plant genes are concerned in these plant-microbe interactions through the production of signal molecules.

10.6.4 HM Ion Uptake from the Soil

The uptake of HMs requires transport across the root cell membrane into the symplast. This process involves specific membrane transporter proteins (Mäser et al., 2001; Axelsen and Palmgren, 2001). About 150 different cation transporters belonging to at least nine different families and are encoded by the genome of *A. thaliana* (Mäser et al., 2001). The abundance of genes implied in metal transport in *A. thaliana*, as well as all other organisms, emphasizes the need for metal homeostasis. At this condition, plants must maintain a fine balance between needful metal availability for metabolism and avoidance of deficiency or toxicity.

Heavy metal concentration within the cell across the plasma membrane can be regulated through transporter proteins that have higher affinity to different binding sites (Nelson, 1999). The electrochemical gradient (proton gradient) across the plasma membrane is liable for uptake of metal ions into the cell. The mechanisms of transport are yet to be fully understood in all cases (Mäser et al., 2001).

These secondary transporters get energy from the fluctuations of membrane potential of the plasma membrane and root epidermal cells (Hirsch et al., 1998). It has been shown that metal transporters act in the primary role for the maintenance of intracellular metal homeostasis (Pilon et al., 2009). Often more than one transport system is responsible for transportation of one metal. In *A. thaliana* a number of transporters of the NRAMP family are involved in transportation of Fe into cells. In addition, the ZIP family member IRT and perhaps a number of members of the 8-member YSL family are also implicited in Fe uptake into cells (Mäser et al., 2001; Curie et al., 2001). The occurrence of several transporters ensures uptake systems with different affinities and capacities. The presence of transporters in internal membranes allows regulation of the storage of HMs in organelles such as vacuoles. Transporters may be cell-type specific. However, more than one metal ion may be transported by a specific transporter. IRT protein mediation of transport of Cd and Fe in roots being the best example.

10.6.4.1 HM Ion Transport through the Plasma Membrane in Roots

Plants possess various families of plasma membrane transporters for HM uptake and homeostasis. HM transporters on the plasma membrane and tonoplast are required to ensure the maintenance of physiological concentrations of HMs at the cellular level. They may also contribute to HM stress responses. These transporters belong to the HM P1B-ATPase, the NRAMP, the CDF (Williams et al., 2000) and the ZIP families (Guerinot, 2000). The biological function, cellular location and metal specificity of most of these transporters in plants are yet to be discovered. At present many plant HM transporters are yet to be identified at the molecular level. The transport function, specificity and cellular location of most of these proteins in plants are unknown.

The ZIP Family – One of the key HM transporter families responsible for HM uptake is the ZIP family. These transporters have been identified in many plant species that are involved in the translocation of divalent cations across membranes. Most ZIP proteins consist of eight transmembrane domains having similar topology. The N- and C-termini of the protein is exposed to the apoplast. A variable cytoplasmic loop between transmembrane domains III and IV of the protein molecule contains a histidine-rich domain. Thus, it can be assumed to be involved in metal binding (Guerinot, 2000) as well as specificity (Nishida et al., 2008).

The first characterized ZIP transporter was the *A. thaliana* IRT1. Subsequently, a total of 15 members of the ZIP family have been identified in the *A. thaliana* genome (Yang et al., 2005). In A. *thaliana,* IRT1 is expressed in root cells and accumulates in response to Fe deficiency, suggesting a role in Fe^{2+} uptake from the soil (Vert et al., 2002). Many metal transporters present low ion selectivity. AtIRT1 can also uptake and transport HM divalent cations such as Mn^{2+}, Cd^{2+} and Zn^{2+} (Cohen et al., 1998; Korshunova et al., 1999). ZIP1 and ZIP3 are expressed principally in the roots and are induced under Zn-limiting conditions.

Genomic sequence analysis of *A. thaliana* detected AtZIP4, a fourth member of the family. It is expressed in roots and shoots and also induced by Zn restriction, suggesting roles of the ZIP family in Zn nutrition. ZIP transporters in plants are also responsible for uptake of Cd from soil into the roots as well as its transport from roots to shoots (Krämer et al., 2007). In hyperaccumulator species, ZIP transporters are necessary (but not sufficient) for the enhanced accumulation of metal ions, and metal accumulating capacity correlates with ZIP expression (Krämer et al., 2007). ZRT3 is another transporter identified in *S. cerevisiae* by functional complementation. It is involved in the mobilization of Zn from vacuoles, along with uptake from the environment (MacDiarmid et al., 2000). Ni may be taken up by Zn transporters, although candidate Ni-specific transporters have also been identified (Assunção et al., 2001; Peer et al., 2003).

The NRAMP Family – NRAMP metal transporters have been revealed to transport a wide range of HMs across plant membranes (Nevo and Nelson, 2006). Their expression in roots and shoots is concerned with transport of metal ions through the plasma membrane and the tonoplast (Krämer et al., 2007). In *A. thaliana* NRAMPs are thought to transport Fe and Cd with NRAMP1 playing a specific role in Fe transport and homeostasis (Thomine et al., 2000). The AtNRAMP1 gene complements the yeast fet3fet4 double mutant, and is induced under limiting Fe conditions (Curie et al., 2000). In transgenic *A. thaliana* AtNrampl over expression is responsible for elevated resistance of the plant to Fe intoxication (Curie et al., 2000)

The Copper Transporters Family (CTR) – The CTR family of transporters was identified later in plants than in yeast and mammals. CTR proteins comprise a putative metal-binding motif in the extracellular domain, three predicted transmembrane domains and a conserved and essential MXXXM motif within the putative second transmembrane domain (Puig and Thiele, 2002). The *A. thaliana* Cu transporter COPT1 was identified by functional complementation of the *S. cerevisiae* mutant ctrl–3, which is defective in Cu uptake (Kampfenkel et al., 1995a). In *A. thaliana*, COPT1 has been shown to transport Cu (Sancenón et al., 2004).

10.6.4.2 Reduced HM Uptake and Efflux Pumping at the Plasma Membrane

The plasma membrane plays an important role in plants in maintaining an optimum concentration of metals within the cell. The plasma membrane either prevents or reduces the uptake of HMs into the cell or by active efflux outside the cell. Ion exclusion or reduced uptakes are less common than active efflux as a sole protective mechanism to control HM accumulation inside the cell. An arsenate-tolerant genotype of *Holcus lanatus* absorbs less arsenate than an equivalent non-tolerant genotype (Meharg and Macnair, 1992). Suppression of the high-affinity arsenate transport system combined with the constitutive synthesis of PCs is responsible (Hartley-Whitaker et al., 2001).

Evidence of plasma membrane efflux transporters activity involved in HM response in plants is scarce. However, the most likely candidate for HM efflux pumps in plants are the P1B-ATPases and the CDF families of transporters. P1B-type ATPases belong to the P-type ATPase superfamily that translocates diverse metal cations across biological membranes using energy derived from ATP hydrolysis (Axelsen and Palmgren, 2001). P1B-type ATPases possess eight predicted transmembrane domains, a CPx (Cys-Pro-Cys/His/Ser) intra-membrane motif is involved in metal translocation with a putative N- or C-terminal metal binding domain (Colangelo and Guerinot, 2006; Ashrafi et al., 2011). P1B-ATPases activate against the electrochemical gradient, leading to pumping out metal ions from the cytoplasm either into the apoplast or the vacuole. P1B-type ATPases were renamed as HM ATPases (HMAs) (Baxter et al., 2003) and can grouped into (1) involved in transport of monovalent cations (Cu/Ag) and (2) involved in the transport of divalent cations (Zn/Co/Cd/Pb) (Baxter et al., 2003). HMAs are more selective than metal uptake transporters. HMA2, HMA3 and HMA4 exclusively export Zn and Cd (Krämer et al., 2007).

Thus, hma2 hma4 double mutants and, to a lesser extent, the hma4single mutant contain low levels of Zn in the shoots, exhibiting Zn-deficiency symptoms but uptake of other micronutrients are unaffected (Hussain et al., 2004). In addition, increased Cd sensitivity and decrease in Cd root-to-shoot translocation was reported (Hussain et al., 2004; Wong and Cobbett, 2009). Both AtHMA2 and AtHMA4 are located on the plasma membrane and heterologous expression of AtHMA4 in yeast induces tolerance to Zn and Cd toxicity (Hussain et al., 2004; Verret et al., 2005), thus advocating activity of this transporter as an efflux pump (Mills et al., 2005).

Involvement of ABC transporters in metal ion efflux from the plasma membrane is also reported. AtPDR8 localized in the plasma membrane of *A. thaliana* root hairs and epidermal cells conferred metal tolerance (Kim et al., 2007). AtPDR8 is induced in the presence of Cd and Pb, and transgenic plants overexpressing the protein do not accumulate Cd in the roots or shoots and are tolerant to normally toxic levels of Cd and Pb. In contrast, mutants accumulate higher levels of Cd and are sensitive to both metals. AtPDR8 probably acts as an efflux pump of these metals at the plasma membrane (Kim et al., 2007).

10.6.5 Root-to-Shoot

Once taken up by the roots, metal ions are loaded into the xylem and transported to the shoots as complexes with various chelators (Figure 10.4). Organic acids, especially citrate, are the major

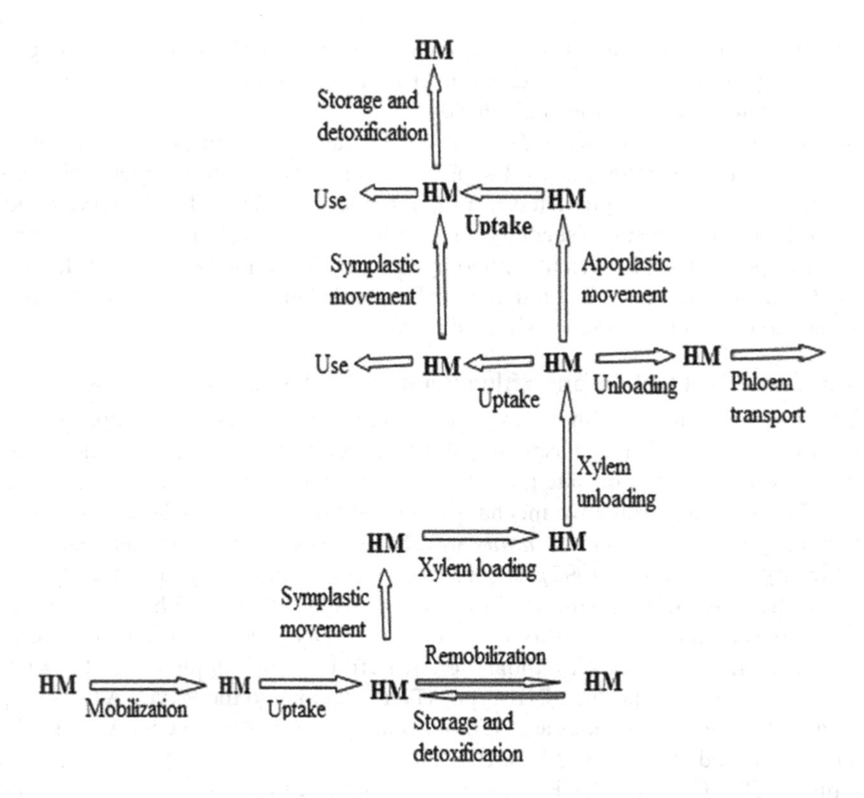

FIGURE 10.4 The path of transition of HM from soil to the site of its use and storage.

chelators for Fe and Ni in the xylem (Tiffin, 1970; Leea et al., 1977). In addition, amino acids are potential metal ligands, for instance Ni may also be chelated by histidine and translocated (Kramer et al., 1996), and the methionine derivative NA is involved in the transport of Cu (Pich and Scholz, 1996). Several types of transporter proteins are involved in the root-to-shoot transport of metals.

Metal ions are also translocated from source to sink tissue via phloem. Therefore, phloem sap contains HMs like Fe, Cu, Zn and Mn arising from source tissue (Stephan et al., 1994). Into the phloem, only NA was identified as a potential metal chelator of Fe, Cu, Zn and Mn (Stephan and Scholz, 1993). NA is involved in the long distance transport of metals inside the xylem and phloem, but other chelators are required for loading. High molecular weight compounds that chelate Ni, Co and Fe are found in the phloem of *Ricinus communis* plants (Wiersma and Van Goor, 1979; Maas et al., 1988), and Zn-chelating peptides are found in *Citrus* spp. (Taylor et al., 1988), but they have yet to be characterized in detail.

10.6.5.1 The HMA Family of Transporters

P-type ATPases reclassified as HMAs transport HM ions. HMA4 is the first gene encoding for the P-type ATPase cloned that characterized in *A. thaliana* (Mills et al., 2003). There are eight HMAs in Arabidopsis, some of them involved in Cu transportation (Andres-Colas et al., 2006; Seigneurin-Berny et al., 2006; Moreno et al., 2008). The rest of them, including HMA2 and HMA4, are indispensable for Zn homeostasis (Hussain et al., 2004). Although HMA2 and HMA4 play essential roles in Zn transport and root-to-shoot translocation, they are also involved in Cd transportation. Overexpressing HMA4, transgenic plants showed increased Cd tolerance and Cd accumulation in shoots (Verret et al., 2004), while increased Cd sensitivity was observed in an hma4 mutant (Verret et al., 2005). A QTL for CdAtHMA4 in Zn and Cd hyperaccumulator *A. halleri* participates in the cytosolic efflux and root-to-shoot translocation of Zn and Cd. It is a plasma membrane transporter of divalent ions responsible for Zn homeostasis and Cd detoxification (Mills et al., 2003; Verret et al., 2004).

Overexpression of the AtHMA4 protein increases Zn and Cd tolerance as well as enhances the root-to-shoot translocation of both metals (Verret et al., 2004). AtHMA5 is expressed constitutively in roots and induced by Cu in other plant organs. However, hma5 mutants are hypersensitive to Cu and accumulate this metal in roots to a greater extent than wild-type plants. It signifies the role of AtHMA5 in root-to-shoot translocation and Cu detoxification (Andrés-Colás et al., 2006). HMA4 is a major determinant of Zn hyperaccumulation and Zn and Cd tolerance in A. halleri (Courbot et al., 2007; Hanikenne et al., 2008)

10.6.5.2 The MATE Family of Efflux Proteins

The multidrug and toxic compound extrusion (MATE) family is the most recently categorized multidrug efflux transporter family. It is a secondary transporter family that couples the translocation of substrates with an electrochemical gradient of cations across the membrane (Kuroda and Tsuchiya, 2009; Shoji, 2014). MATE transporters play important roles in a wide range of biological processes in plants including detoxification of HM toxicity (Li et al., 2002; Diener et al., 2001) and Fe translocation (Inoue et al., 2004; Yokosho et al., 2009). The genes controlling organic anion efflux from roots have been isolated from several crop species (Zhou et al., 2011; Delhaize et al., 2012). MATE transporters have been reported as mediating the citrate efflux to confer plant tolerance to Al toxicity (Magalhaes, 2010). Involvement of the MATE transporters in detoxification of Al was first identified in sorghum (*Sorghum bicolor*, SbMATE) and barley (*Hordeum vulgare*, HvAACT1) respectively through map-based cloning (Magalhaes et al., 2007; Furukawa et al., 2007).

Later study found that the function of HvAACT1 protein is to release citrate to facilitate the translocation of iron from roots to shoots, and the 1-kb insertion in the upstream of the *HvAACT1* coding region in the Al-tolerant barley variety enhances and alters its expression at root tips, which is important in detoxifying Al in barley (Fujii et al., 2012). Overexpression of *HvAACT1* increases citrate efflux and Al tolerance in wheat and barley (Zhou et al., 2013). BoMATE from cabbage (*Brassica oleracea*) requires Al^{3+} to activate citrate efflux, ensuring enhanced Al tolerance in *A. thaliana* (Wu et al., 2014). Many MATE transporters are reported to localize to plasma membranes in the root tips and are related to plant tolerance to Al toxicity. EcMATE1 (*Eucalyptus camaldulensis*), OsFRDL4 (*O. sativa*) and ZmMATE1 (*Zea mays*) are the most studied (Sawaki et al., 2013; Yokosho et al., 2011; Maron et al., 2010).

It is also reported that MATE protein FRD3 takes part in iron-citrate efflux through the loading of Fe^{2+} and citrate into the vascular tissue in the roots. It was found that the xylem exudates from frd3 mutant plants contain less citrate and Fe than its wild counter part. However, transgenic plants overexpressing FRD3 produce more citrate in root exudates. Ferric citrate complexes are required for the translocation of Fe to the leaves because Fe moves through the xylem in its chelated form (Durrett et al., 2007).

10.6.5.3 The Oligopeptide Transporters Family

The oligopeptide transporters (OPTs) are integral membrane proteins. They are responsible for transportation of the substrates-synthesized metal complex of amino acids, including small peptides, secondary amino acids and the modified tripeptide glutathione (Koh et al., 2002; Bogs et al., 2003). Oligipeptides are tentatively 12-transmembrane domains and are distinguish by a number of signature motifs (Wiles et al., 2006). As these transporters are proton-coupled symporters, they are responsible for translocating their substrates in the cytosolic direction (Schaaf et al., 2004; Osawa et al., 2006). They can be divided into two categories: the yellow stripe-like (YSL) and oligopeptide transporter (PT) clades. The YSL gene's function in plants appears to be the transport of metal-chelates (Roberts et al., 2004; Murata et al., 2006) consisting of mugeneic acids or nicotianamine. However, PT genes have only been implicated in the transport of small peptides, glutathione and metal-chelates (Bogs et al., 2003; Vasconcelos et al., 2008). The YSL protein appears to be involved in metal-chelate translocation through intracellular transport, long-distance mobilization and soil scavenging. On the other hand, the PT transporters seem acts as processers for long-distance metal distribution, nitrogen mobilization, HM sequestration, and glutathione transportation (Lubkowitz, 2011).

Eight putative YSL transporters have been identified in *A. thaliana* based on similarity to the maize gene (Colangelo and Guerinot, 2006). AtYSL1 is expressed in the leaf xylem parenchyma, in pollen and young siliques; mutants have a low Fe-NA complex content and cannot germinate normally, indicating a role of this protein in the transport of chelated Fe (Le Jean et al., 2005).

10.6.6 HM Chelation in the Cytosol

In the cell interior, excess HM ions are not immediately required for metabolism. Thus, they may reach toxic concentrations. Plant cells have evolved various mechanisms to store excess metals to prevent their participation in unwanted toxic reactions. At the elevated toxic metal concentration inside the cells, production of chelating compounds is activated. Specific peptides such as PCs and MTs are used to chelate metals in the cytosol to sequester them in specific subcellular compartments. A large number of small molecules including organic acids, amino acids and phosphate derivatives are involved in metal chelation inside the cells (Rauser, 1999).

10.6.6.1 Phytochelatins

Phytochelatins are small peptides with the general structure (γ-Glu-Cys) n-Gly, ($n=2$ to 11, usually 2 to 5), (Figure 10.5) (Jutsz and Gnida, 2015). The structural variants of PCs in plants are (γ-Glu-Cys) n-β-Ala, (γ-Glu-Cys)n-Ser and (γ-Glu- Cys)n-Glu (Cobbett, 2000; Rauser, 2000). Phytochelatins can be produced enzymatically in the presence of many metals (Cobbett, 2000; Rauser, 2000). They are synthesized by phytochelatin synthetase from reduced glutathione through activation by Cd^{2+}, Ag^{+}, Bi^{3+}, Pb^{2+}, Zn^{2+}, Cu^{2+}, Hg^{+} and Au^{+} (Grill et al., 1989). The unusual carboxyamide linkage of phytochelatins in between glutamate and cysteine advocates for their non-ribosomal origin. Actually, PCs are enzymatically synthesized polypeptides, not a direct gene product (Harada et al., 2004). More than 200 plant species are capable of detoxifying HM by phytochelatins (Jutsz and Gnida, 2015).

The range of HMs according to their potency of PCs production is as follows: Cd > Pb > Zn > Ag > Hg > As > Cu (Le Faurcheur et al., 2005). However, HMs such as Ni and Te are not able to trigger PC synthesis (Zenk, 1996). The cysteine thiol groups of PCs react with metals and form chelatable complexes with a molecular weight of 2.5 to 3.6 kDa (Cobbett, 2000). Synthesis involves the chain extension of GS by PCS, a constitutively expressed cytosolic enzyme whose activity is controlled post-translationally because the metal ions chelated by PCs are required for enzyme activity (Grill et al., 1989; Cobbett, 2000). After production in cytosol, PCs are transported as complexes to the vacuoles. PCs seemed to be involved in the homeostasis and trafficking of essential metals such as Cu and Zn (Thumann et al., 1991) and in the detoxification of HMs. On the contrary, they do not seem to be concerned with hyperaccumulation (Ebbs et al., 2002). Contrasting indication accounts

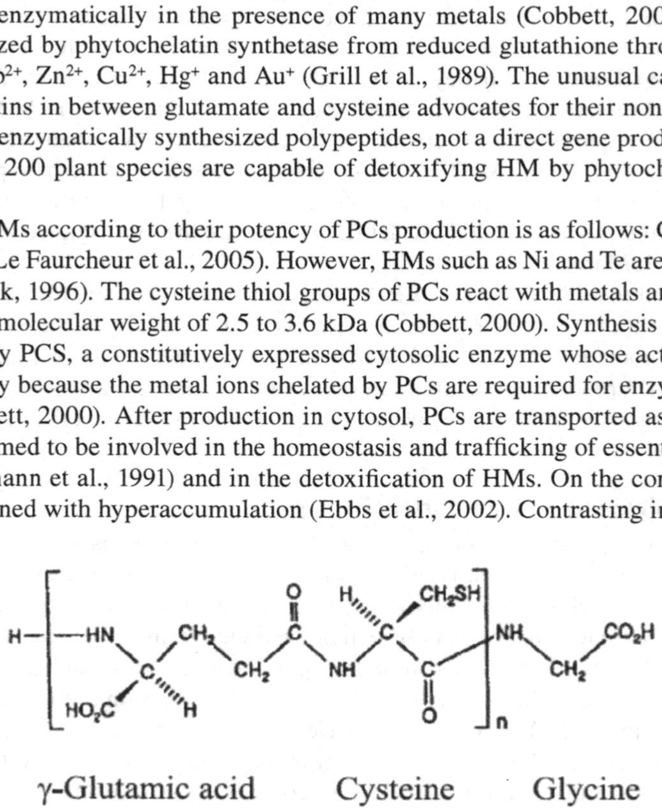

γ-Glutamic acid Cysteine Glycine

FIGURE 10.5 General structure of phytochelatin.

for the role of PCs in HM tolerance. They have a clear role in the response of plants to Cd, e.g. they are induced rapidly in *Brassica juncea* following the intracellular accumulation of Cd, thus protecting the photosynthetic apparatus despite a decline in transpiration and leaf expansion (Haag-Kerwer et al., 1999).

Furthermore, the Cd sensitivity of various *A. thaliana* mutants correlates with their ability to accumulate PCs (Howden et al., 1995). Cd and Cu treatment also induces the transcription of genes related to the synthesis of GS, the precursor of PCs (Xiang and Oliver, 1998). Transgenic *A. thaliana* plants with low GS levels are more sensitive to Cd, whereas, those with elevated GS levels have similar Cd tolerance to wild-type plants (Xiang et al., 2001). Similarly, overexpression of the *Escherichia coli* c-glutamylcysteine synthetase gene in *B. juncea* increased the synthesis of GS and PCs, resulting in greater Cd tolerance (Zhu et al., 1999). The *A. thaliana* gene for PC synthase (CAD1) was identified by using the Cd-sensitive, PC-deficient cad1 mutant (Ha et al., 1999). The Cad1 mutant produces normal levels of GSH but is deficient in PCs and hypersensitive to Cd (Howden et al., 1995). A *Schizosaccharomyces pombe* mutant with the same characteristics has also been identified (Ha et al., 1999). The expression of AtPCS1 from *A. thaliana* and TaPCS1 from wheat in *S. cerevisiae* increases PC synthesis and induces Cd tolerance (Vatamaniuk et al., 1999; Clemens et al., 1999). Furthermore, purified recombinant *A. thaliana* and *S. pombe* PC synthases catalyze the production of PCs from GSH in vitro (Vatamaniuk et al., 1999; Clemens et al., 1999).

To get an idea of the role of PCs in the HM stress response, the sensitivity of cad1 mutants to different metals was analyzed. The *A. thaliana* cad1-3 mutant is more sensitive to arsenate and Cd than wild-type plants. There is no disparity in the case of Zn, selenite and Ni ions (Ha et al., 1999). In opposition to the *S. pombe* pcs mutant, cad1-3 was also insignificantly sensitive to Cu and Hg with intermediate sensitivity to Ag (Ha et al., 1999). The exact role of PCs in Cu tolerance is yet to be determined. Studies of the Cu-tolerant plant *Mimulus guttatus* confirmed a role for PCs in Cu tolerance (Salt et al.,1998). However, Cu-sensitive and Cu-tolerant ecotypes of *S. vulgaris* produced similar amounts of PC during exposure of the root tips to Cu. These findings reflect involvement of other mechanisms for differential tolerance (Schat and Kalff, 1992; De Knecht et al., 1994). Sometimes excessive PC levels in transgenic plants increases the accumulation of HMs without elevating tolerance (Pomponi et al., 2006) and can even confer hypersensitivity to HMs. Indeed, an excessive expression of AtPCS genes confers a hypersensitivity to Cd stress (Lee et al., 2003). This probably advocates for additional significant roles for PCs in the cell. Those may be as S metabolism, antioxidant mechanisms and essential metal homeostasis (Rauser, 1995; Dietz et al., 1999; Cobbett, 2000). The role of PC in HM stress response may perhaps be a side effect of these functions (Steffens, 1990). The ultimate activity of PCs, mainly in response to Cd, involves their accumulation in the vacuoles as complexes (Salt et al., 1998). There they are eventually integrated with S_2^- to form HMW complexes. Cd/H^+ antiporters and ATP-dependent ABC transporters in the tonoplast transported these PC-Cd complexes into the vacuoles (Salt and Wagner, 1993; Salt and Rauser, 1995).

The Cd-sensitive mutant of *S. pombe* can synthesize PCs but not accumulate the Cd-PC-sulfide complexes (Ortiz et al., 1992). Mutation in the gene hmt1 encoding ABC-type transporter, suggests gene-mediated transport and compartmentalization of HMs. PCs also intervene in the root-to-shoot transport of Cd. Expression of wheat TaPCS1 in *A. thaliana* plants accumulates small amounts of Cd in the roots, but Cd transport to the shoots increased. This indicates increased transport efficiency (Gong et al., 2003). PC carries divalent metal from the cytoplasm into the vacuole in the form of a PC-metal complex. The metal ions of this complex undergo dissociation at low pH conditions in the vacuole and combine with organic acids (Clemens et al., 1999; Clemens, 2006). Divalent ions can be released from the complex and used depending on requirements of it, thus ensuring homeostasis of the divalent ions in the cells (Bar-Ness et al., 1991; Cobbett and Goldsbrough, 2002; Hirata et al., 2005; Gasic and Korban, 2007a; Gasic and Korban, 2007b). Phytochelatins are considered to play a major role in HM detoxification (Le Faucheur et al., 2005). Their role in regulating homeostasis of HMs in plants regulates the availability of metal ions in plant cells (Guo et al., 2008).

Clemens et al. (1999) and Ha et al. (1999) have successfully characterized structural genes of this enzyme in Arabidopsis. Two Cd-responsive novel genes coding for the proteins ATMEKKI and a putative famessylated protein with two metal binding motifs have been described in Arabidopsis, and they endow marked tolerance for the plants towards Cd (Suzuki, 2005).

10.6.6.2 Metallothion (MTs)

These are cysteine-rich gene encoded proteins capable of sequestering metals by forming metal-thiol clusters. The expression of MTs in plants growing in absence of metal excess has also been observed (Betrand and Guary, 2002). Like PCs, MTs are a major family of LMW cysteine-rich metal-binding peptides capable of sequestering metals by forming metal-thiolate clusters. They have molecular weight in range of 5 to 20 kDa. Due to the presence of mercaptide groups, they are able to bind metal ions. On the basis of their cysteine content and structure, they can be divided into two different classes that are as follows: (1) Class 1 MTs contain cysteine motifs that align with mammalian MTs and are characterized by the exclusive presence of Cys–X Cys motifs B and (2) Class 2 MTs contain similar cysteine clusters, but they do not align with Classl MTs (Robinson et al., 1993) and both Cys–Cys and Cys–XX–Cys pairs are located in the N-terminal domain (Robinson et al., 1993).

In wheat and rice, MTs are induced by HM ions, such as Cu and Cd (Cobbett and Goldsbrough, 2002). Plant MTs sequester excess metals by coordinating metal ions with the multiple cysteine-thiol groups (Robinson et al., 1993) and have particular affinity for Zn^{2+}, Cu^{+} and Cu^{2+}. Expression of the pea gene PsMTa in *E. coli* has established this (Tommey et al., 1991). *A. thaliana* MT2 is expressed in response to Cu but only marginally in the presence of Cd and Zn (Zhou and Goldsbrough, 1994). Expression of MT1a and MT2a in *A. thaliana* in the trichomes and the phloem, exhibit their involvement both in HM sequestration and transportation (García-Hernández et al., 1998). MTs are responsible both for metal tolerance and metal accumulation, as seen in mtla mutants of *A. thaliana* that are hypersensitive to Cd and accumulate much lower levels of As, Cd and Zn than wild-type plants (Zimeri et al., 2005). Overexpression experiments sustained this finding. *Vicia faba* guard cells overexpressing *A. thaliana AtMT2a* and *AtMT3* can tolerate more elevated levels of Cd than untransformed cells (Lee et al., 2004). Similarly, the overexpression of CcMT2 from *Cajanus cajan* in *A. thaliana* induces both Cd and Cu tolerance and allows both metals to accumulate without disturbing the expression of endogenous transporters (Sekhar et al., 2011). The exact relationship between plant MTs and HMs is not clear to date (Zenk, 1996; Giritch et al., 1998). MTs are known to participate in Cu homeostasis (Cobbett and Goldsbrough, 2002), and *A. thaliana* MT1 and MT2 complement the *S. cerevisiae* MT-defective cupl mutant and confer Cd tolerance (Zhou and Goldsbrough, 1994). MTs biosynthesis is regulated at the transcriptional level. Biosynthesis is induced by several HMs such as Cd, Zn, Hg, Cu, Co and Ni as well as other factors, including hormones and cytotoxic agents (Kagi, 1991; Yang et al., 2005). Regulation of MT genes differs considerably in the plant tissues as well as at developmental stage of the plant (Castiglione et al., 2007). Ahn et al. (2012) demonstrated that the regulation of three *Brassica rapa* MT genes (*BrMT1, BrMT2,* and *BrMT3*) are different under various HM stresses. In Fe-treated seedlings, *BrMT1* and *BrMT2* were not appreciably induced, while elevated expression of *BrMT3* was apparent at 6 and 24 h. In Cu-treated seedlings, *BrMT1* expression increased at 3 h, and it remained at the increased level thereafter, while *BrMT2* expression was down regulated, and *BrMT3* remained unchanged. Similarly, in Zn-treated seedlings, *BrMT1* increased slightly while *BrMT2* was down regulated, and *BrMT3* remained unchanged. In Mn-treated seedlings, *BrMT1* and *BrMT3* genes showed increased expression until 12 h, and then were down regulated. *BrMT2* expression did not change until 12 h, and then was gradually down regulated.

Literature in relation to the role of MTs in HM detoxification and homeostasis is plentiful, but metal-inducibility of plant MTs has not always been demonstrated. Newer information regarding the structures and properties of MTs could clarify their mechanism(s) of action and functions. The use of a model system and a model hyperaccumulator like *Arabidopsis halleri* and some species of

Thlaspi is likely to further elucidate the molecular mechanisms of HM transport, trafficking, tolerance and homeostasis in plants (Jabeen et al., 2009).

10.6.6.3 Ferritins

Ferritins are ubiquitous multimeric proteins that can store up to 4500 iron atoms in a central cavity (Harrison and Arosio, 1996). Plant ferritins are synthesized in response to iron overloading (Murgia et al., 2001, 2002). Ferritin gene expression in plants is regulated by ABA and by antioxidants and serine/threonine phosphatase inhibitors (Savino et al., 1997). Ferritins are therefore a frontline defense mechanism against free iron-induced oxidative stress (Ravet et al., 2009). The major function of plant ferritins is to scavenge free reactive Fe and to prevent the cell from oxidative damage (Ravet et al., 2009).

10.6.6.4 Organic Acids, Amino Acids and Phosphate Derivatives

The binding capability of organic acids and amino acids with HMs involves them in HM toxicity (Rauser, 1999). However, a clear relationship between HM accumulation and the production of these compounds is yet to be established.

Organic acids – Organic acids such as malate, citrate and oxalate confer HM tolerance by transporting metals through the xylem and sequestrating ions in the vacuole, but they have multiple additional roles in the cell (Rauser, 1999). Citrate, which is synthesized in plants by the enzyme citrate synthase, has a higher capacity for metal ions than malate and oxalate, and although its principal role is to chelate Fe^{2+}, it also has a strong affinity for Ni^{2+} and Cd^{2+} (Cataldo et al., 1988). Malate is a cytosolic Zn-chelator in zinc-tolerant plants (Mathys, 1977).

Amino acids and derivatives – These are able to chelate metals and confer to resistance to toxic levels of metal ions. Histidine is considered the most important free amino acid in HM metabolism. Due to the presence of carboxyl, amino and imidazole groups, it is a versatile metal chelator that can confer Ni tolerance and enhance Ni transport in plants when supplied in the growth medium, perhaps reflecting its normal role as a chelator in root exudates (Callahan et al., 2006). It was found that after exposure to Ni, histidine levels also increase in the xylem of *Alyssum lesbiacum*, a Ni hyperaccumulator (Kramer et al., 1996).

The amino acid derivative NA is an aminocarboxylate synthesized by the condensation of three S-adenosyl-L-methionine molecules in a reaction catalyzed by NA synthase (Shojima et al., 1990). NA chelates Fe, Cu and Zn in complexes (Stephan et al., 1996), and then accumulates within vacuoles (Pich et al., 1997); it is not secreted from the roots (Stephan and Scholz, 1993). NA is also involved in the movement of micronutrients in plants (Stephan and Scholz, 1993). The physiological role of NA has been studied extensively in the tomato mutant chloronerva that lacks a functional NA synthase gene and is characterized by the abnormal distribution and accumulation of Fe (Scholz et al., 1985) and Cu (Herbik et al., 1996). NA is also the precursor of the phytosiderophore mugineic acid that binds Zn^{2+}, Cu^{2+} and Fe^{3+} (Treeby et al., 1989). This derivative is synthesized in grasses by the deamination, reduction and hydroxylation of NA (Shojima et al., 1990).

10.6.6.5 Phytate (Myo-Inositol Hexakisphosphate)

Phytate is the principal form of reserve phosphorous in plants and is often localized in the roots and seeds (van Steveninck et al., 1993; Hubel and Beck, 1996). The molecule chelates multiple cations, including Ca^{2+}, Mg^{2+}, and K^{+}, Fe^{2+}, Zn^{2+} and Mn^{2+} (Mikus et al., 1992). The distribution of phytate and its ability to chelate multiple metal species suggest that it could be mobilized as a detoxification strategy for HMs. Addition of Zn to the culture medium leads to the production of Zn^{2+}-containing phytate globoids in the root endoderm and pericycle cells of certain crops (van Steveninck et al., 1993). This sheds light on the involvement of phytate in HM tolerance. Therefore, a controlled synthesis or mobilization of phytate in these cell layers plays a key role in metal ion loading to the aerial parts of plants.

10.6.7 Metal Sequestration in the Vacuole by Tonoplast Transporters

Plants have to expel excess accumulated metal ions from cytoplasm to overcome HM toxicity. Plants either export the ions to the apoplets or compartmentalize them within the cell vacuole by using efflux pumps. The amount of metal ions stored within the vacuole may be ~90% of the cell volume (Vögeli-Lange and Wagner, 1990). Several families of intracellular transporters are involved in this process and they appear to be highly selective.

10.6.7.1 The ABC Transporters

ABC transporters can transport xenobiotics and HMs into the vacuole, and two subfamilies (MRP and PDR) are particularly active in the sequestration of chelated heavy metals. Plant cell vacuoles are, in fact, the major site for accumulation and storage of PC-Cd complexes. PC-Cd complexes are generated in the cytosol and are then translocated by ABC transporters (Vögeli-Lange and Wagner, 1990). In the vacuoles more Cd and sulfide are incorporated to form HMW complexes, the main Cd storage form. The first vacuolar ABC transporter HMT1 was identified by its ability to complement a S. pombe mutant that cannot produce HMW complexes (Ortiz et al., 1992). HMT1 is localized in the tonoplast and transports PC-Cd complexes into the vacuole in a Mg-ATP-dependent manner (Ortiz et al., 1995). A similar protein has been identified in oat roots, but HMT1 homologs are yet to be found in other plants (Salt and Rauser, 1995). In *S. cereviasiae*, the tonoplast ABC pump YCF1 transports Cd into the vacuole as a bis (glutathionato) Cd complex, and confers Cd tolerance (Li et al., 1997).

MRP related sequences like YCF1 have been found in *A. thaliana* (Lu et al., 1997, 1998) and are the most likely candidates for PC-Cd transport across the tonoplast because HMT1 homologs are scarce in plants. In *A. thaliana*, two transporters belonging to the ABC family, AtMRP1 and AtMRP2, have been shown to transport PC-Cd complexes into the vacuole, but the role of AtMRP3 in Cd transport remains unclear (Lu et al., 1997, 1998). AtMRP3 partially complements yeast Dycf1 mutants, but there is no evidence of PC-Cd transport into the vacuole (Tommasini et al., 1998). Moreover, many plants produce Mg-ATP-dependent transporters of GS-S-conjugates (Martinoia et al., 1993).

10.6.7.2 The CDF Transporters

Members of the CDF transporter family (also called MTPs in plants) are involved in the transport of divalent metal cations such as Zn, Cd, Co, Fe, Ni and Mn from the cytoplasm to the vacuole (Krämer et al., 2007; Montanini et al., 2007) and to the apoplast and endoplasmic reticulum (Peiter et al., 2007). Eukaryotic CDF transporters are characterized by six trans- membrane domains, a C-terminal cation efflux domain and a histidine-rich region between trans-membrane domains IV and V (Mäser et al., 2001) that may act as a sensor of metal concentration (Kawachi et al., 2008). The CDF family can be divided into four phylogenetic groups (Mäser et al., 2001), but groups I and III are the most interesting in plants, since these are the ones involved in metal tolerance and accumulation (Krämer et al., 2007). The ZAT1 transporter (later renamed AtMTP1) of *A. thaliana* functions in vacuolar sequestration of Zn (van der Zaal et al., 1999). The gene is constitutively expressed and is not induced by Zn, but its overexpression in transgenic plants exposed to high levels of Zn confers resistance to Zn toxicity and leads to Zn accumulation in the roots without altering Cd sensitivity (van der Zaal et al., 1999). AtMTP1 transports Zn into the vacuole and may represent a Zn tolerance mechanism. Another tonoplast transporter, AtMTP3, is also involved in the transport of Zn into the vacuole (Kramer et al., 2007).

10.6.7.3 The HMA Transporters

HMAs (P1B-ATPases) are involved in the efflux of metal ions from the cytoplasm and the tonoplast (such as AtHMA3). They are thought to contribute to Cd and Zn homeostasis by sequestration into the vacuole (Krämer et al., 2007). However, overexpression of AtHMA3 induces tolerance to Cd,

Pb, Co and Zn. It indicates the involvement of AtHMA3 in the detoxification of a wider range of HMs through storage in the vacuoles (Morel et al., 2009).

10.6.7.4 CaCA Transporters

The CaCA superfamily is responsible for the efflux of Ca^{2+} across membranes against the concentration gradient. A counter electrochemical gradient of other ions, such as H^+, Na^+ or K^+ is involved in this process (Emery et al., 2012). The MHX family, an important member of the CaCA superfamily is a vacuolar Mg^{2+} and Zn^{2+}/H^+ antiport expressed mainly in xylem-associated cells; overexpression in tobacco increases sensitivity to Mg and Zn, although the concentration of these metals in shoots is unchanged (Shaul et al., 1999). The CAX family are Ca^{2+}/H^+ antiports that also recognize Cd^{2+}, suggesting that this is an important route for Cd sequestration in the vacuole (Salt and Wagner, 1993). In *A. thaliana*, only CAX proteins such as AtCAX2 and AtCAX4 seem to be involved in the vacuolar accumulation of Cd. The overexpression of AtCAX2 and AtCAX4 results in the accumulation of more Cd in the vacuoles (Korenkov et al., 2007). AtCAX4 is expressed mainly in root tips and primordia and is induced by Ni and Mn. Root growth in response to Cd and Mn is altered in cax mutants whereas overexpression induces symptoms that are compatible with Cd accumulation (Mei et al., 2009).

10.6.7.5 NRAMP Transporters

NRAMP transporters such as AtNRAMP3 and AtNRAMP4 are localized in the tonoplast and are probably functionally redundant. The nramp3 nramp4 double mutant has a Fe-deficient phenotype in seedlings that can be rescued by providing excess Fe, although the Fe content is the same as in wild-type plants suggesting that AtNRAMP3 and AtNRAMP4 are required to mobilize Fe from the vacuole (Thomine et al., 2003; Lanquar et al., 2005). The overexpression of AtNRAMP3 increases Cd sensitivity (Thomine et al., 2000) and reduces the accumulation of Mn (Thomine et al., 2003), indicating a possible role in the homeostasis of metals other than Fe.

10.6.8 Oxidative Stress Defense Mechanisms and the Repair of Stress-Damaged Proteins

At saturated intracellular metal ions concentration, the defense system of the plant will suffer oxidative stress. The production of ROS and the inhibition of metal-dependent antioxidant enzymes are responsible for this (Schützendübel and Polle, 2002). To overcome this situation plants, activate their antioxidant responses with the induction of enzymes, particularly CAT and SOD, as well as by the production of non-enzymatic free radical scavengers. Exposure to excess Fe induces APX and CAT in the leaves of *Nicotiana plumbaginifolia* (Kampfenkel et al., 1995b). Induction of CAT3 takes place at Cd exposed *B. juncea* plants (Minglin et al., 2005). Its exposure reduces the activity of CAT through oxidation in pea plants. The plant responds by up regulating the transcription of the corresponding gene (Romero-Puertas et al., 2007). The tomato Chloronerva mutant is NA-deficient and contains abnormally high levels of Fe and Cu in leaves resulting in the activation of antioxidant enzymes such as cytosolic APX and Cu/Zn SOD (Pich and Scholz, 1993; Herbick et al., 1996).

It has been found that prolonged treatment with Cd induces SOD activity in tomato seedlings (Dong et al., 2006). In wheat leaves SOD activity increases significantly after exposure to high levels of Cd. This probably reflects the buildup of superoxide (Lin et al., 2007). SOD activity is reduced in pea plants exposed to Cd toxicity (Romero-Puertas et al., 2007). An increase in APX mRNA is also observed in *Brassica napus* cotyledons subjected to toxic levels of Fe (Vansuyt et al., 1997). Several metals are able to induce Fe and Mn-SOD in plants (del Río et al., 1991).

Synthesis of ROS is also counteracted by the activation of the ascorbic acid-GS scavenging system. Cd treatment induces APX in *Phaseolus vulgaris* and *Pisum sativum* (Romero-Puertas et al., 2007). GS plays a key role in metal tolerance due to its ROS scavenging, metal chelating activity and PC biosynthesis capability (Krämer, 2010). It has been found that the expression of the *E. coli* GSS gene gshII in *B. juncea* increased Cd tolerance in seedlings as well as enhanced the accumulating capacity of Cd in adult plants (Zhu et al., 1999). HMs, particularly Cd, reduce the GSH/GSSG ratio and activate antioxidant enzymes such as SOD and GR (Romero-Puertas et al., 2007). GSH (Glu-Cys-Gly), the major intracellular antioxidant inside the cell, is also the precursor of PCs (Cobbett, 2000). It can also form complexes with metal ions, particularly Cd (Wójcik and Tukiendorf, 2011). GS synthesis is activated in response to HM stress.

Heavy metals induce the expression of low molecular weight HSPs in rice (Tseng et al., 1993) as well as in the cell culture of *S. vulgaris* and *Lycopersicon peruvianum* (Wollgiehn and Neumann, 1999). The latter also produces the larger protein HSP70 (Neumann et al., 1994). HSPs could have an important role in the HM response mechanism involving plasma membrane protection and in the repair of stress-damaged proteins.

The SAPs contain either A20/AN1or both zinc finger domains like HSPs and respond to heavy metals stresses in plants. They may function as transcriptional regulators or by direct protein–protein interactions (Dixit and Dhankher, 2011). AtSAP10 is expressed in *A. thaliana* roots and is induced within 30 min by exposure to As, Cd, and Zn (Dixit and Dhankher, 2011). Plants overexpressing AtSAP10 are metal tolerant, they accumulate Ni and Mn in the shoots and roots, but there is no change in Zn content (Dixit and Dhankher, 2011).

REFERENCES

Abeles, F.B., P.W. Morgan, and M.E. Saltveit. 1992. *Ethylene in Plant Biology*. Academic Press Inc., San Diego, CA.

Ahmad M.S., and M. Ashraf. 2011. Essential roles and hazardous effects of nickel in plants. *Rev. Environ. Contam. Toxicol.* 214:125–67.

Ahn, Y.O., S.H. Kim, J. Lee, H.R. Kim, H.-S. Lee, and S.-S. Kwak. 2012. Three Brassica rapa metallothionein genes are differentially regulated under various stress conditions. *Mol. Biol. Rep.* 39:2059–67.

Alam, M.M., K. Nahar, M. Hasanuzzaman, and M. Fujita. 2014. Exogenous jasmonic acid modulates the physiology, antioxidant defense and glyoxalase systems in imparting drought stress tolerance in different *Brassica* species. *Plant Biotechnol. Rep.* 8:279–93.

Andres-Colas, N., V. Sancenon, S. Rodriguez-Navarro, et al., 2006. The Arabidopsis heavy metal P-type ATPase HMA5 interacts with metallochaperones and functions in copper detoxification of roots. *Plant J.* 45:225–36.

Appenroth, K.J. 2009. The definition of heavy metals and their role in biological systems - Chapter 2. In: Varma, A. and I. Sherameti (eds) *Soil Heavy Metals, Soil Biology*, 19. Springer, Berlin, pp. 19–29.

Arazi T, Sunkar R, Kalpan B, Fromm H. 1999. A tobacco plasma membrane calmodulin-binding transporter confers Ni^{2+} tolerance and Pb^{2+} hypersensitivity in transgenic plants. *Plant J.* 20:171–82.

Arazi, T., B. Kaplan, R. Sunkar, and H. Fromm. 2000. Cyclic-nucleotide- and Ca^{2+}/calmodulin-regulated channels in plants: Targets for manipulating heavy-metal tolerance and possible physiological roles. *Biochem. Soc. Trans.* 28:471–5.

Arduini, I., D.L. Godbold, and A. Onnis. 1995. Influence of copper on root growth and morphology of *Pinus pinea* L. and *Pinus pinaster* Ait. seedlings. *Tree Physiol.* 15:411–15.

Arellano, J.B., J.J. Lázaro, J. López-Gorgé, and M. Barón. 1995. The donor side of PSII as the copper-inhibitory binding site. *Photosynth. Res.* 45:127–34.

Aroca, R., R. Porcel, and J.M. Ruiz-Lozano. 2012. Regulation of root water uptake under abiotic stress conditions. *J. Exp. Bot.* 63:43–57.

Arora, A., M.G. Nair, and G.M. Strasburg. 1998. Structure activity relationships for antioxidant activities of a series of flavonoids in a liposomal system. *Free Radic. Biol. Med.* 24:1355–63.

Asgher, M., N.A., Khan, M.I.R., Khan, M. Fatma, and Masood, A. 2014. Ethylene production is associated with alleviation of cadmium-induced oxidative stress by sulfur in mustard types differing in ethylene sensitivity. *Ecotoxicol. Environ. Saf.* 106:54–61.

Ashrafi E., Alemzadeh A., Ebrahimi M., Ebrahimie E., Dadkhodaei N., and Ebrahimi M. 2011. Amino acid features of P1B-ATPase heavy metal transporters enabling small numbers of organisms to cope with heavy metal pollution. *Bioinf. Biol. Insights* 5:59–82.

Assche, F., and H. Clijsters. 1990. Effects of metals on enzyme activity in plants. *Plant Cell Environ.* 24:1–15.

Assunção, A.G.L., P. Da Costa Martins, S. De Folter, R. Vooijs, H. Schat, and M.G.M. Aarts. 2001. Elevated expression of metal transporter genes in three accessions of the metal hyperaccumulator *Thlaspi caerulescens*. *Plant Cell. Environ.* 24:217–26.

Axelsen, K.B., and M.G. Palmgren. 2001. Inventory of the superfamily of P-type ion pumps in Arabidopsis. *Plant Physiol.* 126:696–706.

Bachman, G.R., and W.B. Miller. 1995. Iron chelate inducible iron/manganese toxicity in zonal geranium. *J. Plant. Nutr.* 18:1917–29.

Bae, J., L. D Benoit, and A.K. Watson. 2016. Effect of heavy metals on seed germination and seedling growth of common ragweed and roadside ground cover legumes. *Environ. Pollut.* 213:112–8.

Baetz, U., and E. Martinoia. 2014. Root exudates: The hidden part of plant defense. *Trends Plant Sci.* 19:90–8.

Bajpai, R., D.K. Upreti, and S.K. Dwivedi. 2009. Arsenic accumulation in lichens of Mandav monuments, Dhar district, Madhya Pradesh, India. *Environ. Monit. Assess.* 159:437–42.

Baker, A.J.M. 1987. Metal tolerance. *New Phytol.* 106:93–111.

Bakkaus E., B. Gouget, J.P. Gallien, et al. 2005. Concentration and distribution of cobalt in higher plants: The use of micro-PIXE spectroscopy. *Nucl. Instr. Meth.* B 231:350–6.

Balestrasse, K.B., M.P. Benavides, S.M. Gallego, and M.L. Tomaro, 2003. Effect on cadmium stress on nitrogen metabolism in nodules and roots of soybean plants. *Func. Plant Biol.* 30:57–64.

Barbosa, R.M.T., A.-A.F. Almeida, M.S. Mielke, L.L. Loguercio, P.A.O. Mangabeira, F.P. Gomes. 2007. A physiological analysis of *Genipa americana* L.: A potential phytoremediator tree for chromium polluted watersheds. *Environ. Exp. Bot.* 61:264–71.

Barcelo, J., and C. Poschenrieder. 2004. Structural and ultrastructural changes in heavy metal exposed plants. In: Prasad M.N.V. (ed) *Heavy Metal Stress in Plants*, 3rd edn. Springer, Berlin, pp. 223–48.

Barcelo, J., M.D. Vazquez, and C.H. Poschenrieder. 1988. Cadmium induced structural and ultrastructural changes in the vascular system of bush bean stems. *Bot. Acta* 101:254–61.

Baszynski T., L. Wajda, M. Krol, D. Wolinska, Z. Krupa, and A. Tukendorf. 1980. Photosynthetic activities of cadmium treated tomato plants. *Physiol. Plantarum* 48:365–70.

Baszyński, T., M. Tukendorf, E. Ruszkowska, and E. Skórzyńska. 1988. Characteristics of the photosynthetic apparatus of copper non-tolerance spinach exposed to excess copper. *J. Plant Physiol.* 132:708–13.

Baxter, I., J. Tchieu, M.R. Sussman, et al. 2003. Genomic comparison of P-Type ATPase ion pumps in Arabidopsis and rice. *Plant Physiol.* 132:618–28.

Bazzaz, F.A., G.L. Rolfe and R.W. Carson. 1974. Effects of cadmium on photosynthesis and transpiration of excised leaves of corn and sunflower. *Physiol. Plantarum* 32:373–6.

Biddappa, C.C., and M.G. Bopaiah. 1989. Effect of heavy metals on the distribution of P, K, Ca, Mg and micronutrients in the cellular constituents of coconut leaf. *J. Plant Crops* 17:1–9.

Bishnoi, N.R., L.K. Chugh, and S.K. Sawhney. 1993. Effect of chromium on photosynthesis, respiration and nitrogen fixation in pea (*Pisum sativum* L) seedlings. *J. Plant Physiol.* 142:25–30.

Blaylock, M.J., and J.W. Huang. 2000. *Phytoextraction of Metals, in Phytoremediation of Toxic Metals: Using Plants to Clean Up the Environment.* Raskin, I. and B.D. Ensley (eds). Wiley, New York, pp. 53–70.

Bogs, J., Bourbouloux, A., Cagnac, O., Wachter, A., Rausch, T., and Delrot, S. 2003. Functional characterization and expression analysis of a glutathione transporter, BjGT1, from *Brassica juncea*: evidence for regulation by heavy metal exposure. *Plant Cell Environ.* 26:1703–11.

Borsani, O., J. Zhu, P.E. Verslues, R. Sunkar, and J.K. Zhu, 2005. Endogenous siRNAs derived from a pair of natural cis-antisense transcripts regulate salt tolerance in *Arabidopsis*. *Cell* 123:1279–91.

Bradl, H. 2002. *Heavy Metals in the Environment: Origin, Interaction and Remediation*, Vol. 6. Academic Press, London.

Breckle, S.W. 1989. Growth under stress heavy metals. In: Waisel, Y., U. Kafkafi, A. Eschel (eds) *The Root System The Hidden Half.* Marcel Dekker Inc., New York.

Breckle, S.W. 1991. Growth under stress: Heavy metals. In: Waisel, Y., A. Eshel, and U. Kafnari, (eds) *Plant Roots: The Hidden Half.* Marcel Dekker, New York, pp. 351–373.

Breckle, S.W., and H. Kahle. 1992. Effects of toxic heavy metals (Cd, Pb) on growth and mineral nutrition of beech (*Fagus sylvatica* L.). *Plant Ecol.* 101:43–53.

Bringezu K., O. Lichtenberger, I. Leopold, and D. Neumann. 1999. Heavy metal tolerance of *Silene vulgaris. J. Plant Physiol.* 154:536–46.

Broadley, M.R., P.J. White, J.P. Hammond, I. Zelko, and A. Lux. 2007. Zinc in plants. *New Phytol.* 173:677–702.

Brooks, R.R., R.D. Reeves, R.S. Morrison, and F. Malaisse. 1980. Hyperaccumulation of Copper and Cobalt - A Review. *Bull. Soc. R. Bot. Belg.* 113:166–172.

Brown, J.E., H. Khodr, R.C. Hider, and C.A. Rice-Evans. 1998. Structural dependence of flavonoid interactions with Cu^{2+} ions: Implications for their antioxidant properties. *Biochem. J.* 330:1173--8.

Brune, A., W. Urbach, and K.-J. Dietz. 1995. Differential toxicity of heavy metals is partly related to a loss of preferential extraplasmic compartmentation: a comparison of Cd-, Mo-, Ni- and Zn-stress. *New Phytol.* 129:403–9.

Burda, K., J. Kruk, G.H. Schmid, et al. 2003. Inhibition of oxygen evolution in photosystem II by Cu(II) ions is associated with oxidation of cytochrome b559. *Biochem J.* 371:597–601.

Burton, K.W., E. Moroan, and A. Roig. 1984. The influence of heavy metals upon the growth of sitka-spruce in South Wales forests. I. Greenhouse experiments. *Plant Soil* 78:271–82.

Cakmak, I. 2000. Possible roles of zinc in protecting plant cells from damage by reactive oxygen species. *New Phytol.* 146:185–205.

Callahan, D.L., A.J.M., Baker. S.D. Kolev, and A.G. Wedd. 2006. Metal ion ligands in hyperaccumulating plants. *J. Biol. Inorg. Chem.* 11:2–12.

Cao, F., F. Chen, H. Sun, G. Zhang, Z.H. Chen, and F. Wu. 2014. Genome-wide transcriptome and functional analysis of two contrasting geno types reveals key genes for cadmium tolerance in barley. *BMC Genomics* 15:611.

Cardoso, A.A., E.E.L. Borges, G.A. de Souza, C.J. Silva, R.M.O. Pires, and D.C.F. Dias. 2015. Seed imbibition and germination of *Plathymenia reticulata* Benth. (Fabaceae) affected by mercury: Possible role of aquaporins. *Acta Bot. Bras.* 29:285–91.

Carlson, R.W., and F.A. Bazzaz. 1977. Growth reduction in American sycamore (*Platanus occidentalis*) caused by Pb-Cd interaction. *Environ. Pollut.* 12:243–52.

Carpena, R.O., S. Vazquez, E. Esteban, et al. 2003. Cadmium–stress in white lupin: Effects on nodule structure and functioning. *Plant Physiol. Biochem.* 161:911–6.

Carrió-Seguí, A., A. Garcia-Molina, A. Sanz, and L. Peñarrubia. 2015. Defective copper transport in the *copt5* mutant affects cadmium tolerance. *Plant Cell Physiol.* 56:442–54.

Castiglione, S., C. Franchin, T. Fossati, G. Lingua, P. Torrigiani, and S. Biondi, 2007. High zinc concentrations reduce rooting capacity and alter metallothionein gene expression in white poplar (*Populus alba* L. cv. Villafranca). *Chemosphere* 67:1117–26.

Cataldo, D.A., K.M. McFadden, T.R. Garland, and R.E. Wildung. 1988. Organic constituents and complexation of nickel (II), iron(III), cadmium(II), and plutonium(IV) in soybean xylem exudates. *Plant Physiol.* 86:734–39.

Cedeno-Maldonado, A, and J.A. Swader. 1972. The cupric ions as an inhibitor of photosynthetic electron transport in isolated chloroplast. *Plant Physiol.* 50:698–701.

Chandrashekhar, K.R., and S. Sandhyarani. 1996. Salinity induced chemical changes in *Crotalaria striata* DC. *Indian J. Plant Physiol.* 1:44–8.

Chaoui, A., M.H. Ghorbal and E. El Ferjani. 1997. Effects of cadmium-zinc interactions on hydroponically grown bean (*Phaseolus vulgaris* L.). *Plant Sci.* 126:21–8.

Chatterjee, J., and C. Chatterjee. 2000. Phytotoxicity of cobalt, chromium and copper in cauliflower. *Environ. Pollut.* 109:69–74.

Chen, C., T.H. Chen, K.F. Lo, and C.Y. Chiu. 2004. Effects of proline on copper transport in rice seedlings under excess copper stress. *Plant Sci.* 166:103–11.

Chen, C., H. Dejun and L. Jianquan, 2009. Functions and toxicity of nickel in plants: Recent advances and future prospects. *Clean* 37:304–13.

Chen, J., Yan, Z., and Li, X. 2014a. Effect of methyl jasmonate on cadmium uptake and antioxidative capacity in *Kandelia obovata* seedling sundercadmiumstress. *Ecotoxicol. Environ. Saf.* 104:349–56.

Chen, Y.A., W.C. Chi, N.N. Trinh, et al. 2014b. Transcriptome profiling and physiological studies reveal a major role for aromatic aminoacids in mercury stress tolerance in rice seedlings. *PLOS ONE* 9:e95163.

Cheong, J.J., and Y.D. Choi. 2003. Methyl jasmonate as a vital substance in plants. *Trend Genet.* 19:409–13.

Chmielowska-Bąk, J., I. Lefèvre, S. Lutts, and J. Deckert. 2013. Short term signaling responses in roots of young soybean seedlings exposed tocadmium stress. *J. Plant Physiol.* 170:1585–94.

Cho, U.H., and J.O. Park. 2000. Mercury-induced oxidative stress in tomato seedlings. *Plant Sci.* 156:1–9.

Choi J.M., C.H., Pak and C.W. Lee. 1996. Micronutrient toxicity in French marigold. *J. Plant Nutr.* 19:901–16.

Clarimont, K.B., W.G. Hagar and E.A. Davis. 1986. Manganese toxicity to chlorophyll synthesis in tobacco callus. *Plant. Physiol.* 80:291–93.

Clark, R.B. 1982. Plant response to mineral element toxicity and deficiency. In: Christiansen, M.N. and C.F. Lewis (eds) *Breeding Plants for Less Favourable Environments.* John Wiley & Sons, Hoboken, NJ, pp. 71–142.

Clemens, S. 2001. Molecular mechanisms of plant metal tolerance and homeostasis. *Planta* 212:475–86.

Clemens, S. 2006. Toxic metal accumulation, responses to exposure and mechanisms of tolerance in plants. *Biochimie* 88:1707–19.

Clemens, S., E.J. Kim, D. Neumann, and J.I. Schroeder. 1999. Tolerance to toxic metals by a gene family of phytochelatin synthases from plants and yeast. *EMBO J.* 15:3325–33.

Cobbett, C.S. 2000. Phytochelatin biosynthesis and function in heavy-metal detoxification. *Curr. Opin. Plant Biol.* 3:211–6.

Cobbett, C.S., and P. Goldsbrough. 2002. Phytochelatins and metallothioneins: roles in heavy metal detoxification and homeostasis. *Annu. Rev. Plant. Biol.* 53:159–82.

Cohen, C.K., T.C. Fox, D.F. Garvin, and L.V. Kochian, 1998. The role of iron-deficiency stress responses in stimulating heavy-metal transport in plants. *Plant Physiol.* 116:1063–72.

Colangelo, E.P., and M.L. Guerinot. 2006. Put the metal to the petal: metal uptake and transport throughout plants. *Curr. Opin. Plant. Biol.* 9:322–30.

Colpaert, J., and J. van Assche. 1992. Zinc toxicity in ectomycorrhizal *Pinus sylvestris. Plant Soil* 143:201–11.

Costa, G., and J.L. Morel. 1994. Water relations, gas exchange and amino acid content in Cd-treated lettuce. *Plant Physiol. Biochem.* 32:561–70.

Courbot, M., G. Willems, and P. Motte. 2007. A major quantitative trait locus for cadmium tolerance in *Arabidopsis halleri* colocalizes with HMA4, a gene encoding a heavy metal ATPase. *Plant Physiol.* 144:1052–65.

Crawford, T.W., J.L. Stroehlein, and R.O. Kuehl. 1989. Manganese and rates of growth and mineral accumulation in cucumber. *J. Am. Soc. Hortic. Sci.* 114:300–6.

Creelman, R.A., and J.E. Mullet. 1995. Jasmonic acid distribution and action in plants, regulation during development and response to biotic and abiotic stresses. *Proc. Natl. Acad. Sci. USA* 92:4114–9.

Curie, C., J.M. Alonso, M. Le Jean, J.R. Ecker, and J.F. Briat. 2000. Involvement of NRAMP1 from *Arabidopsis thaliana* in iron transport. *Biochem. J.* 347:749–55.

Curie, C., Z. Panaviene, C. Loulergue, S.L. Dellaporta, J.F. Briat, and E.L. Walker. 2001. Maize yellow stripe1 encodes a membrane protein directly involved in Fe(III) uptake. *Nature* 409:346–9.

DalCorso, G., S. Farinati, and A. Furini. 2010. Regulatory networks of cadmium stress in plants. *Plant Signal Behav.* 5:663–7.

Dalcorso, G., S. Farinati, S. Maistri, and A. Furini. 2008. How plants cope with cadmium: Staking all on metabolism and gene expression. *J. Integr. Plant Biol.* 50:1268–80.

DalCorso, G., A. Manara, and A. Furini. 2013. An overview of heavy metal challenge in plants: From roots to shoots. *Metallomics* 5:1117–32.

Das, P., S. Samantaray, and G.R. Rout. 1997. Studies on cadmium toxicity in plants: A review. *Environ. Pollut.* 98:29–36.

Dat, J.F., S. Vandenabeele, E. Vranova, M. Van Montagu, D. Inze, and F. Van Breusegem. 2000. Dual action of the active oxygen species during plant stress responses. *Cell. Mol. Life Sci.* 57:779--95.

De-Filippis, L.F., R. Hampp, and H. Ziegler. 1981. The effects of sublethal concentrations of zinc, cadmium and mercury on Euglena growth and pigments. *Z. Pflanzenphysiol.* 1:37–47.

De Filippis, L.F., and C.K. Pallaghy. 1976. The effect a sublethal concentration of mercury and zinc on *Chlorella*. I. Growth characteristic and uptake of metals. *Z. Pflanzenphysiol.* 78:197–207.

De Filippis, L.F., and C.K. Pallaghy. 1994. Heavy metals: Sources and biological effect. In: Rai, L.C., J.P. Gaur, and C.J. Soeder (eds) *Algae and Water Pollution*. E. Echweizerbertsche Varlegsbuchhandlung, Stuttgart, Germany, pp. 31–7.

De Filippis, L.F., and H. Ziegler. 1993. Effect of sublethal concentrations of zinc, cadmium and mercury on the photosynthetic carbon reduction cycle of *Euglena. J. Plant. Physiol.* 142:167–72.

De Knecht, J.A., M. van Dillen, P.L.M. Koevoets, H. Schat, J.A.C. Verkleij, and W.H.O. Ernst. 1994. Phytochelatins in cadmium-sensitive and cadmium-tolerant *Silene vulgaris*. Chain length distribution and sulfide incorporation. *Plant Physiol.* 104:255–61.

De, D.N. 2000. *Plant Cell Vacuoles. An Introduction*. CSIRO Publishing, Collingwood, Australia.

del Rio, L.A., L.M. Sandalio, F.J. Corpas, J.M. Palma, and J.B. Barroso. 2006. Reactive oxygen species and reactive nitrogen species in peroxisomes: Production, scavenging, and role in cell signaling. *Plant. Physiol.* 141:330–35.

del Río, L.A., F. Sevilla., L.M. Sandalio, and J.M. Palma. 1991. Nutritional effect and expression of SODs: Induction and gene expression; diagnostics; prospective protection against oxygen toxicity. *Free Radic. Res. Commun.* 2:819–27.

Delhaize, E., J.F. Ma, and P.R Ryan. 2012. Transcriptional regulation of aluminium tolerance genes. *Trends Plant Sci.* 17:341–8.

Devi, S.R., and M.N.V. Prosad. 2005. Antioxident capacity of *Brasicca juncea* plants exposed to elevated levels of copper. *Russ. J. Plant. Physiol.* 50:205–8.

de Souza, M.O., D. Chu, M. Zhao, et al. 1999. Rhizosphere bacteria enhance selenium accumulation and volatilization by Indian Mustard. Plant Physiol. 199:565–73.

Diener, A.C., R.A. Gaxiola, and G.R. Fink. 2001. Arabidopsis ALF5, a multidrug efflux transporter gene family member, confers resistance to toxins. *Plant Cell.* 13:1625–38.

Dietz, K.-J., M. Baier, and U. Krämer. 1999. Free radicals and reactive oxygen species as mediators of heavy metal toxicity in plants. In: Prasad M.N.V. and J. Hagemeyer (eds) *Heavy Metal Stress in Plants: From Molecules to Ecosystems.* Springer-Verlag, Berlin, pp. 73–97.

Dixit, A.R., and O.P. Dhankher. 2011. A novel stress-associated protein 'AtSAP10' from *Arabidopsis thaliana* confers tolerance to nickel, manganese, zinc, and high temperature stress. *PLOS ONE* 6:e20921.

Djingova, R., and I. Kuleff. 2000. Instrumental techniques for trace analysis. In: Vernet, J.P. (eds) *Trace Elements: Their Distribution and Effects in the Environment.* Elsevier Science Ltd., London, UK, p. 146.

Dong, J., F. Wu, and G. Zhang. 2006. Influence of cadmium on antioxidant capacity and four microelement concentrations in tomato seedlings (*Lycopersicon esculentum*). *Chemosphere* 64:1659–66.

Dorling, S.J., S. Leung, C.W.N. Anderson, N.W. Albert, and M.T. McManus. 2011. Changesin1-aminocyclopropane-1-carboxlate (ACC) oxidase expression and enzyme activity in response to excess manganese in white clover (*Trifolium repens* L.). *Plant Physiol. Biochem.* 49:1013–9.

Dubey, D., and A. Pandey. 2011. Effect of nickel (Ni) on chlorophyll, lipid peroxidation and antioxidant enzymes activities in black gram (*Vigna mungo*) leaves. *Int. J. Sci. Nat.* 2:395–401.

Duffus, J.H. 2002. Heavy metals-a meaningless term? *Pure Appl. Chem.* 74:793–807.

Durrett, T.P., W. Gassmann, and E.E. Rogers. 2007. The FRD3-mediated efflux of citrate into the root vasculature is necessary for efficient iron translocation. *Plant Physiol.* 144:197–205.

Ebbs, S.D. and L.V. Kochin. 1997. Toxicity of zinc and copper to Brassica species: Implication for phytoremediation. *J. Environ. Qual.* 26:776–8.

Ebbs, S., I. Lau, B. Ahner, and L. Kochian. 2002. Phytochelatin synthesis is not responsible for Cd tolerance in the Zn/Cd hyperaccumulator *Thlaspi caerulescens* (J. & C. Presl). *Planta* 214:635–40.

Elamin, O.M., and G.E. Wilcox. 1986. Effect of magnesium and manganese nutrition on muskmelon growth and manganese toxicity. *J. Am. Soc. Hortic. Sci.* 111:582–7.

El-Shintinawy, F., and A. El-Ansary. 2000. Differential effect of Cd and Ni on amino acid metabolism in soybean seedling. *Biol. Planta* 43:79–84.

Emery, L., S. Whelan, K.D. Hirschi, and J.K. Pittman. 2012. Phylogenetic analysis of Ca^{2+}/cation antiporter genes and insights into their evolution in plants. *Front. Plant Sci.* 3:1–18.

Ernst, W.H.O., J.A.C. Verkleij, and H. Schat. 1992. Metal tolerance in plants. *Acta Bot. Neerl.* 41:229–48.

Estrabou, C., E. Filippini, J.P. Soria, G. Schelotto, and J.M. Rodriguez. 2011. Air quality monitoring system using lichens as bioindicators in central Argentina. *Environ. Monit. Assess.* 182:375–83.

Farago, M.E., and W.A. Mullen. 1979. Plants which accumulate metals. Part IV. A possible copper-proline complex from the roots of *Armeria maritime. Inorg. Chim. Acta* 32:L93–L94.

Farooqi, Z.R., M. Zafar, M. Iqbal, M. Kabir, and M. Shafiq. 2009. Toxic Effects of lead and cadmium on germination and seedling growth of *Albizia lebbeck* (L.) Benth. Pak. J. Bot. 41:27–33.

Farooqi, Z-U.R., M.Z. Iqbal, M. Kabir, et al. 2011. Tolerance of *Albizia lebbeck* (L) Benth to different levels of lead in natural field condition. *Pak. J. Bot.* 43:445–52.

Fodor, A., A. Szabo-Nagy, and L. Erdei. 1995. The effects of cadmium on the fluidity and H^+-ATPase activity of plasma membrane from sunflower and wheat roots. *J. Plant Physiol.* 14:87–92.

Fodor, F., E. Sarvari, F. Lang, Z. Szigeti, and E. Cseh, 1996. Effects of Pb and Cd on cucumber depending on the Fe-complex in the culture solution. *J. Plant Physiol.* 148:434–9.

Fodor, F. 2002. Physiological responses of vascular plants to heavy metals. In: *Physiology and Biochemistry of Metal Toxicity and Tolerance in Plants.* Prasad M.N.V. and Strzalka K. (eds), Kluwer Academic Publisher, Dortrech, pp. 149–77.

Fontes, R.L.S., and F.R. Cox. 1998. Zinc toxicity in soybean grown at high iron concentration in nutrient solution. *J. Plant Nutr.* 21:1723–30.

Fox, T.C., and M.L. Guerinot. 1998. Molecular biology of cation transport in plants. *Annu. Rev. Plant Physiol.* 49:669–96.

Foy, C.D., R.R. Weil, and C.A. Coradetti. 1995. Differential manganese tolerances of cotton genotypes in nutrient solution. *J. Plant Nutr.* 18:685–706.

Frey, B., K. Zierold, and I. Brunner. 2000. Extracellular complexation of Cd in the Hartig net and cytosolic Zn sequestration in the fungal mantle of *Picea abies–Hebeloma crustuliniforme* ectomycorrhizas. *Plant Cell Environ.* 23:1257–65.

Fujii, M., K. Yokosho, N. Yamaji, et al. 2012. Acquisition of aluminium tolerance by modification of a single gene in barley. *Nat. Commun.* 3:713.

Furukawa, J., N. Yamaji, H. Wang, et al. 2007. An aluminum-activated citrate transporter in barley. *Plant Cell Physiol.* 48:1081–91.

Gajewska, E., and M. Sklodowska. 2007. Effect of nickel on ROS content and antioxidative enzyme activities in wheat leaves. *Biometals* 20:27–36.

Gajewska, E., and M. Skłodowska. 2008. Differential biochemical responses of wheat shoots and roots to nickel stress: Antioxidative reactions and proline accumulation. *Plant Growth Regul.* 54:179–88.

Gallego, S.M., M.P. Benavides, and M.L. Tomaro.1996. Effect of heavy metal ion excess on sunflower leaves: evidence for involvement of oxidative stress. *Plant Science* 121:151–9.

García-Hernăndez, M., A. Murphy, and L. Taiz. 1998. Metallothioneins 1 and 2 have distinct but overlapping expression patterns in Arabidopsis. *Plant Physiol.* 118:387–97.

Gaspar, M. 2011. Aquaporinas: de canais de água a transportadores multifuncionais em plantas. *Rev. Bras. Bot.* 34:481–91.

Gill, S.S., and N. Tuteja. 2010. Reactive oxygen species and antioxidant machinery in abiotic stress tolerance in crop plants. *Rev. Plant Physiol. Biochem.* 48:909–30.

Giritch, A., M. Ganal, U.W. Stephan, and H. Baumlein. 1998. Structure, expression and chromosomal localization of the metallothionein-like gene family of tomato. *Plant Mol. Biol.* 37:701–14.

Godbold, D.L., and C. Kettner. 1991. Lead influences, root growth and mineral nutrition of *Picea abies* seedlings. *J. Plant Physiol.* 139:95–9.

Goldsbrough, P. 2000. Metal tolerance in plants: The role of phytochelatins and metallothioneins. In: Terry N. and G. Banuelos (eds) *Phytoremediation of Contaminated Soil and Water.* CRC Press LLC, pp. 221–33.

Gong, J.M., D.A. Lee, and J.I. Schroeder. 2003. Long-distance root-to-shoot transport of phytochelatins and cadmium in *Arabidopsis. Proc. Natl. Acad. Sci. USA* 100:10118–23.

Grill, E. 1989. Phytochelatins in plants. In: Hamer, D.H. and D.R. Winge (eds) *Metal Ion Homeostasis: Molecular Biology and Chemistry.* Alan R. Liss, New York, pp. 283–300.

Grill, E., S. Loffler, E.L. Winnacker, and M.H. Zenk. 1989. Phytochelatins, the heavy-metal-binding peptides of plants are synthesized from glutathione by a specific c-glutamylcysteine dipeptidyl transpeptidase (phytochelatin synthase). *Proc. Natl. Acad. Sci. USA* 86:6838–42.

Grill, E., E.L. Winnacker, and M.H. Zenk. 1988. Occurrence of heavy metal binding phytochelatins in plants growing in a mining refuse area. *Experientia* 44:539–40.

Groppa, M.D., M.P. Benavides, and M.L. Tomaro. 2003. Polyamine metabolism in sunflower and wheat leaf discs under cadmium or copper stress. *Plant Sci.* 164:293–9.

Guerinot, M.L. 2000. The ZIP family of metal transporters. *Biochim. Biophys. Acta* 1465:190–8.

Guo, X., Y. Wu, Y. Wang, Y. Chen, and C. Chu. 2009. OsMSRA4.1 and OsMSRB1.1, two rice plastidial methionine sulfoxide reductases, are involved in abiotic stress responses. *Planta* 230:227–38.

Ha, S.B., A.P. Smith, R. Howden, et al. 1999. Phytochelatin synthase genes from Arabidopsis and the yeast *Schizosaccharomyces pombe. Plant Cell* 11:1153–63.

Haag-Kerwer, A., H. Schafer, S. Heiss, C. Walter, and T. Rausch. 1999. Cadmium exposure in *Brassica juncea* causes a decline in transpiration rate and leaf expansion without effect on photosynthesis. *J. Exp. Bot.* 50:1827–35.

Hagemayor, J. 1993. Indicators for heavy metals in the terrestrial environment. In: B. Market (Ed.) *Plants as Biomonitors.* V.C.H. Weinheim, New York, 541–56.

Hall, J.L. 2002. Cellular mechanisms for heavy metal detoxification and tolerance. *J. Exp. Bot.* 53:1–11.

Hamer, D.H. 1986. Metallothionein. *Annu. Rev. Biochem.* 55:913–51.

Han, F.X., Y. Su, D.L. Monts, A.C. Waggoner, and J.M. Plodinec. 2006. Binding distribution and plant uptake of mercury in a soil from Oak Ridge, Tennesse, USA. *Sci. Total Env.* 368:753–68.

Han, L., G.-J. Li, K.-Y. Yang, et al. 2010. Mitogen-activated protein kinase3 and 6 regulate *Botrytis cinerea*-induced ethylene production in *Arabidopsis. Plant J.* 64:114–27.

Hanikenne, M., I.N. Talke, and M.J. Haydon. 2008. Evolution of metal hyperaccumulation required cis-regulatory changes and triplication of HMA4. *Nature* 453:391–5.

Harrison, P.M., and P. Arosio. 1996. The ferritins: Molecular properties, iron storage function and cellular regulation. *Biochim. Biophys. Acta* 275:161–203.

Hartley-Whitaker, J., Ainsworth, G., and A.A. Meharg. 2001. Copper- and arsenate-induced oxidative stress in *Holcus lanatus* L. clones with differential sensitivity. *Plant Cell Environ.* 24:713–22.

Hawkes, J.S. 1997. Heavy metals. *J. Chem. Educ.* 74:1369–74.

He, Z.L., X.E. Yang, and P.J. Stoffella. 2005. Trace elements in agroecosystems and impacts on the environment. *J. Trace Elem. Med. Biol.* 19:125–40.

Heale, E.L., and D.P. Ormrod. 1982. Effects of nickel and copper on *Acer rubrum, Cornus stolonifera, Lonicera tatarica* and *Pinus resinosa. Can. J. Bot.* 60:2674–81.

Hegedus, A., S. Erdei, and G. Horvath. 2001. Comparative studies of H_2O_2 detoxifying enzymes in green and greening barley seedings under cadmium stress. *Plant Sci.* 160:1085–93.

Herbette, S., L. Taconnat, V., Hugouvieux, et al. 2006. Genome-wide transcriptome profiling of the early cadmium response of *Arabidopsis* roots and shoots. *Biochimie* 88:1751–65.

Herbik, A., A. Giritch, C. Horstmann, et al. 1996. Iron and copper nutrition-dependent changes in protein expression in a tomato wild type and the nicotianamine-free mutant chloronerva. *Plant Physiol.* 111:533–40.

Hernandez L.E., R. Carpena-Ruiz, and A. Garate. 1996. Alterations in the mineral nutrition of pea seedlings exposed to cadmium. *J. Plant Nutr.* 19:1581–98.

Higuchi, K., K. Kanazawa, N.K. Nishizawa, M. Chino, and S. Mori. 1994. Purification and characterization of nicotianamine synthase from Fe-deficient barley roots. *Plant Soil* 165:173–9.

Hirsch, R.E., B.D. Lewis, E.P. Spalding, and M.R. Sussman. 1998. A role for the AKT1 potassium channel in plant nutrition. *Science* 280:918–21.

Horst, J. 1988. Beschreibung der Gleichgewichtslage des ionenaust-auschs an schwach saoren harzen mit hilfe eines models der oberflachenkomplexbildung. Doctoral thesis, University of Karlsruhe, Kfk report 4464.

Hossain, M.A., M. Hasanuzzaman, and M. Fujita. 2010. Up-regulation of antioxidant and glyoxalase systems by exogenous glycinebetaine and proline in mung bean confer tolerance to cadmium stress. *Physiol. Mol. Biol. Plants* 16:259–72.

Hossain, M.A., M.Z. Hossain, and M. Fujita. 2009. Stress-induced changes of methylglyoxal level and glyoxalase I activity in pumpkin seedlings and cDNA cloning of glyoxalase I gene. *Aust. J. Crop Sci.* 3:53–64.

Howden, R., P.B. Goldsbrough, C.R. Andersen, and C.S. Cobbett. 1995. Cadmium-sensitive, cad1 mutants of Arabidopsis thaliana are phytochelatin deficient. *Plant Physiol.* 107:1059–66.

Hsu, B.D., and J.Y. Lee. 1988. Toxic effects of copper on photosystem II of spinach chloroplasts. *Plant Physiol.* 87:116–9.

Huang, H., and C. Zhang. 1986. The primary studies on the effects of cadmium and lead on the photosynthesis of woody plants. *J. Ecol.* 5:6–9.

Hubel, F., and E. Beck. 1996. Maize root phytase (purification, characterization, and localization of enzyme activity and its putative substrate). *Plant Physiol.* 112:1429–36.

Hussain, D., Haydon, M.J., Y. Wang, et al. 2004. P-Type ATPase heavy metal transporters with roles in essential zinc homeostasis in *Arabidopsis. Plant Cell* 16:1327–39.

Hüttermann, A., I. Arduini, and D.L. Godbold. 1999. Metal pollution and forest decline. In: Prasad, N.M.V. and J. Hagemeyer (eds) *Heavy Metal Stress in Plants: From Molecules to Ecosystems.* Springer-Verlag, Berlin, pp. 253–72.

Iakimova, E.T., E.J. Woltering, V.M. Kapchina-Toteva, F.J.M. Harren, and S.M. Cristescu. 2008. Cadmium toxicity in cultured tomato cells - Role of ethylene, proteases and oxidative stress in cell death signaling. *Cell Biol. Int.* 32:1521–9.

Iglesias, R.G., and M.J. Babiano. 1997. Endogenous abscisic acid during the germination of chick-pea seeds. *Physiol. Plantarum* 100:500–4.

Inoue, H., D. Mizuno, M. Takahashi, H. Nakanishi, S. Mori, and N.K. Nishizawa. 2004. A rice FRD3-like (OsFRDL1) gene is expressed in the cells involved in long-distance transport. *Soil Sci. Plant Nutr.* 50:1133–40.

Israr, M., S. Sahi, R. Datta, and D. Sarkar 2006. Bioaccumulation and physiological effects of mercury in *Sesbania drummonii. Chemosphere* 65:591–8.

Jabeen, R., A. Ahmad, and M. Iqbal. 2009. Phytoremediation of heavy metals: Physiological and molecular mechanisms. *Bot. Rev.* 75:339–64.

Jadia, C.D., and M.H. Fulekar. 1999. Phytoremediation of heavy metals: Recent techniques. *Afr. J. Biotechnol.* 8:921–8.

Jadia, C.D., and M.H. Fulekar. 2009. Phytoremediation of heavy metals: Recent techniques. *Afr. J. Biotechnol.* 8:921–8.

Jain, N., G.D. Ascough, and J.V. Staden. 2008. A smoke-derived butenolide alleviates $HgCl_2$ and $ZnCl_2$ inhibition of water uptake during germination and subsequent growth of tomato – Possible involvement of aquaporins. *J. Plant Physiol.* 165:1422–7.

Jakubowicz, M., H. Gałańska, W. Nowak, and J. Sadowski. 2010. Exogenously induced expression of ethylene biosynthesis, ethylene perception, phospholipase D, and Rboh-oxidase genes in broccoli seedlings. *J. Exp. Bot.* 61:3475–91.

Järup, L. 2003. Hazards of heavy metal contamination. *Br. Med. Bull.* 68:167–82.

Jarvis, M.D., and D.W.M. Leung. 2002. Chelated lead transport in *Pinus radiata*: An ultrastructural study. *Environ. Exp. Bot.* 48:21–32.

Jegerschold, C., J.B. Arellano, W.P. Schroder, P.J.M. van Kan, M. Baron, and S. Styring. 1995. Copper (II) inhibition of electron transfer through photoystem II studied by EPR spectroscopy. *Biochemistry* 34:12747–54.

Jentschke, G., and D.L. Godbold. 2000. Metal toxicity and ectomycorrhizas. *Physiol. Plantarum* 109:107–16.

John, R., P. Ahmad, K. Gadgil, and S. Sharma. 2008. Effect of cadmium and lead on growth, biochemical parameters and uptake in *Lemna polyrrhiza* L. *Plant Soil Environ.* 54:262–70.

Jonak, C., H. Nakagami, and H. Hirt. 2004. Heavy metal stress. Activation of distinct mitogen-activated protein kinase pathways by copper and cadmium. *Plant Physiol.* 136:3276–83.

Kabir, M., M.Z. Iqbal, M. Shafiq, and Z.R. Farooqi. 2008. Reduction in germination and seedling growth of *Thespesia populnea* L. caused by lead and cadmium treatments. *Pak. J. Bot.* 40:2419–26.

Kägi, J.H. 1991. Overview of metallothionein. *Methods Enzymol.* 205:613–26.

Kahle, H. 1993. Response of roots of trees to heavy metals. *Environ. Exp. Bot.* 33:99–119.

Kamal, A.H., and S. Komatsu. 2016. Jasmonic acid induced protein response to biophoton emissions and floodings tress in soybean. *J. Proteomics* 133:33–47.

Kamal, M., A.E. Ghalya, N. Mahmouda, and R. Cote. 2004. Phytoaccumu-lation of heavy metals by aquatic plants. *Environ. Intern.* 29:1029–39.

Kampfenkel, K., S. Kushnir, E. Babiychuk, D. Inzé, and M. Van Montagu. 1995a. Molecular characterization of a putative Arabidopsis thaliana copper transporter and its yeast homologue. *J. Biol. Chem.* 270:28479–86.

Kampfenkel, K., M. Van Montagu, and D. Inze. 1995b. Effects of iron excess on Nicotian plumbaginifolia plants (Implications to oxidative stress). *Plant Physiol.* 107:725–35.

Kasai, Y., M. Kato, J. Aoyama, and H. Hyodo. 1998. Ethylene production and increase in 1-amino-cyclopropane-1-carboxylate oxidase activity during senescence of broccoli florets. *Acta Hort.* 464:153–7.

Kawachi, M., Y. Kobae, T. Mimura, and M. Maeshima. 2008. Deletion of a histidine-rich loop of AtMTP1, a vacuolar Zn^{2+}/H^+ antiporter of Arabidopsis thaliana, stimulates the transport activity. *J. Biol. Chem.* 283:8374–83.

Kehrer, J.P. 2000. The Haber-Weiss reaction and mechanisms of toxicity. *Toxicology* 149:43–50.

Keunen, E., K. Schellingen, J. Vangronsveld, and A. Cuypers. 2016. Ethylene and metal stress: Small molecule, big impact. *Front. Plant Sci.* 6:23.

Khan, M.I.R., and N.A. Khan. 2014. Ethylene reverses photosynthetic inhibition by nickel and zinc in mustard through changes in PSII activity, photosynthetic nitrogenuse efficiency, and antioxidant metabolism. *Protoplasma* 251:1007–19.

Khan, M.I.R., F. Nazir, M. Asgher, T.S., Per, and N.A. Khan. 2015. Selenium and sulfur influence ethylene formation and alleviate cadmium-induced oxidative stress by improving proline and glutathione production in wheat. *J. Plant Physiol.* 173:9–18.

Kim, D.Y., L. Bovet, M. Maeshima, E. Martinoia, and Y. Lee. 2007. The ABC transporter AtPDR8 is a cadmium extrusion pump conferring heavy metal resistance. *Plant J.* 50:207–18.

Kim, Y.S., D. Choi, M.M. Lee, S.H. Lee, and W.T. Kim. 1998. Biotic and abiotic stress-related expression of 1-amino cyclo propane-1-carboxylate oxidase gene family in *Nicotiana glutinosa* L. *Plant Cell Physiol.* 39:565–73.

Kirkham, M.B. 1978. Water relations of Cd-treated plants. *J. Environ. Qual.* 7:334–6.

Kitao, M., T.T. Lei, and T. Koike. 1997. Effects of manganese in solution culture on the growth of five deciduous broad-leaved tree species with different successional characters from northern Japan. *Photosynth.* 36:31–40.

Koeppe, D.E., and R.J. Miller. 1970. Lead effects on corn mitochondrial respiration. *Science* 167:1376–8.

Koh, S., A.M. Wiles, J.S. Sharp, F.R. Naider, J.M. Becker, and G. Stacey. 2002. An oligopeptide transporter gene family in Arabidopsis. *Plant Physiol.* 128:21–9.

Korenkov, V., S.H. Park, N.H. Cheng, et al. 2007. Enhanced Cd^{2+} selective root-tonoplast-transport in tobaccos expressing Arabidopsis cation exchangers. *Planta* 225:403–11.

Korshunova, Y.O., D. Eide, W.G. Clark, M.L. Guerinot, and H.B. Pakrasi. 1999. The IRT1 protein from Arabidopsis thaliana is a metal transporter with a broad substrate range. *Plant Mol. Biol.* 40:37–44.

Kovacs, E., P. Nyitrai, P. Czovek, M. Ovari, and A. Keresztes. 2009. Investigation into the mechanism of stimulation by low-concentration stressors in barley seedlings. *J. Plant Physiol.* 166:72–9.

Kovtun, Y., W.L. Chiu, G. Tena, and J. Sheen. 2000. Functional analysis of oxidative stress-activated mitogen-activated protein kinase cascade in plants. *Proc. Natl. Acad. Sci. USA* 97:2940–5.

Krämer, U. 2010. Metal hyperaccumulation in plants. *Annu. Rev. Plant. Biol.* 61:28.1–28.18
Krämer, U., J.D. Cotter-Howells, J.M. Charnock, A.J.M. Baker, and J.A.C. Smith. 1996. Free histidine as a metal chelator in plants that accumulate nickel. *Nature* 379:635–8.
Krämer, U., I.N. Talke, and M. Hanikenne. 2007. Transition metal transport. *FEBS Lett.* 581:2263–72.
Kukier U., C.A. Peters, R.L. Chaney, J.S. Angle, and R.J. Roseberg. 2004. The effect of pH on metal accumulation in two *Alyssum* species. *J. Environ. Qual.* 33:2090–102.
Kupper, H., I. Setlik, E. Setlikova, N. Ferimazova, M. Spiller, and F.C. Kupper. 2003. Copper induced inhibition of photosynthesis: Limiting steps of in vivo copper chlorophyll formation in *Scenedesmus quadricauda. Funct. Plant Biol.* 30:1187–96.
Kuroda, T., and T. Tsuchiya. 2009. Multidrug efflux transporters in the MATE family. *Biochim. Biophys. Acta* 1794:763–8.
Kuznetsov, V.V., and N.I. Shevyakova. 1997. Stress responses of tobacco cells to high temperature and salinity. Proline accumulation and phosphorylation of polypeptides. *Physiol. Plantarum* 100:320–6.
Lamoreaux, R.J., and W.R. Chaney. 1977. Growth and water movement in silver maple seedlings affected by cadmium. *J. Environ. Qual.* 6:201–5.
Lamoreaux, R.J., and W.R. Chaney. 1978a. Photosynthesis and transpiration of excised silver maple leaves exposed to cadmium and sulphur dioxide. *Environ. Pollut. Plant.*17:259–67.
Lamoreaux, R.J., and W.R. Chaney. 1978b. The effect of cadmium on net photosynthesis transpiration, and dark respiration of excised silver maple leaves. *Physiol. Plantarum* 43:231–6.
Lane, S.D., E.S. Martin and J.P. Garrod. 1978. Lead toxicity effect on indole-3-acetic-induced cell elongation. *Planta* 144:79–84.
Lanquar, V., F. Lelièvre, and S. Bolte, et al. 2005. Mobilization of vacuolar iron by AtNRAMP3 and AtNRAMP4 is essential for seed germination on low iron. *EMBO. J* 24:4041–51.
Lata, S. 1989. Effect of Cadmium on seedling growth, mobilization of food reserves and activity of hydrolytic enzymes in *Phaseolus aureus* L. seeds. *Acta Bot. Indica* 17:290–3.
Lavid, N., A. Schwartz, O. Yarden, and E. Tel-Or. 2001. The involvement of polyphenols and peroxidase activities in heavy metal accumulation by epidermal glands of the waterlily (Nymphaeaceae). *Planta* 212:323–31.
Le Bot, J., E.A. Kirkby, and M.L. Beusichem. 1990. Manganese toxicity in tomato plants: Effects on cation uptake and distribution. *J. Plant Nutr.* 13:513–25.
LeDuc, D.L., M. AbdelSamie, M. Móntes-Bayon, C.P. Wu, S.J. Reisinger, and N. Terry. 2006. Overexpressing both ATPsulfurylase and selenocysteinem ethyltransferase enhances selenium phytoremediation traits in Indian mustard. *Environ. Pollut.* 144:70–6.
Lee, J., R.D. Reevesa, R.R. Brooksa, and T. Jaffréb. 1977. Isolation and identification of a citratocomplex of nickel from nickel-accumulating plants. *Phytochemistry* 16:1503–15.
Lee, J., D. Shim, W.Y. Song., I. Hwang, and Y. Lee. 2004. Arabidopsis metallothioneins 2a and 3 enhance resistance to cadmium when expressed in *Vicia faba* guard cells. *Plant Mol. Biol.* 54:805–15.
Lee, K.C., B.A., Cunningham, G.M. Paulson, et al. 1976. Effects of cadmium on respiration rates and activities of several enzymes in soybean seedlings. *Plant Physiol.* 36:4–6.
Lee, S., J.S. Moon, T.S. Ko, D. Petros, P.B. Goldsbrough, and S.S. Korban. 2003. Overexpression of Arabidopsis phytochelatin synthase paradoxically leads to hypersensitivity to cadmium stress. *Plant Physiol.* 131:656–63.
Lequeux, H., C. Hermans, S. Lutts, and N. Verbruggen. 2010. Response to copper excess in *Arabidopsis thaliana*: Impact on the root system architecture, hormone distribution, lignin accumulation andmineral profile. *Plant Physiol. Biochem.* 48:673–82.
Lewis, S., R.D. Handy, B. Cordi, Z. Billinghurst, and M.H. Depledge. 1999. Stress proteins (HSPs): methods of detection and their use as an environmental biomarker. *Ecotoxicology* 8:351–68.
Li, G., X.R. Meng, G. Wang, et al. 2012. Dual-level regulation of ACC synthase activity by MPK3/MPK6 cascade and its downs tream WRKY transcription factor during ethylene induction in *Arabidopsis. PLoSGenet* 8:e1002767.
Li, L.G., Z.Y. He, G.K. Pandey, T. Tsuchiya, and S. Luan. 2002. Functional cloning and characterization of a plant efflux carrier for multidrug and heavy metal detoxification. *J. Biol. Chem.* 277:5360–8.
Li, Z.-S., Y.-P. Lu, R.-G. Zhen, et al. 1997. A new pathway for vacuolar cadmium sequestration in Saccharomyces cerevisiae: YCF1-catalyzed transport of bis(glutathionato) cadmium. *Proc. Natl. Acad. Sci. USA* 94:42–7.
Liang, W.H., L. Li, F. Zhang, et al. 2013. Effects of abiotic stress, light, phytochromes and phytohormones on the expression of *OsAQP*, a rice aquaporin gene. *Plant Growth Regul.* 69:21–7.

Lin, R., X. Wang, Y. Luo, W. Du, H. Guo, and D. Yin. 2007. Effects of soil cadmium on growth, oxidative stress and antioxidant system in wheat seedlings (*Triticum aestivum* L.). *Chemosphere* 69:89–98.

Lin, Z., S. Zhong, and D.Grierson. 2009. Recent advances in ethylene research. *J. Exp. Bot.* 60:3311–36.

Linger, P., A. Ostwald, and J. Haensler. 2005. *Cannabis sativus* L., growing on heavy metal contaminated soil: Growth, cadmium uptake and photosynthesis. *Biol. Plantarum* 49:567–76.

Liu, C., T. Fukumoto, T. Matsumoto, et al. 2013. Aquaporin OsPIP1;1 promotes rice salt resistance and seed germination. *Plant Physiol. Biochem.* 63:151–8.

Liu, D., W. Jiang, and M. Li. 1992. Effect of trivalent and hexavalent chromium on root growth and cell division of *Allium cepa. Hereditas* 117:23–9.

Liu, H.Y., X. Yu, C.Y. Cui, et al. 2007. The role of water channel proteins and nitric oxide signaling in rice seed germination. *Cell Res.* 17:638–49.

Liu, X.M., K.E. Kim, K.C. Kim, et al. 2010. Cadmium activates *Arabidopsis* MPK3 and MPK6 via accumulation of reactive oxygen species. *Phytochemistry* 71:614–8.

Lösch, R. 2004. Plant mitochondrial respiration under the influence of heavy metals. In: Prasad M.N.V. (ed) *Heavy Metal Stress in Plants*, 3rd edn. Springer, Berlin, pp. 182–200.

Lu, Y.P., Z.S. Li, Y.M. Drozdowicz., S. Hortensteiner, E. Martinoia, and P.A. Rea. 1998. AtMRP2, an Arabidopsis ATP binding cassette transporter able to transport glutathione S-conjugates and chlorophyll catabolites: Functional comparisons with AtMRP1. *Plant Cell* 10:267–82.

Lu, Y.P., Z.S. Li, and P.A. Rea. 1997. AtMRP1 gene of Arabidopsis encodes a glutathione S-conjugatepump: Isolation and functional definition of a plant ATP-binding cassette transporter gene. *Proc. Natl. Acad. Sci. USA* 94:8243–8.

Lubkowitz, M. 2011. The oligopeptide transporters: A small gene family with a diverse group of substrates and functions? *Mol. Plant.* 4:407–15.

Luna, C.M., C.A. Gonzalez, and V.S. Trippi. 1994. Oxidative damage caused by an excess of copper in oat leaves. *Plant Cell Physiol.* 35:11–5.

Hossain, M.A., J.A.T. da Silva, and M. Fujita. 2011. Glyoxalase system and reactive oxygen species detoxification system in plant abiotic stress response and tolerance: an intimate relationship. In: Shanker, A.K., and B. Venkateswarl (eds) *Abiotic Stress in Plants-Mechanisms and Adaptations*. INTECH-Open Access Publisher, Rijeka, Croatia, pp. 235–66.

Ma, J.F., S.J. Zheng, and H. Matsumato. 1997. Detoxifying aluminium with buckwheat. *Nature* 390:569–70.

Maas, F.M., D.A.M. van de Wetering, M.L. van Beusichem, and H.F. Bienfait. 1988. Characterization of phloem iron and its possible role in the regulation of Fe-efficiency reactions. *Plant Physiol.* 87:167–71.

MacDiarmid, C.W., L.A. Gaither, and D. Eide. 2000. Zinc transporters that regulate vacuolar zinc storage in Saccharomyces cerevisiae. *EMBO J.* 19:2845–55.

Macnair, M.R., G.H. Tilstone, and S.E. Smith. 2000. The genetics of metal tolerance and accumulation in higher plants. In: Terry, N. and G. Banuelos (eds) *Phytoremediation of Contaminated Soil and Water.* CRC Press, Boca Raton, FL, pp. 235–50.

Magalhaes, J.V. 2010. How a microbial drug transporter became essential for crop cultivation on acid soils: aluminium tolerance conferred by the multidrug and toxic compound extrusion (MATE) family. *Ann. Bot.* 106:199–203.

Magalhaes, J.V., J. Liu, C.T. Guimaraes, et al. 2007. A gene in the multidrug and toxic compound extrusion (MATE) family confers aluminum tolerance in sorghum. *Nat. Genet.* 39:1156–61.

Maitani, T., H. Kubota., K. Sato, and T. Yamada. 1996. The composition of metals bound to Class III metallothionein (phytochelatin and its desglycyl peptide) induced by various metals in root cultures of *Rubia tinctorum. Plant Physiol.* 110:1145–50.

Maksymiec, W. 2007. Signaling responses in plants to heavy metal stress. *Acta Physiol. Plantarum* 29:177–87.

Maksymiec, W., and Z. Krupa. 2006. The effects of short-term exposition to Cd, excess Cu ions and jasmonate on oxidative stress appearing in *Arabidopsis thaliana. Environ. Exp. Bot.* 57:187–94.

Maksymiec, W., D. Wianowska, A.L., Dawidowicz, S. Radkiewicz, M. Mardarowicz, and Z. Krupa. 2005. The level of jasmonic acid in *Arabidopsis thaliana* and *Phaseolus coccineus* plants under heavy metal stress. *J. Plant Physiol.* 162:1338–46.

Manavalan, L.P., S.K. Guttikonda, L.-S.P. Tran, and H.T. Nguyen. 2009. Physiological and molecular approaches to improve drought resistance in soybean. *Plant Cell Physiol.* 50:1260–76.

Manivasagaperumal, R., S. Balamurugan, G. Thiyagarajan, and J. Sekar. 2011. Effect of zinc on germination, seedling growth and biochemical content of cluster bean (*Cyamopsis tetragonoloba* (L.). *Curr. Bot.* 2:11–5.

Maron, L.G., M.A. Piñeros, C.T. Guimarães, et al., 2010. Two functionally distinct members of the MATE (multi-drug and toxic compound extrusion) family of transporters potentially underlie two major aluminum tolerance QTLs in maize. *Plant J.* 61:728–40.

Marschner, H. 1986. *Mineral Nutrition of Higher Plants*. Academic Press, London, p. 674.

Marschner, H. 1995. *Mineral Nutrition in Higher Plants*, 2nd edn. Academic Press, San Diego, p. 889.

Martinoia, E., E. Grill, R. Tommasini, K. Kreuz, and N. Amrhein. 1993. ATP-dependent glutathione Sconjugate 'export' pump in the vacuolar membrane of plants. *Nature* 364:247–9.

Mäser, P., S. Thomine, and J.I. Schroeder, et al. 2001. Phylogenetic relationships within cation transporter families of Arabidopsis. *Plant Physiol.* 126:1646–67.

Masood, A., Iqbal, N., and N.A. Khan. 2012. Role of ethylene in alleviation of cadmium-induced photosynthetic capacity inhibition by sulphur in mustard. *Plant Cell Environ.* 35:524–33.

Mathys, W. 1977. The role of malate, oxalate, and mustard oil glucosides in the evolution of zinc resistance in herbage plants. *Physiol. Plantarum* 40:130–36.

Matsui, A., J. Ishida, T. Morosawa, et al. 2008. Arabidopsis transcriptome analysis under drought, cold, high-salinity and ABA treatment conditions using timing array. *Plant Cell Physiol.* 49:1135–49.

Meharg, A.A., and M.R. Macnair. 1992. Suppression of the high affinity phosphate uptake system; a mechanism of arsenate tolerance in *Holcus lanatus* L. *J. Exp. Bot.* 43:519–24.

Meharg, A.A. 1993. The role of plasmalemma in metal tolerance in angiosperm. *Physiol. Plantarum* 88:191–8.

Mei, H., N.H. Cheng, J. Zhao, et al. 2009. Root development under metal stress in *Arabidopsis thaliana* requires the H^+/cation antiporter CAX4. *New Phytol.* 183:95–105.

Mertens, J., J. Vangronsveld, D. Van Der Straeten, and M. VanPoucke. 1999. Effects of copper and zinc on the ethylene production of *Arabidopsis thaliana*. In: Kanellis, A.K., C. Chang, H. Klee, A.B. Bleecker, J.C. Pech, and D.Grierson (eds) *Biology and Biotechnology of the Plant Hormone Ethylene II*. Kluwer Academic Publishers, Dordrecht, the Netherlands, pp. 333–8.

Messer, R.L., P.E. Lockwood, W.Y. Tseng, et al. 2005. Mercury (II) alters mitochondrial activity of monocytes at sublethal doses via oxidative stress mechanisms. *J. Biomed. Mater. Res. B* 75:257–63.

Metwally, A., I. Finkemeier, M. Georgi, and K.-J. Dietz. 2003. Salicylic acid alleviates the cadmium toxicity in barley seedlings. *Plant Physiol.* 132:272–81.

Mikus, S.M., M. Bobak, and A. Lux. 1992. Structure of protein bodies and elemental composition of phytin from dry germ of maize (*Zea mays* L.). *Bot. Acta* 105:26–33.

Miller, R.J., J.E. Biuell, and D.E. Koeppe. 1973. The effect of cadmium on electron and energy transfer reactions in corm mitochondria. *Physiol. Plantarum* 28:166–71.

Miller G., V. Shulaev, and R. Mittler. 2008. Reactive oxygen signaling and abiotic stress. *Physiol. Plant* 133:481–9.

Mills, R.F., G.C. Krijger, P.J. Baccarini, J.L. Hall, and L.E. Williams. 2003. Functional expression of AtHMA4, a P1B-type ATPase of the Zn/Co/Cd/Pb subclass. *Plant J.* 35:164–76.

Mills, R.F., A. Francini, P.S.C. Ferreira da Rocha, et al., 2005. The plant P1B-type ATPase AtHMA4 transports Zn and Cd and plays a role in detoxification of transition metals supplied at elevated levels. *FEBS Lett.* 579:783–91.

Moffat, A.S. 1999. Engineering plants to cope with metals. *Science* 285:369–70.

Mohanty, N., I. Vass, and S. Demeter. 1989. Copper toxicity affects photosystem II electron transport at the secondary quinone acceptor, QB. *Plant Physiol.* 90:175–9.

Mohnish, P., and N. Kumar. 2015b. Effect of copper dust on photosynthesis pigments concentrations in plants species. *Int. J. Eng. Res. Manag. (IJERM)* 2:63–6.

Molas, J. 1997. Changes in morphological and anatomical structure of cabbage (*Brassica oleracera* L.) outer leaves and in ultrastructure of their chloroplasts caused by an in vitro excess of nickel. *Photosynthetica* 34:513–22.

Montanini, B., D. Blaudez, S. Jeandroz, D. Sanders, and M.Chalot. 2007. Phylogenetic and functional analysis of the cation diffusion facilitator (CDF) family: improved signature and prediction of substrate specificity. *BMC Genomics* 8:107.

Montero-Palmero, M.B., A. Martín-Barranco, C. Escobar, and L.E. Hernández. 2014. Early transcriptional responses to mercury: A role for ethylene in mercury-inducedstress. *New Phytol.* 201:116–30.

Morel, M., J. Crouzet, A. Gravot, et al. 2009. AtHMA3, a P1B-ATPase allowing Cd/Zn/Co/Pb vacuolar storage in Arabidopsis. *Plant Physiol.* 149:894–904.

Moreno, I., L. Norambuena, D. Maturana, et al. 2008. AtHMA1 is a thapsigargin sensitive Ca^{2+}/heavy metal pump. *J. Biol. Chem.* 283:9633–41.

Muhammad, S., M.I. Zafar, and M. Athar. 2008. Effect of lead and cadmium on germination and seedling growth of *Leucaena leucocephala*. *J. Appl. Sci. Environ. Manag.* 12:61–6.

Murgia, I., J.F. Briat, D. Tarantino and C. Soave. 2001. Plant ferritin accumulates in response to photoinhibition but its ectopic overexpression does not protect against photoinhibition. *Plant Physiol. Biochem.* 39:797–805.

Murgia, I., M. Delledonne, and C. Soave. 2002. Nitric oxide mediates iron-induced ferritin accumulation in Arabidopsis. *Plant J.* 30:521–8.

Murphy, A., and L. Taiz. 1997. Correlation between potassium efflux and copper sensitivity in ten *Arabidopsis* ecotypes. *New Phytol.* 136:211–22.

Murphy, A.S., W.R., Eisenger, J.E., Shaff, L.V., Kochian, and L. Taiz. 1999. Early copper induced leakage of K+ from *Arabidopsis* seedlings is mediated by ion channels and coupled to citrate efflux. *Plant Physiol.* 121:1375–82.

Muthuchelian, K., Bertamini, M., and N. Nedunchezhian. 2001. Triacontanol can protect *Erythrina variegata* from cadmium toxicity. *J. Plant Physiol.* 158:1487–90.

Mysliwa-Kurdziel, B., M.N.V. Prasad and K. Stralka. 2004. Photosynthesis in heavy metal stress plants. In: Prasad, M.N.V. (ed) *Heavy Metal Stress in Plants*, 3rd edn. Springer, Berlin, pp. 146–81.

Nagajyoti, P.C., K.D. Lee, and T.V.M. Sreekanth. 2010. Heavy metals, occurrence and toxicity for plants: a review. *Environ. Chem. Lett.* 8:199–216.

Nagata, T., T. Oobo, and O. Aozasa. 2013. Efficacy of a bacterial siderophore, pyoverdine, to supply iron to *Solanum lycopersicum* plants. *J. Biosci. Bioeng.* 115:1–5.

Nakano, T., K. Suzuki, T. Fujimura, and H. Shinshi. 2006. Genome wide analysis of the ERF gene family in Arabidopsis and rice. *Plant Physiol.* 140:411–32.

Nakashima, K., and K. Yamaguchi-Shinozaki. 2006. Regulons involved in osmotic stress-responsive and cold stress-responsive gene expression in plant. *Physiol. Plant.* 126:62–71.

Nakashima, K., Y. Ito, and K. Yamaguchi-Shinozaki. 2009. Transcriptional regulatory networks in response to abiotic stresse sin *Arabidopsis* and grasses. *Plant Physiol.* 149:88–95.

Naranjo, M.A., C. Romero, J.M. Bellés, C. Montesinos, O. Vicente, and R. Serrano. 2003. Lithium treatment induces a hyper sensitive-like response in tobacco. *Planta* 217:417–24.

Navrot, N., N. Rouhier, E. Gelhaye, and J.P. Jaquotv, 2007. Reactive oxygen species generation and antioxidant systems in plant mitochondria. *Physiol. Plantarum* 129:185–95.

Nelson, N. 1999. Metal ion transporters and homeostasis. *EMBO J.* 18:4361–71.

Neumann, D., O. Lichtenberger, D. Günther, K. Tschiersch, and L. Nover. 1994. Heat-shock proteins induce heavy-metal tolerance in higher plants. *Planta* 194:360–7.

Nevo, Y., and N. Nelson. 2006. The NRAMP family of metal-ion transporters. *Biochim. Biophys. Acta* 1763:609–20.

Nicholls, M., and T.K. Mal. 2003. Effects of lead and copper exposure on growth of an invasive weed, *Lythrum salicaria* L. (Purple Loosestrife). *Ohio J. Sci.* 103:129–33.

Nishida, S., T. Mizuno, and H. Obata. 2008. Involvement of histidine-rich domain of ZIP family transporter TjZNT1 in metal ion specificity. *Plant Physiol. Biochem.* 46:601–6.

Nonogaki, H., G.W. Bassel, and J.D. Bewley. 2010. Germination - Still a mystery. *Plant Sci.* 179:574–81.

Nyitrai, P., M. Mayer, M. Ovari, and A. Keresztes. 2007. Involvement of the phosphoinositide signaling pathway in the anti-senescence effect of low-concentration stressors on detached barley leaves. *Plant Biol.* 9:420–6.

Opdenakker, K., T. Remans, E. Keunen, J. Vangronsveld, and A. Cuypers. 2012. Exposure of *Arabidopsis thaliana* to Cd or Cu excess leads to oxidative stress mediated alterations in MAPKinase transcript levels. *Environ. Exp. Bot.* 83:53–61.

Ortiz, D.F., L. Kreppel, D.M. Speiser, G. Scheel, G. McDonald, and D.W. Ow. 1992. Heavy metal tolerance in the fission yeast requires an ATP-binding cassette-type vacuolar membrane transporter. *EMBO J.* 11:3491–9.

Ortiz, D.F., T. Ruscitti, K.F. McCue, and D.W. Ow. 1995. Transport of metal-binding peptides by HMT1, a fission Yeast ABC-type vacuolar membrane protein. *J. Biol. Chem.* 270:4721–8.

Osawa, H., G. Stacey, and W. Gassmann. 2006. ScOPT1 and AtOPT4 functionasproton-coupledoligopeptide-transporterswithbroad but distinct substrate specificities. *Biochem. J.* 393:267–75.

Pál, M., G. Szalai, E. Horváth, T. Janda, and E. Páldi. 2002. Effect of salicylic acid during heavy metal stress. *Acta Biol. Szeged.* 46:119–20.

Panda, S.K. 2007. Chromium-mediated oxidative stress and ultra structural changes in root cells of developing rice seedlings. *J. Plant Physiol.* 164:1419–28.

Panda, S.K., and S. Choudhury. 2005. Chromium stress in plants. *Braz. J. Plant Physiol.* 17:95–102.

Pandolfini, T., R. Gabbrielli, and C. Comparini.1992. Nickel toxicity and peroxidise activity in seedlings of *Triticum aestivum* L. *Plant Cell Environ.* 15:719–25.

Parys, E., E. Romanowaska, M. Siedlecka, and J. Poskuta 1998. The effect of lead on photosynthesis and respiration in detached leaves and in mesophyll protoplasts of *Pisum sativum*. *Acta Physiol. Plantarum* 20:313–22.

Patra, M., N. Bhowmik, B. Bandopadhyay, and A. Sharma. 2004. Comparison of mercury, lead and arsenic with respect to genotoxic effects on plant systems and the development of genetic tolerance. *Environ. Exp. Bot.* 52:199–223.

Pätsikkä, E., M.F. Kairavuo, E.M. Šerešen, and E. Tyystjärvi. 2002. Excess copper predisposes photosystem II to photoinhibition *in vivo* by outcompeting iron and causing decrease in leaf chlorophyll. *Plant Physiol.* 129:1359–67.

Patterson, W.A., and J. Olsonj. 1983. Effects of heavy metals on radicle growth of selected woody species germinated on filter paper, mineral and organic soil substrates. *Can. J. For. Res.* 13:233–8.

Peer, W.A., M. Mamoudian, B. Lahner, R.D. Reeves, A.S. Murphy, and D.E. Salt. 2003. Identifying model metal hyperaccumulating plants: germplasm analysis of 20 Brassicaceae accessions from a wide geographical area. *New Phytol.* 159:421–30.

Peiter, E., B. Montanini, A. Gobert, et al. 2007. A secretory pathway-localized cation diffusion facilitator confers plant manganese tolerance. *Proc. Natl. Acad. Sci. USA* 104:8532–37.

Peleg, Z., and E. Blumwald. 2011. Hormone balance and abiotic stress tolerance in crop plants. *Curr. Opin. Plant. Biol.* 14:290–5.

Pell, E.J., C.D. Schlagnhaufer, R.N. Arteca. 1997. Ozone-induced oxidative stress: Mechanisms of action and reaction. *Physiol. Plantarum* 100:264–73.

Pesci, P., and R. Reggiani. 1992. The process of abscisic acid induced proline accumulation and the levels of polyamines and quaternary ammonium compounds in hydrated barley leaves. *Physiol. Plantarum* 84:134–9.

Pich, A., and G. Scholz. 1996. Translocation of copper and other micronutrients in tomato plants (*Lycopersicon esculentum* Mill.): Nicotianamine-stimulated copper transport in the xylem. *J. Exp. Bot.* 47:41–7.

Pich, A., S. Hillmer, R. Manteuffel, and G. Scholz. 1997. First immunohistochemical localization of the endogenous Fe^{2+}-chelator nicotianamine. *J. Exp. Bot.* 48:759–67.

Pilon, M., C.M. Cohu, K. Ravet, S.E. Abdel-Ghany, and F. Gaymard. 2009. Essential transition metal homeostasis in plants. *Curr. Opin. Plant. Biol.* 12:347–57.

Polle, A. 1997. Defense against photo-oxidative damage in plants. In: *Oxidative Stress and the Molecular Biology of Antioxidant Defenses*. Scandalios J.G. (ed), Cold Spring Harbor Laboratory Press, Cold Spring Harbor, pp. 623–66.

Popescu, S., G. Popescu, S. Bachan, et al. 2009. MAPK target networks in *Arabidopsis thaliana* revealed using functional protein microarrays. *Genes Dev.* 23:80–92.

Poschenrieder, C.H., B. Gunse, and J. Barcelo.1989. Influence of cadmium on water relations, stomatal resistance and abscisic acid content in expanding bean leaves. *Plant Physiol.* 90:1365–71.

Powell, M.J., M.S. Devis, and D. Francis. 1986. The influence of zinc on the cell cycle in the root meristem of a zinc-tolerant and a non-tolerant cultivar of *Festuca rubra* L. *New Phytol.* 102:419–28.

Prasad, M.N.V. 1997. Trace metals. In: Prasad, M.N.V. (ed) *Plant Ecophysiology*. John Wiley & Sons Inc., New York, pp. 207–49.

Prasad, M.N.V. 1999. Metallothioneins and metal binding complexes in plants. In: Prasad, M.N.V. and J. Hagemeyer (eds) *Heavy Metal Stress in Plants: From Molecules to Ecosystems*. Springer-Verlag, Berlin, pp. 51–72.

Prasad, S.M., R. Dwivedi, and M. Zeeshan. 2005. Growth, photosynthetic electron transport, and antioxidant responses of young soybean seedlings to simultaneous exposure of nickel and UV-B stress. *Photosynthetica* 43:177–85.

Przedpelska-Wasowicz, E.M., and M. Wierzbicka. 2011. Gating of aquaporins by heavy metals in *Allium cepa* L. epidermal cells. *Protoplasma* 248:663–71.

Puig, S., and D.J. Thiele. 2002. Molecular mechanisms of copper uptake and distribution. *Curr. Opin. Chem. Biol.* 6:171–80.

Quartacci, M.F., E. Cosi, and F. Navari-Izzo. 2001. Lipids and NADPH-dependent superoxide production in plasma membrane vesicles from roots of wheat grown under copper deficiency or excess *J. Exp. Bot.* 52:77–84.

Rahoui, S., A. Chaoui, and E. El Ferjani. 2010. Reserve mobilization disorder in germinating seeds of Vicia faba L. exposed to cadmium. *J. Plant Nutr.* 33:809–17.

Rai, V., P. Vaypayee, S.N. Singh, and S. Mehrotra. 2004. Effect of chromium accumulation on photosynthetic pigments, oxidative stress defense system, nitrate reduction, proline level and eugenol content of *Ocimum tenuiflorum*. *Plant Sci.* 167:1159–64.

Rakwal, R., S. Tamogami, and O. Kodama. 1996. Role of jasmonic acid as a signaling molecule in copper chloride-elicited rice phytoalexin production. *Biosci. Biotech. Biochem.* 60:1046–8.

Ramos, J., M.R. Clemente, L. Naya, et al. 2007. Phytochelatin synthases of the model legume *Lotus japonicus*. A small multigene family with different responses to cadmium and alternatively spiced variants. *Plant Physiol.* 143:110–8.

Rao, K.V.M., and T.V. Sresty. 2000. Antioxidative parameters in the seedlings of pigeonpea (*Cajanus cajan* (L.) Millspaugh) in response to Zn and Ni stresses. *Plant Sci.* 157:113–28.

Rao, K.P., G. Vani, K. Kumar, et al. 2011. Arsenic stress activates MAP kinase in rice roots and leaves. *Arch. Biochem. Biophys.* 506:73–82.

Raskin, I., P.B.A.N. Kumar, S. Dushenkov, and D.E. Salt. 1994. Bioconcentration of heavy metals by plants. *Curr. Opin. Biotechnol.* 5:285–90.

Rasmusson, A.G., K.L. Soole, T.E. Elthon. 2007. Alternative NAD(P)H dehydrogenases of plant mitochondria. *Annu. Rev. Plant Biol.* 55:23–39.

Rauser, W.E. 1999. Structure and function of metal chelators produced by plants. The case for organic acids, amino acids, phytin and metallothioneins. *Cell Biochem. Biophys.* 31:19–48.

Rauser, W.E., and E.B. Dumbroff. 1981. Effects of excess cobalt, nickel and zinc on the water relations of *Phaseolus vulgaris. Environ. Exp. Bot.* 21:249–55.

Ravet, K., B. Touraine, J. Boucherez, J.-F. Briat, F. Gaymard, and F. Cellier. 2009. Ferritins control interaction between iron homeostasis and oxidative stress in Arabidopsis. *Plant J.* 57:400–12.

Rengel, Z. 2004. Heavy metals as essential nutrients. In: Prasad, M.N.V. (ed) *Heavy Metal Stress in Plants*, 2nd edn. Springer, Berlin, pp. 271–94.

Rentel, M.C., D. Lecourieux, F. Ouaked, et al. 2004. OXI1 kinase is necessary for oxidative burst-mediated signalling in *Arabidopsis. Nature* 427:858–61.

Revathi, S., and S. Venugopal. 2013. Physiological and biochemical mechanisms of heavy metal tolerance. *Int. J. Environ. Sci.* 3:1339–54.

Roberts, L.A., A.J. Pierson, and Z. Panavise. 2004. Yellow Stripe1 expanded roles for the Maize iron–phytosiderophore transporter. *Plant Physiol.* 135:112–20.

Robinson, N.J., A.M. Tommey, C. Kuske, and P.J. Jackson. 1993. Plant metallothioneins. *Biochem. J.* 295:1–10.

Rodríguez-Serrano, M., M.C. Romero-Puertas, D.M. Pazmiño, et al. 2009. Cellular response of pea plants to cadmium toxicity: Cross talk between reactive oxygen species, nitric oxide, and calcium. *Plant Physiol.* 150:229–43.

Rodríguez-Serrano, M., M.C. Romera-Puertas, A. Zabalza, et al. 2006. Cadmium effect on oxidative metabolism of pea (*Pisum sativum* L.) roots. Imaging of reactive oxygen species and nitric oxide accumulation in vivo. *Plant Cell Environ.* 29:1532–4.

Roelofs, D., M.G.M. Aarts, H. Schat, and N.M. van Straalen. 2008. Functional ecological genomics to demonstrate general and specific responses to abiotic stress. *Funct. Ecol.* 22:8–18.

Romanowska, E., A.U. Igamberdiev, E. Parys, and P. Gargestrom. 2002. Stimulation of respiration by Pb^{2+} in detached leaves and mitochondria of C3 and C4 plants. *Physiol. Plantarum* 116:148–54.

Romero-Puertas, M.C., F.J. Corpas, M. Rodriguez-Serrano, M. Gomez, L.A. del Río, and L.M. Sandalio. 2007. Differential expression and regulation of antioxidative enzymes by cadmium in pea plants. *J. Plant. Physiol.* 164:1346–57.

Ross, S.M. 1994. *Toxic Metals in Soil–Plant Systems.* Wiley, Chichester, p. 469.

Rudus, I., M. Sasiak, and J. Kêpczyñski. 2012. Regulation of ethylene biosynthesis at the level of 1-aminocyclopropane-1-carboxylateoxidase (ACO) gene. *Acta Physiol. Plantarum* 35:295–307.

Rufyikiri, G., S. Declerck, J.E. Dufey, and B. Delvaux. 2000. Arbuscular mycorrhizal fungi might alleviate aluminium toxicity in banana plants. *New Phytol.* 148(2):343–52.

Sagner, S., R. Kneer, G. Wanner, J.P. Cosson, B. Deus-Neumann, and M.H. Zenk. 1998. Hyperaccumulation, complexation and distribution of nickel in *Sebertia acuminata. Phytochemistry* 47:339–47.

Salt, D.E., N. Kato, U. Kramer, R.D. Smith, and I. Raskin. 2000. The role of root exudates in nickel hyperaccumulation and tolerance in accumulator and nonaccumulator species of *Thlaspi*. In: Terry, N. and G. Banuelos (eds) *Phytoremidiation of Contaminated Soil and Water.* CRC Press, Boca Raton, FL, pp. 189–200.

Salt, D.E., and W.E. Rauser. 1995. MgATP-dependent transport of phytochelatins across the tonoplast of oat roots. *Plant Physiol.* 107:1293–301.

Salt, D.E., R.D. Smith, and I. Raskin, 1998. Phytoremediation. *Annu. Rev. Plant Physiol. Plant Mol. Biol.* 49:643–68.

Salt, D.E., and G.J. Wagner. 1993. Cadmium transport across tonoplast of vesicles from oat roots. Evidence for a Cd^{2+}/H^{+} antiport activity. *J. Biol. Chem.* 268:12297–302.

Samuilov, S., B.D. Đunisijević, M. Đukić, and J. Raković. 2014. The effect of elevated Zn concentrations on seed germination and young seedling growth of *Ailanthus altissima* (Mill.) Swingle. *Bull. Faculty Forestry*. 110:145–58.

Sancenón, V., S. Puig, I. Mateu-Andrés, E. Dorcey, D.J. Thiele, and L.Peñarrubia. 2004. The Arabidopsis copper transporter COPT1 functions in root elongation and pollen development. *J. Biol. Chem.* 279:15348–55.

Sanitá di Toppi, L., and R. Gabbrielli. 1999. Response to cadmium in higher plants. *Environ. Exp. Bot.* 41:105–30.

Sanz, A., A. Llamas, and C.I. Ullrich. 2009. Distinctive phytotoxic effects of Cd and Ni on membrane functionality. *Plant Signal. Behav.* 4:980–2.

Savino, G., J.F. Briat, and S. Lobréaux. 1997. Inhibition of the iron-induced ZmFer1 maize ferritin gene expression by antioxidants and serine/threonine phosphatase inhibitors. *J. Biol. Chem.* 272:33319–26.

Sawaki, Y., T. Kihara-Doi, Y. Kobayashi, et al. 2013. Characterization of Al-responsive citrate excretion and citratetransporting MATEs in *Eucalyptus camaldulensis*. *Planta*. 237:979–89.

Schaaf, G., A. Honsbein, A.R. Meda, S. Kirchner, D. Wipf, and N. von Wirén. 2006. AtIREG2 encodes a tonoplast transport protein involved in iron-dependent nickel detoxification in *Arabidopsis thaliana* roots. *J. Biol. Chem.* 281:25532–40.

Schaaf, G., U. Ludewig, B.E. Erenoglu, S. Mori, T. Kitahara, and N. von Wiren. 2004. ZmYS1 functions as a proton-coupled symporter for phytosiderophore- and nicotianamine-chelated metals. *J. Biol. Chem.* 279:9091–6.

Schaller, A., and T. Diez. 1991. Plant specific aspects of heavy metal uptake and comparison with quality standards for food and forage crops. In: Sauerbeck, D. and S. Lubben (eds) *Der Einfluß von festen Abfallen auf Boden, Pflanzen*. KFA, Julich, Germany, pp. 92–125.

Schat, H., and M.M.A. Kalff. 1992. Are phytochelatins involved in differential metal tolerance or do they merely reflect metal-imposed strain? *Plant Physiol.* 99:1475–80.

Schat, H., M. Llugany, and R. Bernhard. 2000. Metal-specific patterns of tolerance, uptake and transport of heavy metals in hyperaccumulating and nonhyperaccumulating metallophytes. In: Terry, N. and G. Banuelos (eds) *Phytoremediation of Contaminated Soil and Water*. CRC Press, Boca Raton, FL, pp. 171–88.

Schellingen, K., D. Van Der Straeten, F. Vandenbussche, et al. 2014. Cadmium-induced ethylene production and responses in *Arabidopsis thaliana* rely on ACS2 and ACS6 gene expression. *BMC Plant Biol.* 14:214.

Scholz, G., Schlesier, G., and K. Seifert. 1985. Effect of nicotianamine on iron uptake by the tomato mutant 'chloronerva'. *Physiol. Plantarum* 63:99–104.

Schützendübel, A., and A. Polle. 2002. Plant response to abiotic stressess: Heavy metal-induced oxidative stress and protection by mycorrhization. *J. Exp. Bot.* 53:1351–65.

Sebastiani I., F. Scebba, and R. Tongttti. 2004. Heavy metal accumulation and growth responses in poplar clones Eridano (Populus deltoides × maximowiczii) and I-214 (P. × euramericana) exposed to industrial waste. *Environ. Exp. Bot.* 52:79–86.

Seigneurin-Berny, D., A. Gravot, P. Auroy, et al. 2006. HMA1, a new Cu-ATPase of the chloroplast envelope, is essential for growth under adverse light conditions. *J. Biol. Chem.* 281:2882–92.

Seregin, I.V., and Ivanov, V.B. 2001. Physiological aspects of cadmium and lead toxic effects on higher plants. *Fiziol. Rast. Moscow* 48:606–63.

Shameer, K., S. Ambika, S.M. Varghese, N. Karaba, M. Udayakumar, and R. Sowdhamini, 2009. STIFDB–*Arabidopsis* stress-responsive transcription factor Data Base. *Int. J. Plant Genomics* 2009:583429.

Shanker, A.K., R. Sudhagar, and G. Pathmanabhan. 2003. Growth, phytochelatin SH and antioxidative response of sunflower as affected by chromium speciation. In: Abstracts of the 2nd International Congress of Plant Physiology on Sustainable Plant Productivity Under Changing Environment, New Delhi, India.

Sharma, S.S., and K.J. Dietz. 2009. The relationship between metal toxicity and cellular redox imbalance. *Trends Plant Sci.* 14:43–50.

Shaul, O., D.W. Hilgemann, J.J. de-Almeida-Engler, M. Van Montagu, D. Inzé, and G. Galili. 1999. Cloning characterization of a novel Mg^{2+}/H^{+} exchanger. *EMBO J.* 18:3973–80.

Shen, Z., X. Li, C. Wang, H. Ghen, and H. Chua. 2002. Lead phyto extraction from contaminated soil with high-biomass plant species. *J. Environ. Qual.* 31:1893–900.

Sheoran, I.S., H.R. Singal, and R. Singh. 1990. Effect of cadmium and nickel on photosynthesis and the enzymes of the photosynthetic carbon reduction cycle in pigeonpea (*Cajanus cajan* L.). *Photosynth. Res.* 23:345–51.

Shiu, S.H., M.C. Shih, and W.H. Li. 2005. Transcription factor families have much higher expansion rates in plants than in animals. *Plant Physiol.* 139:18–26.
Shoji, T. 2014. ATP-binding cassette and multidrug and toxic compound extrusion transporters in plants: A common theme among diverse detoxification mechanisms. *Int. Rev. Cell Mol. Biol.* 309:303–46.
Shojima, S., N.K. Nishizawa, S. Fushiya, S. Nozoe, T. Irifune, and S. Mori. 1990. Biosynthesis of phytosiderophores: in vitro biosynthesis of 20-deoxymugineic acid from L-methionine and nicotianamine. *Plant Physiol.* 93:1497–503.
Siborova, M. 1988. Cd^{2+} ions affect the quaternary structure of ribulose-1,5-bisphosphate carboxylase from barley leaves. *Biochem. Physiol. Pflanzen* 183:371–8.
Siedlecka, A. 1995. Some aspects of interactions between heavy metal and plant mineral nutrient. *Acta Soc. Bot. Pol.* 64:265–72.
Siedlecka, A., and T. Baszynski. 1993. Inhibition of electron flow around photosystem I in chloroplasts of Cd-treated maize plants is due to Cd-induced iron deficiency. *Physiol. Plantarum* 87:199–202.
Silver, S. 1996. Bacterial resistance to toxic metal ions-a review. *Gene* 179:9–19.
Singh, D.P., and S.P. Singh. 1987. Action of heavy metals on Hill activity and O_2 evolution in *Anacystis nidulans. Plant Physiol.* 83:12–4.
Singh, K., Foley, R.C., and L. Oñate-Sánchez. 2002. Transcription factors in plant defense and stress responses. *Curr. Opin. Plant Biol.* 5:430–6.
Singla-Pareek, S.L., S.K. Yadav, A. Pareek, M.K. Reddy, and S.K. Sopory. 2006. Transgenic tobacco overexpressing glyoxalase pathway enzymes grow and set viable seeds in zinc-spiked soils. *Plant. Physiol.* 140:613–23.
Sirhindi, G., M.A. Mir. Abd-Allah, E.F., P. Ahmad, and S. Gucel. 2016. Jasmonic acid modulates the physio-biochemical attributes, antioxidant enzyme activity, and gene expression in glycine max under nickel toxicity. *Front. Plant Sci.* 7:591.
Skórzyńska-Polit, E., A. Tukendorf, E. Selstam, and T. Baszyński. 1998. Calcium modifies Cd effect on runner bean plants. *Environ. Exp. Bot.* 40:275–86.
Skottke, K.R., Yoon, G.M., Kieber, J.J., and A. De Long. 2011. Protein phosphatase 2A controls ethylene biosynthesis by differentially regulating the turn over of ACC synthase iso forms. *PLoS Genet.* 7:e1001370.
Soudek, P., S. Petrova, R. Vankova, J. Song, and T. Vanek. 2014. Accumulation of heavy metals using *Sorghum* sp. *Chemosphere* 104:15–24.
Souza, V.L. 2007. Expressão gênica, respostas morfo-fisioló- gicas e morte cellular induzidas por cádmio em *Genipa americana* L. (Rubiaceae). M.Sc. Dissertation, Universidade Estadual de Santa Cruz, Ilhéus.
Stadtman, E.R., and C.N. Oliver. 1991. Metal catalyzed oxidation of proteins. Physiological consequences. *J. Biol. Chem.* 266:2005–8.
Steffens, J.C. 1990. The heavy metal-binding peptides of plants. *Annu. Rev. Plant. Phys.* 41:553–75.
Stephan, U.W., I. Schmidke, V.W. Stephan, and G. Scholz. 1996. The nicotianamine molecule is made to measure for complexation of metal micronutrients in plants. *Biometals* 9:84–90.
Stephan, U.W., and G. Scholz. 1993. Nicotianamine: Mediator of transport of iron and heavy metals in the phloem? *Physiol. Plantarum* 88:522–9.
Sundaramoorthy P., A. Chidambaram, K.S. Ganesh., P. Unnikannan, and L. Baskaran 2010. Chromium stress in paddy: (i) nutrient status of paddy under chromium stress; (ii) phytoremediation of chromium by aquatic and terrestrial weeds. *C. R. Biol.* 333:597–607.
Suzuki, N. 2005. Alleviation by calcium of cadmium induced root growth inhibition in *Arabidopsis* seedlings. *Plant Biotechnol.* 22:19–25.
Taiz, L., and E. Zeiger. 2002. *Plant Physiology.* Sinauer Associates, Sunderland, MA, USA.
Taylor, K.C., L.G. Albrigo, and C.D. Chase. 1988. Zinc complexation in the phloem of blight-affected citrus. *J. Am. Soc. Hortic. Sci.* 113:407–11.
Thao, N.P., M.I.R. Khan, N. Binh, et al. 2015. Role of ethylene and its cross talk with other signaling molecules in plant responses to heavy metal stress. *Plant Cell Environ.* 169:73–84.
Thomas, F., C. Malick, E.C. Endreszl, and K.S. Davies. 1998. Distinct responses to copper stress in the halophyte, *Mesembryanthemum crystallium. Physiol. Plantarum* 102:360–8.
Thomine, S., F. Lelievre, E. Debarbieux, J.I. Schroeder, and H. Barbier-Brygoo. 2003. AtNRAMP3, a multispecific vacuolar metal transporter involved in plant responses to iron deficiency. *Plant J.* 34:685–95.
Thomine, S., R. Wang, J.M. Ward, N.M. Crawford, and J.I. Schroeder. 2000. Cadmium and iron transport by members of a plant metal transporters family in Arabidopsis with homology to Nramp genes. *Proc. Natl. Acad. Sci. USA* 97:4991–6.
Thumann, J., E. Grill, E.L. Winnacker, and M.H. Zenk. 1991. Reactivation of metal-requiring apoenzymes by phytochelatin-metal complexes. *FEBS Lett.* 284:66–9.

Tiffin, L.O., 1970. Translocation of iron citrate and phosphorus in xylem exudate of soybean. *Physiol.* 45:280–3.
Tomar, M., I. Kaur, A. Neelu, et al. 2000. Effect of enhanced lead in soil on growth and development of Vigna radiate (Linn.) Wilczek. *Indian J. Plant Physiol.* 5:13–8.
Tomaševic, M., S. Rajšic, D. Dorđević, M. Tasić, J. Krstićand, and V. Novaković. 2004. Heavy metals accumulation in tree leaves from urban areas. *Environ. Chem Lett.* 2:151–4.
Tomaševič, M., Z. Vukmirovič, S. Rajšič, M. Tasič, and B. Stevanovič. 2008. Contribution to biomonitoring of some trace metals by deciduous tree leaves in urban areas. *Environ. Monit. Assess.* 137:393–401.
Tommasini, R., E. Vogt, M. Fromenteau, et al. 1998. An ABC-transporter of Arabidopsis thaliana has both glutathione-conjugate and chlorophyll catabolite transport activity. *Plant J.* 13:773–80.
Tommey, A.M., J. Shi, W.P. Lindsay, P.E. Urwin, and N.J. Robinson. 1991. Expression of the pea gene PsMTA in *E. coli*. Metal binding properties of the expressed protein. *FEBS Lett.* 292:48–52.
Tran, L.S.P., R. Nishiyama, K. Yamaguchi-Shinozaki, and K. Shinozaki. 2010. Potential utilization of NAC transcription factors to enhance abiotic stress tolerance in plants by biotechnological approach. *GM Crops* 1:32–9.
Treeby, M., H. Marschner, and V. Römheld. 1989. Mobilization of iron and other micronutrient cations from a calcareous soil by plant-borne, microbial, and synthetic metal chelators. *Plant Soil* 114:217–26.
Trinh, N.N., T.L. Huang, W.C. Chi, S.F. Fu, C.C. Chen, and H.J. Huang. 2014. Chromium stress response effect on signal transduction and expression of signaling genes in rice. *Physiol. Plantarum* 150:205–24.
Tseng, T.S., S.S. Tzeng, C.H. Yeh, F.C. Chang, Y.M. Chen, and C.Y. Lin. 1993. The heat-shock response in rice seedlings: Isolation and expression of cDNAs that encode class-I low-molecular-weight heat-shock proteins. *Plant Cell Physiol.* 34:165–8.
Umezawa, T., Fujita, M., Fujita, Y., Yamaguchi-Shinozaki, K., and Shinozaki, K. 2006. Engineering drought tolerance in plants: Discovering and tailoring genes to unlock the future. *Curr. Opin. Biotechnol.* 17:113–22.
Urano, K., Y. Kurihara, M. Seki, and K. Shinozaki 2010. 'Omics' analyses of regulatory networks in plant abiotic stress responses. *Curr. Opin. Plant Biol.* 13:1–7.
Valliyodan, B., and H.T., Nguyen. 2006. Understanding regulatory networks and engineering for enhanced drought tolerance in plants. *Curr. Opin. Plant Biol.* 9:189–95.
Van Assche, F., and H. Clijsters. 1990. Effects of metals on enzyme activity in plants. *Plant Cell Environ.* 13:195–206.
Van Bentem, S.D., D. Anrather, and I. Dohnal, et al. 2008. Site-specific phosphorylation profiling of *Arabidopsis* proteins by mass spectrometry and peptide chip analysis. *J. Proteome Res.* 7:2458–70.
van der Zaal, B.J., Neuteboom, L.W., J.E. Pinas, et al. 1999. Overexpression of a novel Arabidopsis gene related to putative zinc-transporter genes from animals can lead to enhanced zinc resistance and accumulation. *Plant Physiol.* 119:1047–55.
van Steveninck, R.F.M., A. Babare, D.R. Fernando, and M.E. Steveninck. 1993. The binding of zinc in root cells of crop plants by phytic acid. *Plant Soil* 155–156:525–28.
Van Vliet, C., C.R. Anderson, and C.S. Cobbett. 1995. Copper-sensitive mutant of *Arabidopsis thaliana*. *Plant Physiol.* 109:871–8.
Vangronsveld, J., and H. Clijsters 1994. Toxic effects of metals. In: Farago, M.E. (ed) *Plant and the Chemical Elements: Biochemistry, Uptake, Tolerance and Toxicity.*VCH Verlagsgesellschaft, Weinheim, pp. 150–77.
Vansuyt, G., F. Lopez, D. Inzé, J.F. Briat, and P. Fourcroy 1997. Iron triggers a rapid induction of ascorbate peroxidase gene expression in *Brassica napus*. *FEBS Lett.* 410:195–200.
Varotto, C., D. Maiwald, P. Pesaresi, P. Jahns, F. Salaminian, and D. Leister. 2002. The metal ion transporter IRT1 is necessary for iron homeostasis and efficient photosynthesis in *Arabidopsis thaliana*. *Plant J.* 31:589–99.
Vasconcelos, M.W., G.W. Li, M.A. Lubkowitz, and M.A. Grusak. 2008. Characterization of the PT clade of oligopeptide transporters in rice. *Plant Genome* 1:77–88.
Vassilev, A., F. Lidon, P. Scotti, M. Da Graca, and I. Yordanov. 2004. Cadmium-induced changes in chloroplast lipids and photosystem activities in barleyplants. *Biol. Plant.* 48:153–6.
Vatamaniuk, O.K., Mari, S., Lu, Y.P., and P.A. Rea. 1999. AtPCS1, a phytochelatin synthase from Arabidopsis: Isolation and in vitro reconstitution. *Proc. Natl. Acad. Sci. USA* 96:7110–5.
Velikova, V., T. Tsonev, F. Loreto, and M. Centritto. 2011. Changes in photosynthesis, mesophyll conductance to CO_2, and isoprenoid emissions in *Populus nigra* plants exposed to excess nickel. *Environ. Pollut.* 159:1058–66.
Verkleji, J.A.S. 1993. The effects of heavy metals stress on higher plants and their use as biomonitors. In: Markert, B. (ed) *Plant as Bioindicators: Indicators of Heavy Metals in the Terrestrial Environment.* VCH, New York. 415–24.

Verret, F., A. Gravot, P. Auroy, et al. 2004. Overexpression of AtHMA4 enhances root-to-shoot translocation of zinc and cadmium and plant metal tolerance. *FEBS Lett.* 576:306–12.

Verret, F., A. Gravot, P. Auroy, et al. 2005. Heavy metal transport by AtHMA4 involves the N-terminal degenerated metal binding domain and the C-terminal His11 stretch. *FEBS Lett.* 579:1515–22.

Vert, G., N. Grotz, F. Dédaldéchamp, et al. 2002. IRT1, an *Arabidopsis* transporter essential for iron uptake from the soil and for plant growth. *Plant Cell* 14:1223–33.

Vierke, G., and Stuckmeier, P. 1977. Binding of Copper (II) to proteins of the photosynthetic membrane and its correlation with inhibition of electron transport in class II chloroplasts of spinach. *Z. Naturforsch.* 32:605–10.

Visioli, G., and N. Marmiroli. 2013. The proteomics of heavy metal hyperaccumulation by plants. *J. Proteomics* 79:133–45.

Vögeli-Lange, R., and G.J. Wagner. 1990. Subcellular localization of cadmium and cadmium-binding peptides in tobacco leaves. Implication of a transport functions for cadmium-binding peptides. *Plant Physiol.* 92:1086–93.

Wagner, G.J. 1993. Accumulation of cadmium in crop plants and its consequences to human health. *Adv. Agron.* 51:173–212.

Wainwright, S.J., and H.W. Woolhouse. 1977. Some physiological aspects of copper and zinc tolerance in *Agrostis tenuis* Sibth. cell elongation and membrane damage. *J. Exp. Bot.* 28:1029–36.

Wang, H.Y., M. Klatte, M. Jakoby, H. Bäumlein, B. Weisshaar, and P. Bauer. 2007. Iron deficiency-mediated stress regulation of four subgroup IbBHLH genes in *Arabidopsis thaliana*. *Planta* 226:897–908.

Wang, L.K., J.P. Chen, Y.T. Hung, and N.K. Shammas. 2009. Handbook on heavy metals in the environment. CRC Press, Boca Raton.

Waraich, E.A., R. Ahmad, and M.Y. Ashraf. 2011. Role of mineral nutrition in alleviation of drought stress in plants. *Aust. J. Crop Sci.* 5:764–77.

Wasternack, C. 2014. Action of jasmonates in plant stress responses and development-applied aspects. *Biotechnol. Adv.* 32:31–9.

Watt, M., and J.R. Evans. 1999. Update on root biology proteoid roots. Physiology and development. *Plant Physiol.* 121:317–23.

Weber, M., A. Trampczynska, and S. Clemens. 2006. Comparative transcriptome analysis of toxic metal responses in *Arabidopsis thaliana* and the Cd^{2+}-hypertolerant facultative metallophyte *Arabidopsis halleri*. *Plant Cell Environ.* 29:950–63.

Weis, J.S., and P. Weis. 2004. Metal uptake, transport and release by wetland plants: Implications for phytoremediation and restoration. *Environ. Int.* 30:685–700.

Wiersma, D., and B.J. Van Goor, 1979. Chemical forms of nickel and cobalt in phloem of Ricinus communis. *Physiol. Plantarum* 45:440–2.

Wiles, A.M., F. Naider, and J.M. Becker. 2006. Transmembrane domain prediction and consensus sequence identification of the oligopeptide transport family. *Res. Microbiol.* 4:395–406.

Wilkins, D.A. 1978. The measurement of tolerance to edaphic factors by means of root growth. *New Phytol.* 91:255–61.

Wise, R.R., and A.W. Naylor. 1988. Stress ethylene does not originate directly from lipid peroxidation during chilling-enhanced photooxidation. *J. Plant Physiol.* 133:62–6.

Wisniewski, L., and Dickinson, N.M. 2003. Toxicity of copper to *Quercus robur* (English Oak) seedlings from a copper-rich soil. *Environ. Exp. Bot.* 50:99–107.

Wójcik, M., and A. Tukiendorf. 2011. Glutathione in adaptation of *Arabidopsis thaliana* to cadmium stress. *Biol. Plantarum* 55:125–32.

Wollgiehn, R., and D. Neumann. 1999. Metal stress response and tolerance of cultured cells from *Silene vulgaris* and *Lycopersicon peruvianum*: Role of heat stress proteins. *J. Plant Physiol.* 154:547–5.

Wong, M.K., G.K. Chuan, L.L. Koh, K.P. Ang, and C.S. Hew. 1984. The uptake of cadmium by *Brassica chinensis* and its effect on plant zinc and iron distribution. *Environ. Exp. Bot.* 24:189–95.

Wray, G.A., M.W. Hahn, E. Abouheif, et al. 2003. The evolution of transcriptional regulation in eukaryotes. *Mol. Biol. Evol.* 20:1377–419.

Wu, H., Chen, C., Du, J., et al. 2012. Co-overexpression FIT with AtbHLH38 or AtbHLH39 in *Arabidopsis*-enhanced cadmium tolerance via increased cadmium sequestration in roots and improved iron homeostasis of shoots. *Plant Physiol.* 158:790–800.

Wu, S. 1994. Effect of manganese excess on the soybean plant cultivated under various growth conditions. *J. Plant. Nutri.* 17:993–1003.

Wu, X., R. Li, J. Shi, et al. 2014. *Brassica oleracea* MATE encodes a citrate transporter and enhances aluminum tolerance in *Arabidopsis thaliana*. *Plant Cell Physiol.* 55:1426–36.

Xiang, C., and D.J. Oliver. 1998. Glutathione metabolic genes coordinately respond to heavy metals and jasmonic acid in Arabidopsis. *Plant Cell* 10:1539–50.

Xiang, C., B.L. Werner, E.M. Christensen, and D.J. Oliver. 2001. The biological functions of glutathione revisited in Arabidopsis transgenic plants with altered glutathione levels. *Plant Physiol.* 126:564–74.

Yadav, S.K. 2010. Heavy metals toxicity in plants: an overview on the role of glutathione and phytochelatins in heavy metal stress tolerance of plants. *S. Afr. J. Bot.* 76:167–79.

Yamauchi, M., and X.X. Peng. 1995. Iron toxicity and stress-induced ethylene production in rice leaves. *Plant Soil* 173:21–8.

Yang, T., and B.W. Poovaiah. 2003. Calcium/calmodulin-mediated signal network in plants. *Trends Plant Sci.* 8:505–12.

Yang, X., Y. Feng, Z. He, and P.J. Stoffell. 2005a. Molecular mechanisms of heavy metal hyperaccumulation and phytoremediation. *J. Trace Elem. Med. Biol.* 18:339–53.

Yang, X.E., X.F. Jin, Y. Feng, and E. Islam. 2005b. Molecular mechanisms and genetic basis of heavy metal tolerance/hyperaccumulation in plants. *J. Integr. Plant. Biol.* 47:1025–35.

Yokosho, K., N. Yamaji, and J.F. Ma. 2011. An Al-inducible MATE gene is involved in external detoxification of Al in rice. *Plant J.* 68:1061–9.

Yokosho, K., N. Yamaji, D. Ueno, N. Mitani, and J.F. Ma. 2009. OsFRDL1 is a citrate transporter required for efficient translocation of iron in rice. *Plant Physiol.* 149:297–305.

Yruela, I., G. Gatzen, R. Picorel, and A.R. Holzwarth. 1996. Cu(II)-inhibitory effect on photosystem II from higher plants. A picosecond time-resolved fluorescence study. *Biochemistry* 35:9469–74.

Yuan, Y., H. Wu, N. Wang, et al. 2008. FIT interacts with AtbHLH38 and AtbHLH39 in regulating iron uptake gene expression for iron homeostasis in Arabidopsis. *Cell Res.* 18:385–97.

Zenk, M.H. 1996. Heavy metal detoxification in higher plants-a review. *Gene* 179:21–30.

Zhang, W.H., and S.D. Tyerman. 1999. Inhibition of water channels by $HgCl_2$ in intact wheat root cells. *Plant Physiol.* 120:849–57.

Zhou, G., E. Delhaize, M. Zhou, and P.R. Ryan. 2011. Biotechnological solutions for enhancing the aluminium resistance of crop plants. In: Shanker, A.K. and B. Venkateswarlu (eds) *Abiotic Stress in Plants—Mechanisms and Adaptations*. InTech, Brisbane, pp. 119–42.

Zhou, G., E. Delhaize, M. Zhou, and P.R. Ryan. 2013. The barley MATE gene, HvAACT1 increases citrate efflux and $Al^{(3+)}$ tolerance when expressed in wheat and barley. *Ann. Bot.* 112:603–12.

Zhou, J., and P.B. Goldsbrough. 1994. Functional homologs of fungal metallothionein genes from Arabidopsis. *Plant Cell* 6:875–84.

Zhou, Z.S., S.Q. Huang, K. Guo, S.K. Mehta, P.C. Zhang, and Z.M. Yang. 2007. Metabolic adaptations to mercury-induced oxidative stress in roots of *Medicago sativa* L. *J. Inorg. Biochem.* 101:1–9.

Zhu, Y.L., E.A.H. Pilon-Smits, A.S. Tarun., S.U. Weber, L. Jouanin, and N. Terry 1999. Cadmium tolerance and accumulation in Indian mustard is enhanced by overexpressing gamma-glutamylcysteine synthetase. *Plant Physiol.* 121:1169–77.

Zimeri, A.M., O.P. Dhankher, B. McCaig, and R.B. Meagher. 2005. The plant MT1 metallothioneins are stabilized by binding cadmium and are required for cadmium tolerance and accumulation. *Plant Mol. Biol.* 58:839–55.

11 Plant Response to Multiple Stresses

Da-Gang Hu and Yu-Jin Hao

CONTENTS

11.1 INTRODUCTION

Woody plants, unlike herbaceous annual plants, are more often exposed to various biotic and abiotic stresses due to their perennial characteristics. As sessile organisms, woody plants have evolved a series of intricate mechanisms that allow them to perceive external signals and respond to complicated stress conditions, resulting in minimum damage while conserving valuable resources for growth, development and reproduction. Phytohormones such as abscisic acid (ABA), ethylene (ET), jasmonic acid (JA), and salicylic acid (SA) primarily regulate the protective responses of plants against both biotic and abiotic stresses via synergistic and antagonistic actions that are referred to as signaling crosstalk (Fujita et al., 2006). Research has shown that reactive oxygen species (ROS), nitric oxide (NO) and sugars such as glucose are key signal molecules in responses to multiple stresses (Sheen et al., 1999; Neill et al., 2002). There is increasing evidence that unique combinations of multiple stresses often trigger interactive signaling pathways from various phytohormones and other signal molecules or antioxidants, resulting in distinct transcriptional activation patterns during biotic and abiotic stresses (Shoji and Hashimoto, 2015). Therefore, present techniques of individually imposing each stress in a test to develop stress-tolerant plants may be inadequate.

Large-scale transcriptome analyses strongly support that biotic and abiotic stresses often regulate the expression of multiple overlapping genes in many signaling networks. For instance, the analysis of the correlation between the transcriptional regulation by environmental challenges (e.g. jasmonate application) and by incompatible pathogen infection revealed the expression of considerable overlapping genes in response (Luu et al., 2015). These data suggested that crosstalk between these signaling networks was involved in triggering downstream stress responses. Thus, there is increasing evidence that plant signaling pathways consist of elaborate networks with frequent crosstalk, and, consequently, allow plants to improve both abiotic stress tolerance and disease resistance. In this review, functional genes that are involved in crosstalk and molecular convergence points between biotic and abiotic stress signaling pathways are discussed.

11.2 PHYSIOLOGICAL RESPONSE TO MULTIPLE STRESSES

Abiotic and biotic stresses, such as salinity, drought, oxidative stress, wounding and pathogen attack, are serious threats to plant growth development, causing reduced yield and quality in many plants including woody species. Although intensive research has been done in the field of plant response to stresses, very limited research has focused on plant responses to individual biotic and abiotic factors (Hu et al., 2016a). Due to their perennial nature, woody plants often encounter combinations of diverse stresses arising from various sources (Mahalingam, 2015). The physiological performance of woody plants largely depends on environmental factors including light, temperature, water and air conditions. The effects of these environmental factors on plant functions may often be interactive rather than additive (Sun et al., 2016); therefore, studies of physiological responses of woody plants to multifactorial stresses are very important.

Abiotic and biotic stresses lead to a series of physiological, morphological, biochemical and molecular changes in plants. Such changes significantly affect plant growth and development and productivity (Trivedi et al., 2016; Figure 11.1). In general, salinity, drought and oxidative stresses are often interconnected and constantly induce similar cellular damage. Drought and salt stresses primarily cause osmotic stress, leading to disruption of ion homeostasis and membrane damage in plant cells (Carvalho et al., 2015; Hu et al., 2016a). Oxidative stress that often accompanies extreme temperature, drought or salinity stress can result in denaturation of structural and functional proteins

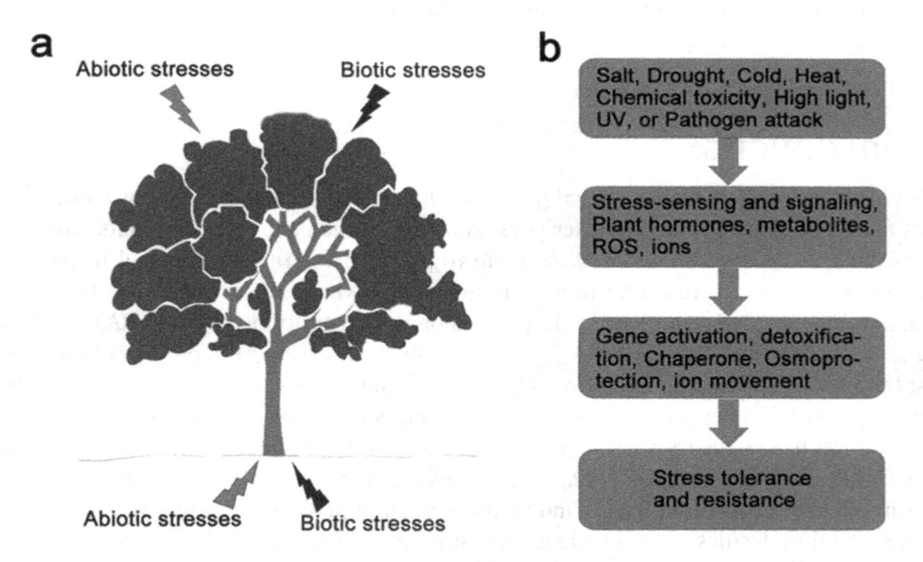

FIGURE 11.1 Schematic illustration of woody plant growth under the environmental stress condition. (a) The root, shoot, leaf, flower and fruit of woody species were continuously subjected to various abiotic and biotic stresses during growth and development. (b) Physiological and molecular responses to different stresses in woody plants. Stresses including drought, salinity, cold, heat, chemical toxicity, high light intensity, UV or pathogen attack cause changes in metabolic reactions, production of reactive oxygen species (ROS) and ion homeostasis, and plant hormone signaling and downstream signaling pathways are then activated by the stress. Expression of the stress responsive genes functions in cellular homeostasis and protects proteins and membrane structures from further damage.

(Wang et al., 2003). To minimize the damage caused by these stresses, woody plants have evolved a series of physiological mechanisms, such as production of stress proteins and accumulation of antioxidants and soluble metabolites (Racchi, 2013; Figure 11.1).

11.2.1 Stress Proteins

Heat shock proteins (HSPs) and heat shock transcription factors (Hsfs), which are proposed to act as ATP-independent molecular chaperones, are ubiquitous proteins that protect cells from the damage of proteotoxic stress. A large amount of studies indicated that the expression of HSPs and Hsfs were up-regulated by heat or other stress stimuli. For example, in peach fruits, the expression level of the *HSP20* and *HSP70* genes was significantly higher after wounding and heat treatment (Tosetti et al., 2014; Spadoni et al., 2014). A recent report with respect to the gene expression in apple showed that the *HSP101* gene was up-regulated in response to *P. expansum*, and the *HSP70* gene was induced by both *P. expansum* and *P. digitatum* (Vilanova et al., 2014). In a wild Chinese grapevine, 10 *VpHsfs* genes were significantly induced by heat stress (Hu et al., 2016d). A genome-wide identification and comparative analysis of the *Hsfs* gene family in Chinese white pear (*Pyrus bretschneideri*) showed that six *PbHsf* genes were up-regulated in the fruit under naturally increased temperature (Qiao et al., 2015). These results suggest that HSPs and Hsfs play a crucial role in abiotic and biotic stress-related responses in woody plants.

Calcium-dependent protein kinases (CDPKs) and calcinuerin B-like (CBL)-interacting protein kinases (CIPKs) were identified as a large class of calcium-mediated family proteins that regulate physiological processes and stress responses in woody plants (Wang et al., 2012; Das et al., 2013; Hu et al., 2016a). Such proteins generally phosphorylate the target proteins such as channel/transporters or transcription factors (TFs), and finally lead to activating the response to diverse stresses. Unlike CDPKs, CIPKs cannot directly bind calcium and are required to interact with CBL proteins to mediate calcium-signaling pathways (Das et al., 2013). Up to this point, several CDPK and CIPK proteins have been discovered. They regulate multiple signaling pathways involved in stress responses in woody plants. For instance, molecular cloning and functional characterization of apple *MdCIPK6L* and *MdSOS2L1* revealed its involvement in multiple abiotic stress tolerances in transgenic apple plants (Wang et al., 2012; Hu et al., 2016a). A recent gene expression analysis of the CDPK family showed that expression of the VaCPK20, 21, 26 and 29 genes was significantly up-regulated by various stresses (Dubrovina et al., 2013, 2015, 2016).

In addition to the aforementioned stress proteins, other proteins, such as MYB, bHLH and NAC family TFs, ABA, SA, JA-related proteins, glucose-regulated proteins (GRPs) and lipid transfer proteins (LTPs) were all involved in various stress responses in woody plant species (Feng et al., 2012; Cao et al., 2013; Jaeckels et al., 2013; Su et al., 2013; Shin et al., 2014; Xu et al., 2015b).

11.2.2 Antioxidants

Low molecular weight antioxidants, such as proanthocyanidins (PAs), glutathione, ascorbrate (vitamin C) and tocopherol, are information-rich redox buffers that are involved in various physiological responses and stress perception. In addition to crucial roles in defense and as enzyme cofactors, cellular antioxidants also influence other biological processes, such as plant growth and development (Tokunaga et al., 2005). The differences in biochemical properties and subcellular localization of the antioxidants and their distinct responses in expression of stress-associated genes resulted in a versatile and agile antioxidative system to maximize the control of the ROS level in plants (Carvalho et al., 2015).

The level of ROS accumulation is largely determined by the antioxidative system. These antioxidants enable woody plants to maintain proteins and other cellular components in an active state for a variety of metabolic processes. Like most of the other aerobic organisms, woody plants maintain most cytoplasmic thiols in a reduced (-SH) state because of the low thioldisulfide redox potential imposed by the glutathione redox buffer. Plant cells in general synthesize a high concentration of ascorbate that provides robust protection against oxidative damage. Redox homeostasis in plants is controlled by a large pool of antioxidants that absorb and buffer oxidants and reductants. Plants also produce proanthocyanidins (PAs) that act as important hydrosoluble and liposoluble redox buffers. Proanthocyanidins are considered to be a major ROS scavenger that is involved in regulation of ABA signaling (Luo et al., 2016; Hu et al., 2016a). Another effective scavenger of ROS is tocopherol. It increases the range of effective superoxide scavenging because the tocopherol redox couple has a more positive midpoint potential than that of the ascorbate pool. The redox buffer function of ascorbate, glutathione, proanthocyanidins and tocopherol pools is one of their most important attributes in responses to stresses in woody plants.

11.2.3 Soluble Metabolites

Woody plants are a nearly unlimited source of phytochemicals. They produce various soluble secondary metabolites that are useful for interaction with the environment, including both abiotic and biotic stresses (Chen et al., 2009). These soluble metabolites contain certain functional molecules, such as soluble sugars, organic acids, hormones and polyamines. As nutrient and metabolite signaling molecules, soluble metabolites in plants are very sensitive to environmental stress. A wide range of external stimuli can trigger a cascade of reactions in plant cells, ultimately leading to formation and accumulation of soluble metabolites that help plants survive various stresses (Chen et al., 2009). This review is oriented towards the effects of soluble sugars, hormones and organic acids on different stress responses in woody plants.

11.2.3.1 Soluble Sugars and Hormones

Soluble sugars, especially sucrose, glucose and fructose, play a distinctly central role in maintaining plant structure and metabolism at the cellular and/or whole plant level. They act as nutrients and metabolite signaling molecules that activate specific or hormone crosstalk transduction pathways, modifying gene expression and proteomic patterns (Smeekens and Hellmann, 2014). In plants, many metabolic reactions and regulations are directly associated with soluble sugars through influencing the production rate of reactive oxygen species (ROS), such as mitochondrial respiration or photosynthesis regulation as well as the anti-oxidative processes, such as the oxidative pentose-phosphate pathway and carotenoid biosynthesis (Couée et al., 2006). Soluble sugars are involved in plant response to many stresses, such as chilling, herbicide injury or pathogen attack, through maintaining the balance of reactive oxygen species. Current transcriptome analyses confirmed the converging or antagonistic relationship between soluble sugar, ROS production and the anti-oxidant process, suggesting that sugar signaling and sugar-modulated gene expression is related to the control of oxidative stress (Bonnefont-Rousselot, 2014). The aforementioned relations place soluble carbohydrates in a crucial role in balancing pro-oxidant and anti-oxidant when plants face the stress challenge.

Soluble sugars are highly sensitive to environmental stresses, directing the supply of carbohydrates from source organs to sink ones (Cho et al., 2010; Yue et al., 2015). Sucrose and hexoses have dual functions in gene regulation as exemplified by the up-regulation of growth-related genes and down-regulation of stress-related genes. Although coordinated and regulated by sugars, both growth- and stress-related genes are up-regulated or down-regulated through Hexokinase (HXK)-dependent and/or HXK-independent pathways (Cho et al., 2010; Yue et al., 2015). HXK is an evolutionarily conserved glucose sensor in many organisms. In plants, HXK has dual functions in glucose metabolism as well as glucose sensing and signaling that modulates many physiological processes by integrating nutrient, light and hormone signals (Cho et al., 2010). HXK1-mediated glucose signaling shows crosstalk with auxin, ABA, ethylene, cytokinin and brassinosteroid signaling and is response to various abiotic and biotic stresses (Figure 11.2; Rolland et al., 2002; Cho et al., 2010; Smeekens and Hellmann, 2014).

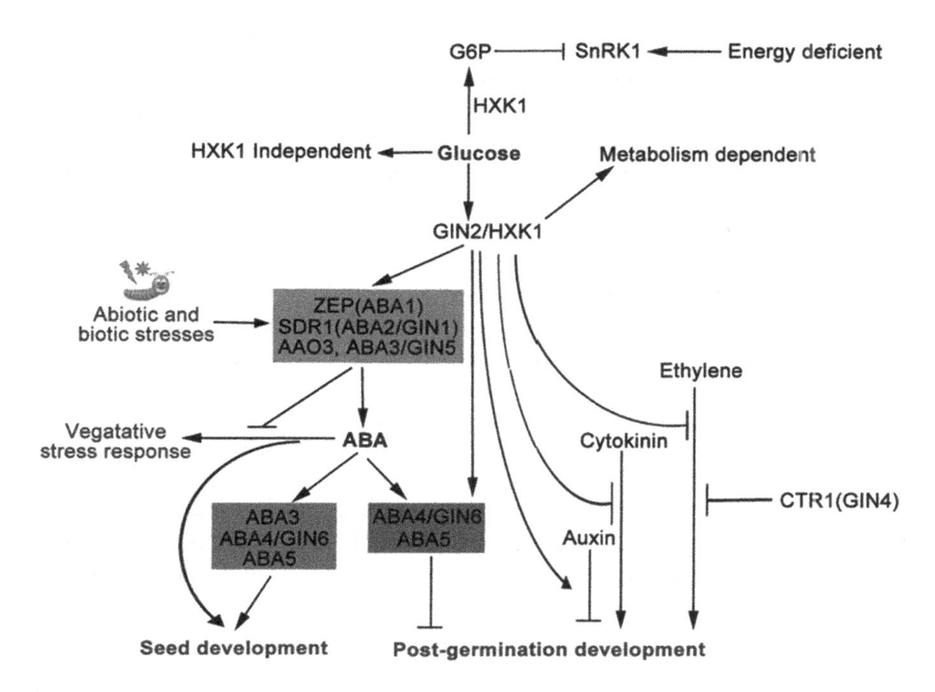

FIGURE 11.2 The hypothetic crosstalk between glucose and hormone signaling in response to various stresses in woody plants.

The sucrose-non-fermenting-1-(SNF1-) related protein pathway, analogue to the protein kinase (SNF-) yeast-signalling pathway, is also involved in sugar sensing and stress-related signal transduction in plants (Figure 11.2; Yang et al., 2015; Yue et al., 2015).

11.2.3.2 Organic Acids

Organic acids have been hypothesized to have functions including root nutrient acquisition, microbial chemotaxis, metal detoxification and numerous life processes in woody plants (Sweetman et al., 2009; Centeno et al., 2011). Organic acids are also involved in energy supply, redox, pH homeostasis and solute potential in plant cells, and promote nutrient absorption, ameliorate aluminum toxicity and help establish microbial communities in rhizospheric soil (Fernie and Martinoia, 2009; Finkemeier and Sweetlove, 2009; Hu et al., 2016c). Malic and citric acids are two main organic acids in most ripe fruits (Seymour et al., 1993). Previous studies suggested that malic and citric acids are responsible for fruit sourness because they release protons to vacuoles and affect the level of sweetness (Bugaud et al., 2011). Studies also showed the close relationship between environmental factors and organic acids. For example, not only cultural practices such as mineral fertilization, irrigation and thinning, but environmental factors such as temperature can modify fruit acidity (Gautier et al., 2005; Burdon et al., 2007; Ramesh Kumar and Kumar, 2007; Thakur and Singh, 2012), indicating the importance of organic acids in plant biology and plant stress responses.

11.3 MOLECULAR RESPONSE TO MULTIPLE STRESSES

In general, multiple stresses can initially reduce source photosynthetic activity by both stomatal and non-stomatal factors in woody plants (Munns and Tester, 2008; Pérez-Alfocea et al., 2010). Regardless of the ionic component of salinity, both stresses can provoke premature senescence due to assimilation and accumulation in source leaves as a result of decreased consumption in sink tissues. This assimilation and accumulation provokes impaired consumption of NADPH by the Calvin Cycle, leading to the so called (feedback) photo-inhibition and photo-oxidation processes, with the subsequent transfer of photosynthetic electrons from over-reduced ferredoxin to oxygen (Mehler Reaction) producing toxic reactive oxygen species that damage cell structures (Keunen et al., 2013; Albacete et al., 2014). There are strong indications from numerous studies that reallocation of assimilates between source and sink tissues also allows the plant to adapt to biotic and abiotic stresses (Keunen et al., 2013; Albacete et al., 2014). Additionally, phytohormone balance can affect plant performance by influencing source-sink relations and metabolism, and thus, stress tolerance and crop yield (Albacete et al., 2014).

The strategies woody plants have adopted to respond to multiple stresses are often coordinated and fine-tuned by regulating molecular activities. Remarkable progress has been made in understanding such molecular mechanisms. In general, responses of plants to stress factors are accompanied by major changes in transcriptome and proteome with both post-transcriptional and post-translational modification in woody plants (Ahuja et al., 2010; Hirayama and Shinozaki, 2010). Recent studies have made efficient use of transcriptome and proteome approaches in identification of transcriptional and proteomic networks associated with stress perception and response in woody species (Gupta et al., 2005; Hu et al., 2016a). A wide range of genes, proteins and enzymes that confer resistance or are regulated in response to environmental stress stimuli have been summarized in Table 11.1. Their descriptions and known or putative functional mechanisms in woody plants are also provided.

11.3.1 Salinity

The widespread increase of soil salinity is a serious environmental problem affecting the growth, productivity and quality of woody plants. During salt stress, the excessive accumulation of Na ions in plant cells induces ionic and osmotic stress. Salinity accumulation changes plasma membrane lipid and protein composition of plant cells, causing ion imbalance and hyperosmotic stress and eventually disturbs normal growth and development of plants (Hu et al., 2012a). The widely studied signaling pathways of plant molecular adaptations to salt stress include salt overly sensitive (SOS) pathway, salt stress-induced production of ROS, and salt-inducible TFs such as MYB, DREB and WRKY (Cao et al., 2013; Shi et al., 2014a,b; Wang et al., 2014b; Yang et al., 2011; Zhou et al., 2008). Recent significant advances are listed in Table 11.1 and Figure 11.3.

In the past two decades, a series of studies demonstrated that woody plants have a functionally conserved SOS pathway in response to salinity stress (Tang et al., 2010; Hu et al., 2016a; Figure 11.3a). Salt stress elicits a rapid increase of the free calcium concentration in cytoplasm, subsequently activates the myristoylated Ca binding protein SOS3. The activated Ca sensor SOS3 physically interacts with SOS2, phosphorylates and activates the Na^+/H^+ exchanger activity of SOS1 at the plasma membrane, resulting in exclusion of excess Na^+ outside the plant cell or sequestering Na^+ into the vacuole and thereby enhancing the salt detoxification of woody plants (Tang et al., 2010; Hu et al., 2016a). Furthermore, SOS2 also modulates salt tolerance by affecting production of ROS (Hu et al., 2016a).

High salinity stress significantly modified the lipid composition of the plasma membrane. Such modifications could influence membrane stability or the activity of plasma membrane proteins such as V-ATPase, HKT and NHX, which provides a mechanism in controlling water permeability and acclimation (Ye et al., 2009; Hu et al., 2016c; Li et al., 2016; Figure 11.3a). Interestingly, Hu et al. (2016c) found that MdSOS2L1 phosphorylated MdVHA-B1 to encode a vacuolar H^+-ATPase subunit at Ser^{396} to modulate malate accumulation in response to salt stress in apple (Hu et al., 2016c). Overexpression of *MdSOS2L1* increased the antioxidant metabolites to enhance salt tolerance in apple and tomato (Hu et al., 2016a; Figure 11.3b).

Recent advance in identification of salt-inducible TFs widens our knowledge of cellular mechanisms in plant adaptation. Genome-wide analysis of the apple MYB TF family led to the

TABLE 11.1
A List of Genes Associated with Different Stresses Cited in this Chapter

Gene Name	Species	Stress Response	References
MdbHLH104	Apple	Iron deficiency	Zhao et al. (2016)
MdCIPK6L	Apple	Multiple abiotic stresses	Wang et al. (2012)
MdCIbHLH	Apple	Cold	Feng et al. (2012)
MdDREB1	Apple	Cold, drought, salt	Yang et al. (2011)
MdMDH	Apple	Cold, salt	Yao et al. (2011)
MdMYB10	Apple	Osmotic stress	Gao et al. (2011b)
MdoMYB121	Apple	Multiple abiotic stresses	Cao et al. (2013)
MdSIMYB1	Apple	Multiple abiotic stresses	Wang et al. (2014b)
MdSOS2L1	Apple	Salt	Hu et al. (2016c)
MdVHA-A	Apple	Drought	Dong et al. (2013)
MdVHA-B1	Apple	Drought, salt	Hu et al. (2012b, 2015)
MdVHP1	Apple	Multiple abiotic stresses	Dong et al. (2011)
MhNCED3	*Malus hupehensis*	Osmotic, Cd^{2+} stress	Zhang et al. (2014b)
MhNPR1	*Malus hupehensis*	Osmotic, salt	Zhang et al. (2014a)
CpP6CS	*Carica papaya*	Temperature	Zhu et al. (2012a)
CpCHT1	*Chimonanthus praecox*	Freezing, Fungus	Zhang et al. (2011)
CpFATB	*Chimonanthus praecox*	Drought	Zhang et al. (2012)
Pbzs-REMORIN	Pear	Fungus	Fu et al. (2012)
PbGST	Pear	Multiple abiotic stresses	Liu et al. (2013)
PbP5CS	Pear	Salt	Silva-Ortega et al. (2008)
PbPGIP	Pear	Fungus	Agueero et al. (2005)
PbSPDS1	Pear	Aluminum	Wen et al. (2009)
PpAKR1	Peach	Multiple abiotic stresses	Kanayama et al. (2014)
PpCBF1	Peach	Cold	Wisniewski et al. (2011)
PpCHLP	Peach	Wounding and fungus	Giannino et al. (2004)
PeCBL6/10	*Populus euphratica*	Multiple abiotic stresses	Li et al. (2012)
PeDHN	*Populus euphratica*	Drought	Wang et al. (2011b)
PeDREB2L	*Populus euphratica*	Multiple abiotic stresses	Chen et al. (2011)
PtaZFP2	Poplar	Multiple abiotic stresses	Martin et al. (2009)
PtCesA	Poplar	Mechanical stress	Wu et al. (2000)
PtGSNOR	Poplar	Chilling	Cheng et al. (2015)
PtADC	*Poncirus trifoliata*	Multiple abiotic stresses	Wang et al. (2011a)
PsHSP70	Tree peony	Heat	Li et al. (2011)
RceIF5A	*Rosa chinensis*	Thermo, oxidative and osmotic	Xu et al. (2011)
FaCBF1	Strawberry	Freezing	Jin et al. (2007)
FaCYF1	Strawberry	Fungus	Martinez et al. (2005)
MusaDHN-1	Banana	Drought and salt	Shekhawat et al. (2011)
VvCBL-1	Grape	Drought	Cuellar et al. (2010)
VvCIPK23	Grape	Drought	Cuellar et al. (2010)
VvWRKY1	Grape	Fungus	Marchive et al. (2007)

identification of the *MdoMYB121* gene conferring abiotic stress tolerance, including salt tolerance in plants (Cao et al., 2013). Characterization of a R2R3 MYB gene (*MdSIMYB1*) showed its role in apple salt tolerance (Wang et al., 2014b). The dramatic abundance of DREB protein in apple in response to salt stress also highlighted its significant role in salt tolerance (Yang et al., 2011). The enhancement of tolerance to salt stress by modifying the redox state and salicylic acid content via the cytosolic malate dehydrogenase gene in transgenic apple plants was evident (Wang et al., 2016;

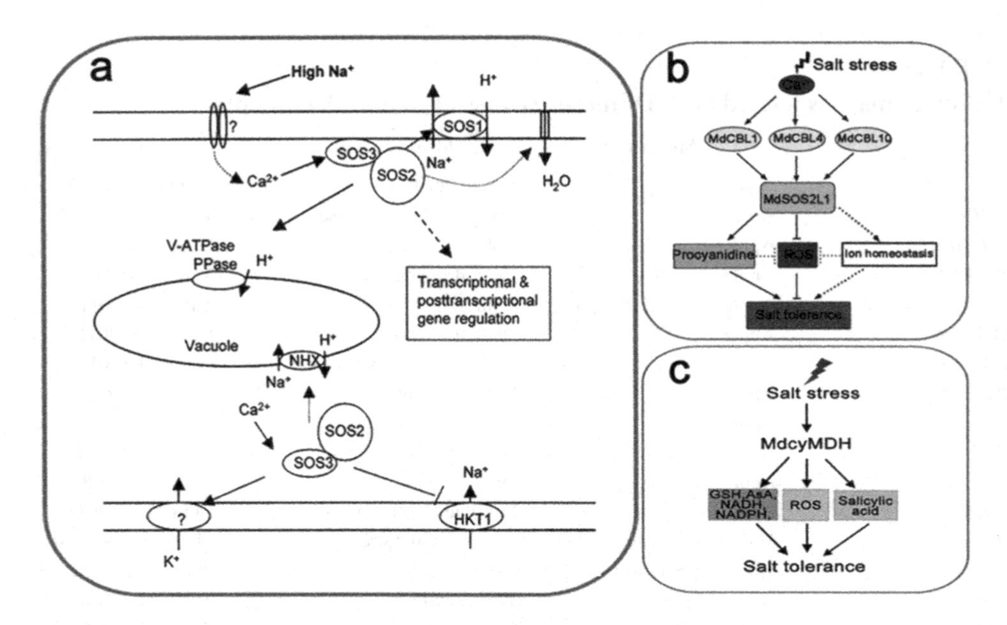

FIGURE 11.3 Diagrammatic representation of plant adaptations to salinity stress factors. (a) Regulation of ion (e.g. Na^+ and K^+) homeostasis by the SOS pathway. High Na^+ stress initiates a Ca signal that activates the SOS3-SOS2 protein kinase complex, which then stimulates the Na^+/H^+ exchange activity of SOS1 and regulates the expression of some genes transcriptionally and post-transcriptionally. SOS3-SOS2 may also stimulate or suppress the activities of other transporters involved in ion homeostasis under salt stress, such as vacuolar H^+-ATPases and pyrophosphatases (PPase), vacuolar Na^+/H^+ exchanger (NHX), and plasma membrane K^+ and Na^+ transporters (Zhu, 2002). (b) The mechanism of MdSOS2L1 improves the antioxidant metabolites via a CBL-dependent signal pathway to enhance salt tolerance in apple. (c) The enhancement of tolerance to salt stress by modifying the redox state and salicylic acid content via the cytosolic malate dehydrogenase gene in transgenic apple plants.

Figure 11.3c). Attempts to enhance salinity tolerance are currently being applied in cell type-specific manipulation of transport processes in commercially relevant woody plants.

11.3.2 Drought

Continuous water deficit or drought stress is one of the most important factors affecting plant growth, development, survival and crop productivity (Todaka et al., 2015). Woody plants are generally able to make physiological responses to drought, for instance, decrease photosynthetic activity, alter cell wall elasticity, close stomata or even generate toxic metabolites causing plant death. Acclimation of woody plants to drought stress is controlled by molecular cascades and is involved in cross-talk with other stresses and plant hormone signaling mechanisms. Here, we discussed key molecular programs proposed to confer tolerance to drought stress along with their envisaged modes of action (Figure 11.4; Table 11.1). Specific focus was given to ABA dependent, ABA-independent, and DREBs-related mechanisms (Figure 11.4).

ABA is a signal intermediate that influences the expression of many genes. ABA plays an important role in response to drought stress by regulating stomatal closure (Figure 11.4). Many abiotic stress-inducible genes are activated or inhibited by ABA, while other genes do not respond to ABA, suggesting the involvement of both ABA-dependent and independent regulatory systems (Yang et al., 2011). Several *cis*-elements including the dehydration responsive element/C-repeat (DRE/CRT) and the ABA-responsive element (ABRE) have been shown to be crucial to expression of ABA-dependent and ABA-independent genes in plants responding to drought stress (Aquino, 2014).

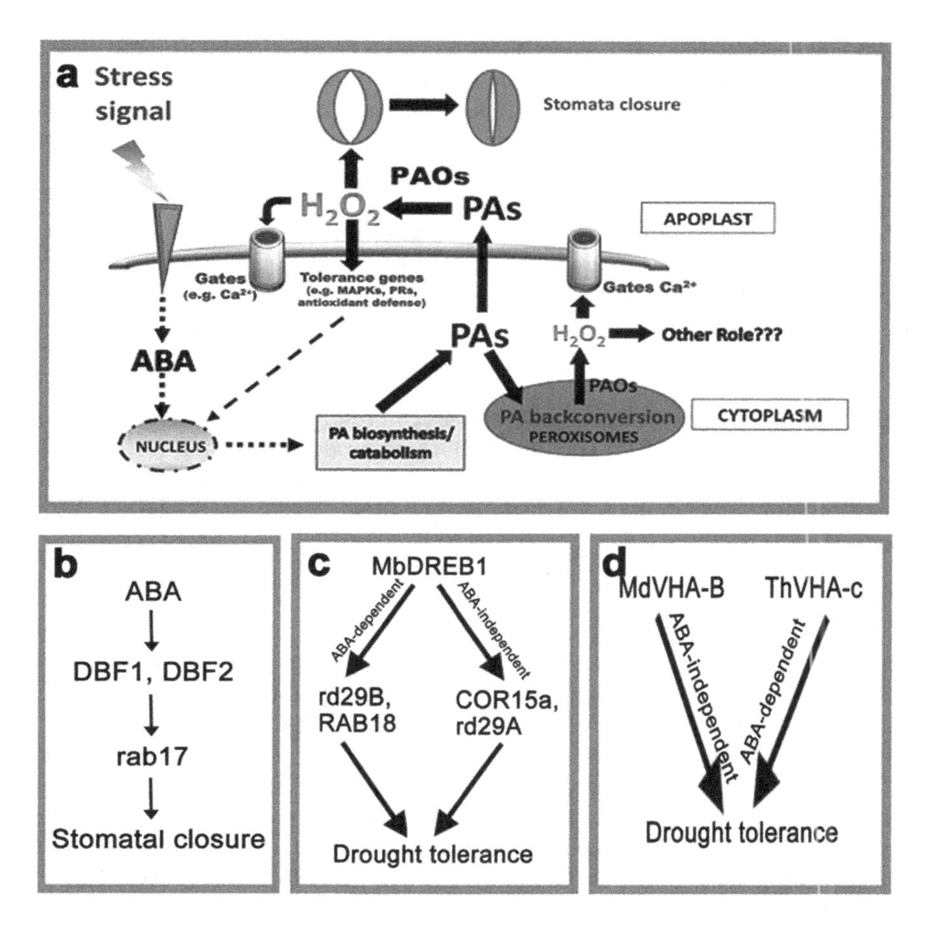

FIGURE 11.4 Diagrammatic representation of plant adaptations to drought stress. (a) A proposed model for PAO-derived H_2O_2 participation in stress and development (Konstantinos et al., 2010). During development, intrinsic clues induce PA catabolism and generation of H_2O_2. This molecule seems to act as signal to activate Ca^{2+} channels, and stomata closure (herein). Higher Spd titers or increased PAO activity results in high levels of H_2O_2 that activate Ca^{2+} channels, but the Ca^{2+} influx is either not restricted or the influx is exceedingly high leading to deleterious effects. Under stress, Paexodus into the apoplast is signaled where it is oxidized by PAO. The generated H_2O_2 induces either expression of tolerance effector genes or PCD syndrome. The result depends both on the size of the H_2O_2 and on the restoration of intracellular PA homeostasis, which is brought about by the simultaneous induction of the PA biosynthetic genes. All of these factors reinforce the view that only optimal oxidation of Spd by PAO can lead to normal molecular responses. Moreover, in the case of the control of cation channels the back-conversion pathway seems to have a significant role, but a role for this pathway in other responses, like effector-genes induction remains to be established. (b) The mechanism of maize DRE-binding proteins DBF1 and DBF2 are involved in rab17 regulation through the drought-responsive element in an ABA-dependent pathway. (c) *MbDREB1* functions as a transcription factor and increases plant tolerance to drought stress via both ABA-dependent and ABA-independent pathways. (d) The vacuolar H^+-ATPase subunit genes *ThVHAc1* and *MdVHA-B* confer tolerance to drought stress via ABA-dependent and ABA-independent signaling pathway.

In grape, ABA-dependent amine oxidases-derived H_2O_2 affected regulated stomatal conductance (Konstantinos et al., 2010; Figure 11.4a). In maize, DRE-binding proteins DBF1 and DBF2 that are AP2/EREBP TF family members were involved in *rab17* regulation through the drought-responsive element in an ABA-dependent pathway (Kizis and Pages, 2002; Figure 11.4b). Yang et al. (2011) reported that *MbDREB1* functioned as a TF and increased plant tolerance to drought stress via both ABA-dependent and ABA-independent pathways in dwarf apple (Figure 11.4c). A novel vacuolar membrane H^+-ATPase c subunit gene (*ThVHAc1*) from *Tamarix hispida* conferred tolerance to several abiotic stresses via an ABA-dependent stress signaling pathway in *Saccharomyces cerevisiae*, while ectopic expression of an apple vacuolar membrane H^+-ATPase B subunit gene *MdVHA-B* conferred tolerance to drought by the ABA-independent signaling pathway in transgenic tomato (Gao et al., 2011a; Hu et al., 2012b; Figure 11.4d). Other studies demonstrating the complexity of transcription, post-transcription and post-translational modifications and different action modes in the stomatal guard cell response added new information on the role of ABA in mediating drought responses (Zhang et al., 2006).

11.4 INTERACTIONS BETWEEN BIOTIC AND ABIOTIC STRESSES

The signal transduction of biotic and abiotic stresses can result in a complex arrangement of interacting factors (Suzuki et al., 2014) (Figure 11.5). Gene products resulting from both biotic and abiotic stresses may control the response specificity of the plant to multiple stresses (Santino et al., 2013). Transcriptomic, proteomic and genetic analyses have greatly increased our knowledge of these processes.

11.4.1 Transcription Factors (TFs)

TFs are of great importance in generating specificity in response to biotic and abiotic stresses. Their regulation, particularly for the downstream events, provides numerous opportunities for conferring biotic and abiotic stress tolerance in woody plants (Kizis and Pages, 2002; Marchive et al., 2007; Chen et al., 2011; Hu et al., 2016d).

The MYB TFs, a large functionally diverse group, are considered to be associated with both biotic and abiotic stress responses, especially for regulation of the phenylpropanoid biosynthesis pathway (Dubos et al., 2010; Xu et al., 2015a). Genome-wide classification and evolutionary and expression analysis of citrus MYB TF families found that *CsMYB71/114/173* were induced by both NaCl and mannitol; however, other members such as *CsMYB33/37/55/129* were induced by NaCl and repressed by mannitol. These results support their specific roles in differentiating responses to individual stresses (Hou et al., 2014). Research also found that *MdMYB10* induced by osmotic stress promoted anthocyanin biosynthesis in apple (Gao et al., 2011b; Hu et al., 2016b). Anthocyanins are very important in plant resistance against pathogenic bacteria and fungi by quenching ROS that are generated from the pathogen and the infected plant (Mierziak et al., 2013), indicating that there is a link between anthocyanin and ROS in governing a broad spectrum of stress tolerances. Mellway et al. (2009) reported that the *MYB134* gene encoding an R2R3 MYB TF regulated proanthocyanidin synthesis in poplar was induced by mechanical wounding, *M. medusae* infection and elevated ultraviolet B. These data indicate that MYB TFs are the key regulators in crosstalking between biotic and abiotic stress responses.

Two other TFs, NAC and AP2/ERF (ethylene response factor), were reported to be associated with stress signaling. Recent discoveries suggested that they might be excellent targets for improving broad-spectrum tolerance in woody plants through genetic engineering (Le Hénanff et al., 2013; Su et al., 2013; Qin et al., 2014). Genome-wide analysis found that the *NAC* (*NAM*, *ATAF1,2*, and *CUC2*) TFs were responsive to multiple stresses in apple (Su et al., 2013). In a new model woody plant *Jatropha curcas*, overexpressing *JcNAC1* showed enhanced tolerance to drought and increased susceptibility to pathogens (Qin et al., 2014). Similarly, grapevine *NAC1* TF acted as a node of convergence to regulate developmental processes, abiotic stresses and necrotrophic/biotrophic pathogen tolerance (Le Hénanff et al., 2013). In addition, three AP2/VpERFs from Chinese wild *Vitis pseudoreticulata* showed different roles in either preventing disease progression via regulating the expression of relevant defense genes or directly involving abiotic stress responsive pathways (Zhu et al., 2013).

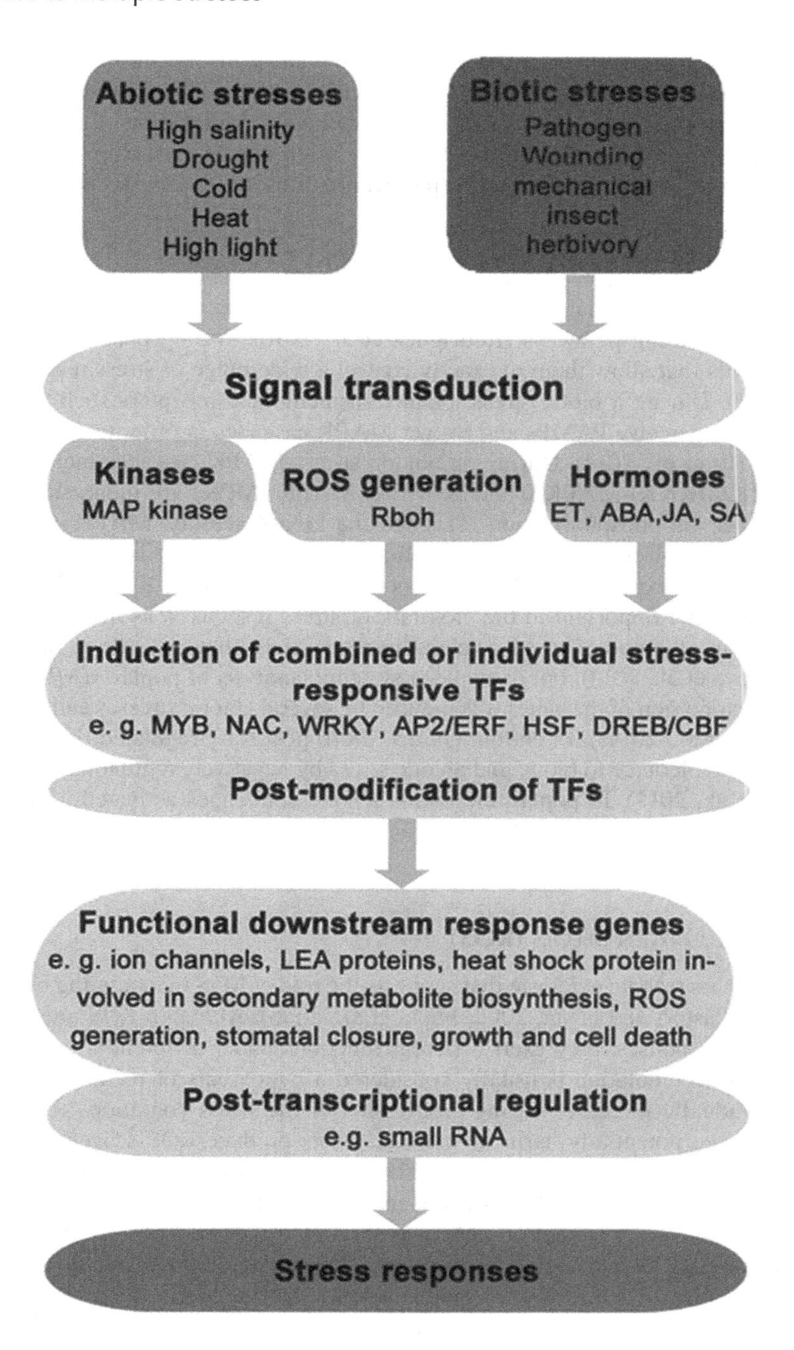

FIGURE 11.5 Convergence points in abiotic and biotic stress signaling networks.

Other TFs acting in both biotic and abiotic stress pathways have recently been identified. For example, the grapevine VpWRKY3 is involved in ABA signal pathway and salt stress as well as activating defence-related genes (Zhu et al., 2012b); the *Muscadinia rotundifolia* MrWRKY30 enhanced plant resistance to downy mildew pathogen (*Peronospora parasitica*) and cold stress, yet had a negative effect on salt tolerance (Jiang et al., 2015); overexpression of the *MrCBF2* gene enhanced both biotic and abiotic stress tolerance in *Arabidopsis* (Wu et al., 2016) and transcriptional profiling of apple fruit showed that the heat shock protein (*HSP*) and *HSP*-related genes were involved in induced heat and pathogen resistance (Spadoni et al., 2015). Many TFs conferred a growth impediment when overexpressed indicating a shift in balance between allocation of resources into stress protection rather than growth. Therefore, transgenes often need to be expressed under a stress-inducible promoter in order to overcome this problem (Atkinson and Urwin, 2012).

11.4.2 Mitogen-Activated Protein Kinase (MAPK) Cascades

The MAPK cascades are crucial in woody plants for transducing the perception of stress stimuli into intracellular signaling pathways (Rodriguez et al., 2010). Woody plants have multitudinous MAPK components that allow them to exactly control a wide range of stress responses (Atkinson and Urwin, 2012). During a biotic stress, plant transmembrane receptors such as FLAGELLIN SENSING2 (FLS2) perceive PAMPs and trigger MAPK cascades in order to establish a pathogen response signal (Li et al., 2014). During an abiotic stress, MAPK cascades including MAPKKK (MAP3K/MEKK/MKKK)-MAPKK (MKK/MEK)-MAPK (MPK) are induced and sequentially activate each other via phosphorylation. Through the kinase activity, MAPK cascades translate incoming environmental cues into post-translational modifications of the target proteins to ultimately reorganize gene expression and stress responses (Danquah et al., 2014). Additionally, MAPK cascades are especially important in the crosstalk of stress responses, as many MAPKs are activated by more than one type of stress or plant hormone; thus, they are able to integrate different signals (Smékalová et al., 2014). For example, a systemic analysis of poplar *MAPKK* family genes showed that the expression of the genes was induced by several abiotic stresses and stress-associated hormones (Wang et al., 2014a). In cotton, *GhMAP3K40* positively regulated the defense response but caused reduced tolerance to biotic and abiotic stress by negatively regulating growth and development (Chen et al., 2015). In poplar, biotic and abiotic challenges activated MAPKs that functioned as a convergence point for pathogen defense and oxidant stress signaling cascades (Hamel et al., 2005).

11.4.3 Reactive Oxygen Species (ROS)

There is gradually increasing evidence that supports the role of ROS in mediating biotic and abiotic stress responses (Pastori and Foyer, 2002; Fujita et al., 2006). Although ROS are critical to biotic and abiotic stress responses, they exert different functions under different stresses. In general, a low level of ROS accumulation is usually considered a byproducts of metabolism; however, the concentration gradually increases during progression of the stress condition. To minimize damage originating from potentially harmful stresses, plants produce ROS-scavenging enzymes and antioxidants (Carvalho et al., 2015; Luo et al., 2016; Hu et al., 2016a). Woody plants can vigorously produce ROS in response to pathogen infection, and this is known as the oxidative burst. Production of ROS confines the further spread of pathogens through hypersensitive response and cell death, and then coordinates ROS-scavenging mechanisms (Fujita et al., 2006; Carvalho et al., 2015).

Evidence also supports that ROS are signal molecules. During biotic and abiotic stress, the ROS responsive genes were specifically induced (Pastori and Foyer, 2002; Fujita et al., 2006). Research found that ROS was immediately produced by membrane-bound NADPH oxidases during pathogen infection or wounding, subsequently diffused into plant cells and activated diverse plant defenses (Fujita et al., 2006; Carvalho et al., 2015). Production of ROS is also required for ABA and ethylene-driven stomatal closure. The process of ROS production may contribute towards the positive effect of ABA and ethylene during early pathogen response (Beguerisse-Dıaz et al., 2012).

It is reported that some TFs and protein kinases could integrate ROS-scavenging mechanisms in response to stresses. In *Tamarix hispida*, *ThWRKY4* functioning as a TF to modulate reactive oxygen species positively regulated biotic and abiotic stress tolerances (Zheng et al., 2013). Similarly, overexpression of *GhWRKY39* positively regulated the plant response against pathogens and salinity by regulating the reactive oxygen species system via multiple signaling pathways (Shi et al., 2014). As mentioned above, biotic and abiotic challenges activated MAPKs in poplar as well as in herbaceous species, which then functioned as a convergence point for pathogen defense and oxidant stress signaling cascades (Hamel et al., 2005).

As biotic and abiotic stresses often occur in combination, the relationship between ROS-scavenging mechanisms and the stress response in plants is complicated. Therefore, identification of master regulators of ROS metabolisms may provide new candidates for generating a broad-spectrum of stress tolerance (Fujita et al., 2006; Carvalho et al., 2015).

11.5 PERSPECTIVES: CONSIDERATION OF PLANT STRESS RESEARCH IN A NEW VIEW

Responses of woody plants to major biotic and abiotic stresses are being extensively studied mostly in studies on individual stresses. Studies focusing on combinations of multiple stresses that influence woody plants have only recently been attempted by very few research groups. Those studies of individual and combinatorial stresses allow us to scrape together the regulatory network of molecular interactions that are responsible for different stresses in woody plants. Generally, woody plants activate specific or non-specific stress responses when they encounter adverse environmental conditions to protect their normal growth and development and reproduction. Current studies support that signal specificity can be achieved through the exact interaction with components of each pathway, especially the TFs, ROS, HSFs, small RNAs and hormones (ET, ABA, SA, and JA) (Fujita et al., 2006). Traditionally, an individual stress on a woody plant has been studied as a single stimulus that triggers linear signal transduction pathways. However, single factor research no longer seems sufficient because signaling pathways of both biotic and abiotic stresses are closely connected in a wide molecular interaction network.

An important goal of stress research in woody plants is to improve their tolerance or resistance to different stresses. With the changing climate, novel stress combinations such as rising CO_2 and UV-B light are a growing or even great threat to sustainable agriculture. Such threats pose the new challenges in developing plant materials with tolerances to multiple stresses (Newton et al., 2011). There is no doubt that better understanding of the key stress-regulatory networks and the underlying effects of different combinations of adverse conditions will help address the aforementioned threats and challenges. Research advances in stress physiology in model plant species, such as *Arabidopsis* and rice, have provided a possible way for exploration in other species including woody plants. Research on major regulatory genes involved in biotic and abiotic stress responses in the model species provide excellent candidates for manipulating stress tolerance in woody plants (You et al., 2014; Zhao et al., 2016). For example, apple *MdDRB1* functioning in the miRNA biogenesis in a conserved way is a master regulator in controlling adventitious rooting, leaf curvature and tree architecture (You et al., 2014). The transgenic apples with *MdDRB1* showed significantly increased tolerance to salinity, drought and cold (You et al., 2014; Ren et al., 2016). Transgenic analysis found that MdbHLH104 acted together with other apple bHLH TFs to regulate iron uptake via modulating the expression of the *MdAHA8* gene and PM H^+-ATPase activity in apple (Zhao et al., 2016). Such genes may therefore provide more opportunities for creating broad-spectrum stress-tolerant woody plants. Moreover, overexpression of *MdSOS2L1* significantly improved salt tolerance via a conserved SOS pathway in apple, suggesting some signal pathways in woody plants are similar to the model plants *Arabidopsis* and rice (Hu et al., 2012a, 2016a). Therefore, the conserved signal regulatory network can be referred to model plants such as *Arabidopsis* and rice.

Many research projects on major plant stresses of woody plants are conducted in a laboratory setting in which the experimental conditions cannot fully reflect conditions in the field; therefore, such lab-based research may lead to completely different results when plants perceive and respond to each unique set of stress conditions (Bray, 2004). Studies also showed that changes in growth parameters, such as temperature or sugar levels, obviously affected the responsive priorities of plants under the stress condition (Luna et al., 2011). Therefore, the experimental conditions (timing and severity of the stress, the condition of plant growth and nutrient availability) should be set up to as closely as possible resemble the field condition to achieve a meaningful result.

Furthermore, the new genome editing technology CRISPR-Cas is attracting a growing cadre of devotees. Applications of CRISPR-Cas, such as gene activation and silencing, visualization of genome dynamics and genome-wide functional screens show far greater efficiency than traditional genetic engineering strategies in a few organisms (fungal, bacterial, plant and animal species); therefore, it will be easily adopted in the study of stress responses of woody plants in future. In a word, one of the greatest challenges and aims for plant scientists in the twenty-first century is to breed stable multiple stress tolerance traits in plants including woody species to increase yields and qualities, especially in the field with adverse environmental conditions.

REFERENCES

Agueero, C. B., Uratsu, S. L., Greve, C., et al. 2005. Evaluation of tolerance to Pierce's disease and Botrytis in transgenic plants of *Vitis vinifera L.* expressing the pear *PGIP* gene. *Molecular Plant Pathology* 6(1):43–51.

Ahuja, I., de Vos, R. C., Bones, A. M., Hall, R. D. 2010. Plant molecular stress responses face climate change. *Trends in Plant Science* 15(12):664–674.

Albacete, A. A., Martínez-Andújar, C., Pérez-Alfocea, F. 2014. Hormonal and metabolic regulation of source-sink relations under salinity and drought: From plant survival to crop yield stability. *Biotechnology Advances* 32(1):12–30.

Aquino, S. O. D. 2014. Análise funcional da região promotora de haplótipos do gene *CcDREB1D* de *Coffea canephora* via transformação genética de *Nicotiana tacabum cv. SRI. Embrapa Recursos Genéticos e Biotecnologia-Tese/dissertação (ALICE).*

Atkinson, N. J., Urwin, P. E. 2012. The interaction of plant biotic and abiotic stresses: From genes to the field. *Journal of Experimental Botany* 63(10):3523–3543.

Beguerisse-Dıaz, M., Hernández-Gómez, M., Lizzul, A., Barahona, M., Desikan, R. 2012. Compound stress response in stomatal closure: A mathematical model of ABA and ethylene interaction in guard cells. *BMC Systems Biology* 6(1):1.

Bonnefont-Rousselot, D. 2014. Obésité et stress oxydant. *Obésité* 9(1):8–13.

Bray, E. A. 2004. Genes commonly regulated by water-deficit stress in *Arabidopsis thaliana. Journal of Experimental Botany* 55(407):2331–2341.

Bugaud, C., Deverge, E., Daribo, M. O., Ribeyre, F., Fils-Lycaon, B., Mbéguié-A-Mbéguié, D. 2011. Sensory characterisation enabled the first classification of dessert bananas. *Journal of the Science of Food and Agriculture* 91(6):992–1000.

Burdon, J., Lallu, N., Yearsley, C., Osman, S., Billing, D., Boldingh, H. 2007. Postharvest conditioning of Satsuma mandarins for reduction of acidity and skin puffiness. *Postharvest Biology and Technology* 43(1):102–114.

Cao, Z. H., Zhang, S. Z., Wang, R. K., Zhang, R. F., Hao, Y. J. 2013. Genome wide analysis of the apple MYB transcription factor family allows the identification of *MdoMYB121* gene confering abiotic stress tolerance in plants. *PLOS ONE* 8(7):e69955.

Carvalho, L. C., Vidigal, P., Amâncio, S. 2015. Oxidative stress homeostasis in grapevine (*Vitis vinifera* L.). *Frontiers of Environmental Science* 3:20.

Centeno, D. C., Osorio, S., Nunes-Nesi, A., et al. 2011. Malate plays a crucial role in starch metabolism, ripening, and soluble solid content of tomato fruit and affects postharvest softening. *Plant Cell* 109:1–23.

Chen, F., Liu, C. J., Tschaplinski, T. J., Zhao, N. 2009. Genomics of secondary metabolism in *Populus*: Interactions with biotic and abiotic environments. *Critical Reviews in Plant Sciences* 28(5):375–392.

Chen, J., Xia, X., Yin, W. 2011. A poplar DRE-binding protein gene, *PeDREB2L*, is involved in regulation of defense response against abiotic stress. *Gene* 483(1):36–42.

Chen, X., Wang, J., Zhu, M., et al. 2015. A cotton Raf-like *MAP3K* gene, *GhMAP3K40*, mediates reduced tolerance to biotic and abiotic stress in *Nicotiana benthamiana* by negatively regulating growth and development. *Plant Science* 240:10–24.

Cheng, T. Chen, J., Abd Allah, E. F., Wang, P., Wang, G., Hu, X. and Shi, J. 2015. Quantitative proteomics analysis reveals that S-nitrosoglutathione reductase (GSNOR) and nitric oxide signaling enhance poplar defense against chilling stress. *Planta* 242:1361–1390. doi: 10.1007/s00425-015-2374-5.

Cho, Y. H., Sheen, J., Yoo, S. D. 2010. Low glucose uncouples hexokinase 1-dependent sugar signaling from stress and defense hormone abscisic acid and C2H4 responses in *Arabidopsis. Plant Physiology* 152(3):1180–1182.

Couée, I., Sulmon, C., Gouesbet, G., El Amrani, A. 2006. Involvement of soluble sugars in reactive oxygen species balance and responses to oxidative stress in plants. *Journal of Experimental Botany* 57(3):449–459.

Cuéllar, T., Pascaud, F., Verdeil, J. L., et al. 2010. A grapevine Shaker inward K^+ channel activated by the calcineurin B-like calcium sensor 1-protein kinase CIPK23 network is expressed in grape berries under drought stress conditions. *The Plant Journal* 61(1):58–69.

Danquah, A., de Zelicourt, A., Colcombet, J., Hirt, H. 2014. The role of ABA and MAPK signaling pathways in plant abiotic stress responses. *Biotechnology Advances* 32(1):40–52.

Das, R., Pandey, A., Pandey, G. K. 2013. Role of calcium regulated kinases: Expressional and functional analysis in abiotic stress signaling. *Changing Views on Living Organisms* 49:377–425.

Dong, Q. L., Liu, D. D., An, X. H., Hu, D., Yao, Y., Hao, Y. 2011. *MdVHP1* encodes an apple vacuolar H^+-PPase and enhances stress tolerance in transgenic apple callus and tomato. *Journal of Plant Physiology* 168(17):2124–2133.

Dong, Q. L., Wang, C. R., Liu, D. D., Hu, D. G., Fang, M. J., You, Y. X., Yao, C. X., and Hao Y. J. 2013. MdVHA-A encodes an apple subunit A of vacuolar H+-ATPase and enhances drought tolerance in transgenic tobacco seedlings. *J. Plant Physiol.* 170:601–609. doi: org/10.1016/j.jplph.2012.12.014.

Dubos, C., Stracke, R., Grotewold, E., Weisshaar, B., Martin, C., Lepiniec, L. 2010. MYB transcription factors in *Arabidopsis. Trends in Plant Science* 15(10):573–581.

Dubrovina, A. S., Kiselev, K. V., Khristenko, V. S. 2013. Expression of calcium-dependent protein kinase (*CDPK*) genes under abiotic stress conditions in wild-growing grapevine *Vitis amurensis. Journal of Plant Physiology* 170(17):1491–1500.

Dubrovina, A. S., Kiselev, K. V., Khristenko, V. S., Aleynova, O. A. 2015. *VaCPK20*, a calcium-dependent protein kinase gene of wild grapevine *Vitis amurensis* Rupr mediates cold and drought stress tolerance. *Journal of Plant Physiology* 185:1–12.

Dubrovina, A. S., Kiselev, K. V., Khristenko, V. S., Aleynova, O. A. 2016. *VaCPK21*, a calcium-dependent protein kinase gene of wild grapevine *Vitis amurensis* Rupr., is involved in grape response to salt stress. *Plant Cell, Tissue and Organ Culture (PCTOC)* 1(124):137–150.

Feng, X. M., Zhao, Q., Zhao, L. L., et al. 2012. The cold-induced basic helix-loop-helix transcription factor gene *MdCIbHLH1* encodes an ICE-like protein in apple. *BMC Plant Biology* 12(1):22.

Fernie, A. R., Martinoia, E. 2009. Malate. Jack of all trades or master of a few? *Phytochemistry* 70(7):828–832.

Finkemeier, I., Sweetlove, L. J. 2009. The role of malate in plant homeostasis. *F1000 Biology Reports* 1:47.

Fu, Z. F., Yao, C. C., Zhang, C. H., Wang, Y. J. 2012. Cloning and functional analysis of *PbzsREMORIN* gene related to *Venturia nashicola* resistance in 'Zaosuli'pear. *Acta Horticulturae Sinica* 1:004.

Fujita, M., Fujita, Y., Noutoshi, Y., et al. 2006. Crosstalk between abiotic and biotic stress responses: A current view from the points of convergence in the stress signaling networks. *Current Opinion in Plant Biology* 9(4):436–442.

Gao, C., Wang, Y., Jiang, B., et al. 2011a. A novel vacuolar membrane H^+-ATPase c subunit gene (*ThVHAc1*) from *Tamarix hispida* confers tolerance to several abiotic stresses in *Saccharomyces cerevisiae. Molecular Biology Reports* 38(2):957–963.

Gao, J. J., Zhang, Z., Peng, R. H., et al. 2011b. Forced expression of *Mdmyb10*, a myb transcription factor gene from apple, enhances tolerance to osmotic stress in transgenic *Arabidopsis. Molecular Biology Reports* 38(1):205–211.

Gautier, H., Rocci, A., Buret, M., Grassely, D., Causse, M. 2005. Fruit load or fruit position alters response to temperature and subsequently cherry tomato quality. *Journal of the Science of Food and Agriculture* 85(6):1009–1016.

Giannino, D., Condello, E., Bruno, L. 2004. The gene geranylgeranyl reductase of peach (*Prunus persica [L.] Batsch*) is regulated during leaf development and responds differentially to distinct stress factors. *Journal of Experimental Botany* 55(405):2063–2073.

Gupta, P., Duplessis, S., White, H., Karnosky, D. F., Martin, F., Podila, G. K. 2005. Gene expression patterns of trembling aspen trees following long-term exposure to interacting elevated CO_2 and tropospheric O_3. *New Phytologist* 167(1):129–142.

Hamel, L. P., Miles, G. P., Samuel, M. A., Ellis, B. E., Seguin, A., Beaudoin, N. 2005. Activation of stress-responsive mitogen-activated protein kinase pathways in hybrid poplar (*Populus trichocarpa*× *Populus deltoides*). *Tree Physiology* 25(3):277–288.

Hirayama, T., Shinozaki, K. 2010. Research on plant abiotic stress responses in the post-genome era: Past, present and future. *The Plant Journal* 61(6):1041–1052.

Hou, X. J., Li, S. B., Liu, S. R., Hu, C. G., Zhang, J. Z. 2014. Genome-wide classification and evolutionary and expression analyses of citrus *MYB* transcription factor families in sweet orange. *PLOS ONE* 9(11):e112375.

Hu, D. G., Li, M., Luo, H., et al. 2012a. Molecular cloning and functional characterization of *MdSOS2* reveals its involvement in salt tolerance in apple callus and *Arabidopsis*. *Plant Cell Reports* 31(4):713–722.

Hu, D. G., Ma, Q. J., Sun, C. H., Sun, M., You, C., Hao, Y. 2016a. Overexpression of *MdSOS2L1*, a CIPK protein kinase, increases the antioxidant metabolites to enhance salt tolerance in apple and tomato. *Physiologia Plantarum* 156(2):201–214.

Hu, D. G., Sun, C. H., Ma, Q. J., et al. 2016b. MdMYB1 regulates anthocyanin and malate accumulation by directly facilitating their transport into vacuoles in apples. *Plant Physiology* 170(3):1315–1330.

Hu, D. G., Sun, C. H., Sun, M. H., Hao, Y. J. 2016c. MdSOS2L1 phosphorylates MdVHA-B1 to modulate malate accumulation in response to salinity in apple. *Plant Cell Reports* 35(3):705–718.

Hu, D. G., Sun, M. H., Sun, C. H., et al. 2015. Conserved vacuolar H^+-ATPase subunit B1 improves salt stress tolerance in apple calli and tomato plants. *Scientia Horticulturae* 197:107–116.

Hu, D. G., Wang, S. H., Luo, H., et al. 2012b. Overexpression of *MdVHA-B*, a V-ATPase gene from apple, confers tolerance to drought in transgenic tomato. *Scientia Horticulturae* 145:94–101.

Hu, Y., Han, Y. T., Zhang, K., et al. 2016d. Identification and expression analysis of heat shock transcription factors in the wild Chinese grapevine (*Vitis pseudoreticulata*). *Plant Physiology and Biochemistry* 99:1–10.

Jaeckels, N., Tenzer, S., Rosfa, S., Schild, H., Decker, H., Wigand, P. 2013. Purification and structural characterisation of lipid transfer protein from red wine and grapes. *Food Chemistry* 138(1):263–269.

Jiang, W., Wu, J., Zhang, Y., Yin, L., Lu, J. 2015. Isolation of a *WRKY30* gene from *Muscadinia rotundifolia* (Michx) and validation of its function under biotic and abiotic stresses. *Protoplasma* 252(5):1361–1374.

Jin, W. M., Dong, J., Yin, S. P., Yan, A. L., Chen, M. X. 2007. *CBF1* gene transgenic strawberry and increase freezing tolerance. *Acta Botanica Boreali-Occidentalia Sinica* 27(2):223.

Kanayama, Y., Mizutani, R., Yaguchi, S., et al. 2014. Characterization of an uncharacterized aldo-keto reductase gene from peach and its role in abiotic stress tolerance. *Phytochemistry* 104:30–36.

Keunen, E. L. S., Peshev, D., Vangronsveld, J., Van Den Ende, W. I. M., Cuypers, A. N. N. 2013. Plant sugars are crucial players in the oxidative challenge during abiotic stress: Extending the traditional concept. *Plant, Cell and Environment* 36(7):1242–1255.

Kizis, D., Pages, M. 2002. Maize DRE-binding proteins DBF1 and DBF2 are involved in *rab17* regulation through the drought-responsive element in an ABA-dependent pathway. *The Plant Journal* 30(6):679–689.

Konstantinos, P. A., Imene, T., Panagiotis, M. N., Roubelakis-Angelakis, K. A. 2010. ABA-dependent amine oxidases-derived H_2O_2 affects stomata conductance. *Plant Signaling and Behavior* 5(9):1153–1156.

Le Hénanff, G., Profizi, C., Courteaux, B., et al. 2013. Grapevine *NAC1* transcription factor as a convergent node in developmental processes, abiotic stresses, and necrotrophic/biotrophic pathogen tolerance *Journal of Experimental Botany* 64(16):4877–4893.

Li, C., Liang, B., Chang, C., Wei, Z., Zhou, S., Ma, F. 2016. Exogenous melatonin improved potassium content in *Malus* under different stress conditions. *Journal of Pineal Research* 61(2):218–229.

Li, C. Z., Zhao, D. Q., Tao, J. 2011. Molecular cloning and sequence analysis of a heat shock protein gene *PsHSP70* from tree peony (*Paeonia suffruticosa* Andr.). *Journal of Yangzhou University (Agricultural and Life Science Edition)* 4:011.

Li, D., Song, S., Xia, X., Yin, W. 2012. Two CBL genes from *Populus euphratica* confer multiple stress tolerance in transgenic triploid white poplar. *Plant Cell, Tissue and Organ Culture (PCTOC)* 109(3):477–489.

Li, L., Li, M., Yu, L., et al. 2014. The FLS2-associated kinase BIK1 directly phosphorylates the NADPH oxidase RbohD to control plant immunity. *Cell Host and Microbe* 15(3):329–338.

Liu, D., Liu, Y., Rao, J., et al. 2013. Overexpression of the glutathione S-transferase gene from *Pyrus pyrifolia* fruit improves tolerance to abiotic stress in transgenic tobacco plants. *Molecular Biology* 47(4):515–523.

Luna, E., Pastor, V., Robert, J., Flors, V., Mauch-Mani, B., Ton, J. 2011. Callose deposition: A multifaceted plant defense response. *Molecular Plant–Microbe Interactions* 24(2):183–193.

Luo, P., Shen, Y., Jin, S., et al. 2016. Overexpression of *Rosa rugosa* anthocyanidin reductase enhances tobacco tolerance to abiotic stress through increased ROS scavenging and modulation of ABA signaling. *Plant Science* 245:35–49.

Luu, V. T., Schuck, S., KIM, S. G., Weinhold, A., Baldwin, I. T. 2015. Jasmonic acid signalling mediates resistance of the wild tobacco Nicotiana attenuata to its native Fusarium, but not Alternaria, fungal pathogens. *Plant, Cell and Environment* 38(3):572–584.

Mahalingam, R. 2015. *Consideration of Combined Stress: A Crucial Paradigm for Improving Multiple Stress Tolerance in Plants*. Springer International Publishing:1–25.

Marchive, C., Mzid, R., Deluc, L., et al. 2007. Isolation and characterization of a *Vitis vinifera* transcription factor, *VvWRKY1*, and its effect on responses to fungal pathogens in transgenic tobacco plants. *Journal of Experimental Botany* 58(8):1999–2010.

Martin, L., Leblanc-Fournier, N., Azri, W., et al. 2009. Characterization and expression analysis under bending and other abiotic factors of *PtaZFP2*, a poplar gene encoding a Cys2/His2 zinc finger protein. *Tree Physiology* 29(1):125–136.

Martinez, M., Abraham, Z., Gambardella, M., Echaide, M., Carbonero, P., Diaz, I. 2005. The strawberry gene *Cyf1* encodes a phytocystatin with antifungal properties. *Journal of Experimental Botany* 56(417):1821–1829.

Mellway, R. D., Tran, L. T., Prouse, M. B., Campbell, M. M., Constabel, C. P. 2009. The wound-, pathogen-, and ultraviolet B-responsive *MYB134* gene encodes an R2R3 MYB transcription factor that regulates proanthocyanidin synthesis in poplar. *Plant Physiology* 150(2):924–941.

Mierziak, J., Kostyn, K., Kulma, A. 2013. Flavonoids as important molecules of plant interactions with the environment. *Molecules (Basel, Switzerland)* 19(10):16240–16265.

Munns, R., Tester, M. 2008. Mechanisms of salinity tolerance. *Annual Review of Plant Biology* 59(1):651–681.

Neill, S. J., Desikan, R., Clarke, A., Hurst, R. D., Hancock, J. T. 2002. Hydrogen peroxide and nitric oxide as signalling molecules in plants. *Journal of Experimental Botany* 53(372):1237–1247.

Newton, A. C., Johnson, S. N., Gregory, P. J. 2011. Implications of climate change for diseases, crop yields and food security. *Euphytica* 179(1):3–18.

Pastori, G. M., Foyer, C. H. 2002. Common components, networks, and pathways of cross-tolerance to stress. The central role of 'redox' and abscisic acid-mediated controls. *Plant Physiology* 129(2):460–468.

Pérez-Alfocea, F., Albacete, A., Ghanem, M. E., Dodd, I. C. 2010. Hormonal regulation of source-sink relations to maintain crop productivity under salinity: A case study of root-to-shoot signalling in tomato. *Functional Plant Biology* 37(7):592–603.

Qin, X., Zheng, X., Huang, X., et al. 2014. A novel transcription factor *JcNAC1* response to stress in new model woody plant *Jatropha curcas*. *Planta* 239(2):511–520.

Qiao, X., Li, M., Li, L., Yin, H., Wu, J., Zhang, S. 2015. Genome-wide identification and comparative analysis of the heat shock transcription factor family in Chinese white pear (*Pyrus bretschneideri*) and five other Rosaceae species. *BMC Plant Biology* 15(1):1.

Racchi, M. L. 2013. Antioxidant defenses in plants with attention to *Prunus* and *Citrus* spp. *Antioxidants* 2(4):340–369.

Ramesh Kumar, A., Kumar, N. 2007. Sulfate of potash foliar spray effects on yield, quality, and post-harvest life of banana. *Better Crops* 91:22–24.

Ren, Y., Zhao, Q., Zhao, X., Hao, Y., You, C. 2016. Expression and function analysis of apple *MdDRB1* Gene. *Acta Horticulturae Sinica* 43(6):1033–1043.

Rodriguez, M. C. S., Petersen, M., Mundy, J. 2010. Mitogen-activated protein kinase signaling in plants. *Annual Review of Plant Biology* 61(1):621–649.

Rolland, F., Moore, B., Sheen, J. 2002. Sugar sensing and signaling in plants. *The Plant Cell* 14(suppl 1): S185–S205.

Santino, A., Taurino, M., De Domenico, S., et al. 2013. Jasmonate signaling in plant development and defense response to multiple (a) biotic stresses. *Plant Cell Reports* 32(7):1085–1098.

Seymour, G. B., Taylor, J., Tucker, G. A. 1993. *Biochemistry of Fruit Ripening*. London: Chapman & Hall.

Sheen, J., Zhou, L., Jang, J. C. 1999. Sugars as signaling molecules. *Current Opinion in Plant Biology* 2(5):410–418.

Shekhawat, U. K. S., Srinivas, L., Ganapathi, T. R. 2011. *MusaDHN-1*, a novel multiple stress-inducible SK3-type dehydrin gene, contributes affirmatively to drought-and salt-stress tolerance in banana. *Planta* 234(5):915–932.

Shi, W., Hao, L., Li, J., Liu, D., Guo, X., Li, H. 2014a. The Gossypium hirsutum *WRKY* gene *GhWRKY39-1* promotes pathogen infection defense responses and mediates salt stress tolerance in transgenic *Nicotiana benthamiana*. *Plant Cell Reports* 33(3):483–498.

Shi, W., Liu, D., Hao, L., Wu, C., Guo, X., Li, H. 2014b. *GhWRKY39*, a member of the *WRKY* transcription factor family in cotton, has a positive role in disease resistance and salt stress tolerance. *Plant Cell, Tissue and Organ Culture (PCTOC)* 118(1):17–32.

Shin, S., Lv, J., Fazio, G., Mazzola, M., Zhu, Y. 2014. Transcriptional regulation of ethylene and jasmonate mediated defense response in apple (*Malus domestica*) root during *Pythium ultimum* infection. *Horticulture Research* 1(1):14053.

Shoji, T., Hashimoto, T. 2015. Stress-induced expression of NICOTINE2-locus genes and their homologs encoding ethylene Response Factor transcription factors in tobacco. *Phytochemistry* 113:41–49.

Silva-Ortega, C. O., Ochoa-Alfaro, A. E., Reyes-Agüero, J. A., Aguado-Santacruz, G. A., Jiménez-Bremont, J. F. 2008. Salt stress increases the expression of *p5cs* gene and induces proline accumulation in cactus pear. *Plant Physiology and Biochemistry* 46(1):82–92.

Smeekens, S., Hellmann, H. A. 2014. Sugar sensing and signaling in plants. *Frontiers in Plant Science* 5.

Smékalová, V., Doskočilová, A., Komis, G., Šamaj, J. 2014. Crosstalk between secondary messengers, hormones and MAPK modules during abiotic stress signalling in plants. *Biotechnology Advances* 32(1):2–11.

Spadoni, A., Guidarelli, M., Phillips, J., Mari, M., Wisniewski, M. 2015. Transcriptional profiling of apple fruit in response to heat treatment: Involvement of a defense response during *Penicillium expansum* infection. *Postharvest Biology and Technology* 101:37–48.

Spadoni, A., Guidarelli, M., Sanzani, S. M., Ippolito, A., Mari, M. 2014. Influence of hot water treatment on brown rot of peach and rapid fruit response to heat stress. *Postharvest Biology and Technology* 94:66–73.

Su, H., Zhang, S., Yuan, X., Chen, C., Wang, X., Hao, Y. 2013. Genome-wide analysis and identification of stress-responsive genes of the *NAM-ATAF1, 2-CUC2* transcription factor family in apple. *Plant Physiology and Biochemistry* 71:11–21.

Sun, C. X., Li, M. Q., Gao, X. X., Liu, L. N., Wu, X. F., Zhou, J. H. 2016. Metabolic response of maize plants to multi-factorial abiotic stresses. *Plant Biology* 18(S1):120–129.

Suzuki, N., Rivero, R. M., Shulaev, V., Blumwald, E., Mittler, R. 2014. Abiotic and biotic stress combinations. *New Phytologist* 203(1):32–43.

Sweetman, C., Deluc, L. G., Cramer, G. R., Ford, C. M., Soole, K. L. 2009. Regulation of malate metabolism in grape berry and other developing fruits. *Phytochemistry* 70(11–12):1329–1344.

Tang, R. J., Liu, H., Bao, Y., Lv, Q., Yang, L., Zhang, H. 2010. The woody plant poplar has a functionally conserved salt overly sensitive pathway in response to salinity stress. *Plant Molecular Biology* 74(4–5):367–380.

Thakur, A., Singh, Z. 2012. Responses of 'Spring Bright' and 'Summer Bright' nectarines to deficit irrigation: Fruit growth and concentration of sugars and organic acids. *Scientia Horticulturae* 135:112–119.

Todaka, D., Shinozaki, K., Yamaguchi-Shinozaki, K. 2015. Recent advances in the dissection of drought-stress regulatory networks and strategies for development of drought-tolerant transgenic rice plants. *Frontiers in Plant Science* 6:84.

Tokunaga, T., Miyahara, K., Tabata, K., Esaka, M. 2005. Generation and properties of ascorbic a cid-overproducing transgenic tobacco cells expressing sense RNA for L-galactono-1,4-lactone dehydrogenase. *Planta* 220(6):854–863.

Tosetti, R., Tardelli, F., Tadiello, A., et al. 2014. Molecular and biochemical responses to wounding in mesocarp of ripe peach (*Prunus persica* L. Batsch) fruit. *Postharvest Biology and Technology* 90:40–51.

Trivedi, D. K., Huda, K. M. K., Gill, S. S., Tuteja, N. 2016. Molecular chaperone: Structure, function, and role in plant abiotic stress tolerance. *Abiotic Stress Response in Plants* 2016:135–154.

Vilanova, L., Wisniewski, M., Norelli, J., et al. 2014. Transcriptomic profiling of apple in response to inoculation with a pathogen (*Penicillium expansum*) and a non-pathogen (*Penicillium digitatum*). *Plant Molecular Biology Reporter* 32(2):566–583.

Wang, J., Sun, P. P., Chen, C. L., Wang, Y., Fu, X., Liu, J. 2011a. An arginine decarboxylase gene *PtADC* from *Poncirus trifoliata* confers abiotic stress tolerance and promotes primary root growth in *Arabidopsis*. *Journal of Experimental Botany* 62(8):2899–2914.

Wang, L., Su, H., Han, L., Wang, C., Sun, Y., Liu, F. 2014a. Differential expression profiles of poplar *MAP* kinase kinases in response to abiotic stresses and plant hormones, and overexpression of *PtMKK4* improves the drought tolerance of poplar. *Gene* 545(1):141–148.

Wang, Q. J., Sun, H., Dong, Q. L., et al. 2016. The enhancement of tolerance to salt and cold stresses by modifying the redox state and salicylic acid content via the cytosolic malate dehydrogenase gene in transgenic apple plants. *Plant Biotechnology Journal* 14(10):1986–1997.

Wang, R. K., Cao, Z. H., Hao, Y. J. 2014b. Overexpression of a R2R3 MYB gene *MdSIMYB1* increases tolerance to multiple stresses in transgenic tobacco and apples. *Physiologia Plantarum* 150(1):76–87.

Wang, R. K., Li, L. L., Cao, Z. H., et al. 2012. Molecular cloning and functional characterization of a novel apple *MdCIPK6L* gene reveals its involvement in multiple abiotic stress tolerance in transgenic plants. *Plant Molecular Biology* 79(1–2):123–135.

Wang, W., Vinocur, B., Altman, A. 2003. Plant responses to drought, salinity and extreme temperatures: Towards genetic engineering for stress tolerance. *Planta* 218(1):1–14.

Wang, Y., Wang, H., Li, R., Ma, Y., Wei, J. 2011b. Expression of a SK2-type dehydrin gene from *Populus euphratica* in a *Populus tremula* × *Populus alba* hybrid increased drought tolerance. *African Journal of Biotechnology* 10(46):9225–9232.

Wen, X. P., Ban, Y., Inoue, H., Matsuda, N., Moriguchi, T. 2009. Aluminum tolerance in a spermidine synthase-overexpressing transgenic European pear is correlated with the enhanced level of spermidine via alleviating oxidative status. *Environmental and Experimental Botany* 66(3):471–478.

Wisniewski, M., Norelli, J., Bassett, C., Artlip, T., Macarisin, D. 2011. Ectopic expression of a novel peach (*Prunus persica*) CBF transcription factor in apple (*Malus×domestica*) results in short-day induced dormancy and increased cold hardiness. *Planta* 233(5):971–983.

Wu, J., Folta, K. M., Xie, Y., et al. 2016. Overexpression of *Muscadinia rotundifolia CBF2* gene enhances biotic and abiotic stress tolerance in *Arabidopsis. Protoplasma* 1:1–13.

Wu, L., Joshi, C. P., Chiang, V. L. 2000. A xylem-specific cellulose synthase gene from aspen (*Populus tremuloides*) is responsive to mechanical stress. *The Plant Journal* 22(6):495–502.

Xu, J., Zhang, B., Jiang, C., Ming, F. 2011. *RceIF5A*, encoding an eukaryotic translation initiation factor 5A in *Rosa chinensis*, can enhance thermotolerance, oxidative and osmotic stress resistance of *Arabidopsis thaliana. Plant Molecular Biology* 75(1–2):167–178.

Xu, W., Dubos, C., Lepiniec, L. 2015a. Transcriptional control of flavonoid biosynthesis by MYB-bHLH-WDR complexes. *Trends in Plant Science* 20(3):176–185.

Xu, Z. C., Yin, J., Zhou, B. 2015b. Grape seed proanthocyanidin protects liver against ischemia/reperfusion injury by attenuating endoplasmic reticulum stress. *World Journal of Gastroenterology: WJG* 21(24):7468.

Yang, W., Liu, X. D., Chi, X. J., et al. 2011. Dwarf apple *MbDREB1* enhances plant tolerance to low temperature, drought, and salt stress via both ABA-dependent and ABA-independent pathways. *Planta* 233(2):219–229.

Yang, Y., Tang, N., Xian, Z., Li, Z. 2015. Two SnRK2 protein kinases genes play a negative regulatory role in the osmotic stress response in tomato. *Plant Cell, Tissue and Organ Culture (PCTOC)* 122(2):421–434.

Yao, Y. X., Dong, Q. L., Zhai, H., You, C. X., and Hao Y. J. 2011. The functions of an apple cytosolic malate dehydrogenase gene in growth and tolerance to cold and salt stresses. *Plant Physiol. Biochem.* 49:257–264. doi: 10.1016/j.plaphy.2010.12.009.

Ye, C. Y., Zhang, H. C., Chen, J. H., Xia, X. L., Yin, W. L. 2009. Molecular characterization of putative vacuolar NHX-type Na^+/H^+ exchanger genes from the salt-resistant tree *Populus euphratica. Physiologia Plantarum* 137(2):166–174.

You, C. X., Zhao, Q., Wang, X. F., et al. 2014. A dsRNA-binding protein MdDRB1 associated with miRNA biogenesis modifies adventitious rooting and tree architecture in apple. *Plant Biotechnology Journal* 12(2):183–192.

Yue, C., Cao, H. L., Wang, L., et al. 2015. Effects of cold acclimation on sugar metabolism and sugar-related gene expression in tea plant during the winter season. *Plant Molecular Biology* 88(6):591–608.

Zhang, L. H., Jia, B., Zhuo, R. Y., et al. 2012. An acyl-acyl carrier protein thioesterase gene isolated from wintersweet (*Chimonanthus praecox*), *CpFATB*, enhances drought tolerance in transgenic tobacco (*Nicotiana tobaccum*). *Plant Molecular Biology Reporter* 30(2):433–442.

Zhang, J., Jia, W., Yang, J., Ismail, A. M. 2006. Role of ABA in integrating plant responses to drought and salt stresses. *Field Crops Research* 97(1):111–119.

Zhang, J. Y., Qu, S. C., Qiao, Y. S., Zhang, Z., Guo, Z. R. 2014a. Overexpression of the *Malus hupehensis MhNPR1* gene increased tolerance to salt and osmotic stress in transgenic tobacco. *Molecular Biology Reports* 41(3):1553–1561.

Zhang, S. H., Wei, Y., Liu, J. L., et al. 2011. An apoplastic chitinase *CpCHT1* isolated from the corolla of wintersweet exhibits both antifreeze and antifungal activities. *Biologia Plantarum* 55(1):141–148.

Zhang, W., Yang, H., You, S., Xu, Y., Ran, K., Fan, S. 2014b. Cloning, characterization and functional analysis of the role *MhNCED3*, a gene encoding 9-cis-epoxycarotenoid dioxygenase in *Malus hupehensis Rehd.*, plays in plant tolerance to osmotic and Cd^{2+} stresses. *Plant and Soil* 381(1–2):143–160.

Zhao, Q., Ren, Y. R., Wang, Q. J., Yao, Y., You, C., Hao, Y. 2016. Overexpression of *MdbHLH104* gene enhances the tolerance to iron deficiency in apple. *Plant Biotechnology Journal* 14(7):1633–1645.

Zheng, L., Liu, G., Meng, X., et al. 2013. A *WRKY* gene from *Tamarix hispida*, *ThWRKY4*, mediates abiotic stress responses by modulating reactive oxygen species and expression of stress-responsive genes. *Plant Molecular Biology* 82(4–5):303–320.

Zhou, Q. Y., Tian, A. G., Zou, H. F., et al. 2008. Soybean WRKY-type transcription factor genes, *GmWRKY13*, *GmWRKY21*, and *GmWRKY54*, confer differential tolerance to abiotic stresses in transgenic Arabidopsis plants. *Plant Biotechnology Journal* 6(5):486–503.

Zhu, J. K. 2002. Salt and drought stress signal transduction in plants. *Annual Review of Plant Biology* 53(1):247–273.

Zhu, X., Li, X., Zou, Y., Chen, W., Lu, W. 2012a. Cloning, characterization and expression analysis of Δ^1-pyrroline-5-carboxylate synthetase (*P5CS*) gene in harvested papaya (*Carica papaya*) fruit under temperature stress. *Food Research International* 49(1):272–279.

Zhu, Z., Shi, J., Cao, J., He, M., Wang, Y. 2012b. *VpWRKY3*, a biotic and abiotic stress-related transcription factor from the Chinese wild *Vitis pseudoreticulata*. *Plant Cell Reports* 31(11):2109–2120.

Zhu, Z., Shi, J., Xu, W., et al. 2013. Three ERF transcription factors from Chinese wild grapevine *Vitis pseudoreticulata* participate in different biotic and abiotic stress-responsive pathways. *Journal of Plant Physiology* 170(10):923–933.

Index

I

J

L

M

R

S

T

W

X

Z

W

X

Z

Printed in the United States
by Baker & Taylor Publisher Services